THE ENVIROPAEDIA®

Environmental Encyclopaedia
&
Networking Directory
for
Southern Africa

Presented to: _____

_____

From: _____

Message: _____

_____

_____

_____

_____

Published – May 2004

The ENVIROPAEDIA

ISBN 0-620-31893-7 © COPYRIGHT

To find solutions and create a better future we must first know and understand the nature of the issues.

WARNING – The *Enviropaedia* will not provide a "sanitised" perspective of environmental issues.

There are many opposing and contradicting views on any environmental topic – depending on whether you look at the issue from a scientific, economic, social, political or moral perspective, all of which are legitimate. In order to build bridges of understanding and promote constructive dialogue, The *Enviropaedia* provides a platform and invites all sectors to present their viewpoints. When reading the Topics and Guest Essays you will therefore find a great diversity of style and content. Whilst the authors of Topics have been asked to present a "balanced" perspective of the subject, in contrast, the authors of Guest Essays have been invited to express their personal views and promote their own perspective. It is therefore important for readers to note who the authors are and what organisation they represent as this may influence the nature and content of their topic or essay. In all instances of conflicting viewpoints, instead of taking sides, let us look for common ground and solutions.

THE ENVIROPAEDIA WISHES TO THANK:

FOUNDING AUTHOR: Arend Hoogervorst BSc. (Hons), MPhil, Pr.Sci.Nat, MIEnv.Sc, MIWM. Wrote the entire "Encyclopaedic" content of the first edition of The *Enviropaedia* in 2000. This material is intended to provide short, concise explanations of a range of topics, terms and issues. All of these (except those specifically identified as being written by guest authors) were written in a deliberately neutral, balanced and factual manner. This material continues to be used and provides the foundation for all subsequent editions. With some 20 years experience in the "environmental business", Arend is a consulting environmental scientist and advisor in private practice, working for Eagle Environmental. He is also editor and publisher of the *Eagle Bulletin*, and writes widely on environmental and industrial issues in other publications including *European Chemical News* and *Business Day*.

SPECIALIST ENVIRONMENTAL EDUCATION AUTHOR – Dr Eureta Rosenburg (nee Janse van Rensburg)
Looking at individual topics in isolation would not provide a coherent and holistic picture of what is happening to our environment. We therefore asked Dr Eureta Rosenburg to tackle the immensely difficult task of writing a series of balanced yet challenging "overviews" on a range of key environmental issues and controversial topics of current importance. Dr Rosenburg has a background in the medical and social sciences, and a PhD in environmental education. She was previously Associate Professor in the Murray & Roberts Chair of Environmental Education at Rhodes University. Dr Rosenburg now consults on environmental education, training and research for clients that include the City of Cape Town, South African National Parks and National Government. Dr Rosenburg was recently commissioned by Prof Kader Asmal – Department of Education, to write the handbook on Environmental Education for Teachers in Schools.

GUEST AUTHORS:

Dr Kelvin Kemm, MD of Stratek Business Strategy Consultants and Director of the "Green & Gold Forum." Dr Kemm is a respected industrial scientist and influential Author, Lecturer, Strategic Planner, Consultant and Advisor to the Industrial community. We thank Dr Kemm for his contribution of over 20 wide-ranging environmental topics.

The following **Guest Authors** who have taken a great deal of time, care and effort in writing their individual specialist topics :

Randall Adriaans – City of Cape Town; **Dr John Anderson** – Gondwana Alive; **Glenn Ashton** – Ekogaia Foundation and Safe Age; **Prof. Kader Asmal** – Department of Education; **Dr W.R. Bainbridge** – Bainbridge Resource Management and Wilderness Action Group; **Mark Borchers** – Sustainable Living Centre; **Peter Borchert** – Africa Birds & Birding and Africa Geographic Magazines; **Børge Brende** – UN Commission on Sustainable Development and Norwegian Minister for Environment; **Dr Gwen Breedlove** – University of Pretoria; **Therese Brincate** – WWF (SA); **Etienne Bruwer** – Greenhaus Architects; **Cheryl Carolus** – SA Tourism; **Tessa Chamberlain** – Pick 'n Pay; **Cormac Cullinan** – EnAct International; **Dr Jenny Day** – University of Cape Town; **Prof Maarten de Wit** – University of Cape Town; **Tony Dixon** – Institute of Directors SA; **Patrick Dowling** – WESSA; **Heather Dugmore** – Environmental Writer; **Earthlife Africa**, **Environmental Evaluation Unit** – University of Cape Town; **Enviro Facts** Information Leaflets; **Sandra Fowkes** – Santam Cape Argus Ukuvuka; **Lauren Gardener** – WESSA; **Prof J.P. Hattingh** – University of Stellenbosch; **Dr Sibbele Hietkamp** – CSIR; **Karen Ireton** – Anglo American; **Alison Kelly** – WESSA; **Ruan Kruger** –Development Bank of Southern Africa; **Jeff Le Roux** – Enviroserv; **Dr Heather Malan** – University of Cape Town; **Lester Malgas** – SouthSouthNorth; **Prof P. Marjanovic** – University of Witwatersrand; **Karen Marx** – Birdlife South Africa; **Dr Ian McCallum** – Author and Psychologist; **David McDonald** – BOTSOC; **Executive Mayor Nomaindia Mfeketo** – Mayor, City of Cape Town; **Guy Midgley** – National Botanical Institute; **Mohamed Vali Moosa** – DEA&T; **Glyn Morris** – Agama Energy; **Mokhethi Moshoeshoe** – African Institute for Corporate Citizenship; **Andrew Muir** – Wilderness Foundation; **Credo Mutwa** – African Sangoma; **Deon Nel** – WWF(SA); **Professor Johan Nel** – North-West University (Potchefstroom Campus); **David Newton** – TRAFFIC East/Southern Africa; **Nicky Newton-King** – JSE Securities Exchange SA; **Piet Odendaal** – NACA; **Mark Obree** – City of Cape Town; **Dr Herman Oelsner** – Oelsner Group; **Tony Phillips** – Barloworld; **Professor Gordon Pirie** – University of Western Cape; **John Richards** – SAFM; **Trevor Sandwith** – C.A.P.E; **Chamilla Sanua** – Weleda Pharmacy; **Lynne Shannon** – Department of Marine & Coastal Management (DEA&T); **Robert Small** – Abalimi Bezekhaya; **Dr Merle Sowman** – University of Cape Town; **Jonathan Spencer Jones** – ESI Africa Magazine; **Richard Starke** – Global Ocean; **Willem Steenkamp** – Frandevco; **Thys Strydom** – Allganix Holdings; **Ken Stucke** – ERA Architects; **Ronel Suthers** – African Gabions ; **Robert Swan** (O.B.E) – Inspire!; **P.F. Theron** – Institute of Waste Management SA; **Dr Jennifer Thomson** – University of Cape Town; **Lt Col Etienne F van Blerk** – Department of Defence; **Prof Rudy van der Elst** – Oceanographic Research Institute, Durban; **Danie Van der Walt** – 50/50 SABC2; **L.D. van Essen** – University of Pretoria; **Prof Willem van Riet** – Peace Parks Foundation; **Braam van Wyk** – University of Pretoria; **Llewellyn van Wyk** – CSIR; **Professor Gerard H Verdoorn** – Endangered Wildlife Trust; **Wayne Visser** – KPMG and SA New Economics Network; **Terry Winstanley** – **Winstanley, Smith & Cullinan Inc**; **Rachel Wynberg** – Biowatch SA.

EDITOR AND PROJECT DIRECTOR: David Parry-Davies

After 20 years international management experience in the corporate banking, finance and insurance sector, David turned his attention to pursue his lifelong interests in nature and the environment. After working with various environmental organisations including Wildlife Aid and Friends of the Earth in the UK, David established Eco-Logic Development (Pty) Ltd in London – to work with corporate and local government institutions in implementing Agenda 21 initiatives. David published several magazines in the UK – designed to motivate public participation and commitment to ecological sustainability. After returning to South Africa, David conceived and designed The Enviropaedia as a practical tool to facilitate environmental education and motivate business and public participation.

This publication, he dedicates to his children Richard and Laura Parry-Davies – may they and their generation value wisdom before knowledge and beauty above possession.

ENVIRONMENTAL NETWORKING MANAGER: Penny Sparrowhawk

With 15 years experience in the corporate marketing and IT sectors, Penny has brought a wealth of knowledge and expertise to The Enviropaedia. This in turn fulfills her interests in nature and the environment and continues her lifelong involvement in social responsibility projects. Her humour and friendly support for NGOs and businesses alike in the process of building "The Network" has been invaluable.

PROOF READER: Peter Lague

After many years in the teaching profession and the educational publishing industry, Peter now runs his own publishing project management business. His primary concern has always been with social justice, a crucial underpinning of which is ensuring sustainable and equitable development, with a focus on environmental integrity.

COVER PAINTING by: Joe Joubert

The cover painting has evolved as an extension of the original Enviropaedia logo and reflects the collaborative efforts of and between the editor's desire to stimulate a 're-view' of our place and role within cosmology and the masterful skills of one of South Africa's most innovative and deep-thinking new artists, **Johan Joubert.**

As a key to some of the symbolism in the cover painting, consider the following:

◆ We, as human beings, are made of the same basic, organic "stuff" as a plant, a mouse or an elephant.
  There is not much difference between us.
  We all come from the same source and are all the "fruit" of this wonderful Earth.

◆ All species on Earth share the need for a healthy "environment" in order to survive.
  However one significant difference between us and other species is that, as a result of our thoughts and the consequent decisions we make, we humans effectively "hold the future of the entire planet within our minds."

◆ By our thoughts and our actions today we can literally create a "desert" of the future – or we can restore and regenerate the wonderful abundance and diversity of life on this planet, through our considerate and intelligent decisions and actions. Given just a little nourishment, nature, and the human spirit, can overcome seemingly insurmountable obstacles – we can grow, flourish and achieve the seemingly impossible.

"Your every thought, decision and action has an impact on the Earth and makes a significant difference.
So consider, what are you going to do? Your every footstep counts!"

See http://joejoubert.crweb.net/ or www.joejoubert.com for other examples of his exciting and thought-provoking work.

PHOTOGRAPHS: We would like to acknowledge and thank the many individuals and organisations who have contributed photographs to The Enviropaedia, in particular; Earthyear Magazine; Bruce Sutherland of City of Cape Town; Guy Stubbs – Photographer; WESSA – Africa Wildlife Magazine; and Dr John Ledger.

PAPER: We have used "Educator" RECYCLED paper supplied by Sappi Fine Papers, to support the principle of recycling. We wish to illustrate that the quality of today's recycled paper has improved greatly and is appropriate for use in many publications that currently use virgin paper. We also wish to highlight that the more recycled paper is used – the greater the cost efficiencies will be achieved, thus lowering the cost of production of recycled paper.

PUBLISHED by: ECO-LOGIC PUBLISHING CC – CK 1999/017347/23

PO Box 425, Simonstown, 7995, W Cape, Tel: (021) 786 4311, Email: enviropaedia@iafrica.com, Web: www.enviropaedia.com

REPRODUCTION by: CASTLE GRAPHICS PTY (LTD)   **PRINTED by:** LOGO PRINT

THE ENVIROPAEDIA - AT WORK

THE ENVIROPAEDIA IS COMMITTED TO SUPPORT:

THE ENVIRONMENTAL COMMUNITY • POVERTY ALLEVIATION • ENVIRONMENTAL EDUCATION

The Publishers of The Enviropaedia would like this publication to generate a direct and tangible benefit to organisations that have worked • To protect our natural environment • To assist people in uplifting themselves • To assist in creating a greater awareness and understanding of the principles of "Sustainable Development." We have therefore allocated **R58 000** from the sale of both the first and second editions of The Enviropaedia to selected organisations that we believe have an active "hands-on" approach and are making a successful and measurable contribution toward "sustainability."

ECOLINK

Ecolink is a training and development NGO, founded in 1985 by Dr Sue Hart, Executive Director. Ecolink's mission is, "To enhance the quality of life for people in their own environment in a sensitive manner and create an awareness of how they interact with their natural resources." EcoLink is committed to offering new oportunities for sustainable self-development through environmental education and skills training. **To date, The Enviropaedia has allocated R16 000 from book sales to Eco-Link.**

CHEETAH OUTREACH

Cheetah Outreach campaigns to bring the plight of the free-ranging cheetah into full focus within the communities of South Africa through financial partnerships with the Cheetah Conservation Fund and the National Wild Cheetah Management Programme, to support in-situ conservation environmental education programmes and workshops, and training of cheetah ambassadors.

To date, The Enviropaedia has allocated R16 000 from book sales to Cheetah Outreach and has donated books to the value of R29 250, which have been distributed by Cheetah Outreach to underprivileged schools. The funds have enabled them to visit many more underprivileged schools and the copies have assisted in developing teacher skills in planning and creating Environmental Education lesson plans. The cash donation also paid to send one teacher to the United States on a two month Fellowship in Environmental Education.

Photo: Annie Beckhelling, Shadow and David Parry-Davies

FOOD AND TREES FOR AFRICA

Food and Trees for Africa, established by Jeunesse Park in 1990, is a national greening NGO facilitating a healthy and sustainable quality of life for disadvantaged communities through the promotion of tree planting, permaculture, urban greening and environmental awareness. **To date, The Enviropaedia has allocated R16 000 from book sales to Food and Trees for Africa.**

Photo: Penny Sparrowhawk & Nan Rice

DOLPHIN ACTION AND PROTECTION GROUP

Founded by Nan Rice in 1977, with the express aim of campaigning for the protection and conservation of Dolphins and Whales, DAPG has since then broadened its role and activities and has run many successful National educational and fundraising campaigns. **To date, The Enviropaedia has allocated R10 000 from book sales to DAPG.**

+ OVER R456 000 WORTH OF COPIES OF THE ENVIROPAEDIA HAVE BEEN DISTRIBUTED FREE OF CHARGE to underprivileged schools and communities for environmental education purposes and also to a variety of environmental NGOs, to raise funds for their work.

Photo: Mr Windsor Shuenyane (of SAB) – distributing some of the books that were sponsored by SA Breweries – to many underprivileged schools in the Western Cape.

PAGE REFERENCE GUIDE

Photo: WESSA - Africa Wildlife Magazine

ACTIVISM

The Concise Oxford Dictionary defines activism as "a policy of vigorous action in a cause, especially in politics." The opposite of active is passive or inactive. But as far as environmental protection is concerned, just what does activism entail? As in the above definition it means standing up to be counted. It means that instead of resigning ourselves to the idea that nothing can be done about halting the precipitous decline of planetary ecosystems, intelligent and concerned people can turn the situation around by active participation or intervention in defence of our global resources. What forms can activism take? At one end of the scale, there are the confrontational types of organisations such as Greenpeace and Earth First! whose members put their lives and money on the line to raise public awareness of important environmental issues in often spectacular and sometimes illegal actions. On the other end of the scale, there are increasing numbers of ordinary people who are willing to participate in recycling and reducing their footprint on the earth in many simple and non-confrontational ways. In between these two extremes, are many shades of activism that allow for everyone to participate actively, at their own level and in their own way. However, as the dominant species, it becomes ever more apparent that humans must recognise our responsibility and become far more active in halting and undoing the damage we have inflicted on this fragile planet. It is now essential that we individually participate and embrace a degree of activism in whatever forms we are capable and comfortable with, in order to ensure a better legacy for our children and for all of the organisms with which we share this planet.

Guest Author: Glenn Ashton ~ Ekogaia Foundation

Key Associated Topics:

Consumerism; Development of Land; Public Participation; Stewardship; Green Consumer Guide; Eco-Logical Lifestyle Guide.

Associated Organisations:

Eco-Access; Ekogaia Foundation; SA Climate Action Network (SACAN); Southern Health & Ecology Institute.

AEROSOL

The word "aerosol" describes the dispersion, through the air, of minute solid or liquid particles that are so light that they fall to the earth very slowly. The most common natural aerosols in the atmosphere are haze and cloud. However, in everyday use, "aerosol" describes a pressurised spray can containing a propellant. When forced from a spray can, aerosols have more movement horizontally and can be directed easily at someone or something. At one stage, the most common pressurised propellants in aerosol cans were CFC (chlorofluorocarbon) gases because of their chemical "unreactiveness" to the product they sprayed. However, due to environmental problems with CFCs, no aerosol spray cans have used CFCs as a propellant (with the exception of a small number of specialist medical uses such as asthma pumps) in South Africa since 1992. Aerosol sprays can also be created using pump action bottles, which can be refilled. These do not need gas propellants such as butane (which is not an ozone-depleting gas but is a "greenhouse gas"). The most common example of these is ladies perfume spray bottles. Pressurised aerosol cans can, and should, be recycled.

Key Associated Topics:

Air Quality; Atmospheric Pollution; Global Warming; Greenhouse Gases; Montreal Protocol; Ozone.

Associated Organisations:

Aerosol Manufacturers' Association.

AESTHETICS

Aesthetics is a discipline which describes how attractive, beautiful or pleasing to the senses (sight, smell, touch, hearing and taste) a particular scene, building, painting, or image may be to the observer. As a discipline, aesthetics has become more formalised in recent years with the development of visual assessments and analyses as a part of landscape, topography and planning study. The work of the landscape architect includes consideration of aesthetics from a design, function and practicality point of view. There is no "right" or "wrong" in aesthetics because every person has a different opinion or preference. Aesthetics practitioners, such as landscape architects, will study characteristics, preferences and patterns and try to achieve consensus amongst users and interested and affected parties. In projects and developments, careful attention to colour, shape, form and texture of buildings can have a significant effect on the acceptance of such structures by neighbours and other interested and affected parties. Aesthetics and landscape design are used to enhance structures in game reserves to ensure that they blend in with their surroundings.

Key Associated Topics:

Environmental Management Planning; Integrated Environmental Management; Landscape Architecture: Sustainable Development Section – The Built Environment.

Associated Organisations:

ERA Architects; Greenhaus Architects; Institute of Landscape Architects.

AFFORESTATION

Afforestation refers to the establishment of commercial tree crops by purposeful planting on land previously not used for tree crops, e.g. the establishment of monocultures of pines, eucalypts or wattles in areas of primary grassland in South Africa. The resultant tree stand usually consists of a single species (monoculture), referred to as a plantation. In contrast, reforestation is the replanting of trees on previously forested land. In practice, however, the terms afforestation and reforestation are often used interchangeably, resulting in considerable confusion as to the precise nature of a tree-planting activity. This is particularly the case in areas where natural forest is being harvested commercially, as in many temperate parts of the northern hemisphere, followed by selective replanting with one or a few (but not all) of the species originally present in the forest.

The impoverished tree stand thus established through so-called reforestation approaches a plantation rather than a true natural forest. Reforestation also refers to the establishment of a new commercial tree crop on land from which a planted tree crop had been harvested, regardless of the type of natural vegetation that originally existed in the area. Because of the widespread misuse of these terms to hide environmentally de-structive activities, it is essential to establish whether the particular tree planting activity is for commercial purposes (establishment of a plantation) or the restoration of a natural forest for non-commercial purposes following damage or degradation (rather rarely attempted because of its non-commercial nature). Deforestation, on the other hand, is the destructive process of clearing a natural forest.

Commercial afforestation rather than deforestation is a cause of major environmental degradation and social problems in many parts of the world. Effects of large-scale monoculture tree plantations, especially on grassland biodiversity can be disastrous. Negative impacts are aggravated by the uncontrolled spread of alien trees beyond the plantation, by bush encroachment due to inappropriate management of grassland enclaves within plantation areas and by changes in the local hydrology. Although tree plantations are often promoted as environmentally friendly under the pretence that they are "forests", in many parts of the world these plantings are seen as green wastelands from a biodiversity point of view. A clear distinction should therefore be made between the concepts "forest" and "plantation". Forests are complex, self-regenerating natural ecosystems rich in biodiversity; plantations are artificial plantings of a tree crop.

South Africa already has more than 1,5 million hectares of alien tree plantations, mostly composed of eucalypts, pines and wattles. Timber companies are, however, increasingly acknowledging the negative effects of afforestation and are attempting to lessen the negative impact by setting aside land (still negligible it must be said) for conservation purposes.

Guest Author: A.E. van Wyk ~ Department of Botany, University of Pretoria

Key Associated Topics:

Agroforestry; Forestry; Tropical Rain Forest.

Associated Organisations:

Development Bank of SA; Sappi.

Photo: DWAF (Earthyear Magazine)

AFRICAN ENVIRONMENTAL TRADITION

Words of wisdom from Credo Mutwa
~ African Sangoma

Photo: Ingrid Penderis

The ailing future of living things

Respected friends, when most people talk about the extinction of wildlife as well as plant life on this planet of ours, they tend to make the mistake of talking about something, which happened in the remote past. These people miss the important and horrifying fact that extinction has now become a yearly, monthly and even daily process, which is going on even as I speak. The extinction of animals, the vanishing of valuable plants in our time, is accompanied by other things of an extremely disturbing nature. These ongoing extinctions are accompanied by worrying phenomena such as drastic and radical changes in the weather patterns of our country. They are also accompanied by attitudes within the minds of many people, that bode ill for the future of living things upon this Earth.

There was once a time in Africa

There was once a time in Africa, and in the lands of people such as the Celts of Europe, when the protection of living things was part and parcel of the religion of the people.

In the land of the Xhosa people there is a sacred crane, a graceful long-legged bird, whose feathers were only worn by warriors who had proven their bravery in battle and their loyalty to the tribe many times. This was the blue crane, the "Indwa", a bird symbolic of selfless courage and loyalty. A bird whose feathers were regarded by the Xhosa people and the Zulu people in the same way that British soldiers regard the Victoria Cross. If you injured this bird, if you broke one of its legs, your own leg was broken with a heavy stone; and if you were a wealthy man, ten head of cattle were taken from you afterwards as a fine. And the injured bird was finished off and its remains were cremated inside one of your huts, which was afterwards burned to the ground.

Amongst the trees that enjoyed protection was a large acacia, which was especially protected and revered by the Batswana people. This tree is known as the "musu", which means "dead person tree", and a branch of this tree was only cut off from the tree when the tribal chieftain had died, and that branch was used in the sacred fire which was lighted for two nights next to the home of the dead man to light his soul back to the after world and then to light it again so that it would quickly return through reincarnation to the home of the dead one's family. Even today you will find such trees standing in Batswana territory long, long after other trees of other kinds have been burned as fuel wood. And if you look closely at these trees, which the Batswana people protect so fiercely, you'll find that they are the kind of trees upon whose branches migratory birds use to rest on during their long flights from we knew not where towards we knew not which destination.

Today it is quite common to see children urinating and defecating into rivers and streams in South Africa. In the olden days those children were often punished in a terrible way. They were buried up to their necks in river sand and then left there. If the child survived the horrible punishment, it was regarded as being a sign that the gods had forgiven him or her, but should he or she die, it was believed that the gods had taken the child's soul as a sacrifice.

Africans were not the only people who protected animals and plant life with horrific laws such as these. The ancient Celtic people in Ireland, Scotland, Brittany and Wales had laws protecting nature just as draconian as those that you found in Africa. Anyone caught cutting down an Oak tree for pleasure in ancient Ireland, suffered exactly the same punishment that was inflicted upon African offenders. Their stomachs were cut open and their entrails tied around the tree in order to appease the fearsome goddess "Danu".

Religion and nature; are we master or servant?

The ancient Greeks also revered nature and had laws against its destruction and debasement. But, as the winged centuries went by, there rose religions in the Middle East and later in Europe, which viewed nature as an enemy, as something outside man's orbit, something to be mastered, conquered and exploited. The Christian Bible says that man was given mastership over the Earth and everything living that there was upon it, be it plant or animal. But, we in Africa were not taught that kind of thing. We were not told that man was the master of creation, we were not told that man could do whatever he wanted to the animals, the fish and the foul and the trees of the Earth, on which he found himself. No, we were taught that man was the caretaker of all living things, and far from being superior to the animals and to the birds and the fish, far from being the master of trees and grasses, man was the servant, and a very weak servant at that, of life on earth.

It was no accident that when the white man came to South Africa, he found the land teeming with animals of all kinds. He found millions of springbok and wildebeest, of zebras and elephants, swarming upon the face of the African veld. Africans protected animals. Africans regarded the existence of animals on this earth as ensuring the continuing fertility of this earth. Even those great destroyers of African crops in olden days, locusts, were viewed as a necessary part of existence with the human or the animals on the earth. I remember when white farmers cursed and wept at what the locusts were doing to their crops. I remember my grandfather saying to me in a beehive hut in Natal: "Son of my daughter, these grasshoppers are known by the Zulu people as the "Izinkunbi", the fertilizers of the land. The white men should not kill these grasshoppers; rather they should allow them to

Printed on Sappi recycled paper

come and go, because when the grasshoppers leave after eating almost all our crops, they will leave the land more fertile than ever before, because the grasshoppers, the "Izinkunbi", defecate and their droppings are in every nook and cranny of the country and their dung will make the green maize plants of next year carry heavy cobs of maize". My grandfather's words proved true. Those parts of the country over which the locusts had swarmed in 1937 enjoyed good rain as well as good harvests.

Sinking of medicinal plants into the hole of extinction

I first became a sangoma in 1937 and in those days there were healing plants, which were plentiful all over Natal and other parts of South Africa. I could list well over 50 different kinds of plants, which we used in the healing of sickness amongst our people as traditional healers. But many of these plants have vanished, never to return, because extinction has now become an accelerating and ongoing process. And as the extinction of valuable plants goes on in South Africa, our people suffer horribly. As a direct result of this, as traditional, safe, herbal medicines vanish, our people are growing more and more to seek refuge in highly dangerous chemical substances in their battle against sickness. I have tried in vain to fight against this thing, against these alien, un-African chemicals and my fight has repeatedly failed.

One of the things that brings about the extinction of medicinal plants in South Africa is the fact that since the late 1920s, there has existed in South Africa what are called "muti-shops", shops run by Indian, white as well as African business men. These shops literally sell tons of herbal plants and bark of trees all over the main cities of our country. It is this huge "muti" industry and not the traditional healers, which is responsible for the depletion of many precious herbal medicines. Anyone who blames traditional healers for this extinction should ask himself: "Is it traditional healers who are exporting tons of South African herbal plants and other natural resources to nations in the far east in container ships?" Because this is exactly what is happening! Tons of African plants are being ripped out of the land and exported to India, to Hong Kong and other places far away and no traditional healer is involved in this criminal industry, but enterprising business men, to whom money is everything, business men who do not care about the extinction of our country's precious plant life. The people who export African herbs to the Far East are the same people who are exporting rhino horn to Hong Kong. Unscrupulous, merciless people who blind themselves to their responsibility as human beings, in the glare of gold and greed.

I strongly believe that the continued extinction of animal as well as plant species in our country is behind the deterioration in health amongst all human beings that we see in Africa today. Because all the things are interlinked and the destruction of one is the destruction of all. I sometimes suspect that the terrible disease AIDS resulted, amongst other things, from the continuing extinction of living things in our country. An extinction, which has thrown the entire girdle of nature out of balance.

Preventing Gondwana's demise

The prevention of Gondwanaland's demise must not be made into something that belongs only to scientists; it must be the national duty of every one of us, no matter how high in the ladder of modern life or how low in that ladder one happens to be standing. The preservation of our planet is our duty, and no politician or whoever should stand in our way in the performance of this important duty. One thing that amazes me is that there are many white men and women amongst the ranks of South Africa's animal and plant conservation crusaders who wrongly believe that only they and they alone should fight for the conservation of the green life of our country. These people are wrong, arrogantly, stupidly wrong! If conservation is to have any meaning whatsoever, it should be restored back to the hearts and the minds of grassroots level. The white man must not think that only he can save this country. He is failing. We want to fight for our country, we want to ensure that a hundred years from now this country will at least still have some of the sacred animals and trees that we used to see in the past. Africans must be encouraged to join this crusade in large numbers. Africa must be encouraged to once more become her own protector as she was, before colonialist guns, canons and muskets enslaved her. We revered nature, at a time when Victorian era Englishmen, Frenchmen and Germans regarded nature as something to be raped, to be enslaved, to be tamed and destroyed. I once read a poem written by an Englishman of those times, a poem, which stated and I quote: "interrogate nature with power". As if Mother Nature was a captive, chained hand and foot in some dark and wet dungeon where she would be tortured with electrical power for the benefit of some vampire lord in some nameless European country. Interrogate nature with power indeed! Look where it has got us, gentlemen and ladies! Our people did not see humanity as beings apart from the rest of nature, our people did not see humankind as a race of "uber menchen", super men over other living things; our people were fully aware of the slavish dependence that man had on nature. They knew that if nature is destroyed, man will die.

And here ends my story

Many years ago, so goes a story, there was a very cruel chieftain, in the land of the Batswana. A mad man who used to kill animals for pleasure. One day his people got tired of what this tyrant was doing. So they seized him, bound him and placed him upon a sled and they pulled him over many days to a far away desert plain, where there was no water, no grass, no animals, save snakes and lizards. And out there in the heart of the desert, the angry people set their chief free. They gave him no weapon, no tool, no vessel, and left him with these words: "Great king, we leave you now to rule over the type of country that you appear to prefer." It is said by the storytellers that the man only survived a few days before he perished and was found as a heap of bleached bones at the very foot of a tall sand dune. He who had wanted to turn the land into a desert, died in the heart of a desert. That was the Africa that was our mother. Those were the laws that were the religion, and here ends my story.

This Article consists of extracts from an interview with Credo Mutwa by the Gondwana Alive Society. The full interview can be read in their publication Towards Gondwana Alive Vol. 2 – A set of 100 strategies towards stemming the Sixth Extinction. This 330 page, full colour book is available from: Dr. John M Anderson, Managing Editor, "Gondwana Alive", National Botanical Institute, Pretoria, Private Bag X101, Pretoria, 0001, South Africa.

DE BEERS

A DIAMOND IS FOREVER

From carats to conservation and communities

Over the past few years, De Beers has undergone significant transformation to meet the global challenges and opportunities of business. An essential part of this is our focus on sustainable development, which evolves dynamically to take into account the changing needs of all our stakeholders. De Beers is committed to the principles of sustainability in terms of ethical accountability and investing appropriately in all capital stocks - natural, human, social and manufactured and financial – to achieve the flow of benefits for the company and beyond. Particularly in the context of South Africa, De Beers is committed to realise the objectives of the new mining legislation and, in the African context, the success of the New Partnership for Africa's Development (NEPAD).

De Beers has wide coverage of ISO 14001 certified environmental management systems at its operations in both the marine and terrestrial environments. Aside from on-site management and initiatives, De Beers also invests heavily in technology research and development to enhance eco-efficiency. As an offset to its mining footprint resulting from the various operations, De Beers has set aside reserves, several times the size of the mining footprint, for conservation of biodiversity. One such reserve has been incorporated as a core area of the Limpopo / Shashe Transfrontier Conservation Area. In recognition of this contribution and of De Beers collaborative relationship with South African National Parks and the Peace Parks Foundation, De Beers received an award at the World Summit for Sustainable Development in 2002.

In the past few years, the diamond industry has worked with governments and NGOs to overcome a range of difficult and challenging issues. With diamonds being a symbol of enduring values, 'conflict diamonds' brought together two unacceptably opposed concepts. As the leader in the global diamond industry, De Beers has been a major player in the formulation and introduction of the Kimberley Process Certification Scheme for rough diamond exports to end trade in 'conflict diamonds'.

A key initiative currently underway through De Beers' Supplier of Choice strategy is the development of the Diamond Best Practice Principles. This will provide a management framework and assurance programme for the De Beers Group and its associated business partners to a series of ethical standards, particularly related to human rights.

De Beers is the first mining company to have provided free access to HIV/AIDS treatment for employees and a spouse or life partner. A number of the company's operations have been awarded the highest ratings in terms of safety and integrated safety, health and environmental systems by the National Occupational Safety Association (NOSA).

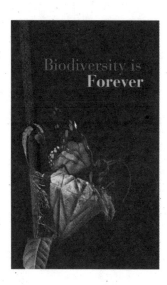

Biodiversity is **Forever**

De Beers is committed to upholding fundamental human rights and respecting cultures, customs and values of indigenous peoples especially in protected areas. Among the issues of concern for De Beers is the future of the San of the Central Kalahari Game Reserve in Botswana and it is vital that a collaborative solution is found to this. In Canada, where we are looking at bringing new mines on line, the surrounding communities are involved in planning aspects to address possible concerns and to build long term relationships through the opportunities presented for development.

Through its various corporate social investment programmes, De Beers is involved in supporting people with purpose and addressing community needs in education, welfare, health, HIV/Aids, skills development and community upliftment. The essential objective with each of these programmes is in giving a hand and not in giving a handout.

De Beers is fully committed to transformation and black economic empowerment. In order to put these into practice, the company has implemented several initiatives including employment equity, empowerment contracting and outsourcing and the development of ownership through joint ventures and historically disadvantaged businesses.

Throughout Africa and elsewhere in the world, the way De Beers operates is to bring lasting benefits to communities and countries as well as to the diamond industry. De Beers' commitment to sustainable development was ingrained more than 50 years ago by Sir Ernest Oppenheimer, the grandfather of Nicky Oppenheimer, our current Chairman, when he said the purpose of the company was to make profits for its shareholders, but in a way that would make a real and lasting contribution to the countries and the communities in which it operates.

Gary M Ralfe
Managing Director

For more information, log on to
www.debeersgroup.com

AGENDA 21

"Agenda 21", so named because of its position in the meeting agenda, is a global plan of action for sustainable development agreed to by most United Nations member nations at the United Nations Conference on Environment and Development (also called the Earth Summit or UNCED) held in Rio de Janeiro, Brazil, in 1992. The member nations agreed to aim for more balanced development, in order to minimises any negative environmental impact on the earth. The Agenda 21 document contains some forty separate sections of concerns and outlines a total of over 2 500 recommendations. Agenda 21 focuses on partnerships involving the public and all relevant stakeholders to resolve developmental problems and to plan strategically for the future. It also tries to address the practicalities of applying sustainable development principles in human activity and development. The global Commission on Sustainable Development (CSD) was created in December 1992 to ensure effective follow-up of UNCED, and to monitor and report on the implementation of the agreements at local, national, regional and international levels. The South African "custodian" for Agenda 21 is the Department of Environmental Affairs and Tourism (DEAT). Chapter 28 of Agenda 21 sets out targets and an approach that local authorities in each country should undertake a consultative process with their populations and achieve a consensus on a local Agenda 21 for the community. At the nineteenth special session of the UN General Assembly (UNGASS), popularly known as Earth Summit + 5, held in June 1997, Thabo Mbeki (the then vice-president of South Africa) stated that Agenda 21 remained the fundamental programme of action for achieving sustainable development for South Africa. The 55th General Assembly session decided in December 2000 that the CSD would serve as the central organising body for the 2002 World Summit on Sustainable Development (WSSD), which was held in Johannesburg, South Africa in August 2002.

Key Associated Topics:

Agenda 21 is a central tenet of Sustainable Development – and relates to practically every other Topic within *The Enviropaedia*, but in particular: Local Agenda 21; Sustainable Development; World Summit on Sustainable Development.

Associated Organisations:

Cape Town Environmental Management & Solid Waste; Dept. Environmental Affairs & Tourism; Development Bank of SA.

AGRICULTURE

Agriculture is the formalised and structured means by which humankind grows food to feed the growing population. As more people move from rural areas to cities, it has become more necessary to use intensive farming methods to increase crop yields. Early agricultural systems (agro ecosystems) only needed to provide sufficient food to feed the farmer and his or her family. Historical "slash and burn" programmes would clear land, grow crops and as soon as yields dropped, the farmer would move on to another area. The farmer would return to the original plot but only after a number of decades, which allowed the fallow land to recover. Increasing populations of people and their animals caused more inten-sive land use. The fallow periods become shorter and so the need to develop more sophisticated systems arose. These included irrigation, fertilisers, pesticides and herbicides, and, more recently, genetically manipulated crops. Agriculture modifies the environment and needs to be undertaken with proper consideration for the ecosystem carrying capacity and wider environmental impacts. Agro ecosystems cover more than 25% of global land area but almost 75% of that area has poor soil fertility and approximately 50% has steep slopes, which limit production capacity. About 66% of agricultural land has been degraded in the past 50 years by erosion, salinisation (accumulation of salts in soil which limit plant growth), compaction, nutrient depletion, biological degradation or pollution. About 40% of land has been strongly or very strongly degraded. With this decline in availability and effectiveness of land to grow food, agro ecosystems will face the challenge over the next 20 years of having to feed an additional 1.7 billion people.

Key Associated Topics:

Agroforestry; Carrying Capacity; Game Farming; Genetically Modified Organisms (GMOs); Integrated Pest Control; Land Degradation; Organic Farming; Permaculture; Pesticide Treadmill; Pesticides; Soil; Trench Gardening; Urban Agriculture.

Associated Organisations:

Agricultural Research Council; Barlofco; Biowatch South Africa; Cedara College of Agriculture; Elsenburg College of Agriculture; Nursery Association of KwaZulu-Natal; Permacore; S A Freeze Alliance on Genetic Engineering; S A Sugar.

AGROFORESTRY

Agroforestry is the practise of combining agriculture and forestry technologies to optimise the use of land, to prevent land degradation and to produce more sustainable land-use systems. Typical agroforestry practises in use are Alley Cropping and Riparian Forest Buffers. In Alley Cropping, an agricultural crop is grown simultaneously with a long-term tree crop to provide an income while the tree crop matures. In North America fine hardwoods like walnut, oak and ash are favoured tree species, while a nut crop is a useful intermediate species. Riparian Forest Buffers are natural or re-established forests along riverbanks. These forests can be composed of trees, shrubs and grasses and they reduce riverbank erosion, protect aquatic environments and increase biodiversity. Riverbank buffers also serve to reduce the damage caused by flooding, by acting as brakes on fast flowing water when it overflows its banks, and also by making riverbanks more absorbent to the flood water. Agroforestry is typically practised by the more advanced farmers and land-use authorities, and considerable work still needs to be carried out to develop even more innovative techniques and applications.

Guest Author: Dr Kelvin Kemm

Associated Topics:

Agriculture; Eco-efficiency; Forestry; Soil; Permaculture.

Associated Organisations:

Ecoguard Distributors; Permacore; Sappi.

AIR QUALITY

Guest Essay by Pieter du Toit
~ National Association for Clean Air (NACA)

South Africa is an industrialised country and is very dependent on coal as a primary energy source. The burning of coal however causes significant air pollution. Electricity in South Africa is primarily generated by means of coal burning power stations. Coal is also burned and processed in our major industries and a large proportion of our residential sector uses coal for heating and cooking. People are therefore exposed to air pollution from many sources. The deteriorating air quality in turn will have a negative affect on the health and wellbeing of our population. This will also directly affect the environment as a whole.

An estimated 20-30 million South Africans regularly inhale smoke from burning coal or wood in poorly ventilated homes. This can cause acute and chronic lung diseases, including bronchitis and lung cancer. On the streets, one runs into air pollution from traffic emissions, from badly maintained vehicles, which emit carbon monoxide fumes, particulates, lead and various hydrocarbons. The vehicle density in our cities is increasing at an alarming rate, thereby aggravating urban pollution. Lead in car fumes contaminates the soil and body tissues. Benzene in our fuel

is a well known carcinogen. Petrol station attendants, motor mechanics and people living near busy roads face an increased risk of kidney, lung and brain damage from leaded petrol, and precautionary measures should therefore be taken. Government restrictions require that all lead be removed from fuels as from 2006.

At work one is exposed to many sources of air pollution. In the industrial and mining environment workers may be exposed to chemical and metallurgical fumes, dust and solvent emissions and radiation. Even in the office, workers may inhale ozone released from photocopiers, which can cause headaches and eye, nose and throat irritations, as well as tissue damage. The so-called 'sick building syndrome' can occur where windows are seldom opened and air conditioners fail to filter stale air effectively; thus circulating germs cause infections and poor health in general. Farm and municipal workers can breathe in noxious fumes when they spray crops or pavements to kill pests or weeds. In the building industry some paints, solvents and carpet glues release some of the most noxious gases known.

Our environment is influenced by air emissions at a local, regional and global scale.

Our large-scale emissions of acid gases such as sulphur and nitrogen oxides is a particularly worrying problem in southern Africa as this can cause acid rain and the resulting acid deposition then leads to acidification of our soil and water sources.

Whilst it is possible to minimise or eliminate emissions from many sources, such measures can be very costly and it is therefore important to prioritise expenditure in the most cost effective manner possible. Protective measures can also be taken in the workplace, however, many workers are uninformed or too scared to complain, or they lack the motivation to wear masks and safeguard their health.

The effects of air pollution on people and the environment are often indirect, such as the depletion of the ozone layer.

Air pollution and ozone

Synthetic chemicals used as coolants, such as chloro-fluoro-carbons (CFCs) react with ozone molecules high up in the atmosphere. Here, a layer of ozone protects the earth from some of the sun's more harmful rays. But it seems that the ozone layer is becoming thinner in places because of such chemical reactions. This results in an increase in solar radiation reaching the earth, which can cause skin cancer and eye cataracts, and damage plant life. In the sea below the Antarctic 'ozone hole' there has been a 12% reduction in the activity of phytoplankton, (the tiny plants which form the basis of the marine food web) as a direct result of the damaged ozone layer.

The effects of air pollution on human health are sometimes direct and acute. For example, the leak of noxious gases from a factory near Chatsworth in Durban caused ear, nose and throat irritations, which people could immediately link to the smell in the air. In Bhopal, India, a gas leak in a Union Carbide plant killed and maimed thousands of people overnight.

More often, however, the effects of air pollution are less obvious and more long-term, making it difficult to identify and prove who is responsible for causing the damage.

In Durban, for example, where the airport and heavy industries are close to residential areas, communities have been concerned for years about the impact on their health. Research in the United Kingdom and Australia has shown strong links between children who develop cancer and those who live in the vicinity of major roads, airports, or the producers, refiners and industrial users of petroleum products.

Poor monitoring of air pollution is just one factor hindering effective environmental governance in South Africa. The responsible government agency, the Department of Environmental Affairs and Tourism (DEA&T), has limited capacity and struggles to enforce even our existing laws. New air legislation delegates powers to provincial and local authorities but the same capacity problems are present at those levels. We are therefore reliant on industry to be self-disciplined and to take a responsible approach to air quality monitoring and control.

Some industries, particularly those companies that trade internationally, are keen to comply with international standards. This is the result of the corporate governance approach that drives them. However many smaller companies lack the international links to motivate or enable them to reduce their noxious emissions. The most cost effective approach to air quality monitoring and control in South Africa may lie in the pooling of resources between the government, NGOs and industry.

The latest State of the Environment Report says that government also needs the help of vigilant citizens who are prepared to take action, using their Constitutional right to an environment that is not detrimental to their health and well-being. (See the DEA&T's User Guide to the National Environmental Management Act, Guide 1. No.107 of 1998: Your Right to Take Action to Prevent Environmental Damage in Terms of NEMA.) However citizens cannot take on the responsibility alone. For example, whilst we can encourage one another to reduce vehicle emissions by using buses and trains, we also need an effective public transport system. This requires action from local, provincial and national government. Similarly, it takes a strong political will to enforce and improve vehicle emission laws.

Accepting pollution-producing industries into South Africa could consign this country to a box labelled 'third world'.

We hope that the new Air Quality Management Act, currently being considered by Parliament, will not allow this to happen.

A cost-benefit analysis must weigh up the advantages of jobs and investment against the disadvantages of pollution. This should take account of the resulting ill health of workers and their families (thus increasing national medical costs) and environmental damage, which threatens tourism potential and long-term development prospects.

Cleaner production

In the International marketplace, the ability to show that your product has been produced using "cleaner production" techniques will, according to Eskom, become a competitive advantage. Soon this may also be the case locally, as environmental awareness increases, resulting in a greater number of South Africans coming to see pollution as a human rights issue.

The onus rests on every citizen to play his or her part in reducing energy usage and air pollution in South Africa. Reduction of electricity wastage at home and at work will reduce air polluting emissions and will help preserve our natural resources. Setting firm and clear objectives and targets and effective collaboration between the major stakeholders (governmental departments, NGOs, civic societies and industrial organisations) can lead to more effective implementation of legislation and reduction of emissions, if we are all willing to accept our individual responsibility and take the appropriate ACTION!

Key Associated Topics:

Atmospheric Pollution; Carbon Dioxide; Carbon Monoxide; Catalytic Converter; Climate Change; Eco-taxes; Energy; Environmental Health; Fossil Fuels; Global Warming; Methane; Oxygen; Ozone; Particulates; Photochemical Smog; Pollution; Polluter Pays Principle.

Associated Organisations:

Ceatec; Groundwork; Holgate & Associates; Museum Park Enviro Centre; National Air Pollution Services; Richards Bay Clean Air Association; SI Analytics.

Photo: Chris van der Merwe

ALIEN INVASIVE PLANTS IN SOUTH AFRICA

Many plant species have a strong ability to grow in similar situations but away from their native habitats. The result is that many plants are now found in places where they did not originate. This process of global distribution of plants has been happening for millions of years. In the past 1 000 years – as people have dispersed across the globe – this has speeded up, escalating still further over the past 300 years as modern world travel has developed. Plants have been distributed as crop plants and ornamentals to foreign lands, often displacing the local flora with negative consequences. In many, if not most, cases invasive alien plants that have originated from horticulture are plants selected by gardeners for the same qualities that make them potentially invasive. Some of these characteristics are rapid growth, early maturity, large quantities of seeds that are easily dispersed, the ability to out-compete other plants and disease, and pest resistance. Like many other parts of the world, southern Africa has also been affected by the global distribution of plants. With its diverse natural environment, southern Africa provides habitats suitable for many species ranging in origin from the tropics to mediterranean type environments and deserts. In South Africa 198 plant species have been declared weeds and invaders. These plants are termed "invaders" because they spread and displace the indigenous plants. The question then is "Why are invasive alien plants such a problem?" Apart from displacing the natural flora and therefore impacting negatively on biodiversity they also use more water than the better-adapted natural flora. They also intensify wild fires should these occur. These negative impacts call for concerted action for the control of these invasive alien plants.

What is being done to combat the spread of invasive alien plants? The major group of offending plant species in southern Africa has been identified, and the *Working for Water* Programme is active throughout South Africa in clearing alien plant species. Legislation has now been enacted to combat the problem of invasive plant species that threaten the natural flora of the country and, in turn, valuable water resources. Some of the most widespread offending species in random order are *Acacia mearnsii* (Black Wattle), *Saligna* (Port Jackson Willow), *Cyclops* (Rooikrantz), *Melanoxylon* (Blackwood), *Lantana camara* (Lantana), *Chromolaena odorata* (Triffid weed), *Solanum mauritianum* (Bugweed), *Hakea sericea* (Silky hakea), *Pinus pinaster* (Cluster pine), and *Melia azedarach* (Syringa or Persian lilac). The problem of invasive plants is large, and it requires active public and private participation to combat this 'growing' threat. Agricultural landowners need to familiarise themselves with those species that pose a threat on their own land and eradicate them. The gardening public, in turn, should be aware of those invasive alien plant species that they may have on their suburban properties and remove them. Local hack-groups exist whose purpose is to clear invasive alien plants from public land for the benefit of local communities and their environment. This activity should be greatly encouraged to increase awareness of the increasing threat of invasive plant species and to ensure that action is taken against this threat at the local level.

Guest Author: David J. McDonald ~ Botanical Society of SA

Key Associated Topics:

Fire; Integrated Pest Control; Land Degradation; Water Quality & Availability.

Associated Organisations:

Bergvliet School; Botanical Society of SA; Conservation Support Services; Enact International; Global Invasive Species Programme; Hilland & Associates; Hout Bay & Llandudno Heritage Trust; Santam-Cape Argus Ukuvuka; Thorntree Conservancy; W Cape Nature Conservation Board; Working for Water.

ANIMALS AND ANIMAL WELFARE

Animal welfare covers a wide range of interests and concerns ranging from specific welfare and cruelty-related issues such as bear baiting, bull fighting, and pâté de foie gras (goose liver pate) to minks and fur coats, and cows and shoe leather and avian rehabilitation. Human beings have established close emotional and work ties with many animals over the centuries e.g. domesticated dogs, the horse, the mule and the falcon. Animals evoke a wide range of emotions in humans, especially when human beings inflict cruel treatment on animals. One of the principles raised is that animals do not have the intellectual and physical abilities to defend themselves against humans and therefore humans have a moral and spiritual obligation to defend and protect animals from harm or injury. Tremendous effort is put into helping whales or dolphins who are beached through pollution, confusion or other human-related interference because of the strong emotional feelings evoked from seeing dolphins or whales in their natural surroundings. Certain fundamentalists argue that nature is cruel and human beings are part of nature – therefore actions to minimise cruelty, death or loss are interfering with the natural order and balance of nature. It must be emphasised that that no one can argue against those who gain personal satisfaction from helping animals or defending animals against cruel and barbarous acts. Animal welfare groups form an important part of environmental management structures and systems, even if not everyone agrees with the wide range of philosophical views and attitudes that exist amongst the various interest groups.

Key Associated Topics:

Avian Rehabilitation; Conservation; Ecological Intelligence; Biodiversity; Birds; Stewardship; Wildlife.

Associated Organisations:

The Africat Foundation; Africa Geographic; Bio-Experience; Animal Anti-Cruelty League; Beauty Without Cruelty; Carthorse Protection Association; International Fund for Animal Welfare; People's Dispensary for Sick Animals; Society for Animals in Distress; South African Association Against Painful Experiments on Animals; SPCA; World of Birds Wildlife Sanctuary.

Photo: Earthyear Magazine

ANTARCTICA

Guest Essay by Robert Swan OBE ~ INSPIRE!

Actions speak louder than words

The paradox of the seemingly timeless, yet rapidly changing wilderness is what draws INSPIRE! as an organisation back into Antarctica's brutal embrace, time and time again.

Well over a decade ago, I was the first person in history stupid enough to walk to both the South and the North Poles. We later spearheaded one of the most unusual clean-ups in Antarctic history – Mission Antarctica – raising enough sponsorship over a decade to undertake the titanic task of removing over 1 000 tons of solid waste from the continent and successfully recycling it in Montevideo, Uruguay.

Extraordinary, life-threatening conditions on our North and South Pole expeditions, caused by the hole in the ozone layer and global warming – and even litter in these remote regions – convinced me that in the new millennium, the true heroes are the real people in industry, business, government and communities who don't sit around talking. They are the people who are getting together and actually going out there and doing something proactively, however seemingly small and insignificant. The world is changing before our eyes and we need to do something about it.

There are no simple answers to Antarctica and what is happening to it. But, while it may be located at the end of the earth, it is not as remote as many might imagine.

Almost 70% of the world's fresh water is locked in its increasingly fragile embrace. Antarctica is the Earth's ecological dynamo – the icy engine sustaining vital global systems – and the last great wilderness area left on the planet.

At the dawn of the new millennium it is clear that this seemingly remote part of our planet is under serious threat. Illegal fishing has the potential to undermine the entire Antarctic marine ecosystem, while scientific research, military bases and in particular, rapidly growing eco-tourism, pose potentially serious threats to the environment.

These challenges have a common root: they are governmental and commercial activities operating in a part of the world where the regulation of such activities is, to put it euphemistically, rather benevolent.

And like a delicate umbrella poised above all this, the Antarctic Treaty valiantly attempts to prescribe proper conduct to its signatories south of the 60 degrees South Latitude.

The review process of the Antarctic Treaty and the 1991 Environmental Protocol that resulted in a 50-year ban on all mineral exploitation, begins in 2041, which is the name we gave our Antarctic yacht, '2041'.

This last great wilderness on earth will remain free from commercial exploitation as long as the Antarctic Treaty and the Environmental Protocol are in place and as long the signatories continue to respect it and abide by its provisions. INSPIRE! is collecting signatures from Environment Ministers and Heads of State around the world, endorsing the terms of the Treaty and Protocol. After all, Antarctica is the axis of the turning world. It is the only continent nobody owns. It has never known war. And it is dedicated to science, exploration and global harmony. We draw our Inspiration from Antarctica. It provides a potential model for our global future

Every year since the Mission Antarctica clean-up, INSPIRE! returns to Antarctica to promote personal leadership and teamwork through positive participation on real missions in which we tackle real environmental challenges.

Whether it is dismantling fuel tanks that are relics of the former Soviet Fleet's Cold War activities, or establishing an inspiration station driven by wind and solar energy or to inspire young people to use renewable energy resources, the INSPIRE! Antarctic Mission aims to ensure that our children, and our children's children, will make a responsible and informed choice in the year 2041, for an environmentally sound and sustainable future in Antarctica.

On our expeditions to Antarctica, drawing teams of people from across the globe, INSPIRE! takes small, positive steps to make a difference and provides an inspiring and responsible example that people can use, to promote a sustainable future in their own lives, in their workplaces, in their communities, in their environments, in their nations.

You can do it too. You can join us on a real mission, on one of our Antarctic expeditions.

Or, in your own environment, you can do the same thing. Through personal leadership, teamwork and positive participation, you can tackle a real environmental mission in your area, thus finding your own unique story and realising your own vision for a sustainable future.

Key Associated Topics:

Bioprospecting; Global Warming; Wilderness Area.

Associated Organisations:

Antarctic & Southern Ocean Coalition (ASOC); Inspire!

AQUACULTURE

Guest Essay by Richard Starke
~ Global Ocean (Pty) Ltd.

"A new way to feed the world"

A total of 70% of the world's conventional commercial fish species are now fully exploited, overexploited, depleted or recovering from depletion (Halvorsen, 1999).

Aquaculture is the production of aquatic organisms in fresh, brackish and salt water. Mariculture is the production of brackish water and marine species. Aquaculture has been the world's fastest growing food production system over the past decade rising from 3.6m tonnes in 1970 to 45m tonnes in 2000 (from 5.3% to 32.2% of total production).

When South Africa's Minister of Environment recently reduced the number of licenses issued for line fishing, we witnessed the huge impact that dwindling fish stocks can have on coastal communities. The reduction of the fishing effort was in response to scientific advice, which suggested that the stocks of several important linefish species have collapsed. The furore that accompanied the reduced allocation of line fish permits provided further proof of the urgent need to kick start aquaculture development in South Africa.

Photo: Earthyear Magazine

Still there is apathy and a general lack of response to these signals, particularly from Government. We see fishing communities out of work, with no alternatives provided by either Government or the corporations that once thrived in their towns. The large fishing companies are slow to react and switch to aquaculture. We see budgets for applied aquaculture research and Government assistance at paltry levels. And we see futile competition and a lack of co-operation in fledgling aquaculture industries.

Aquaculture in South Africa today is dogged by a lack of foresight and coherence.

Five points to consider

1. Community – South Africa is a developing nation and a significant emphasis should be placed on the provision of jobs through empowering coastal and inland communities to thrive in the aquaculture industry;

2. Diversity – we should not focus on a single saviour species. Risk should be alleviated through research and development, and the co-operative development of new culture species;

3. Environment – aquaculture should follow sustainable utilisation criteria and not be at the expense of our beautiful country;

4. High-tech – the days of low-tech aquaculture are over. Technology drives efficiency and productivity, and in the end, profits;

5. Legislation should be designed to promote investment and encourage sustainable development in aquaculture.

Eco-challenge

Modern aquaculture must be a sustainable development in an industry that pleases the environmentalist, animal welfare concerns, economists and industrialists. It wasn't always environmentally friendly for the early intensive industry. Pollution from salmon and shrimp farming was particularly worrying.

The South African challenge is to grow fish on the shore in competitively priced systems that recycle most of the water needed to provide an 'Aquanomically' suitable environment for the species. Recycling more than 95% of the circulated water is essential and such processes are able to deal with the accumulated waste in Eco friendly ways. The bio-technology is available and up and running with an all-biological non-chemical input that meets the strictest EIA requirements.

This is achieved with small inflows of seawater that have little or no impact on the surrounding area. We now need industry wide research with cross-pollination between competing companies and departments to ensure that the breeding loop is closed for as many species as practical. This will open options for our ex-fisherman to grow fish that eat low value input food for maximum value output

In Europe research into the production of new candidate species includes Cod, Plaice and Haddock, which are currently major contributors to their capture fisheries. In France and Italy pilot farms exist producing meagre which is similar to the cob (kob, or kabeljou) species found off our coast (Argyrosomus sp – Mild meagre). Experimental work starting with the Two Oceans Aquarium and involving MCM, Global Ocean and I&J is already underway to growout the faster growing "Dusky Cob".

The production of High Value species for export makes economic sense but to feed our people lower value species must also be grown in the same recirculating systems.

We are standing on the threshold of an era of growth in aquaculture and we hope that our combined knowledge, wisdom and drive will fuel the fires of the aquaculture engine. Let the industry grow, provide jobs, generate profits and reduce the pressure on our oceans.

Key Associated Topics:

Fisheries; Marine Environment; Marine life of Southern Africa.

Associated Organisations:

Unilever Centre For Environmental Water Quality; Ecosense; Irvin & Johnson; Global Ocean; Rand Afrikaans University (RAU).

"CARING FOR THE ENVIRONMENT SHOULD BE AS NATURAL AS BREATHING"

At I&J, protecting the environment is no special task. It's as natural as making sure that our trawlers are operating, our suppliers are paid and our staff is content. It's integrated into our everyday planning and management systems. Protecting the environment is an integral part of our business.

Our activities impact on the sea, land and air. We acknowledge this. We've built into our work ethic a recognition of the fragility of the environment and a striving to protect it by whatever means possible.

AT SEA

I&J has one of the largest fishing fleets in the Southern Hemisphere operating in over 500 000 square miles of ocean. To prevent marine pollution each I&J trawler has a 'waste package area' for non-biodegradable waste which is returned to shore for proper disposal. Each of our trawler crew members is trained in marine resource management and, via our Environmental Award Scheme, is incentivised to return all non-biodegradable waste at the end of each voyage. We use virtually every gram of fish caught. I&J is the only deep sea trawling company that makes a practice of saving fresh fish waste to produce fishmeal either on board its vessels or in its shore-based plant. Together with international experts I&J has developed fish net technology that avoids the catching of undersized fish. We work very closely with Government in the research of the biology of trawl fish species and we are host to numerous scientific researchers on board our trawlers.

ON LAND

Fresh water is precious. Our largest fish processing plant, employing some 1300 people, has implemented a progressive water reduction programme that has reduced the use of fresh water by 40%. Water purity is also a priority, at I&J's abalone farm on the Cape south coast a locally designed filtration system is installed. This ensures that all the farm's water going back into the sea is pure and the area's environment remains pristine. I&J pioneered the culturing and farming of abalone and our farm is the largest in South Africa. Aware of the ever decreasing abalone stocks through poaching, I&J is funding a R&D experiment with the University of Cape Town to evaluate the potential for reseeding the South African coastline with abalone.

IN THE AIR

Air pollution, noise, human stress levels, lighting, waste disposal, stagnant water and dust are some of the factors that I&J has been monitoring for many years as part of the Company's Loss Control system.

IN THE MIND

I&J sponsors many projects aimed at creating public awareness of the fragility of the marine environment and the need to protect it. These continuing sponsorships range from 14 years of sponsorship of Government's National Marine Day to feeding the marine life at The Two Oceans Aquarium for 9 years. Children are special to I&J and we sponsor a number of lively educational materials aimed at primary school children.

CONSERVATION THROUGH INNOVATION

We believe that unless thoughtfully monitored and cared for, our country's finely balanced marine environment and resources will gradually diminish in quality and quantity.

The marine conservation policies of new entrants to South Africa's fishing industry must be carefully considered before access is granted. New, innovative ways to control fishing fleets must be researched and developed. For example, I&J was one of the first companies in the world to develop a vessel monitoring system which gives technical, medical and fishing advice to the entire I&J fleet 24-hours a day. We do not pretend to have all the answers. However, our deep concern and our active, ongoing commitment is, hopefully, more than just a drop in the ocean.

Prof. B Figaji
CHAIRPERSON: I&J BOARD

IRVIN & JOHNSON LIMITED

ASBESTOS

Asbestos is the fibrous form of several silicate minerals. Its physical characteristics of being a good insulator, fire resistant and chemically inert have made it invaluable to many areas of commerce and industry. It is used for building and construction purposes (roofing, fencing, wall panels) and as insulation lagging. However, asbestos proves a serious health threat. The danger comes from the fine fibres breaking loose and being inhaled into the lungs. Prolonged exposure to these forms of asbestos causes mesothelioma (a cancer of the lung and abdominal lining) or asbestosis (a chronic lung condition that eventually makes breathing almost impossible). The most dangerous forms of asbestos are crocidolite or blue asbestos and amosite or brown asbestos.

The dangers posed by using asbestos lagging have meant that most companies have begun programmes to replace the lagging with safer forms of insulation. In many instances, there are alternative materials, which are not more expensive and have the same properties as asbestos. Most companies using asbestos in building materials use chrysolite or white asbestos, which is encapsulated in the cement matrix and is safe, if managed with care. Asbestos is very dangerous if full and complete precautions are not taken to protect workers and the environment. It can take between 15 and 30 years after exposure for the asbestos fibres to cause mesothelioma or asbestosis. There are currently a number of court cases going on around the world, in which asbestos processing companies are being sued for damages by their ex-employees and surrounding communities for not protecting employees and communities from asbestos exposure.

Key Associated Topics:

Accumulation; Environmental Justice; Environmental Liability for Remediation; Environmental Risk and Liability; Hazardous Wastes; Pollution; Polluter Pays Principle.

Associated Organisations:

Dial Environmental Services; South African Institute of Ecologists and Environmental Scientists.

ATMOSPHERE

The atmosphere is the thin layer around the globe upon which all livings things on earth depend for their survival. As a comparison, the thickness of the atmosphere around the earth is the same as the thickness of the skin of an apple compared to its diameter. The atmosphere consists of two main layers: the troposphere, which extends from sea level to about 17 kilometres up (about 95% of the mass of all the gases and 90% of water vapour are in this layer) and the stratosphere, which extends from 17 kilometres up to about 48 kilometres upwards. The ozone layer is in the upper part of the strato-sphere. Above the stratosphere are two further layers called the ionosphere (which reflects radio waves between distant ground radio stations) and the exosphere (a zone of rarefied helium and hydrogen gases, which merges with "outer space"). A controlled environment is created by the atmos-phere, which enables living beings of all kinds to survive.

Photo: Mark Shuttleworth

The atmosphere acts as a shield against dangerous ultra-violet radiation and also forms an insulating layer, which raises the temperature to levels where life can comfortably exist. Without the insulating layer of the atmosphere, the temperature on earth would fall to –240°C at night.

Key Associated Topics:

Air Quality; Atmospheric Pollution; Ecosphere & Ecosystems; Global Warming & Climate Change – The World Response; Photochemical Smog; Scrubber; Ozone.

Associated Organisations:

Aerosol Manufacturers' Association; Dial Environmental Services; Enviroleg; Hillside Aluminium; National Air Pollution Services; National Association for Clean Air; Resource Recovery Systems; Richards Bay Clean Air Association; SI Analytics; Softchem.

ATMOSPHERIC POLLUTION

Pollutants released into the atmosphere cause many disturbances to the natural balance of atmospheric systems. These disturbances include: ozone depletion (caused by CFCs), acid rain (caused by excessive sulphur oxides), global warming (caused by increased carbon dioxide from fuel combustion), particulates (caused by combustion and desertification and resulting in climate change through reflection of incoming solar radiation by particulates) and climate change (temperature changes). The atmosphere is the earth's life support system and needs to be treated carefully. Pollutants come from sources such as car exhausts, industrial emissions, fires and chemical processes in industries. We need to reduce atmospheric pollution as a part of our drive to become more sustainable because continued pollution of our planet's life support systems will have severe impacts on the quality of life and environment of present and future generations. Atmospheric pollution causes imbalances in the atmosphere. Whilst there appears to be the capacity to manage those imbalances within the atmosphere, it is not known how long this capacity will last. It is important to ensure that atmospheric pollution is reduced through better management of wastes and the development of cleaner production programmes in industry.

Key Associated Topics:

Air Quality; Atmosphere; Carbon Tax; Carbon Trading; Ecosphere & Ecosystems; Energy; Global Warming & Climate Change; Methane; Ozone; Particulates; Photochemical Smog; Pollution; Polluter Pays Principle; Scrubber; Transboundary Pollution; Sustainable Development Section: – Accounting for Nature's Services.

Associated Organisations:

Aerosol Manufacturers' Association; Dial Environmental Services; Edward Nathan & Friedland; Enviroleg; Hillside Aluminium; National Air Pollution Services; National Association for Clean Air; Richards Bay Clean Air Association; SI Analytics; Wales Environmental Partnerships.

AVIAN REHABILITATION

Avian rehabilitation describes the work that is undertaken – mostly by NGOs – to take injured, polluted or distressed birds, nurse them back to health and release them back into the wild. A recent high profile example of this work is the 20 000 African penguins – some 50% of the already endangered South African population – that were threatened by oil pollution from the sunken iron ore carrier, *Treasure*. While the oil-soaked birds were cleaned up, the other birds were airlifted from Cape Town to Port Elizabeth to temporarily remove them from the threat of oil contamination. The cleaning of each oil-soaked penguin cost R2000 and the costs of cleaning the birds and the coastline exceeded R40 million. Most of the clean up work on the penguins was carried out by volunteers. There are also focused groups who cater for the rehabilitation of specific bird species, e.g. vultures, other raptors and blue cranes. A key aspect of avian rehabilitation is to rehabilitate the birds in such a manner that they can be returned to

Photo: *Wildlife Aid - UK*

the wild. In this way, the rehabilitation can contribute to the protection and sustainability of the species.

Key Associated Topics:

Animals & Animal Welfare; Birds; Ecological Intelligence; Stewardship.

Associated Organisations:

African Birds & Birding; Bateleur Rehabilitation Centre; Centre for Rehabilitation of Wildlife; Eagle Encounters; EWT; Southern African Foundation for the Conservation of Coastal Birds; WESSA; World of Birds Wildlife Sanctuary.

BACTERIA

Bacteria are single cell organisms. They are prokaryotic (which means their genetic material is not enclosed by a nuclear membrane and they do not have mitochondria or plastids). Many are decomposers, which get the nutrients they need to survive by breaking down complex organic compounds in the tissues of living or dead organisms into simpler inorganic nutrient compounds. This is why wounds must be cleaned – to prevent these organisms from living off the damaged cells of the wound and turning it "septic" or infected. If wounds are left unattended long enough, gangrene (which is an advanced form of infection) may set in and can result in limb amputation or even death of the patient.

Bacteria can also seriously affect humans and domestic animals. The disease, tetanus, is caused by the bacterium, *Clostridium tetani*, and is characterised by muscles becoming rigid and inflexible. Anthrax, which affects cattle and sheep, is a bacterial disease. It spreads through spores and can result in Septicaemia (blood poisoning) and death in animals and humans.

Some bacteria, such as cyano bacteria (previously known as blue-green algae), can photosynthesise: they combine inorganic chemicals, using sunlight, to make the organic nutrient compounds they need to live. Whilst in the direct human context bacteria are often seen as a problem, they form a key part of the functioning of natural ecosystems. For example, of direct human value are bacteria that have been isolated, which live on oil and hydrocarbons. These have a use in dealing with oil spill clean ups and situations where oil clean-up detergents cannot be used for fear of causing other environmental damage. Bacteria occur everywhere in the biosphere and are responsible for activities ranging from the souring of milk to the decay of dead animals.

Key Associated Topics:

Bio-remediation; Composting; Decomposition; Eutrophication.

Associated Organisations:

IBA Environnement; South African Institute of Ecologists and Environmental Scientists.

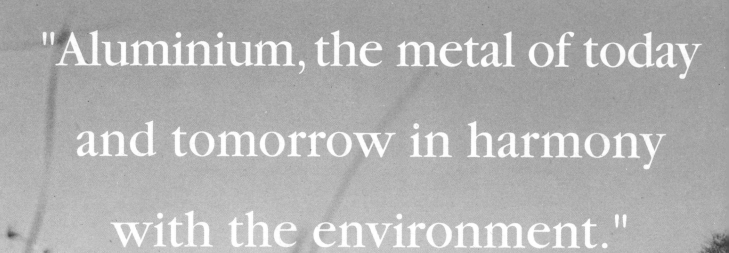

"Aluminium, the metal of today and tomorrow in harmony with the environment."

Our vision is to be the world's leading aluminium smelter. Yet it's not an ambition we'll realise without taking responsibility. The kind of responsibility that recognises we are part of a bigger world, and how we treat our corner of the globe today will affect the environment we'll all share tomorrow.

And so while current operations and development must contribute to economic value, we recognise our obligation to conduct our business safely to minimise environmental risk and contribute in a manner sensitive to the needs and social values of the people affected.

Hillside
ALUMINIUM

ISO 9002   ISO 14001
SABS

BASEL CONVENTION

The Basel Convention or the "Basel Convention on the Control of Transboundary Movements of Hazardous Wastes and their Disposal", is a United Nations Convention administered by a secretariat based in Geneva, Switzerland. The convention was first signed in May 1992 and South Africa became a signatory to the convention in May 1994. The main objectives of the Convention are to encourage the reduction of production of hazardous waste and to minimise the movement of such waste between countries. It also seeks to encourage the disposal of hazardous waste in an environmentally safe and responsible way. The convention requires that certain classified hazardous wastes may only be moved between signatory countries under cover of a permit. There have been various discussions regarding the classification of substances. Some industries, for example the metal recycling industry, have had problems with the manner in which their recyclable products have been classified. The underlying value of a convention such as this is that it encourages companies and countries to look at ways and means of reducing the production of hazardous waste at source. South African legislation is moving more towards compliance with the Convention and the *White Paper on Integrated Pollution and Waste Management for South Africa* includes specific provisions for new legislation, which will align South African waste management practice to the Basel Convention.

Key Associated Topics:

Hazardous Wastes; Integrated Pollution and Waste Manage-ment; Transboundary or Transfrontier Pollution.

Associated Organisations:

Institute of Waste Management; Jarrod Ball Associates.

BIODEGRADABLE

Biodegradable means that a substance can be broken down by living organisms such as microbes and bacteria and absorbed into the environment without leaving toxic or long lived residual substances. Waste products that are biodegradable cause less harm in the environment because, after they have biodegraded, their constituent parts are used by other organisms and processes in nature. For example, when a creature dies, it decomposes into the soil and the nutrients that result provides food and sustenance for other organisms. Another key aspect to biodegradability is the question of recognising the carrying capacity of an ecosystem. It is necessary to ensure that biodegradable products are not added to the environment in such large quantities that they overwhelm the ability of the system to absorb the materials. There is currently a drive to try and link factories where the waste product of one business can provide the feedstock of another. For example, a chemical plant that produces carbon dioxide as a by-product could be located close to a company that produced carbonated soft drinks. Humankind produces waste products that can, in some cases, remain unchanged in the environment for many decades. Substances that are not biodegradable should be recycled or reused to promote sustainability.

Key Associated Topics:

Bacteria; Bio-remediation; Composting; Decomposition.

Associated Organisations:

Enchantrix; Enviroserv Waste Management; Recycling Forum (The National); Recycling Projects.

BIODIVERSITY

Biodiversity, or biological diversity, describes the extraordinary diversity of plant, animal and insect species that exist on earth. Each grouping of species has a different genetic make-up to cope with a specific range of circumstances such as climate, food supply, habitat, defence and movement.

Biodiversity is made up of three related concepts: genetic diversity, species diversity and ecological diversity. Genetic diversity is the variability of genes within a single species (for example, there are African and Indian elephants, which are the same species, but have a different genetic make-up). Species diversity is the variety of species on earth and in different parts of the planet (e.g. forests, lakes, oceans). Ecological diversity is the variety of biological communities that interact with one another and with their environments.

To date, scientists have classified almost 1.5 million different species on earth. However, it is suspected that the actual number may exceed 10 million, with insects accounting for as much as three quarters of that total. Diversity of species allows a maintenance of ecological stability. Any reduction of this diversity directly threatens and weakens ecosystems. This is particularly true in monoculture systems of agriculture where the loss of species diversity opens crops to a greater risk of disease and pest infestations. Loss of biodiversity represents one of the greatest environmental threats to the future of humankind. Whilst it may not be evident for current generations, future generations could be wiped out by disease and pestilence because of their inability to use the earth's natural resources to defend themselves and the organisms on which they depend for their survival.

Key Associated Topics:

Biodiversity & The 6th Extinction; Birds; Cape Floral Kingdom; Ecological Intelligence: Endangered Species; Environmental Ethics; Extinction; Fisheries; GAIA; Marine Life of Southern Africa; Wildlife.

Associated Organisations:

Bainbridge Resource Management; Biowatch; BirdLife SA; Botanical Society of South Africa; Cape Action for People and the Environment; Cape Town (City of) Environmental Management; Conservation International; Development Consultants of SA; Doug Jeffery Environmental Consultant; EWT; Ezemvelo Galago Ventures; Global Invasive Species Programme; Gondwana Alive; IUCN SA; KZN Wildlife; NaDeet; Sustainability United; South African Association for Marine Biological Research; WESSA; WWF; Wilderness Foundation.

There's only one giant this gentle.

TEAMWORKS 21049/E

BIODIVERSITY AND THE SIXTH EXTINCTION

Guest Essay by Dr John M Anderson ~ Gondwana Alive

"The SIXTH GREAT extinction spasm of life is upon us, grace of mankind" – Wilson (1992)

"For each of the Big Five [extinctions] there are theories of what caused them, some are compelling, but none proven. For the sixth extinction, however, we do know the culprit. We are." – Leakey & Lewin (1995)

"Imagine an asteroid the diameter of central Sydney slamming into the Earth. We humans are that asteroid … We are forging the Sixth Global Extinction." – Anderson (1999)

Ask the first 100 persons you encounter at your local shopping mall, on a Saturday morning in Pretoria, what this 'Sixth Extinction' might be, and the chances are 99 of them will shrug their shoulders in vacant response. Do the same in London or Tokyo and the response will be the same. That is our greatest problem and its greatest cause. Democracy is arguably the smartest form of government humanity has devised till now, yet almost no member of these spreading democracies has any notion of the most overwhelming, indeed catastrophic, threat to their existence. All humanity, past and present, are the cause of the Extinction, yet virtually all of humanity remains unaware of its throttling embrace.

So what is the Sixth Extinction, when did it begin, what is its extent, and what is its cause? How long has the term been around?

Over the past 500 million years, since complex life colonised the seas, then the continents, there have been five major global extinction events. For one or other reason – asteroid hits, vast volcanic activity, global CO_2 poisoning, abrupt climatic change, oceanic stagnation – perhaps 90 percent of all life forms (species) have gone extinct in a geological moment. The last such event, the fifth, was when the dinosaurs died out 65 million years ago.

The Sixth Extinction of life is now "upon us" – it is of our time. We are in its remorseless grip. It is not hovering out there on the horizon threatening: it is happening. It began around 140 000 years ago and has been increasing in magnitude, exponentially, ever since. We are within the tsunami, perhaps halfway overwhelmed. And we, every one of us – through our ignorance, our divisions, our political, philosophical and economic systems, our science and industry, our inventive genius and our exploding population – are its cause.

The notion of asteroids hitting the Earth and causing mass global extinctions was first raised by Luis Alvarez and colleagues in Science in 1980. It ignited something of a revolution in thinking about the history of life. Through the rest of the 1980s, extinctions and their causes became very lively science. The previous Big Five extinctions were dissected and debated at great length; and slowly the Sixth – ours – has become more tangibly defined. Only within the last decade, with the pair of books by Edward O. Wilson, "*The Diversity of Life*" (1992) and Richard Leakey "*The Sixth Extinction:*

Biodiversity and its Survival" (1995), has the term "the Sixth Extinction" begun to emerge in academic circles. It hasn't yet surfaced in meaningful measure beyond.

The exponentially increasing Sixth Extinction can be shown pretty convincingly to parallel humankind's headlong expansion in numbers from literally 1 of a kind some 140 000 years back to over six billion individuals today (Fig.1). And this population explosion can be inseparably tied to three successive, seminal communication revolutions: language, writing and printing. It is ironic that it is our sheer genius that is propelling us towards our pending demise.

CAUSING THE EXTINCTION

1. Mitochondrial Eve and the origin of language (hunter-gatherers, the 1st phase of extinction)

- Global population – from 1 individual to 5 million persons
- Centre of dispersal – eastern or southern Africa

Somewhere in Africa south of the Sahara, Mitochondrial Eve and her kin, around 140 000 BP, developed some elemental ascendancy over their fellows. It was a dramatic time in Earth history. A huge swing of some 10 degrees C in global climate, from peak glacial to peak interglacial, occurred between 145 000 and 135 000 BP. What adaptation other than a quantum leap forward in the capacity to use language could have conferred on her clan such titanic advantage that they weathered the climate change, populated the region, the continent and finally the world to the total exclusion of all less endowed archaic Homo sapiens rivals, including Homo neanderthalensis. Within 130 000 years Eve's descendents had colonised virtually all the world and reached saturation population for hunter-gatherers of perhaps 5 million.

As we moved further afield from our mother continent, Africa, we became more lethally feral, not unlike cats or mice on Marion Island or rabbits criss-crossing Australia. On reaching Australia (>62 000 BP), the Americas (ca 12 000), Madagascar (ca 1 500 BP) and New Zealand (ca 1 000 BP), we hunted to extinction the prolific megafauna in our path: 80-90% of the species of larger mammals and flightless land birds succumbed.

2. Invention of writing (farmers and city dwellers, the 2nd phase of extinction)

- Global population – from 5 million to 500 million persons
- Centre of dispersal – Middle East, Persia; ca 5 500 BP

Between the Tigris and Euphrates Rivers, in modern day Iraq, some 5 500 BP, the curtain between prehistory and history was drawn aside: writing was invented. The first texts, stylized picture images were etched into wet clay tablets. They had simple pragmatic purpose, reflecting no great mystical portent or dynastic ego, but rather a tally of cattle or sacks of grain. It soon gave rise to a succession of civilizations in the Middle East with their legal codes, sophisticated farming records, administrative bureaucracy, political hierarchies, enabled by the wedge-shaped characters of cuneiform and the pictorial symbols of hieroglyphics. Writing was the key to advanced agriculture, city-states, nations and civilizations; and a new quantum in population increase.

Over the following 5 000 years, humanity raced to a global population of 500 million – 100 times that of their prehistoric/pre-writing forebears. Yet that was the merest hint of what was to come. With farming, cities and roads, humankind set alight the second phase of extinction: habitat transformation. Floral diversity – species, communities, ecozones – now became severely hit. Primary nature gave way to monoculture, to domestication.

3. Invention of the printing press (scientific & industrial revolutions, the 3rd phase of extinction)

◆ Global population – 500 million to 5 000 million

◆ Centre of dispersal – Western Europe, Germany; 1454 AD

In Mainz, Germany, in 1454 AD, Johan Gutenburg invented the printing press – or rather he was the first to develop and combine movable type, an oil-based printing ink and a suitable press into an effective system for the mass production of books and other documents. It was the mother of revolutions.

STEMMING THE EXTINCTION

4. The electronics revolution (World Wide Web)

◆ Global population – 5 000 000 to 6 000 000 million

◆ Centre of dispersal – Global; 1990s

In Cambridge, England, in 1897, Sir Joseph John Thompson discovered the electron, opening up the astonishing new world of subatomic particles and providing a basis for understanding electricity. Within 50 years (1946) appeared ENIAC, the world's first electronic calculating machine; and within 100 years (1991), the World Wide Web, fashioned by Tim Berners-Lee. Order immediately reshaped the chaos of cyberspace. In just five years, the 600,000 users of the internet had burgeoned to 40 million. By 1999, 150 million people logged on to the Internet weekly. Now, suddenly, our global family truly lives in a global village. We can use those uniquely human attributes – speech, language and writing – to communicate with each other anywhere, anytime, instantly.

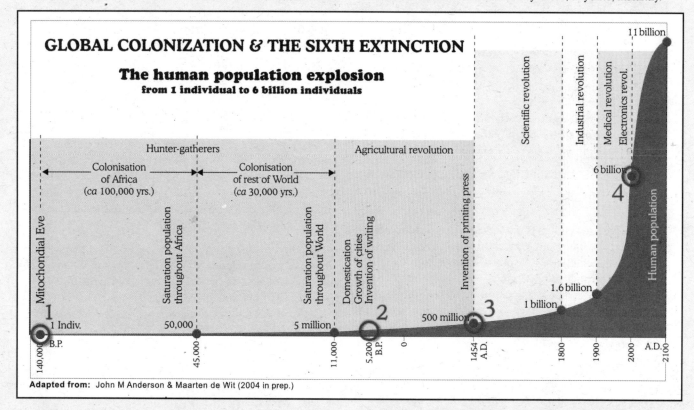

GLOBAL COLONIZATION & THE SIXTH EXTINCTION

The human population explosion
from 1 individual to 6 billion individuals

Adapted from: John M Anderson & Maarten de Wit (2004 in prep.)

It spawned the age of global exploration, it ignited Humanism and the Reformation, it made possible the Scientific Revolution followed by the Industrial Revolution and ultimately the medical revolution and the electronics revolution.

In 500 years, Eve's human family recolonised the world, this time in ships, trains, cars and airplanes, and exploded in numbers by a further 10-fold to over 5 billion – and decimated primary nature in the process. Gutenburg could not have suspected what he was unleashing. In direct proportion to the sophistication of our discoveries, inventions and growth of knowledge, pristine nature has been erased around us – like penicillin's effect on mould. We have already eliminated 70% of the world's most diverse hotspots and 50% of the tropical rainforests.

With the world's knowledge at our fingertips and linkable and accessible, we can negotiate a new balance of nature, we can negotiate cessation of our population explosion and of the Sixth Extinction. Whilst the first three communication revolutions spawned our explosive rise, the fourth can and must take us to a new century of understanding.

During this very early stage of the fourth communications revolution, we have a clear choice: we either allow the fourth and final phase of the extinction to blaze out of control and engulf us; or we unite as the one global family that we are and stem this extinction. I continue – in driving the Gondwana Alive initiative – in the firm conviction that we will take the second route. There could be no possible point otherwise. And the positive spin off is prodigious.

BIOLOGICAL PEST CONTROL

Biological pest control is man's use of a specifically chosen living organism to control a particular pest. The chosen organism may be a predator, parasite or disease, which will attack the harmful insect. A complete biological control programme is a form of manipulating nature to obtain a desired effect, which will be least harmful to beneficial insects, but will attack the target insect almost like a 'living insecticide'.

Biological control methods can be used as part of an overall integrated pest management programme. Biological control can be more economical than some pesticides, and it may have less legal and environmental implications than chemical insecticides.

Unlike most insecticides, biological controls are very specific for a particular pest. Other helpful insects and animals are unaffected and there is less danger of impact on water quality. However, some disadvantages of biological controls are that they usually need more intensive management and planning than pesticide use, and usually require more education and training on the part of the implementation team. Successful use of biological control requires a significant understanding of the biology of both the pest and its enemies. Another drawback is that the results of using biological control are not as dramatic or quick as using pesticides.

The three main approaches to biological control are:

1. Classical biological control, which is the importation of an enemy of the pest. Simply it is a case of travelling to the country of origin of the pest and finding a natural enemy.
2. The Augmentation approach is a case of increasing the population of a natural enemy of a pest. This can be done by breeding them in a laboratory, or by making the natural environment more favourable to the natural enemy.
3. The Conservation approach is the process of cultivating the natural enemies of the pest by identifying any factors that limit their effectiveness, and then controlling them or enhancing them to optimise their ability to attack the pest.

Guest Author: Dr Kelvin Kemm

Key Associated Topics:

Alien Invasive Plants; Bacteria; Bio-remediation; Entomology; Eutrophication; Integrated Pest Control; Organic Farming; Pesticide Treadmill; Pesticides.

Associated Organisations:

Plaaskem; University of Cape Town – Zoology Department.

BIOMASS

Biomass is used to denote large quantities of biological material, usually in a dead state, that can be used for some useful purpose. For example, the burning of wood or grass is a case of 'burning biomass to produce heat'.

Biomass can also be processed to obtain certain chemicals, such as extracting vitamins or useful oils, from the biological material. The term *Biomass* is usually reserved for waste material left over from some other activity. For example, leaves and branches left over after trees have been harvested, or sugarcane after the sucrose-bearing liquid has been squeezed out of it, is biomass.

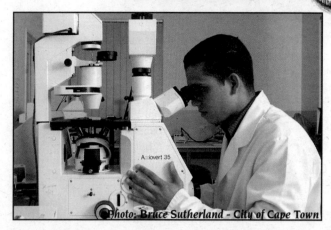

Photo: Bruce Sutherland - City of Cape Town

The use of biomass for some useful purpose is seen as a positive move towards recycling and towards promoting sustainable development.

Guest Author: Dr Kelvin Kemm

Key Associated Topics:

Biodegradable; Composting; Methane; Waste Management.

Associated Organisations:

Agama Energy; Botanical Society of SA; Gondwana Alive; Mondi Wetlands Project.

BIOME

Biomes are the world's major environmental communities classified according to the predominant vegetation and characterised by adaptations of organisms to that particular environment.

The importance of biomes cannot be underestimated. Biomes have changed and moved many times during the history of life on Earth. More recently human activities have also impacted on the biomes. The Earth's biomes are classified in various ways. For example the major divisions are: Aquatic, Deserts, Forests, Grasslands and Tundra.

In turn, these biome divisions are further subdivided. The Aquatic biome is composed of two basic regions: Freshwater (ponds, dams and rivers) and Marine (oceans and estuaries). The Desert biome is divided into four major types: Hot and Dry, Semiarid, Coastal, and Cold. The Forest biome is divided into three major types: Tropical, Temperate, and Boreal. The Grassland biome is divided into two subdivisions: Temperate grasslands, and Tropical or Savannah grasslands. The Tundra biome consists of two subdivisions: Arctic tundra and Alpine tundra.

Within these subdivisions there are further classifications depending on factors such as temperature ranges and seasonal variations.

Guest Author: Dr Kelvin Kemm

Key Associated Topics:

Biosphere; Communities; Ecosphere and Ecosystems; GAIA.

Associated Organisations:

Gondwana Alive; EWT; WESSA; WWF.

BIOPROSPECTING

'Biodiversity prospecting', sometimes shortened to 'bio-prospecting', is the exploration of biodiversity for commercially valuable genetic resources and biochemicals. It describes a search for resources, and the collection of resources with the intention to commercialise them. Bioprospecting can also include the collection of traditional knowledge relating to the use of these resources from local communities. When biodiversity or knowledge about biodiversity is collected without permission from the owners of these resources – and then patented – it may be referred to as 'biopiracy'. Bioprospecting does not include all research on biodiversity; in particular it does not include academic or conservation research (although these may have commercial applications in the future). It also does not include any commercial use of natural resources – for example it does not include the trade in commodities, even if they are medicinal plants. It does not include logging or mining or commercial agriculture, or even the local collection and sale of non-timber forest or veld products. Bioprospecting therefore only refers to a small group of activities undertaken by a small number of commercial sectors. As a result, and because bio-prospecting usually involves taking small samples of material, its impact on the environment is usually minimal. However, it is important to ensure that bioprospecting is done in a sustainable and ethical manner and results in fair benefits for the country and local people from which the genetic resources are prospected. This is an important objective of the International Convention on Biological Diversity. Because South Africa is rich in biodiversity and traditional knowledge, and also has well-developed research capacity and institutions, we are an important player in bioprospecting, and there are several projects underway to investigate the potential of our local plants and animals. One of the best-known examples is the development by the CSIR and the UK-based company Phytopharm of an anti-obesity drug, based on the San people's traditional knowledge of a Kalahari plant called Hoodia. Another example is the development of South African plants for ornamental horticulture products by the US-based company, Ball Horticulture. Up until recently there was no legislation to control bioprospecting, but the new Biodiversity Act requires users of biodiversity first to obtain permission to commercialise local biodiversity and related knowledge. Users are also obliged to ensure that they share benefits fairly with holders of knowledge and those providing the biological resources. Whilst these developments are certainly an improvement on the previous lack of control over bioprospecting, it is important to recognise that bioprospecting brings only limited benefits, and that these are very seldom financial. Countries wanting to commercialise their biodiversity need to do so as part of an overarching and multi-faceted strategy, which considers bioprospecting as only one of many different options to reap benefits from biological resources.

Guest Author: Rachel Wynberg ~ Biowatch South Africa

Key Associated Topics:

Antarctica; Cape Floral Kingdom; Cultural Resources; Medicinal Herbs.

Associated Organisations:

Agricultural Research Council; Biowatch SA; National Botanical Institute; School of Botany and Zoology – University of Natal; Department of Plant, Sciences and Genetics – University of Orange Free State.

BIO-REMEDIATION

Bio-remediation is the method of using living organisms as a means of cleaning up or removing pollution from soil or water. Usually, this is done using special types of micro-organisms, such as natural or bio-engineered bacteria, which convert the pollution or hazardous waste into harm-less substances. For example, bio-remediation is being used to treat contaminated groundwater where the water is pumped up from underground, treated with the micro-organisms, which remove the pollution at the surface, and then pumped back underground again. Another example is a new type of bacteria, which has been especially "bred" to breakdown oil spills into harmless bio-products such as water, carbon, carbon dioxide and oxygen. This is a new way of dealing with waste and pollution and will prob-ably be developed more extensively in the future. Wetlands are a natural bio-remediation process, which filter out pollutants from water. Progress has been made in developing artificial wetlands, which have been used for cleaning up industrial pollutants as well as providing an alternative form of sewage treatment in remote areas. Creative research and development is focussing on the "copying" of nature's clean-up systems and processes in order to develop efficient and effective pollution control and reduction methods.

Key Associated Topics:

Bacteria; Biodegradable; Decomposition; Integrated Pollution & Waste Management; Pollution; Polluter Pays Principle.

Associated Organisations:

Groundwater Association of KwaZulu-Natal; IBA Environnement; Lombard and Associates; Peninsula Permits.

BIOSPHERE

The biosphere is the envelope around the Earth, containing the life supporting systems of the planet. So all the soil, inland water, and the sea form part of the biosphere. The air space in which the birds fly is definitely part of the biosphere and so is some of the atmosphere above 'bird-height' because that part of the atmosphere is linked to weather patterns, which affect life. The part of the high upper atmosphere near where satellites orbit is not part of the biosphere, but there is no hard and fast rule as to how high the biosphere extends. Similarly, on land, the biosphere includes the soil area, and also the depth to which the deepest roots of large trees penetrate. The seabed, even in the deepest places, is also part of the biosphere. The biosphere is an area of complex interactions, which all collectively function to keep the Earth in the state that Mother Nature intended.

Guest Author: Dr Kelvin Kemm

Key Associated Topics:

Atmosphere; Biome; Ecosphere and Ecosystems; GAIA.

Associated Organisations:

Space for Elephants; Gondwana Alive; Mondi Wetlands Project; Settlement Planning Services; Wilderness Action Group; WWF.

BIRDING

Guest Essay by Peter Borchert
~ Africa Birds and Birding Magazine

One often hears comments such as: 'isn't birding a rather elitist pastime – what real contribution does it make to conservation, and how relevant is it in the context of the endless list of pressing environmental issues facing our planet?'

At one level this perception of the world of birding, and birders in particular, is understandable. Regrettably birders can be rather snooty towards the uninitiated, giving the impression of being members of an exclusive club. Not helping matters, they are also popularly and all too often lampooned as strangely dressed eccentrics, 'nutters' in fact, festooned with notebooks, field guides and binoculars as they tramp through swamps and jungles in search of some rarity. The reality is something quite different as the hobby of bird watching is one of the fastest-growing leisure activities in the world, drawing its aficionados from many walks of life, young and old. An interest in birding has become an expression of lifestyle and part of a multi-million dollar industry built on the sales of binoculars, cameras, books and magazines, garden feeders, bird-attracting garden plants, and nature-based travel.

Yet it has to be acknowledged that the millions of bird-watching enthusiasts around the world, still represent a very small slice of privileged human beings. Field guides, though good value for money, are not high on the list of priorities for the millions of Africans struggling to support themselves and their families; binoculars, even an inexpensive pair, are way beyond the budgets of many; and yes, travel to far-off wilderness areas to add species to one's life-list, is not cheap.

But this is not where it all has to start. One only has to peer out of a home or office window, or make the short journey to an urban park, to begin a lifelong adventure with birds. Birds are the most accessible of all the wild creatures on our planet. And southern Africa, with some 950 recorded species, has more than its fair share. They are everywhere. And if they are not, then you can be sure that all is not well in that particular neck of the woods. Birds are extraordinarily good indicators of the general environmental state of the area and I am sure that most conservationists would agree that if we can only 'save the birds' then our planet will be able to take care of itself. It follows, therefore, that if a love, fascination and respect for bird life can be kindled among the greatest number of people, then a major, positive step in the conservation of our planet will have been taken. Yes, this is an over-simplification, maybe even naïve, but in essence it is not wrong. The question is: 'How can the pleasures and value of birds and birding be communicated and nurtured on what seems to be such an impossibly wide basis?'

Any conversation along these lines inevitably leads to statements such as: 'What we need to do is educate people'. Apart from the perhaps unintended, but nevertheless implicit condescension in such comments, what follows is sadly, but so often the case, complete defeat. To educate one needs funds, infrastructure, and manpower that all too often just are not available. But if we are sincere, then there is something that each one of us can do.

The obvious step is to join an organisation such as BirdLife South Africa, the local affiliate of BirdLife International, a worldwide association that has as one of its stated objectives the widening of an awareness of birds and their conservation needs. In South Africa, BirdLife, which supports a rapidly growing number of conservation projects that draw corporate funders, scientists and local communities into partnership, has been instrumental in motivating, establishing and running internationally important bird areas such as Wakkerstroom in the Free State, and it acts as an umbrella organisation for some 28 regional bird clubs throughout the country.

While wholeheartedly encouraging and endorsing involvement with BirdLife and one of its local branches, the call to all amateur birders should be to take their commitment a step further. Cast your mind back to that time when you started and were unashamedly chuffed with every positive identification you could make, and when your list of lifers was growing with just about every bird you saw – even if it was only a European Starling or a House Sparrow. If you were to succeed in generating that same sense of wonder in only one other person who has as yet had no exposure to the intense and consuming interest of birding, then I would argue that you will have done more for birds, conservation, and the general quality of our planet than you could have imagined.

Photo: Richard du Toit – Londolozi

Key Associated Topics:

Avian Rehabilitation; Biodiversity; Birds; Communities; Eco-tourism; Endangered Species; Ecological Intelligence; Extinction; Stewardship.

Associated Organisations:

Africa Birds & Birding; BirdLife South Africa; Riviera Wetlands and African Bird Sanctuary; Rondevlei Nature Reserve; EWT:– African Wattled Crane Working Group – Blue Swallow Working Group – Vulture Study Group – The Poison Working Group – The Raptor Conservation Group.

BIRDS

Birds are excellent indicators for most forms of biodiversity and sustainability.

They can tell us a great deal about the state of our environment. For example, studying birds continues to be one of the best ways of monitoring environmental pollution. Worldwide, birds are now understood to be key indicators of changing land use and long-term sustainability. Birds play a significant role in ecosystems in that they are dispersers of seeds and their decline could have an impact on the success rates of many plants. These in turn supply food to higher organisms whose survival also becomes threatened.

Over a fifth of all known bird species are giving some cause for concern in terms of global extinction risk. 1 111 species (11%) have been identified as threatened, 11 (0.1%) are conservation dependent, 66 (1%) are data deficient, and 875 (9%) are near threatened. Many common species are also declining in numbers. One bird in eight could join the extinction list this century. The decline of bird species can be attributed to a wide range of reasons, most of which are related to human activities or encroachment. These include habitat destruction, introduced species and exploitation. Practices such as urbanisation, loss of wetlands, physical interference with nests, poisoning from pesticides and pollution threaten our birds and therefore the biodiversity and sustainability of our planet.

Although intensive conservation action for individual species can be successful, it is more effective to protect the areas in which birds live. The indications are that the distribution of non-avian endemics coincides well with endemic bird distribution. If we can save the birds, we have a good possibility of saving many of the earth's other creatures at the same time.

Guest Author: Karen Marx ~ BirdLife South Africa

Key Associated Topics:

Avian Rehabilitation; Biodiversity; Birding; Communities; Eco-tourism; Endangered Species; Extinction: Stewardship.

Associated Organisations:

Africa Birds & Birding; Bateleur Rehabilitation Centre; BirdLife South Africa; EWT:– African Wattled Crane Working Group – Blue Swallow Working Group – Vulture Study Group – The Poison Working Group – The Raptor Conservation Group; Southern African Foundation for the Conservation of Coastal Birds; World of Birds Wildlife Sanctuary.

CAPE FLORAL KINGDOM

The plant life of planet Earth is regarded by bio-geographers as falling into six floral "kingdoms", each characterised by a core set of species. On the south-western tip of Africa is the smallest of these, the Cape Floral Kingdom. It is the only one in the world contained within a single national boundary, and is the richest in that it has the most species, and even then represents a third of South Africa's plant species in only 6% of its area. This kingdom includes several types of vegetation, including forests, the succulent and Nama karoos, and extends as far as the thicket biome in the Eastern Cape.

At the heart of the CFK, however, is the fynbos vegetation of the Western Cape. Fynbos, loosely regarded as the South African equivalent to the vegetation of the Mediterranean Basin, is phenomenally rich in plant biodiversity. Of the approximately 8 700 species found here, 68% are found nowhere else in the world. As in other parts of the world, the treasures of the region are under threat. Urban expansion, increased environmental demands by industry and agriculture, invasion by exotic weed species, and even uncontrolled development in response to the growth of tourism, have all contributed to the CFK becoming one of the "hottest of hotspots" globally in the global fight to conserve biodiversity. More than 1 700 plant species here are regarded as under the threat of extinction in the wild. This is three-quarters of the SA national list. Protection and conservation management in the region are of extremely high priority.

Photo: Earthyear Magazine

Join the Botanical Society of South Africa!!!

Bring your family and join the Botanical Society of South Africa! Why? Because the Botanical Society is the organisation focused on appreciating and conserving the remarkable floral heritage of South Africa. We do this by offering members free entry to all eight of the National Botanical Gardens in the country. But this is not the only benefit. Our 16 branches offer members opportunities to be involved with local volunteer initiatives and exciting walks and talks that are not only educational but fun too! If you join the Botanical Society you will also receive the quarterly magazine, Veld & Flora, packed with interesting articles about our floral heritage. The conservation actions of the Society are making a real difference and by supporting these efforts with your membership and donations you are part of a growing group of people who have realised that our planet is worth preserving. Are you prepared to only be a couch potato or are you willing to make a real difference and get involved with a society that has more than just flower power?

For further information and membership application contact:
The Botanical Society of South Africa
Private Bag X 10 Claremont 7735

Tel. (021) 797-2090 • Fax (021) 797-2376
e-mail: info@botanicalsociety.org.za • Website: www.botanicalsociety.org.za

With support from the Global Environmental Facility, South Africa has prepared a strategy and action plan to address the threats to the Cape Floral Kingdom and adjacent marine environment. Termed Cape Action for People and the Environment, it is one of the world's most progressive region-wide eco initiatives to involve government, NGOs and community-based organizations in co-operative action.

Photo: BOTSOC

Guest Author: Trevor Sandwith ~ Cape Action for People and the Environment (C.A.P.E.)

Key Associated Topics:

Biodiversity; Biome; Bio-prospecting; Endangered Species; Heritage; Indigenous Plants; Stewardship.

Associated Organisations:

Botanical Society of South Africa; Cape Action for People and the Environment; Cape Environmental Trust; Cape Town City Council – Environmental Management Department; Coastec; Conservation International; National Botanical Institute; National Parks; Santam-Cape Argus Ukuvuka; Western Cape Nature Conservation Board.

CARBON

Carbon is a building block of the molecules, which make up living cells and is an essential element for life on earth. It is a basic ingredient of bones, carbohydrates, connective tissue, hormones, nerve cells and proteins. As an important component of carbohydrates such as sugar and starch, carbon is a key component of photosynthesis and respiration, the mechanisms used to convert the sun's energy into forms that can be used by life on earth. Fossil fuels (natural gas, oil and coal) are all carbonaceous compounds, which at a previous time in the earth's evolution were built up by living cells through photosynthesis.

Carbon, in gas, liquid or solid form, moves between the atmosphere, oceans, rivers and landmass via chemical processes and through geophysical means. It moves through the atmosphere as carbon dioxide and carbonic acid, forms bicarbonates and flows into the oceans. The oceans absorb carbon dioxide from the atmosphere and organisms absorb carbon from the atmosphere. There are complex balances that are continually changing as a result of short, medium and long-term actions, both natural and man-made. A clear example of how excessive changes in carbon in the biosphere can impact on the environment is that of global warming. Excess releases of carbon from fossil fuel burning create a build up of carbon dioxide in the atmosphere, which results in increased global temperatures.

Key Associated Topics:

Carbon Dioxide; Carbon Monoxide; Carbon Tax; Carbon Trading; Climate Change; Fossil Fuels; Global Warming & Climate Change; Kyoto Agreement.

Associated Organisations:

Department of Geography & Environmental Studies – University of Stellenbosch; Department of Geography & Environmental Studies – University of Western Cape; SouthSouthNorth.

CARBON DIOXIDE

Carbon dioxide is a gas that occurs naturally in the air and is produced when animals breathe, vegetation rots and when material containing carbon is burnt or broken down. Carbon dioxide is a key component in photosynthesis, which is the major base source of food for organisms on earth. In photosynthesis, carbon dioxide combines with water and sunlight (energy) to produce sugars which, in turn, provide energy for plants from which they live and grow. Millions of years ago, the earth's atmosphere consisted mainly of carbon dioxide until plants gradually removed the gas and, through decomposition, "fixed" the carbon in the fossil fuels, coal, oil and gas. Humankind is now reversing that process by releasing the carbon again through excessive burning of fossil fuels. Carbon dioxide is one of the so-called "greenhouse gases" which absorb heat in the atmosphere. The release of carbon dioxide through excessive burning of fossil fuels adds to global warming. The carbon dioxide is absorbed back into the system physically through plants and trees but this can be arrested if humankind cuts down too many trees and reduces that absorption capacity.

Key Associated Topics:

Biomass; Carbon; Carbon Monoxide; Carbon Tax; Carbon Trading; Climate Change; Fossil Fuels; Global Warming & Climate Change; Kyoto Agreement.

Associated Organisations:

Department of Geography & Environmental Studies – University of Stellenbosch; Department of Geography & Environmental Studies – University of Western Cape; SouthSouthNorth.

CARBON MONOXIDE

Carbon monoxide is a colourless gas produced by incomplete combustion, mainly from motor vehicles and cigarettes. It is found naturally in the atmosphere in minute quantities. Carbon monoxide is toxic because it combines with haemoglobin (the red blood pigment) to form pink carboxyhaemoglobin, which prevents the blood from performing the function of transporting oxygen to body tissues. The natural levels of carbon monoxide in the blood are around 0.5%. Levels in a driver sitting in dense traffic may rise to 2%, smokers may have levels of between 4 and 8% and chain smokers may achieve 9 to10%. At these levels, no permanent harm is known of, except in people with heart or lung complaints: However, higher concentrations can be fatal.

Key Associated Topics:

Air Quality; Atmospheric Pollution; Catalytic Converter; Fossil Fuels; Global Warming & Climate Change; Greenhouse Gases; Kyoto Agreement.

Associated Organisations:

Department of Geography & Environmental Studies – University of Stellenbosch; Department of Geography & Environmental Studies – University of Western Cape; SouthSouthNorth.

CARBON TAX

Carbon Tax is a tariff charged by governments on business, industries and energy sources that emit greenhouse gases, particularly Carbon Dioxide (CO_2), into the atmosphere, and in doing so contribute to pollution, global warming and climate change. The charge is typically levied per tonne of CO_2 emitted.

The idea behind Carbon Tax is to oblige emitters to increasingly factor environmental costs into the production of goods and services, as these costs have traditionally been seen as external (not accountable for or relevant) to the activities of these product and service providers.

Carbon Tax therefore makes business and industry players, and subsequently the consumer, accountable for environmental degradation caused by their activities and consumption patterns. It also provides an incentive for business, industry and the consumer to promote the implementation of technologies and practices that are less carbon-intensive.

Carbon Tax may also be administered in the form of a levy, which can then be used to provide a source of funding for the transition to cleaner technologies or to address health impacts related to carbon emissions.

Guest Author: Lester Malgas ~ Environmental Writer, SouthSouthNorth (SSN)

Key Associated Topics:

Agenda 21; Atmospheric Pollution; Carbon; Carbon Dioxide; Carbon Monoxide; Carbon Trading; Eco-Tax; Fossil Fuels; Global Warming & Climate Change; Kyoto Agreement; Polluter Pays Principle.

Photo: Cape Metropolitan Council

Associated Organisations:

Renewable Energy And Energy Efficiency Partnership (Reeep); SouthSouthNorth; Department of Environmental Affairs & Tourism.

CARBON TRADING

Carbon trading is a type of emissions trading, and is the trading of the rights to emit greenhouse gases. The unit traded is a tonne of carbon dioxide equivalent and is termed a 'carbon credit' or 'carbon allowance'.

Trading of carbon credits occurs between governments and corporations, often facilitated by a broker. Emissions trading schemes typically set legally backed targets for missions reductions and impose financial penalties for non-compliance. Entities that are able to reduce their emissions cheaply are thus able to sell their surplus allowances to entities that are unable to bring about cheap reductions.

Emissions allowances are thus transferred to the site of high use value, creating a situation where emissions reductions are achieved as cost-effectively as possible for all parties involved in the market. Since trading began in 1996, the price of carbon credits has ranged from $1.50 to $60 per tonne.

Carbon trading represents an innovative international market approach to the environmental response to climate change and it is claimed that double the targeted emissions reductions have been achieved through such trading schemes.

Guest Author: Lester Malgas ~ Environmental Writer, SouthSouthNorth (SSN)

Key Associated Topics:

Agenda 21; Atmospheric Pollution; Carbon; Carbon Dioxide; Carbon Monoxide; Carbon Tax; Eco-Tax; Global Warming & Climate Change; Kyoto Agreement; Polluter Pays Principle.

Associated Organisations:

Department of Environmental Affairs & Tourism; Renewable Energy And Energy Efficiency Partnership (Reeep); SouthSouthNorth.

CARCINOGEN

A carcinogen is any substance that is capable of causing cancer in animal tissues. Doctors and scientists do not fully understand how carcinogens cause cancer but much is known about the kinds of substances that cause the cancers. The first carcinogen substance to be discovered was arsenic but many, many more have been added to the list. Substances such as beryllium, cadmium, cobalt, chromium and asbestos have been isolated as carcinogens and the list has grown to include over 300 different major substances. The importance of communication between workers, interested and affected parties and industry is to ensure that full information is freely available on chemical substances being used in the workplace so that the risks of dealing with carcinogenic substances can be discussed and clearly understood. Ideally, carcinogenic substances should be eliminated from the workplace. However, where these substances are necessary and have no immediate substitutes, they can be used provided that the users wear correct and complete protective clothing.

Key Associated Topics:

Accumulation; Asbestos; Basel Convention; Environmental Risk and Liability; Hazardous Wastes; Plutonium; Pollution; Polluter Pays Principle; Precautionary Principle.

Associated Organisations:

Enchantrix; Industrial Health Research Group – University of Cape Town; South African Institute of Ecologists and Environmental Scientists.

CARRYING CAPACITY

Carrying capacity, in the case of organisms, is the maximum number of organisms that can be supported, fed or is able to survive in any specific habitat or ecosystem without causing the breakdown of the habitat or ecosystem. For example, the carrying capacity of a lift might be five people. That would be governed by the weight of those five people. Similarly, carrying capacity might be affected by the capacity of chemicals within a system to react with substances added. For example, excessive phosphates could exceed the carrying capacity of an ecosystem and cause it to function in an imbalanced mode. Understanding the limits of an ecosystem is important when considering sustainable development, because one needs to know the maximum carrying capacity so that one can manage the number of organisms and keep those numbers below the maximum to prevent the system breaking down. It is not, for example, known what the carrying capacity of the earth is with regard to human population, but it is known that we are stretching the limits of some of the resources that the earth needs to supply the requirements of humankind.

Key Associated Topics:

Biodiversity; Consumerism; Ecosphere and Ecosystems; Game Farming; Population; Poverty Alleviation; Sustainable Development.

Photo: Earthyear Magazine

Associated Organisations:

Brousse-James and Associates; Conservation Management Services; Elephant Management and Owners' Association; Game Rangers' Association; Wildlife Management – University of Pretoria.

CATALYTIC CONVERTER

Certain metals have the property to act as catalysts. This is a characteristic, which enables them to promote, speed up or make more efficient, certain chemical reactions without being permanently changed themselves. The platinum group of metals can convert or "catalyse" the breakdown of noxious exhaust gases of petrol driven cars (nitrogen oxides, carbon monoxide and hydrocarbons) to safer nitrogen, carbon dioxide and water. A catalytic converter, or auto catalyst, fitted correctly to a car's exhaust system, can reduce the effect of the noxious gases on the environment. However, auto catalysts will only work if the car uses unleaded petrol. This is one of the reasons unleaded petrol is being introduced in South Africa. A growing concern regarding the effects of vehicle exhaust emissions on human health and natural ecosystems is causing serious consideration to be given to requiring all new vehicles to have catalytic converters fitted. If this were to be introduced, its effects would be gradual whilst the existing stock of vehicles without converters came to the end of their lives. This could take between 10 and 25 years to complete.

Key Associated Topics:

Air Quality; Atmospheric Pollution; Carbon Monoxide; Fossil Fuels; Industrial Ecology; Petrochemicals; Photochemical Smog; Pollution; Synthetic Fuels.

Associated Organisations:

Audi; Ford Motor Company of Southern Africa; Toyota; Wales Environmental Partnerships.

CHERNOBYL

Chernobyl is a Russian nuclear power station, some 80km from Kiev in the Ukraine, where a nuclear accident occurred on 26th April 1986. The accident was caused by the deliberate switching off of safety systems, which resulted in the deaths of at least 31 people at the time of the event and many more in the subsequent clean up process. Over 200 others suffered acute radiation sickness and it is believed that over 200 000 people had to be evacuated. Fifteen years after the event, those people have still not all been permanently relocated. Over 150 000 square kilometres in Belarus, Russia and the Ukraine are contaminated and an "exclusion zone", 30 km in radius around the site, is totally uninhabited. In July 1987, six former senior officials and technicians from the station were sentenced to terms of imprisonment in labour camps for blatant violation of safety procedures. The accident resulted in radioactive fallout over Europe, which will increase the number of people likely to develop cancer in that area over the next 50 years. The accident caused the nuclear industry to re-examine all its emergency planning and safety procedures. The political and economic situation in the former Soviet Union is causing many international nuclear scientists to worry about the state of nuclear power stations and the possibility that other events like Chernobyl could occur because of the shortage of skilled manpower and money to operate the power station in a safe manner. International agencies will be spending over US$765 million over the next eight to nine years to reinforce the shelter covering the controversial fourth reactor on the site, to shore up the deteriorating "Sarcophagus" cover put over to contain future escapes and releases.

Key Associated Topics:

Contaminated Land; Eco-catastrophe; Energy; Nuclear Energy – The Counter Debate; Radiation.

Associated Organisations:

Koeberg Nuclear Power Station; Department of Environmental Affairs & Tourism; Department of Minerals & Energy.

CITES

The Convention on International Trade in Endangered Species of Wild Fauna and Flora (CITES) is an international agreement that was signed by 21 countries in Washington, DC in 1973, and came into force in 1975. Over 160 countries are currently party to this Convention, which is the largest wildlife conservation agreement in existence.

The CITES provisions assist member countries to regulate the international commercial trade in fauna and flora as well as parts and derivatives thereof. Member countries regulate trade using a system of permits and certificates issued in accordance with the decisions and resolutions taken at the Conference of the Parties, which is held roughly every two years.

CITES accords varying degrees of protection to wild animal and plant species, depending on their biological status and the effect international trade has or could have on them. These degrees of protection are afforded through the following Appendices:

◆ Appendix I contains species which are in danger of extinction and that may be negatively affected by trade. Such species cannot be traded among member countries except under exceptional circumstances such as scientific purposes. Appendix I specimens cannot be traded for commercial purposes, although specific exemptions are permitted.

◆ Appendix II contains species, which are not necessarily currently threatened with extinction but may become so unless trade is strictly regulated so as to ensure sustainability. Appendix II further contains so-called look-alike species, which, due to their similarity in appearance to certain regulated species, must be managed so as to ensure effective control of both species.

◆ Appendix III contains species that are subject to regulation within the jurisdiction of a Party and for which the co-operation of other Parties is needed to prevent or restrict their exploitation.

Upon accession to CITES, a country designates one or more governmental departments as its Management Authority. The Management Authority is responsible for issuing permits and compiling annual trade reports. A designated Scientific Authority provides the scientific expertise to the Management Authority on which wildlife import and export approvals are based. The Scientific Authority also determines when to limit or prohibit trade that may be detrimental to a species' survival.

Guest Author: David Newton ~ National Representative, TRAFFIC East/Southern Africa

Key Associated Topics:

Conservation; Endangered Species; Extinction; Wildlife.

Associated Organisations:

EWT; WESSA; WWF; Conservation International; Department of Environmental Affairs & Tourism.

Photo: WESSA National Photographic Salon

The National Cleaner Production Centre

The South African National Cleaner Production Centre (NCPC) aims to enhance the competitiveness and productive capacity of the national industry through the adoption of Cleaner Production (CP) techniques, and the transfer and development of environmentally sound technologies. The NCPC was launched during the World Summit on Sustainable Development in Johannesburg in 2002 and marks one of the tangible results of this event. This is a cooperation programme between South Africa and the United Nations Industrial Development Organisation (UNIDO) over a three-year period with financial assistance from the dti and the governments of Austria and Switzerland. The NCPC is hosted at the Council for Scientific and Industrial Research (CSIR) in the Manufacturing and Materials Division, Pretoria, with Regional Focal Points in Cape Town and Durban. UNIDO/UNEP long-term partnership has contributed to the creation of a strong network of Cleaner Production Centres and programmes in more than 30 countries around the globe.

The NCPC acts as a launch pad for the implementation of activities relating to international environment conventions such as the Kyoto Protocol of the United Nations Framework Convention on Climate Change or the Stockholm Convention on the Reduction and Elimination of Persistent Organic Pollutants.

The NCPC will play an important catalytic role and coordinating role in fostering the promotion of sustainable industrial development in South Africa by building national capacity in cleaner production, fostering policy dialogues between industry, government and academia and promoting investments to develop and transfer environmentally sound technologies. Through the NCPC and its regional Cleaner Production Focal Points an active national Cleaner Production infrastructure will be established that will facilitate the sustainable application of cleaner production in South Africa. In order to enhance the sustainable application of the concept, the programme is based on a multi-stake holders approach and aims at involving the different levels of industry, government, academia and the financial sector.

CP is an internationally-recognised tool to simultaneously meet the challenges of both competitiveness and environmental protection. Consumers and major international trading partners increasingly demand products that not only meet their quality requirements but which are also produced under environmentally and socially acceptable conditions. CP looks at production processes from the design phase, through manufacture, use/ operation and disposal. It seeks to eliminate or at least minimize the production of waste/ pollution, and to optimise the consumption of resources (energy, water) during both product manufacture and use. CP activities generally improve an organisation's bottom line and invariably improve productivity and competitiveness.

CP can be achieved in a number of different ways, of which the three most important are

- Changing attitude,
- Applying know-how, and
- Improving technology.

The work of the NCPC is guided by an Executive Board, involving the main national and international stake holders of the Centre. A representative Advisory Board provides an industry perspective. The main activities of the NCPC cover in-plant assessments; promotion of CP technologies and investments; training; information dissemination and CP policy advice to the Government.

The NCPC mission supports South African industries to be more sustainable and globally competitive.

The policy of the NCPC is based on the national government's mandate for:

- SME priority, in line with the government's drive for industrial growth
- Regional and sectoral operation
- Coordinated national activities in CP. Because of a large number of initiatives and the cross cutting nature of CP projects, there is a need to ensure awareness and cooperation between key stake holders to maximise benefit for the country and avoid duplication

A network of local and international partnerships strengthens the NCPC's ability to meet the needs of industry and ensures that best practices are developed. The NCPC has signed a memorandum of understanding with the Kingdom of Denmark for the operation of the Cleaner Textile Linkage Centre (CTELC) in Cape Town. The NCPC and CTELC share the same project steering committee so as to optimise use of resources.

NCPC Structure

Advisory Board	NCPC	Executive Board
Executive board, industry associations, DEAT, DWAF, organised labour, WRC, SABS, Mintek, BEE, academia, stake holders, etc.	Directors, Deputy Director	UNIDO, **dti**, Austria, Switzerland, CSIR, 3 industry reps, NCPC, financial institution

Economist

Regional Sectoral Coordinator and Academic Coordinator	Regional Sectoral Coordinator and Academic Coordinator	Regional Sectoral Coordinator and Academic Coordinator

Enquiries

Tel: +27 12 841 3338
Fax: +27 12 841 2135
E-mail: ncpcinfo@ncpc.co.za
www.ncpcsa.co.za

NATIONAL CLEANER PRODUCTION CENTRE

ncpc

SOUTH AFRICA

CLEANER PRODUCTION

Cleaner production describes a philosophy that is the continuous use of industrial processes and products to prevent the pollution of air, water and land, to reduce wastes at source, and to minimise risks to the human population and the environment. Some of the ways of promoting cleaner production include improving house-keeping in factories (cleaning up and tidying up wastes), training workers, reducing wastage by analysing production processes to make more efficient use of raw materials, minimising spills and losses, and improving methods and systems of waste management and pollution control. Cleaner production was an initiative started by the Industry and Environment section of the United Nations Environment Programme (UNEP). The Industry and Environment section has produced a range of publications as well as case studies and contacts to help those who wish to explore Cleaner Production in more detail. For more information, refer to the International Cleaner Production Information Clearing House (ICPIC) at http://www.unepie.org The chemical industry in South Africa is implementing cleaner production approaches and more information on cleaner production can also be obtained by writing to the Director of the Chemical and Allied Industries' Association (CAIA).

Key Associated Topics:

Air Quality; Eco-efficiency; Global Compact; Industrial Ecology; Integrated Environmental Management; Integrated Pollution and Waste Management; Life Cycle Assessment; Safety Health & Environment; Waste Management.

Associated Organisations:

Beco Institute for Sustainable Development; BMW; Cape Town (City of) Solid Waste; Clothing Textile Environmental Linkage Centre; Chemical and Allied Industries' Association; Development Bank of SA; Engen; Hewlett-Packard; IBA Environnement; Institute of Waste Management; Liz Anderson and Associates; National Association for Clean Air; National Cleaner Production Centre; Responsible Container Management Association of SA.

CLIMATE CHANGE

There is great controversy over the state of global climate at the present time. However, the weight of evidence seems to suggest that the ambient global temperature has risen, probably as a result of human industrial activities. National and international authorities have begun to institute actions and programmes to modify industrial activities (such as carbon dioxide release) which impact on climate. Accordingly, South Africa is a signatory to the United Nations Framework Convention on Climate Change (For more information on the Convention, go to www.unfccc.de/) and is committed to the implementation of programmes, which will reduce carbon dioxide emissions.

There are many theories that suggest that the earth is warming whilst others suggest that the next ice age is very close. Both sets of theories tend to agree that human activities are affecting the world's climate.

One important aspect to bear in mind is that heating and cooling in different parts of the globe vary according to aspect and attitude, so it is very difficult to generalise about the whole world. There is significant evidence that the increase of carbon dioxide, caused by the large scale burning by people of fossil fuels (such as coal and oil), is having an effect on climatic conditions. The problem is in deciding whether the changes currently occurring are significant in relation to the patterns of the past. Humankind's time on earth and the ability to record and measure only stretch back a few thousand years and the scientific accuracy of this is difficult to confirm. Global climate is something that has existed for millions and millions of years. We have no accurate way of knowing whether what we are seeing now is just a temporary fluctuation or is evidence of a longer-term change in weather patterns. The earth's climate has naturally gone through changes throughout its history. However, with the onset of increased atmospheric pollution from humankind's activities and the resulting disturbance of natural balances in the atmosphere, scientists fear that the earth's climate will change rapidly with widespread consequences for human life.

Key Associated Topics:

Air Quality; Atmospheric Pollution; Carbon Tax; Carbon Trading; Energy; Global Warming & Climate Change; Kyoto Agreement.

Associated Organisations:

Cape Town (City of) Environmental Management; Centre for Environmental Studies – University of Pretoria; Imbewu Enviro; IUCN SA; Minerals & Energy Education & Training Institute; Price WaterhouseCooper; SA Climate Action Network (SACAN); SouthSouthNorth; Sustainable Energy Africa; WWF.

Photo: Inspire!

COASTAL ZONE MANAGEMENT (INTEGRATED)

Photo: Bruce Sutherland - City of Cape Town

The South African coast

The coast is a distinct, but limited spatial area that derives its character mainly from the direct interactions between the land and the sea. South Africa's coast extends for approximately 3 000 km from the Namibian border in the west, to the border with Mozambique in the east.

Nearly a third of our population live at the coast, mainly in the cities of Durban, Cape Town and Port Elizabeth, as well as centres such as East London, Saldanha and Richards Bay.

Our coast provides food and sustenance for many people living in both rural and urban coastal areas, and our ports are a gateway to international trade. The coast is also an area of tremendous importance for recreation and tourism, whilst supporting a range of coastally-dependent businesses and commercial ventures. Furthermore, the cultural, educational, religious, scientific and spiritual importance of the coast and its resources, are invaluable. The coast is a national asset, belonging to all people of South Africa, and its products account for around 35% of our national Gross Domestic Product.

Given the enormous value of coastal resources, as well as the mounting human pressure on coastal ecosystems worldwide, increasing attention is being given to improving the management of all coastal regions.

Efforts to improve coastal management

International concern for the coast, and efforts to enhance coastal management, are reflected in a number of international recommendations and agreements such as Agenda 21 and the various conventions, which deal with coastal resources.

Furthermore, guidelines for integrated coastal management (ICM) have been developed by, for example, the World Bank, the Organisation for Economic Co-operation and Development, the United Nations Environment Programme, and the World Conservation Union.

In Africa, coastal nations are focusing efforts in a co-operative manner through two United Nations Environment Programme conventions, namely:

(i) The Abidjan Convention for the Co-operation in the Protection and Development of the Marine and Coastal Environment in the West and Central African Region; and

(ii) The Nairobi Convention on the Protection, Management and Development of the Marine and Coastal Environment in the East Africa Region.

Coastal management in South Africa is in the process of being transformed by the introduction of the integrated coastal management (ICM) approach. The White Paper for Sustainable Coastal Development in South Africa, released in April 2000, ushered in this new era for coastal management, and is a product of an extensive public participation, research and analysis process. The White Paper is driven by the need to realize the opportunities our coast provides to build our nation and transform our economy and society. It deliberately seeks to improve the quality of life for current and future generations of South Africans. It recognizes that, in order to do so, we must maintain the diversity, health and productivity of coastal ecosystems. The challenge, posed by the White Paper, is to achieve sustainable coastal development through a dedicated and integrated coastal management approach, in partnership with all South Africans.

The White Paper noted that existing South African legislation affecting coastal management has been fragmented, administered by diverse government departments, and, in some cases, is outdated or inappropriate. It recommended that legislative changes be made to facilitate integrated coastal management. Consequently, new national legislation, the National Environmental Management Coastal Zone Bill is presently being drafted to give effect to the coastal policy commitments.

In view of the multiplicity of activities that take place in the coastal environment, most government departments, at all spheres of governance, are responsible for aspects of coastal management in South Africa. Marine and Coastal Management (MCM) Directorate, within the Department of Environmental Affairs and Tourism (DEA&T), is ultimately responsible for providing national leadership for promoting sustainable coastal development. This is largely achieved through the CoastCARE partnership programme, which involves both the private and public sector, and seeks to enhance opportunities and address barriers to achieving sustainable livelihoods for poor coastal communities. Other CoastCARE initiatives include Working for the Coast, Local Demonstration Projects, and the Blue Flag programme.

Guest Author: Environmental Evaluation Unit, UCT

COGENERATION

Cogeneration is when two or more forms of energy are utilised at the same time, from the same source. The most common form of cogeneration is the production of heat and electricity. In North America the term 'cogeneration' is usually used, whereas in Europe the term 'combined heat and power' (CHP) is frequently used.

In most industrialised countries the largest single wastage of energy occurs at power plants when fossil fuels are burnt to produce electricity. Generally thermal power plants only convert about one third of the fossil fuel energy to electricity, the rest is discarded as waste heat.

Cogeneration is possible when this waste heat can be harnessed for some useful purpose such as heating water, or space heating in buildings. By effectively using cogeneration, energy efficiencies can be raised from about 30% to 70% or more. However, electricity can be transmitted over long distances without losing much energy, whereas heat usually needs to be used close to the source of production. As a result, an important factor in effective cogeneration is the innovative thinking needed to develop applications for the various energy forms.

If cogeneration is designed in as a critical consideration at the stage of initial project planning, then an energy production facility can be designed with cogeneration in mind, from inception. There are many cogeneration possibilities in a variety of industries. For example, waste heat from cement kilns, and bagasse from the sugar industry that can be burned to produce heat.

The concept of cogeneration is a challenge to the ingenuity of designers and planners wherever energy forms are available that would otherwise be wasted.

Guest Author: Dr Kelvin Kemm

COMPOSTING

Composting is the decomposition of plant and other organic remains to make an earthy, dark, crumbly substance that will enrich soil. Composting is one of the natural alternatives to using synthetic fertilisers. Compost needs air, moisture, worms and microbes. It is common practice to mix in "browns" (dry dead plant materials such a straw leaves, and wood chips or sawdust) and "greens" (fresh green materials such as green weeds, fruit and vegetable scraps and peelings, coffee grounds and tea bags, some horse manure, etc.). The "browns" and "greens" should be mixed together, and slightly moistened, making sure that there is plenty of air circulating. Within a couple of weeks fresh compost will have formed, which can either be dug into the soil, used as mulch or top dressing, or it can be combined in equal parts with water and used as a "compost tea" as a quick boost for indoor plants.

Photo: Earthyear Magazine

Photo: Bruce Sutherland - City of Cape Town

CONNEP (P)

CONNEP stands for "**CON**sultative **N**ational **E**nvironmental **P**olicy (**P**rocess)." In August 1995, the CONNEP Conference was held in Johannesburg to decide how to proceed in developing a national environmental policy, which would be fair and complete, and discussed and developed with all interested and affected parties from all the different sectors of South African society. A draft discussion document was widely circulated and various provincial workshops were held to discuss and debate its contents. In addition to the workshops, meetings were held throughout the country to discuss the document and gather ideas. Meetings were held in civic halls, schools, churches and even under the shade of large trees. A second national meeting was held to discuss a Green Paper (formal discussion document), which was produced from the results of the discussions that had been held. After some two years of consultations, discussions, debates and submissions, a White Paper (Government Policy document) entitled A *White Paper on Environmental Management Policy for South Africa* was published in May 1998. This resulted, later in 1998, in the promulgation of the National Environmental Management Act, South Africa's first democratic, consultative, multi-disciplinary environmental management Act. The CONNEP exercise represented one of the most thoroughly debated and discussed policy exercises ever undertaken in the democratic world. It was an exercise in consultation and an exercise in participation. The trends developing to date seem to suggest that those involved in the consultations have taken ownership of the policies and legislation and are beginning to put their environmental preferences into practice through the implementation of the resultant legislation.

Key Associated Topics:

Environmental Governance; Environmental Law; Environmental Justice; NEMA.

Associated Organisations:

Contact Trust; Jarrod Ball and Associates; Department of Environmental Affairs and Tourism.

CONSERVATION

Conservation used to be the term used to describe the management of game reserves and parks for the purpose of recreation and protection of flora and fauna.

The modern-day understanding of conservation relates to the wise management of natural resources to ensure the credibility and survival of diverse species within their natural habitat. It is gradually extending to cover the industrial and commercial context of environmental concerns as well. Conservation and protection of the vital bio-diversity of the species on earth is a vital function. The loss of species represents an unacceptable drop in the available gene pool and the potentially irreplaceable loss of species that could provide the bases for the curing of AIDS or the drug that provides the answer to prevent the human body rejecting a donor heart.

Modern users of the term, "conservation" see it as covering a very broad area of interest that ranges from managing animals, right up to reviewing Environmental Impact Assessments as an interested and affected party in a stakeholder meeting. Conservation is one of the key action strategies that form a part of the philosophical and practical approach to implementing sustainable development.

Key Associated Topics:

African Environmental Tradition; Agenda 21; Antarctica; Biodiversity & the Sixth Extinction; Birds; CITES; Deforestation; Desertification; Ecological Intelligence; Endangered Species; Environmental Ethics; Extinction; Fisheries; Gondwana Alive Corridors; Grasslands; Marine Life of Southern Africa; Mountains; National Parks; Rivers and Wetlands; Soil; Stewardship; Sustainable Development; Transfrontier Parks; Tropical Rain Forest; Water Quality and Availability; Wilderness Area; Wildlife; Wildlife Management; World Parks Congress.

Associated Organisations:

Africa Birds & Birding; Africa Geographic; Africat Foundation (The); Bio-Experience; BirdLife Africa; Cape Action for People and the Environment (CAPE); Centre for Conservation Education; Cheetah Outreach; De Wildt Cheetah and Wildlife Centre; Conservancy Association; Conservation Support Services; Delta Environmental Centre; Durban Parks Board; Dolphin Action Protection Group; EWT; Eco Africa Travel; Eco Systems; Elephant Management & Owners' Association; Ezemvelo African Wilderness; Ezemvelo KZN Wildlife; Forum for Economics & the Environment; Green Clippings; Green Trust; IUCN SA; Mondi Wetlands Project; NaDeet; Open Africa; Peninsula Permits; SA Wildlife College; SA National Parks; SA Foundation for the Conservation of Coastal Birds; SA Wildlife College; Space for Elephants Foundation; Vervet Monkey Foundation; White Elephant; Wilderness Leadership School Trust; Wilderness Foundation; WESSA; WWF.

TRUST IN OUR FUTURE

For more than seventy years, South African consumers have trusted Woolworths to give them the products that they wanted. Quality, goodness, safety, value for money and attention to detail have been at the heart of the Woolworths brand. In the context of the 21st Century crisis of sustainability, these merits have expanded beyond products on shelves to encompass communities, society and the environment.

At, the core of Woolworths operations, there has always been a strong set of principles and an unyielding commitment to being a responsible corporate citizen. Woolworths employment practices, supplier sourcing, supplier development programmes and social investments are underpinned by the firm belief that as a truly South African business, they have a responsibility to make a positive difference.

Part of this responsibility is evident in the efforts that Woolworths makes in working towards a sustainable environment. They have introduced sustainability reporting that puts them in line with world best practices. There is a dedicated Woolworths sustainability forum comprised of senior people who implement strategies that ensure that every aspect of Woolworths' business is aligned to the principle of delivering sustainable products and services with care and concern. This includes product safety, reduction of waste, animal welfare and the responsible use of energy. There is a commitment to work hard to use natural resources efficiently and reduce the use of substances, materials and methods that are harmful to people and the environment.

Woolworths is acutely aware of the impact that an organisation has on the communities around it, and that an organisation can only survive and thrive with the support and endorsement of these communities. They recognise their interdependence and the great need to commit to a sustainable future for all.

In the past Woolworths has quietly developed and implemented sustainable empowerment projects through successful partnerships with communities and their stakeholders. In 2003, the Woolworths Trust was launched to provide a new focussed channel for their substantial investments and efforts.

The Woolworths Trust Eduplant programme is currently Woolworths' flagship investment. It is managed by Food & Trees for Africa, South Africa's leading greening organisation. The Trust's programme reaches thousands of schools throughout South Africa motivating schools and their communities to grow their own organic foods, to learn about good nutrition, to provide fresh, healthy food for those living with HIV/Aids, to alleviate poverty and develop skills. The programme reflects many of Woolworths brand values – good food, good nutrition, care for the environment and community development. In a country where more than half of our children are malnourished, the Trust's programme is meeting a real and urgent need.

Woolworths is committed to seeking out this kind of actual effectiveness as an end-result of its corporate policies, activities and investments. There is a keen awareness that talk is not enough; appropriate, significant action is required.

A practical example of this is Woolworths' involvement in the provision of "badger-friendly" honey – a South African first. Woolworths played a pivotal role in this movement, contributing to raising consumer awareness of the issue as well as working closely with conservation organisations and beekeepers to establish a national badger-friendly initiative. This declaration has been accepted and promulgated by various beekeeping associations, the South African Federation of Bee Industry Executive, Cape Nature Conservation, the World Wildlife Fund (WWF), The Wildlife & Environment Society of SA (WESSA), The Endangered Wildlife Trust (EWT) and other NGO's.

Two years ago, the indiscriminate killing of the Honey Badger (Mellivora capensis) by bee farmers was brought to light through the efforts of researchers Keith and Colleen Begg. The Honey Badger is relatively small predator with a varied diet that includes rodents, reptiles, scorpions, spiders and the larvae of bees. As a top predator in many of its ranges, the Honey Badger plays an important role in maintaining a healthy balance in an eco-system.

This attractive and intrepid mammal is sparsely distributed from South Africa, through Africa to the Middle East. In this country it is listed in the Red Data Book of Mammals as a vulnerable species, indicating that they may become endangered in the near future if they are not properly protected. Despite a reputation for being surprisingly fierce, the Honey Badger is actually shy, reclusive and mostly nocturnal. They are found in various habitats, often outside of protected conservation areas. Honey Badgers are naturally attracted to beehives, seeking out the nutritious larvae of bees, rather than the actual honey.

Beekeepers play a vital role in the South African farming economy, and not just for the honey that they produce. Bees are also specifically farmed so that they will pollinate orchards, and they are thus, very important to the South African fruit-growing industry. The Honey Badger's natural predilection for bee larvae brings them into conflict with beekeepers as they can cause substantial damage and loss to the farmers. It has been common practice for beekeepers to indiscriminately set traps or deliberately poison the Honey Badger, contributing to its vulnerable status as a species.

Woolworths began to work closely with Keith and Colleen Begg and it was discovered that there are better solutions to this conflict - such as raising the hives out of reach of Honey Badgers. They initiated a project to raise awareness, to develop a badger-friendly code of practice and actively encourage beekeepers to use alternative methods to protect their hives that did not impact negatively on the Honey Badger population in South Africa.

Woolworths adopted a "badger-friendly" honey policy, ensuring that all honey sold by Woolworths nationwide has been produced without harm to Honey Badgers. They have been steadfast in their commitment to this policy, even when adverse weather conditions in 2003 affected their supply of "badger-friendly" honey.

The badger-friendly project is just one example of the kind of emphasis we need business to place on long-term sustainability over short-term profit.

Woolworths' most recent commitment to animal welfare is their landmark statement that from March 2004, all eggs available at Woolworths will be eggs that are laid by non battery chickens – that is, eggs produced by hens that roam freely in large barns and outside areas. Woolworths is the first retailer in South Africa to adopt such a strong animal welfare stance.

Woolworths has made similar commitments that range from an animal welfare policy that embraces global standards for the ethical and humane treatment of animals, and is approved by the S.P.C.A. - to being the leading supplier of the widest range of organic and free range food in South Africa.

Woolworths has a powerful vision of a prosperous and healthy future for all South Africans. This vision is the inspiration to commit to and implement policies and practices right now that will ensure a sustainable and vibrant environment for generations to come.

WOOLWORTHS
the difference

CONSUMERISM

TOWARDS SUSTAINABLE CONSUMPTION AND PRODUCTION

Guest Essay by Dr Eureta Rosenberg

What's it about?

You are a consumer – whether you only spend pocket money, get the groceries, order office stationery, or do the procurement for a whole department. Consumption is the purchase and use of goods and services, and consumerism is the 'shop-till-you-drop' syndrome – that ever-increasing spiral of over-consumption which characterises modern society.

I love shopping. Not the weekly chore, but those fun expeditions when the new season's goodies arrive and the mall is mesmerizing with colour and creative design. I enjoy getting gifts that match the recipient and occasion, and I've discovered the delight of dressing Baby in clothes that show how special he is ... For let's face it, we consume not only for survival and comfort, but also for variety, stimulation and to express relations, status, identity …

So, I'm guilty of over-consumption. Guilty, because I know that the word 'consume' also means to 'use up' and 'destroy' and that over-consumption is environmentally destructive and enjoyed at the expense of poor people – one of whom waits outside the mall at the first traffic lights.

To reduce my ecological footprint I try to be a green consumer. In the supermarket I go for the 'green' paper towels (which are white) and the 'eco-friendly' detergents. Then I check the label to see what makes this product more environmentally-friendly than the next one, and the information is seldom enough. Words like bio-degradable, surfactant and recyclable are used, but seem to obscure more than they reveal – especially because other products don't always tell me what they contain. Usually I buy the green goods anyway, even if the information is not convincing. Sometimes their price puts me off, for they are seldom the cheapest. But we should buy them anyway, I reason, so as to send a message to the producer.

MEASURE YOUR PERSONAL ECOLOGICAL FOOTPRINT

Choose the description that best matches your lifestyle. Circle its score. Then add up.

Food: You consume plenty and pay little attention to where your food is produced. Score: 100	**Food:** You eat locally grown, vegetarian food and have little food waste. Score: 32
Waste: You produce very little waste and reuse/recycle everything possible. Score: 30	**Waste:** You recycle little or none of your waste. Score: 100
Water: You take mostly showers, and don't have a dishwasher or hosepipe. Score: 1	**Water:** You take lots of baths, have a dishwasher, hosepipe, etc. Score: 50
Holiday: You take at least one long distance flight per year. Score: 65	**Holiday:** You usually holiday close to home. Score: 10
Electricity: You conserve energy and get your electricity from renewable sources. Score: 2	**Electricity:** You use many appliances, often leaving them on or on standby. Score: 50
Heating: You use electric heating sparingly, have excellent insulation and low bills. Score: 10	**Heating:** You keep your home warm using electricity, have poor insulation and high electricity bills. Score: 45
Transport: You travel mostly by car. Score: 75	**Transport:** You travel mostly by public transport, cycling or walking. Score: 10
Paper: You regularly buy newspapers and new books. Score: 10	**Paper:** You share newspapers and usually borrow books rather than buy them. Score: 5

Your total score gives an idea of the area of land that would be needed to sustainably support your lifestyle – your ecological footprint. The chart shows the number of planets that would be needed to sustainably support global consumption, if everyone lived like you.

Your Score   Number of Planet Earths Needed

100   ●

200   ● ●

300   ● ● ●

400   ● ● ● ●

© *Best Foot Forward* 2000 (see www.bestfootforward.com)

That's what green consumers do.

But hang on, what message are we sending? That customers will buy a product labelled 'environmentally-friendly' even without clear information about what makes it so? That we are willing to pay more for green goods? Mmm …

Green consumerism could be undermined by corporate opportunism. Corporates drive consumption, for traditionally this has been the way to grow wealth. (It is not yet clear to many businesses that in the long term and in indirect ways, it will be in their own interests to reduce the level of consumption.) So if I ran a business based on goods, I would be tempted to let green consumerism work for me, by supplying "green" products that would encourage customers to consume more, not less. And most people fit in with this, for consumerism is an almost universal core value. It is a central process of our social and economic life. Whatever the occasion – a funeral or a matric farewell – we spend as much as possible to show its importance. We measure the growth of an economy in terms of how much the nation consumes. We justify expensive extras on our President's jet by linking it to his status.

Consumption is "the major cause of the continued deterioration of the global environment" (Agenda 21, United Nations Conference on Environment and Development, 1992).

Now, the President doesn't really need expensive trimmings to indicate that he must be respected; no more than my baby needs cute clothes to be adored. He already is. Does an economy need ever-increasing consumption in order to be sustaining, that is, to sustain the health and well-being of all people?

Consumption-filled lifestyles are actually unhealthy. In 2000 a World Watch Institute report warned that people are consuming more meat, coffee, cars and pills and getting more obese. The World Health Organisation describes obesity as "today's principal neglected public health problem". More than one billion people are now overweight and numbers are increasing everywhere.

Now, in economic terms spending on cholesterol tablets and heart-bypass operations are forms of consumption and are simply added 'to the Gross Domestic Product, which measures the size of the economy.

So over-consumption makes the economy look healthier, even if it is not sustaining the health and well-being of people.

These are my depressing thoughts as I leave the supermarket and drive off, noting the sign at the traffic lights: "No job, no money, please help".

A matter of equity and ecology

At least 18 millions South Africans need to consume more – more nourishing food, clean water, energy, health care. There is a way of reasoning – which underpins economic policies – that if some of us work, earn and spend more, the growing economy will also benefit the poor who do not consume enough. There is unfortunately little evidence to support this belief. In the past years our economy has grown and consumption has increased, but mainly among those already well-off, and in the period of this economic growth we also shed 800 000 jobs. We are still number three among the most unequal nations of the world. A mere 10% of households still consume almost 50% of the available goods and services in South Africa, while the poorest 10% consume just over 1%.

Wealthy and middle class South Africans form part of the global 'consumer class', roughly the 20% of the world's citizens who have direct access to a car. The global consumer class is found not only in wealthy countries – there are consuming elites in poorer countries too. It is these plus minus 20% of the world's citizens who eat 45% of all the meat and fish consumed, own 87% of all the cars, use 84% of all the paper and 75% of all energy including 68% of all electricity – in the process generating 75% of the annual global pollution.

This inequality is neither acceptable nor sustainable, in terms of social justice and ecological impact. Often, these considerations are intertwined. The over-consumption of the consumer class squanders natural resources which poorer people directly depend on, and creates pollution and health hazards which affect the poorest people most.
(See POVERTY ALLEVIATION.)

What did you have for breakfast? Bacon and eggs, coffee? Did you know that the world population of pigs, chickens and cows almost tripled between 1961 and 2000? This is threatening water quality and adding large volumes of methane to a growing load of greenhouse gases. The consumption of coffee is linked to deforestation and farmers giving up food security to grow the cash crop.

It is not possible to extend the over-consumption patterns of the consumer class to the rest of the world, even if this is said to be a form of 'democratization'. The earth simply cannot provide the necessary resources or absorb the resulting increased pollution. In fact we are already consuming more than the Earth's biological capacity can support. According to the WWF Living Planet Report 2002 – using UN projected scenarios, which assume slowed population growth, steady economic development, and more resource-efficient technologies – the world's ecological footprint will in the next 50 years continue to increase, from a level 20 per cent above the Earth's biological capacity to a level between 80 and 120 percent above its capacity. Clearly, our current patterns of consumption are un-sustainable.

To develop equitably and sustainably, we should aim for contraction and convergence. The consumer classes must contract (reduce) consumption while the poor increase their consumption to a point of convergence, where both meet. It is not so much about re-distribution, as about restraint. Rough calculations suggest that the consumer class needs to bring down their overall use of the environmental resources by 80-90% during the next 50 years.

> There will be no equity, and no environmental sustainability, unless the consumer classes become able and willing to live comfortably at a drastically reduced level of resource demand. Such a transformation of wealth is the central challenge of sustainability.

What can be done?

Will a reduction in consumption not harm the economy? No – bear in mind that many proposals for tackling environmental destruction and poverty are based on developing a new kind of economy that does not require harmful and unfair consumption patterns (see ECONOMICS).

I. INFORMED CONSUMER CHOICES

Green consumption can be a powerful tool to encourage business to provide more environmentally-friendly products and services. To make better choices, consumers need accurate, adequate and accessible environmental and health information about ALL products and services (not just those labelled green). Producers, retailers and public authorities must provide this information, through extensive product labelling, customer call services, user-friendly research reports, etc.

In the supermarket I would, for example, need information not only on the eco-friendly paper towels, but on the other paper towels too, and on those cotton kitchen cloths which could do the same job. I need information on the materials used in making them (paper, cotton, bleaches, dyes); the environmental impact of the manufacturing process (how much water was used to grow the trees and the cotton; what pollution is cause by bleaching, dyeing); the impact of use (which product is more hygienic, for example); and disposal – which would cause the least harm when dumped. A useful tool for informing consumer decision-making is the life cycle analysis. It examines the impacts of a product or service and the processes involved in producing it, from production through use to disposal.

> Using Life Cycle Analysis, the Wuppertal Institute in Germany found that producing one litre of orange juice used 25 kg of raw materials and 12m² of land, mostly in Brazil. This did not include the infrastructure to produce the raw materials.

2. TRANSFORM PRODUCTION PATTERNS

Recognising that increasing numbers of consumers are choosing the 'environmentally friendly' option, some producers have put their efforts into changing perceptions of their product (with huge advertising budgets) rather than changing their product or the way it is produced. A far more constructive, effective and responsible approach would be to re-think the production process.

As a rule of thumb: For every ton of waste produced at the consumer end, five tons are produced during manufacturing, and a massive 20 tons of waste at the site(s) of origin.

Some of the many possibilities include:

◆ Radically increasing resource productivity – Producers use existing resources and energy more effectively and efficiently, or find different materials and processes altogether. Examples include resource-light architecture, regional food markets, hydrogen engines, smart cars and making appliances such as photocopiers more recyclable.

◆ Re-modelling production on ecosystems – 'Cleaner production' strategies involve the re-use of materials in continuous closed cycles, as happens in nature. Examples include wind power and permaculture.

◆ Applying the precautionary principle – Processes, materials, chemicals and products must be proven to be safe prior to their release on the market.

◆ Restoration of living systems – through deliberate investment in rivers, mountain slopes, soils, etc. to restore, sustain and expand our natural capital. Examples include afforestation, township gardens, tackling erosion and de-contaminating soil.

◆ Emphasising 'real wealth' – thus reducing the importance of goods for both producer and consumer, as we re-value those forms of wealth which cannot be bought. One can make money without adding more things to the world, as companies who 'sell results rather than things' can testify.

While one can find fault with any of these individual strategies (for example, attempts to increase resource productivity could make products less durable; in their early developmental stages the benefit of new energy-efficient products could be cancelled out by the energy used to produce them), we should use them as part of a 'whole system' approach, which includes industrial ecology innovations such as IWEX.

Photo: Guy Stubbs Photographer

Pick 'n Pay's own environmental commitment & initiatives

Pick 'n Pay has always upheld a policy of social and environmental consciousness, but in recent years has looked more critically at its obligations towards staff, customers, shareholders, communities and the environment in which it operates.

Our role as a good corporate citizen is not just about acting on these obligations by applying the principles of sustainable development, but to play a significant role in creating environmental awareness within the company and the general public.

In 1989, the Company's first campaign was created with several million Rands worth of donations to NGOs and community-based organisations involved in environmental projects. Stores set up depots for collection of consumer waste and campaigned for the reduction and re-use of packaging.

Communicating with staff

Communicating environmental issues to our employees is an integral part of the company's on-going education programme and all staff are exposed to the company's philosophy in their induction.

Staff environmental groups (Green Teams) work together to set goals for their work and home environments with emphasis on creating a positive impact on communities in need of upliftment.

Pick 'n Pay launched the Environmental Awards competition in 1997 to stimulate and reward employees for their involvement in caring for the environment both in the workplace and at home.

Pick 'n Pay employees, Lewis Phiri and Louw Heydenreich, were chosen to assist Robert Swan (the first man to walk both the North and South Poles) with his project to clean up Antarctica.

Communicating with customers and the general public

National Environment Days

Pick 'n Pay employees participate actively in all four nationally-designated Environment Days with Green Teams enlisting members of the community for help.

Talkabout

Pick 'n Pay launched the Talkabout series of booklets in 1989 which covers both ecological and social issues. The booklets are distributed free of charge in all Pick 'n Pay stores.

Pick 'n Pay staff and the children of the Riding for the Disabled school in Constantia, Cape Town, participating in the Trees for Life project.

friendly packaging. Pick 'n Pay launched the Enviro Facts project in 1990. These information leaflets cover a range of environmental issues, providing an overview of each topic or issue. The booklets are distributed free of charge in all Pick 'n Pay stores.

Pick 'n Pay continues to be committed to the environment

Raymond Ackerman, founder and Chairman of Pick 'n Pay Stores Limited, once said: "Besides being a useful predicator of performance, I believe that values are the very lifeblood of an organisation." Indeed Pick 'n Pay has made protection of the environment central to its value system, and this attitude will continue to be reflected in the way the company conducts its business in the years ahead.

Pick 'n Pay's Green Team's hard at work!

Carrier Bags

Pick 'n Pay embarked on a "Go Green" communication exercise to advise consumers of the need to switch to more environmentally

Pick'n Pay
We're on your side
www.picknpay.co.za

Y&R Hedley Byrne 34084

Inspiring Example – The IWEX (Industrial Waste Exchange) project links 'waste material generators' to 'waste material users'. Among its success stories is Warner Lambert SA, whose factory in Retreat, Cape Town, listed its waste in the IWEX catalogue. Now it exchanges a range of wastes, for example plastic waste with Atlantic Plastics – 200-300 kg every three weeks – and cardboard waste with a community outreach programme that makes furniture.
(See www.capetown.gov.za/iwex)

3. ECONOMIC REFORMS THAT RE-VALUE NATURE

Technology for greener production already exists. To make sure it is used, we also need economic reforms. These should be based on an economic framework in which natural resources are valued at their replacement costs. For economic innovations that could inspire more sustainable production and consumption see ECONOMICS. They include:

◆ Tax reform – Taxes would shift from labour to resource consumption and the pollution and waste that result from consumption.

◆ Subsidies – Some government subsidies support environmental destruction, foster inefficiency and consumption, and prevent innovation and conservation.

◆ User fees for the commons – User fees can help protect common goods – such as the ocean or atmosphere – by raising the price of using them. Take the example of aviation, which is a growing cause of air pollution and is currently exempt from the Kyoto Protocol. In the USA a national railway went bankrupt, whilst numerous domestic airlines survive. If an appropriate price was asked for the true cost of air travel, the railway, a more environmentally-friendly means of transport, could be financially viable.

◆ Price incentives – Full cost accounting would reflect the total costs of products and services, including the cost of their environmental impacts, in prices. This may mean that energy generated from fossil fuels would be more expensive than energy generated from renewable sources. This would encourage businesses to compete to produce goods and services with the lowest impact on the natural resource base.

This, like the others above, is a controversial strategy. How would price increases affect the poor? Remember we only want to reduce the consumption of the consumer class. There are ways of pricing which do not harm the poor (viz. South Africa's Water Act). Cheaper renewable energy could benefit those living in remote areas not reached by the main grid.

Inspiring Example: A local government in Korea introduced a pay-per-volume garbage removal fee. This encouraged consumers to recycle more, transformed the packaging industry, and in three years reduced the size of the waste stream by 20%.

◆ Legal incentives – In the current skewed and competitive market economy, these reforms could only take root if we introduced a legal framework of rights and responsibilities. To illustrate the need for this: When the Korean government introduced a tax on engine size to encourage the use of smaller cars, they were pressured to withdraw the tax by the USA, whose interest was to protect its export of luxury cars. The counter-productive effect of vested interests calls for universally enforceable regulations that protect government, corporate, consumer and citizen rights and at the same time reflect our growing understanding of our social and environmental responsibilities.

Towards a post-consumerist world

Where does all this leave us? If we gain a better understanding of the size of our ecological footprint, and the impact of the products we love to consume, we may end up buying less. Before my mind's eye the colourful mall disappears and a new structure arises, with large vats from which we re-fill our household detergent … It could be a rather bleak picture – but it need not be! The 'eco-mall' could be bursting with stalls where we buy (or exchange) the tools and materials to make our own gifts, and crafts which sing the praises of our children … There could be lively studios selling skills and designs to help us do this. Our core value would be creativity, rather than consumption; instead of cash, we would give time; our status would be based on the services we offer, rather than things we show off …

Sounds 'far out'? Yes! However, Einstein pointed out that "we will not solve the problems of today – by using the same thinking that created these problems in the first place". We will not achieve equity and ecological sustainability by continuing to over-consume, with a bit of green-label buying on the side. And changes in end-consumption are only one part of the solution. We need a whole system approach, where micro-level changes in the home or business are matched by macro-changes in the economy, systematically tackling the drivers of over-consumption. We do have the power to influence and work towards these changes.

Useful resources

Adbusters: www.adbusters.org

Centre for Sustainable Consumption and Production: www.grip.no

Clearing house for Applied Futures: www.agenda-transfer.de

Consumers International: www.consumersinternational.org

International Institute for Environment and Development: www.iied.org

World Resources Institute: www.wri.org

United Nations Working Group on Sustainable Production: www.unep.frw.uva.nl

South African New Economics: www.sane.org.za

Key Associated Topics:

Ecological Intelligence; Economics; Population; Poverty; Sustainable Development Section: – The Retail Industry – Investing in the JSE SRI Index; Resource Section – Green Consumer Guide – Eco-Logical Lifestyle Guide.

Associated Organisations:

Centre for Environmental Studies – University of Pretoria; Earthlife Africa; Enchantrix; Sustainable Living Centre.

CONTAMINATED LAND

One of the by-products of industrial society is an increase in contaminated land resulting from releases of chemicals, illegal dumping, seepages and disposal of off-specification products in the ground. Unless careful record is kept of the activities on particular pieces of land, it becomes very difficult to keep track of the status and state of health of that land. A prime example of this is the "US Love Canal" case, in which an abandoned toxic waste dump was allowed to become a play area for a residential housing estate. This resulted in attendant illnesses, which shocked the American nation. There was evidence of birth deformities, abortions, child sickness, increased rates of cancer and direct poisonings. Incidents of this nature have not yet occurred in South Africa to the same extent as Love Canal but there have been examples of companies having to deal with serious land contamination, which, had they been in industrial areas, could have had significant health consequences for communities. Contaminated land also carries serious implications for company employees who may have to work in the area on a daily basis and may even unknowingly handle the contaminated substances.

Key Associated Topics:

Bio Remediation; Carcinogen; Chernobyl; Eco-catastrophe; Environmental Liability for Remediation; Environmental Risk and Liability; Hazardous Wastes; Leachate; Landfill; Pollution; Polluter Pays Principle; Rehabilitation of Land; Used Oil; Waste Management.

Associated Organisations:

Barlofco; Chemical Emergency Co-ordinating Services; Envirocover; Kantey & Templer; Morris Environmental & Groundwater Alliances; Wales Environmental Partnerships.

CULTURAL RESOURCES

Cultural resources are those lessons that we have learned and carried forward from our history, which show us how we developed as human beings and how our society evolved. These resources could be prehistoric pots, old buildings, folk songs, traditional practices, languages, ancient stories, books and papers, paintings, rock art, battlefields, relics from bygone ages, old photographs or films, and so on.

Photo: Guy Stubbs Photographer

They are all records of what previous generations did, said, thought and created. We learn from the past so that we can plan and develop for the future. It is important that we recognise the importance of the past and preserve examples to help future generations understand how we developed and perhaps to save them from making the same mistakes. South Africa is a diverse society with many different components to it. The different components and ethnic groupings within our society have differing views on culture as it affects them. A society's laws, morals, ethics and standards are based on the cultural principles that have developed as the society has matured and grown.

Key Associated Topics:

African Environmental Tradition; Agenda 21; Heritage; Indigenous Peoples; Women and the Environment.

Associated Organisations:

Agency for Cultural Resource Management; Albany Museum; Cape Town Heritage Trust; Centre for Conservation Education.

Drain — Historical spills — Toxic waste

Basement

UST

Hazardous vapours in soil, buildings and utilities

Contamination absorbed into soil

Separate-phase product

Water table

Groundwater contaminated with dissolved product

Flow

Diagram by Morris Environmental (MEGA)

A complicated site showing contaminant sources, phases, plume and effects

DEFORESTATION

Photo: Copyright South African Tourism

The world's forests, both temperate and tropical, are under enormous pressures from disease (often as a result of air pollution, notably from acid rain), felling for timber needs, clearing for agriculture, or loss under reservoirs through the damming of rivers. According to the Food and Agriculture Organisation of the United Nations (FAO), deforestation was concentrated in the developing world, with a net loss of some 180 million hectares between 1980 and 1995, or an average of 12 million hectares per annum. The main causes of deforestation are poverty (extension of subsistence farming), overpopulation and ignorance. Sustainable forestry practices involve the re-planting at least as many, if not more, trees than are removed by logging, which offsets at least some of the logging losses. Deforestation is also harmful to the earth and its inhabitants because trees absorb large amounts of carbon dioxide and, in turn, release oxygen. Consequently, the loss of forests is affecting the earth's ability to cleanse and freshen the air. Loss of indigenous forests also causes loss of biodiversity of species within the forest ecosystem. A recent (during the year 2000) World Resources Institute Study suggests that the current loss of natural forests is approximately 160 000 square kilometres per year, an area the size of England and Wales combined. In addition to the direct effects of logging, disturbing trends have been noted in the quality of forests. A 1995 survey of forest conditions in Europe indicated that over 25% of the trees assessed were suffering from significant defoliation. Annual European surveys showed the numbers of completely healthy trees falling from 69% in 1988 to 39% in 1995.

Key Associated Topics:

Aforestation; Forestry; Poverty; Tropical Rain Forest.

Associated Organisations:

Department of Water Affairs and Forestry; Ken Smith Environmental; SAPPI.

DESALINATION

Desalination describes the process of removing dissolved salts from salt water or brackish (slightly salty) water to make it fit for consumption by humans or for use for agricultural and other human activities. Common types of desalination include distillation and reverse osmosis.

Distillation involves the use of energy to condense the liquids and cool them rapidly to convert them into water condensate. Reverse osmosis is a process using membranes or thin "skins" through which water will pass, leaving the salts behind.

Some countries in the Middle East have established desalination plants to produce fresh water for human consumption because of the desperate shortage of fres water. There are about 7 500 desalination plants in 120 countries in the world, which provide about 0.1% of all freshwater used by humankind.

South Africa is fast running out of potential new sources of water and desalination is likely to become one of the options to be considered to augment current water supplies after 2015. Desalination's high costs mean that it only becomes a feasible option when all other practical options such as reduction, recycling, or re-use are exhausted, and when the high costs are no longer such a significant factor in decision-making.

Key Associated Topics:

Agriculture; Groundwater; Water Quality and Availability.

Associated Organisations:

Freshwater Research Unit – University of Cape Town; Water Institute of Southern Africa.

Photo: Copyright South African Tourism

DESERTIFICATION

Guest Essay by Communications Unit ~ Desert Research Foundation of Namibia

Desertification is defined as "land degradation in arid, semi-arid and dry sub-humid areas resulting from various factors, including climatic variations and human activities". The important point is that there is loss of productivity of the land. Land that originally could support many people with, for example, crops, pastures for livestock and wild foods, becomes more difficult to farm and less able to provide the necessary resources for people and animals.

A desertified area may look more barren and desert-like, but this is not always the case. Unfortunately, desertification is often misinterpreted as encroachment of deserts (especially of sand dunes) on previously fertile land. The true physical manifestations of desertification include deforestation, deterioration of rangelands, soil erosion, bush encroachment and salinisation of soils.

Issues Three essential components of desertification should be noted:

◆ Human influence is almost always dominant, but climate – particularly potential long-term climate change – may also have an influence;

◆ The processes are not the result of normal rainfall variation or drought, but may be initiated or worsened by these short-term factors;

◆ The processes are more or less irreversible, except possibly over a long time period and usually at considerable cost.

Human influence comprises actions such as permanent settlement in areas that were previously used on a seasonal basis, and excessive pressure on land from large numbers of livestock, which do not give any chance for recovery of favoured grasses and young trees. The most palatable grasses, including the perennials, diminish and do not leave a seed-bank as they are grazed heavily. In addition, the physical action of the hooves of many livestock destroys grass clumps and prevents new ones from starting. Similarly, young trees do not germinate and there is no replacement of old ones.

These processes are exacerbated by years that are drier than normal, and in drought years degraded land is likely to look more barren and desert-like, especially when livestock are also in poor condition. But the process of degradation does also continue during periods of normal or good rainfall, if the causal factors persist.

It is difficult to reverse the pattern of degradation and to restore the productivity of land, especially if it has been degraded for many years. Soil that is lost by erosion takes a very long time to recover – at least hundreds of years. Once lost, the seed bank of grasses may take years or decades to build up again. Areas that have become impenetrable because of overgrowth of invasive trees, such as bushy thorn-trees, must be physically cleared before they can be utilised by livestock. Soils that have become salinised by inappropriate irrigation will be infertile until the heavy salt load is washed out by natural rainfall, a process that is obviously slow, especially in areas where rainfall is low already.

A failure in policies and planning to accommodate the variability of climatic conditions in tropical and subtropical dry lands is a main contributing factor. For instance, planning for drought conditions in a systematic way is rarely carried out, with the result that emergency measures are engaged unnecessarily. The emergency measures may even become the norm, which carries the likely impact of over-exploitation of scarce resources such as groundwater and pastures, and permanent settlement on lands, which are better suited to seasonal use.

Photo: Desert Research Foundation of Namibia

Scope In 1994 it was estimated that about 30% of the agriculturally usable land in Africa was degraded, and about 23% of the usable land worldwide. Strong and extreme degradation amounted to about 8% in Africa. Converting these figures into monetary terms is even more difficult than making the estimates in the first place. The total figure for Africa probably runs into hundreds of millions of dollars.

Solutions Improved management of land that recognises natural variability of rainfall and finds alternative ways of making a living, especially in drought conditions, has had successful results. Many policies and laws that govern farming and land tenure are the root cause of desertification. Correction of these policies so that they promote sustainable management of natural resources is necessary for long-lasting solutions. In parallel with this, restoration of degraded land can only be sustained if it carries the support and dedication of the farmers on the ground.

Countries in southern Africa are developing and implementing National Action Plans to combat desertification. The work has to be carried out on many fronts, from the level of farmers and communities, to government and non-governmental agencies that provide them with support, up to the level of national and international policies. The International Convention to Combat Desertification, implemented by the UN and member countries that have ratified the Convention, is the main organising body.

Key Associated Topics:

Deforestation; Global Warming; Groundwater; Land Degradation; Soil; Water Quality & Availability.

Associated Organisations:

Desert Research Foundation of Namibia; Environmental Monitoring Group; NaDeet.

DEVELOPMENT OF LAND

(The EIA, Public Participation and Rezoning Process)

Guest Essay by Patrick Dowling ~ WESSA

The rezoning of land from one land-use to another is one of the things that can "trigger" an environmental impact assessment (EIA). As this often involves the conversion of healthy agricultural land (in a country where this is scarce) or pristine natural tracts to another less generally beneficial purposes, which will put numerous pressures on the environment, it is important that rezoning is dealt with very carefully.

Photo: Bruce Sutherland - City of Cape Town

Proponents (developers), local authorities, interested and affected parties, provincial authorities, consultants, specialists and non-governmental organizations can all play roles in an EIA, often making it a complex process even without disputes. When these do occur lawyers and the courts become involved.

The officials should scrutinise applications, assess Plans of Study for Scoping and EIA processes, evaluate Scoping and Environmental Impact Reports and then make a decision. Officials, the public and the consultants need to have a clear understanding of the roles and responsibilities of the officials – and the rules of the game. The provincial officials should be available to provide information during public processes. The public perception that officials are far more ready to advise and assist developers and their consultants must be corrected through constructive action.

No two EIA processes are the same, but in any particular development proposal, there is wide consensus about the minimum required in terms of the process. The Plan of Study for the EIA process should become a key focus area in which the public are also involved. It must serve as a blueprint for a process that everyone feels that they can live with.

Public participation takes various forms – open house meetings, public meetings, workshops with various stakeholder groups or meetings with specific I&APs. A public process that is defined prior to any public involvement more often than not incorrectly anticipates the level of public concern and opens the door to accusations of manipulation and lack of transparency. The public also needs to be included in the establishment of time frames, to which consultants can then be held. In almost every process, the public complains about the process being rushed and that insufficient time has been allowed to digest the documents, understand the issues and then comment. Processes are dogged by the "hurry up and wait syndrome" where each step appears unreasonably rushed and then there is a long period of waiting between steps, where the public have no idea what is going on, or why they are waiting.

It should be clearly understood that no specialists should be appointed, or specialist studies undertaken prior to the completion of the Scoping Process. It is becoming increasingly common for a process to be far advanced by the time it reaches the public arena. This too arouses suspicions of manipulation.

The inclusion of the public in the design phase of the EIA process has positive implications for the developer and the public participation consultant too. It would serve the purpose of smoothing the process and provide a mechanism for dealing with those who are guilty of stringing out the process unnecessarily, possibly to fulfill a hidden agenda. A disgruntled public is easily manipulated by vociferous and unreasonable arguments against the development.

The key issue is that the public knows when they are being meaningfully included in a process and when they are being fobbed off. Public participation is unwieldy, untidy and often unpredictable, but this is not reason enough to try to short-circuit it.

A transparent and adequate public process assists in resolving conflict and minimises the odds that the decision will be challenged in court. Appeals and High Court reviews hold up a development application process for much longer periods of time than does a comprehensive public process – and are more costly too. The public should therefore be involved early in the process and should be part of the design of the process and the adoption of rules. Transparency might have become an overused word, but this does not make it less important.

The public should be guaranteed access to any information that is relevant to the issue under discussion There should be equal opportunity to participate in proceedings, advertisements need to be obtrusively and timeously placed in publications and, throughout the life of the process, adequate time must be allowed for comment. Reports must be accessible to members of the public and not loaded with difficult technical jargon. Meaningful alternatives to the proposed development must be discussed and evaluated during the process.

Realistic development alternatives are often not put forward or adequately assessed. This aspect of the EIA process can be circumvented by putting on the table alternatives that are either obviously uneconomic to the developer or so extensive that the environmental impacts are ludicrous. The process will consequently automatically home in on a developer's serious proposal, without adequate consideration of alternatives. Alternative that are put forward for consideration must all be reasonable, feasible and viable. It is also incumbent on the developer to demonstrate the need and desirability of the proposed developments. This is seldom done. This includes the developer providing clear evidence of the profitability of the development proposal to back up the stated minimum level of development to "break even".

In addition, there should be some mechanism by which it is possible to reject out of hand a development proposal that is obviously outrageous in terms of its impact on the environment and its conflict with existing Structure Plans and other planning legislation. Once an EIA process has been started, it is almost inevitable that some sort of development will occur. This perception needs to be changed. No go areas must be clearly established. No property owner has the right to expect to be granted enhanced property rights.

Decision-makers have fallen into the habit of approving developments, but slapping on a hefty number of complex conditions on the approval. Some of these "conditions", by rights, should have been fully investigated by specialists prior to the approval of the development. The effect of some of the attached conditions is to transfer decision-making responsibility to the Environmental Monitoring Committee. This responsibility eventually comes to rest with the Residents' Association that is formed. Under these circumstances, there is little prospect of the conditions being enforced.

The solution is simple. Decision-making responsibility and enforcement of compliance rests fairly and squarely on the relevant tiers of government. It is not something that can be conveniently "privatised". If necessary, the EIA process must be revisited over and over again, until almost all conditions have been met, so that a simple and enforceable Record of Decision can be issued. If decision-makers are doubtful about their capacity to ensure compliance with the conditions in the Record of Decision, then the development should not be approved. As citizens of South Africa, we need to take a firm stand against authorities abrogating their decision-making responsibilities.

Key Associated Topics:

Environmental Ethics; Environmental Impact Assessment (EIA); Interested & Affected Parties (IAPs); Land Use Planning; Public Participation; Scoping; Social Impact Assessment; Stakeholders; Urbanisation.

Associated Organisations:

Cebo Environmental Consultants; Duard Barnard; Settlement Planning Services; Web Environmental Consultancy; WESSA.

Photo: Earthyear Magazine

DILUTION

Dilution in the context of environmental management describes a strategy to deal with pollution. For example, certain liquid effluents are disposed of down sewers by being heavily diluted with clean or relatively clean water. This used to be a cost effective waste management solution when water costs were very low (i.e. a sub-economic price which did not relate to the cost of supplying the quantities and qualities of water). However, the new National Water Act has dictated that water prices will be more economically related and, as a result, the use of potable water as a dilutant becomes prohibitive and alternatives must be sought. This strategy forces industry to look for alternatives, which often include either reducing or eliminating the effluent in the first place. It is also possible to "dilute" chimneystack emissions by adding air or burning gases to dilute concentrated emissions. The current weak South African air pollution legislation means that there are no strong controls or pressures regarding the dilution of air emissions. (It could be argued that in some cases dilution of particular air emissions makes little difference to the health or environmental impacts of the gases and particulates.) However, the current air pollution legislation is in the process of being re-written and the new regulations will be stricter, echoing the strategies of the water legislation.

Key Associated Topics:

Air Quality; Atmospheric Pollution; Freshwater Ecology; Integrated Pollution & Waste Management; Rivers and Wetlands; Pollution; Particulates; Water Quality & Availability.

Associated Organisations:

Enviroserv Waste Management; Groundwater Association of KZN; Groundwater Consulting Services; Water Institute of South Africa; Institute of Waste Management.

DIOXINS

Dioxin refers to a group of chemical compounds that share certain similar chemical structures and biological characteristics. There are several hundred of these toxic compounds that are members of related families. The families are complex chemicals with complex names such as chlorinated dibenzo-p-dioxins (CDDs), chlorinated dibenzofurans (CDFs) and polychlorinated biphenyls (PCBs). The most well known dioxin is 2,3,7,8 - tetrachlorodibenzo-p-dioxin (TCDD) and sometimes the term "dioxin" is used to refer to TCDD only. Dioxins such as CDDs, CDFs, and PCBs were created intentionally for industrial purposes, but others are created unintentionally from processes such as the incineration of waste, the burning of fossil fuels, and the chlorine bleaching of pulp and paper. Cigarette smoking also contains small amounts of dioxins. Dioxins are also produced in nature in forest fires and volcanoes.

Dioxins are extremely persistent compounds. When released into water they tend to settle and can then be further ingested by fish or aquatic organisms. Dioxins can also be deposited on plants and be taken up by animals and other organisms.

Different dioxins have different toxicities, but mostly they are found in a mix rather than as single compounds. The most noted health effect in people exposed to large amounts of dioxin is chloracne. This is a skin disease with acne-like lesions. Other effects of exposure to large amounts of dioxin include skin rashes and skin discolouration.

Another concern is that exposure to small amounts of dioxin could cause cancer in adults who were exposed to dioxins in the work place for many years. Most of the population has low level exposure to dioxins, and it has been estimated that over 95% of such exposure comes through the dietary intake of animal fats. Dioxin is also found in small quantities in the soil.

First world countries have moved to reduce the amount of dioxins in the environment by controlling industrial use (e.g. PCBs have been banned), and ensuring high temperature incineration of certain wastes to optimise complete burning, thereby ensuring that dioxins are not inadvertently produced as a by-product of incomplete incineration.

Guest Author: Dr Kelvin Kemm

Key Associated Topics:

Atmospheric Pollution; Carcinogen; Environmental Liability for Remediation; Fossil Fuels; Hazardous Wastes; Incineration; Integrated Pollution Control and Waste Management; Petrochemicals; Polluter Pays Principle; POPS & POPS Convention; Precautionary Principle.

Associated Organisations:

Institute of Waste Management; Chemical and Allied Industries' Association.

DISSOLVED OXYGEN CONTENT

Dissolved oxygen content (DOC) is an important means of measuring the "health" of a stretch of river or quantity of water by checking the concentration of oxygen. The healthier the water body is, the higher the dissolved oxygen content will be. A level of 8-9 mg/l (milligrams of oxygen per litre of water) could be expected in a normal stream, whereas 4 mg/l and below are situations where it is very difficult for organisms to survive and usually means that there are some forms of pollution present. Dissolved oxygen content is usually measured in mg/l (milligrams per litre) or mg/m^3 (micrograms per cubic metre). The relationship between dissolved oxygen content within water and the biological oxygen demand (BOD) of the organisms within a body of water helps us to understand the carrying capacity of organisms for that particular stretch of water. Studies of aquatic ecosystems consider BOD and DOC when trying to determine management strategies for these systems. The addition of high BOD wastes to an aquatic ecosystem can have disastrous effects upon the system and its ability to support life.

Key Associated Topics:

Carrying Capacity; Freshwater Ecology; Rivers and Wetlands; Water Quality & Availability.

Associated Organisations:

Freshwater Research Unit – University of Cape Town; Water Institute of Southern Africa.

DRIP IRRIGATION

Drip irrigation is a very economical irrigation system, which delivers small quantities of water to crops close to their roots with minimal loss to evaporation or runoff. It consists of thin pipes, which are punctured beneath the shadow of each plant and deliver drops of water rather than a stream or flow of water. The Israelis who began to grow oranges in the desert with limited water supplies pioneered drip irrigation. Their success in growing the world renown "Jaffa" oranges has encouraged many others to develop irrigation systems, which deliver minimal quantities of water exactly where they are needed. South Africa is a "water deficient" country, meaning that water is in short supply. Agriculture and market gardening enterprises need to make more use of drip irrigation both to save money and also to reduce overall water consumption. Some 65% of total water usage in South Africa is consumed by agriculture and conservative estimates suggest that up to 40% of that water may be lost by evaporation due to wasteful field spray irrigation methods. Increased water costs in South Africa over the next two years are likely to put pressure on farmers to change their irrigation methods to more economical and efficient methods.

Key Associated Topics:

Agriculture; Eco-efficiency; Permaculture; Water Quality and Availability.

Associated Organisations:

Agricultural Research Council; Agricultural Resource Consultants; Plaaskem.

Photo: African Wildlife Magazine

ECO-CATASTROPHE

This describes any major catastrophe which results in a major decline or drop in biodiversity or environmental quality. An eco-catastrophe can be both natural and man-made. An example of a natural eco-catastrophe could be a major volcanic eruption resulting in lava flows, ash, clouds of dust and smoke causing serious air, water and soil pollution, and the destruction of vegetation and farming land. An example of a human-induced eco-catastrophe would be the loss of the Lapp reindeer herds in Lapland as a result of the nuclear accident at Chernobyl in 1986, or the ecological destruction of the Alaskan coastline caused by the oil spill from the super-tanker, Exxon Valdez in 1989. The volcanic eruption of Mount St Helens in the State of Washington, USA, was a natural eco-catastrophe, which resulted in the largest natural ash cloud of the 20th Century, and the biggest landslide known to man, releasing an estimated 4.6 billion cubic metres of rock, almost a third of the mountain's visible surface mass.

Key Associated Topics:

Chernobyl; Environmental Liability for Remediation; Environmental Risk and Liability; Exxon Valdez Disaster; Fire.

Associated Organisations:

Department of Environmental Affairs & Tourism; Earthlife Africa; South African Institute of Ecologists and Environmental Scientists; WESSA; WWF.

ECO-EFFICIENCY

Eco-efficiency describes the ecological efficiency of goods and services by measuring their economic price (that includes their consumer demand and their monetary cost) and checking it against its production or manufacturing successes in reducing environmental impact, improving quality of life and lessening overall negative environmental impact on the earth. The higher or better the eco-efficiency ratio of the goods and services, the less the negative environmental impact and the lower the use and abuse of natural resources.

Manufacturers must be encouraged to improve their eco-efficiency by using Life Cycle Assessment and other environmental evaluation tools such as environmental auditing and environmental impact assessment and sustainable design theory (See the Centre of Sustainable Design's website at: www.cfsd.org.uk).

Key Associated Topics:

Cleaner Production; Industrial Ecology; Life Cycle Assess-ment; Organic Farming; Permaculture; Renewable Energy; Solar Power; Wave Power; Wind Farming.

Associated Organisations:

African Wind Energy Association; Agama Energy; Beco Institute for Sustainable Development; Energy Efficient Options; Oelsner Group; Renewable Energy And Energy Efficiency Partnership; Sustainable Energy Society of SA.

ECO-LABELLING

Eco-labelling is a concept that ensures that products are labelled with information that informs the consumer or the user exactly what ingredients are in the product, whether or not they are "environment friendly" and whether the packaging is bio-degradable or recyclable.

A current example of the importance of eco-labelling can be found by looking at the importance of eco-labelling in genetically modified products. Sound and accurate eco-labelling is important because those who may be allergic to particular products need to know if those products or genetic components of those products are in the food they eat. Eco-labelling would form a part of a company's environmental management system and would indicate how serious a company is about ensuring that its products and packaging do as little harm as possible to the environment. The South African Bureau of Standards operates an "environment mark" in conjunction with its system of agreed and approved standards. The ISO 14000 environmental management series includes two standards, which specifically refer to eco-labelling, namely ISO 14020 and ISO 14021.

Key Associated Topics:

Consumerism; Genetically Modified Organisms; Green Consumer Guide; ISO 14000 Series; Organic Foods.

Associated Organisations:

Biolytix; Biowatch SA; Enchantrix; Stratek; Clothing Textile Environmental Linkage Centre; SA Freeze Alliance on Genetic Engineering.

ECOLOGICAL FOOTPRINTS

An "Ecological footprint" is a description of the ecological impact a company or group of people have on the earth. The bigger the footprint the worse the impact. It is a complicated measurement, which includes information such as population numbers, technology used, energy consumed, water used, use of natural resources, etc. For example, a group of San Bushmen in southern Africa who live off the earth have a very small ecological footprint, whereas a large manufacturing plant in an industrial area in Europe would have a much bigger ecological footprint. The concept of the ecological footprint can be used as a sustainable development indicator to show how people or industries are working towards minimising their negative impacts on the environment. The principles of sustainable development require us to make our individual and collective ecological footprints as small as possible.

Key Associated Topics:

Carrying Capacity; Cleaner Production; Consumerism; Eco-efficiency; Environmental Impact Assessment; Industrial Ecology; Life Cycle Assessment.

Associated Organisations:

Centre for Environmental Studies – University of Pretoria; Delta Environmental Centre; Enviroleg; Natural Step The; Richards Bay Minerals; Ezemvelo African Wilderness; WWF; WESSA.

Photo: Department Minerals & Energy

ECOLOGICAL INTELLIGENCE

Guest Essay by Dr Ian McCallum

(This essay is an extract on "Spirit and Soul" from Dr McCallum's book, Ecological Intelligence. The poem is from his book of poetry Wild Gifts.)

The environmental crises of our day, ranging from the denuding of tropical forests, acid rain, air and water pollution, diminishing wilderness areas, the introduction of alien vegetation and green house warming, all have one thing in common – the human factor. A sobering thought. It is difficult to counter the argument that human beings are the most dominant and, as seen through human eyes, the most successful species on earth. Indeed, there is hardly a place on the face of our planet that we have not explored, settled, and altered in some way to satisfy our own ends and, as the writer and scientist, E. O. Wilson puts it, "we have become a geographical force more destructive than storms and droughts." It is a fact that death and extinction is on the cards for all of the Earth's species, but, prior to the emergence of the human animal, nowhere in the evolutionary narrative does it show any one species having driven any other into extinction.

Notwithstanding the many tiresome predictions of ecological doom already proved wrong, we would be naïve in the extreme to deny that the alarm bells regarding the human equation on this planet are ringing and that they have been doing so for a long time. The big question remains – is it too late? Something in me says yes, the downward spiral of human co-existence with this planet has already begun and that it is too late to make amends; but something in me also says no. It is that something that allows me to continue my work as a psychiatrist, that affirms the belief that people can learn to see themselves differently and that if we can relinquish the arrogant notion of the human animal at the genetic apex of creation in favour of the overwhelming evidence that we are a part of a web of life, then I believe we have good reason to be optimistic about our relationship with the Earth. Let us be clear about this relationship, however. The Earth doesn't need healing. We do. We are the ones who need to do the reaching out, not to save Nature or to save the Earth, but to rediscover ourselves in it. Only then can we heal the split.

To effectively rediscover ourselves in Nature it is crucial that we learn a lot more about our own nature, about our past, especially our wild side and why, as a species, we are blind to what we are doing to the Earth. We are often told that conservation is doomed without public participation and I agree with this perception. But there is more. I believe it is doomed unless we can become a lot more evolutionary and psychologically minded.

Psychology is the science that studies behaviour and mental processes. It includes the study of relationships – the way we relate to people, places and events, and to the way we relate to ourselves. It is also the study of the psychodynamics of survival, which includes an awareness of those two great, yet poorly understood, dimensions of psychology – spirit and soul.

Psychology is not an exact science and probably never will be, which is why for many scientists it is regarded as being too abstract or too theoretical to be relevant to their work. It is essential however, that this attitude be changed for not only are we all naturalists of sorts, all of us scientifically curious, we are all philosophers and psychologists also, even in a small way. And what is more, we can't help it! It is in our nature to be objective, to explore and to define our outer world, but this is only a part of our nature.

Human nature is powerfully subjective also; it is both abstract and abstracting, never entirely satisfied with what can be measured, which is why, for everything wonderful about science, it seldom answers the deep, existential questions in our lives. Mostly indescribable outside of the language of poetry, how can we ever measure those great imponderables of non-science like spirit and soul? They refuse to be measured and yet, it would seem, we cannot live without them. Usually dismissed by science as the "untouchables", these dimensions of human psychology are the catalysts in the human quest for meaning. The words are at the tips of our tongues in our descriptions of kinship, belonging, connection and continuity and any attempt to define the human animal without them will be one that is both dry and incomplete. They are beyond the explanatory level of the intellect, but this does not mean that they are irrelevant. Instead, they need to be understood as natural, psychologically significant and therefore valid. It is important however, that the words be rescued from their conventional religious restrictions, because unless we do so, they will continue to be regarded as being of no practical use to ecological science.

To me, it is important that the words spirit and soul be seen as complimentary opposites, that they are inextricably linked, but essentially different. For example, if spirit is anticipatory then soul is reflective. If spirit is toward continuity, discovery and the future, then soul is historical. Soul points to our origins, to our biology and ultimately toward our humanity. It is about seasons and subjectivity, of the need to belong, to share, to remember and to celebrate where we have come from. If spirit is individual, then soul is collective. If spirit is cool, pointed and soaring, and where the experiences of it are peaked, then soul is earth-bound and warm. Soul is the music of the valleys and the experiences of it are rounded. If to be spiritual is to have one's eyes turned toward the sky, towards life after death, then to be soulful is to love the Earth and everything that has come out of it. If spirit gives us wings, then soul gives us roots. To be soulful then, is to be mindful of the wild parts of our nature, our cold and warm-blooded history and to embrace it.

Finally, I believe that the presence of the soul places of the world – the wild places and the animals that live there, are important for our overall sense of identity. It fits into a notion that a sense of soul and with it, our sense of self, is deeply linked to an ancient, genetic memory of where we have come from. It is a reminder of the privilege of being a human animal. Let this poem speak for itself.

WILDERNESS

Have we forgotten
that wilderness is not a place,
but a pattern of soul
where every tree, every bird and beast
is a soul maker?

Have we forgotten
that wilderness is not a place
but a moving feast of stars,
footprints, scales and beginnings?

Since when
did we become afraid of the night
and that only the bright stars count?
Or that our moon is not a moon
unless it is full?

By whose command
were the animals
through groping fingers,
one for each hand,
reduced to the big and little five?

Have we forgotten
that every creature is within us
carried by tides
of earthly blood
and that we named them?

Have we forgotten
that wilderness is not a place,
but a season
and that we are in its
final hour?

Key Associated Topics:

African Environmental Tradition; Conservation; Consumerism; Cultural Resources; Environmental Ethics.

Associated Organisations:

Botanical Society of SA; Earthlife Africa; Eco Link; Ezemvelo African Wilderness; WESSA; Wilderness Leadership School; WWF.

ECO-MANAGEMENT AND AUDIT SCHEME (EMAS)

EMAS is a voluntary European Community programme to promote continuous environmental performance improvements of industries by committing company site management teams to evaluate and improve their environmental performance and provide relevant information to the public. The scheme does not replace existing European Community or national environmental legislative requirements. Companies who want to register their sites must set targets for improving their environmental performance (e.g. reducing pollution, saving energy or recycling wastes) and must make the management changes necessary to achieve the improvements.

The EMAS requirements are demanding and require serious commitment on the part of companies. In years to come, South African companies who carry out a great deal of trade with European companies will probably have to adhere to EMAS requirements in order to be able to compete with European businesses. For further information, see the EMAS Frequently Asked Questions page (FAQ) at www.europa.eu.int/comm/environment/emas.

Key Associated Topics:

Cleaner Production; Environmental Auditing; Environmental Management Planning Systems; Industrial Ecology; Integrated Environmental Management; Integrated Pollution & Waste Management.

Associated Organisations:

Wales Environmental Partnerships.

Photo: Bruce Sutherland - City of Cape Town

ECONOMICS

An Environmental Perspective — Unmasking the myths of the predatory lion economy

Guest Essay by Wayne Visser ~ KPMG

It is not uncommon to hear our present global economic system being compared to a predatory natural environment. We might imagine the market as a great African plain where competition for scarce resources dominates the life of every species. We can see successful companies, as the supreme hunters in this eat-or-be-eaten world, like the awesome lions of the bushveld.

It is certainly a compelling analogy and one that is perpetuated in boardrooms and business schools around the world. We need only look at our economic and business language to realise that predatory behaviour is believed to be imperative to survive and thrive in the marketplace. We must 'target' our customers, 'chase' higher growth and better profits, 'hunt' for potential merger or acquisitions partners, and 'go for the kill' in our sales pitches.

But is this lion-king economy really the kind of place that we want to live in? Of course, it's great if you're at the top of the food chain, but what about the other, more vulnerable species? Is the free market really working 'for the common good', as it should according to Adam Smith's 'invisible hand'? Or are communities and the environment being sacrificed to keep companies well fed?

This article questions many of the beliefs which dominate economic thinking today, and asks whether we can design an economy 'as if people and the planet mattered' – a sustainability-driven elephant economy.

Myth 1: The bushveld exists solely to serve the lion-king

One hundred years ago, nobody questioned the role of the economy in our lives. Business and the economy existed to serve the welfare of society – to provide the goods and services we needed in order to maximise our quality of life.

Certainly, it was not the other way round. People did not exist to serve the economy.

And yet, some time in the past century, the tables have turned. Personal, community and even country survival are increasingly dependent on working within the formal, money economy. Furthermore, companies are able to justify all kinds of unethical practices in the name of profits or job creation, whether it is restricting the accessibility of lifesaving drugs, or causing wholesale destruction of the natural environment.

Partly this is a systemic problem in the way our economy works. It encourages dependency on money, institutionalises growth and incentivises short-sighted thinking. It is also about balance of power and accountability. Today, many companies are more influential than whole countries, yet they remain accountable only to their shareholders, whose sole criterion is dividends.

This entrenched situation is the same as saying that the entire bushveld exists only to serve the lion-king. Which, of course, is neither desirable nor sustainable, even in a 'survival of the fittest' context. In actual fact, nature is dominated by cooperative, symbiotic relationships that weave together into a complex, dynamic balance.

Myth 2: The bushveld contains unlimited food for the lion

We all know that economies depend on the natural environment, both to supply resources and act as a sink for our wastes. But the rate at which most modern economies are gobbling up raw materials and pumping out pollution far exceeds the ecosystem's ability to regenerate or assimilate these materials. Our economies, especially from the first world, are acting like lions that entertain the misguided belief that their appetite and food source is unlimited. A few statistics illustrate the point.

We have lost over ten percent of the species that were living a few hundred years ago. Conservation biologists are predicting that half of the diversity of life will be lost in the next century if the present rates of habitat destruction and disturbance continue. Already, according to the World Resources Institute, in the last 50 years we have lost, destroyed or seriously depleted two thirds of the world's agricultural land, half of the freshwater wetlands, mangrove swamps and rivers, a quarter of the marine fish stocks and 20 percent of the forests.

Some scientists are talking about humanity bringing about the 6th mass extinction in the history of the Earth. This is a matter for grave concern bearing in mind that, on the evidence of the previous mass extinctions, it takes between 10 million and 100 million years to recover former levels of biological diversity.

Unfortunately, as the election of George W Bush in the USA has shown, plugging short-term economic growth and business interests at the expense of the environment still wins political votes. So, until enough social pressure can be exerted and today's perverse economic incentives corrected, our modern economies will continue to function like planet guzzling monsters engaged in an unsustainable feeding frenzy.

Myth 3: Growth of the pride is good for the bushveld

This myth relates to a widely held belief that economic growth is always good and should be continuously strived for. At the heart of this assumption is the idea that if the economy is growing, everyone is becoming progressively better off. Wealth that is generated supposedly "trickles down" through the society and the general standard of living is raised. For this reason, growth in gross domestic product (GDP) has come to be regarded as an indictor of a country's level of development and quality of life.

The evidence, however, is beginning to challenge this myth. The United Nations' Human Development Index concludes that "the link between economic prosperity and human development is neither automatic nor obvious". In a similar vein, the World Economic Forum's Pilot Environmental Sustainability Index concludes that "there is no clear relationship between a country's observed economic growth rate and its environmental sustainability".

What we do know is that over the past 50 years, while the global economy has steadily grown, income inequality has increased (i.e., the rich have become richer and the poor have become poorer). Furthermore, several indicators that adjust GDP for negative factors such as environmental degradation, poverty and health (e.g., the Index for Sustainable Economic Welfare) show that, since the 1970s, our quality of life has been declining despite the increase in GDP.

This is a fundamental challenge to one of the biggest myths of our time and one that pervades all business and economic thinking – that growth is good and bigger is better. Now we have to face the fact that economic growth does not automatically benefit either society or the environment. And in the age of sustainability, where economic, social and environmental performance are linked, business will need to examine these relationships and impacts more carefully. When the lion pride grows, it may well be at the expense of other species and the environment.

Myth 4: The lion ignores the state of the bushveld

No one would be naive enough to say that lions are unaware of the state of their environment – whether affected by drought or rain or disease. Lions constantly pick up on signals from the bushveld, which tell them how they need to modify their own behaviour to survive and thrive in the changing conditions. Unfortunately, our modern economies often either do not receive this feedback effectively or choose to ignore it.

At the moment, if there is a tragic accident that kills hundreds of people, or an environmental catastrophe that requires extensive clean-up, GDP goes up, the economy grows and we are all meant to be better off. Which is, of course, nonsense! This is because, in the economists' own jargon, 'externalities' are not internalised. Pollution is a typical negative externality. It imposes costs on others (e.g., health costs, loss of aesthetic value, ecological damage), but these costs are not paid for by the polluter.

Unless our economies can begin to 'internalise' these externalities (e.g., through pollution or social 'sin' taxes), companies and the public will continue to ignore the steady decline in the state of the environment or health of the community. In fact, our economies not only fail to tax resource use and pollution, they still subsidise resource-intensive, polluting activities, such as the fossil fuels industry.

So the challenge is to begin to generate the correct signals in the economy. Feedback on negative social and environmental impacts needs to be translated into economic terms, through higher costs, lost revenues, taxes, consumer boycotts and the like. This will allow players in the economy to adjust their behaviour to maximise the welfare of society and the planet.

Myth 5: All other species are indebted to the lion

Money is the lubricant of the economy and sometimes it feels like it really *does* make the world go round. But we should not forget that money is meant to be a tool in our hands, a simple mechanism to help us meet our needs more efficiently and effectively. In this respect, money is like a public service – something that is meant to serve the common good of society.

Few people know, however, that most of the money in circulation is created through commercial debt (i.e. through loans made by commercial banks). According to the fractional reserve banking system, commercial banks can lend out (and charge us interest on) more than five times the amount of money they physically have (i.e., they create money 'out of nothing' and charge people for it). What's more, one of the largest expenditure items in governments' budgets is interest payable to these commercial banks; meanwhile other public services like health, education and environment are crying out for more government funds.

The interest rate mechanism that our national economic system uses is problematic for a number of other reasons as well. For instance, there is evidence to suggest that interest is one of the chief causes of the 'trickle up' effect we are seeing today (i.e., why the rich are getting richer and the poor poorer). It's obvious really: the wealthy receive net interest, since their investments generally exceed their debts and they get preferential rates, while the opposite is true for the poor, who end up paying out net interest.

Another insidious effect of interest is linked to the widely applied practice of discounting, whereby a given amount of money today is worth more than the same amount in the future. According to environmental economist Herman Daly, this can results in "economically rational" extinction of species. For example, if the interest rate exceeds the growth rate of a natural forest, it makes *economic* sense to cut down the forest, sell the wood and invest the money in an interest-bearing account rather than to allow the forest to continue growing. Daly calls this phenomenon "killing the goose that lays the golden egg". "The fact that individual capitalists are made better off by killing the goose and putting their money in a faster growing asset", he says, "does not alter the fact that society has lost a perpetual stream of golden eggs."

Myth 6: Fatter lions do not necessarily increase the health of the bushveld

A lot of the calls for increasing free trade in the globalising economy come from the acquisitive appetites of first world multi-national companies that have saturated their own domestic markets. Under the World Trade Organisation, even some basic social and environmental standards are being challenged as restrictions on free trade that should not be allowed.

Critics of the globalised free trade regime talk about the 'race to the bottom' in terms of these standards. In other words, they fear that economies and companies will simply trade with and invest in countries that have the lowest social and environmental standards. This would make narrow business sense, since higher standards always cost money and, in the short term, reduce profits.

Many are beginning to believe that Adam Smith's invisible hand of the market, which is meant to automatically result in the 'common good', is probably more a conjuring trick than the truth. There are too many examples of countries and companies sacrificing the morally 'right' option in favour of economic growth and profits. For example, one of US President Bush's reasons for not backing the global action plan on climate change (the Kyoto Protocol) was that it would hurt the US economy. Likewise, oil companies have until recently resisted being proactive on reducing their greenhouse gas emissions because their profits may suffer.

So, we should not judge the health of our society or the planet by how large or wealthy our economies or businesses are. At the same time, it is simplistic to just brand companies as villains, without taking into account the economic system that shapes their behaviour. The relationship between how well a lion is eating and the health of its surroundings is neither simple nor direct.

Myth 7: The lion is eternal king by design and birthright

It is clear that our modern, globalising economy is dominated by predatory tendencies and that in many cases, our cultural integrity, community health and natural ecosystems are being sacrificed in the hunt. But this is not to say that the economy cannot be transformed to become a more cooperative, compassionate force in society. We made it, we can change it.

It is all a question of emphasis. Until now, economics has chosen to highlight and dramatise the predatory aspects in society, whereas these are the exception rather than the rule. Even in nature, the underlying characteristic is that of interdependent relationships and symbiotic cooperation. In a sustainability era, economies' success will depend on being able to create win-win outcomes for communities and the planet.

To extend the metaphor, the predatory lion will lose its throne to the sustainability elephant. An elephant is a powerful symbol for the economy of the future. Elephants are gentle animals, with an extremely rich and complex social life. They are matriarchal societies and display a wide range of behaviours that show their compassionate nature. They demonstrate a remarkable capacity for learning and sharing their knowledge over long distances using infrasound.

Elephants are also a role model for sustainability. They have been around in various forms for 50 million years and are supreme adaptors, having survived an ice age and today still living on high mountain slopes, in tropical forests, savannah bushveld and sandy deserts. And they leave in their wake conditions that are highly favourable for other species, such as water holes and previously inaccessible vegetation.

A vision of the sustainability elephant economy

So what can we do to transform our economies into ones that serve not only business interests, but communities and the environment as well? The South African New Economics (SANE) Foundation, whose mission is the promotion of a more just, sustainable economy, describes what steps we can begin to take:

1. *Tax bads and reward endeavour* – At present we do the opposite. We need to tax pollution, non-renewable resources, currency and land speculation; and to reduce income tax.
2. *Promote local* – We need to move from export-led growth to consumption-led enterprise, especially for subsistence sectors such as food, fibre, fuel, and furniture.
3. *Restrict international currency movement* – We need to control speculative 'footloose' capital and begin to restore national sovereignty, using a Tobin Tax, and other controls. These will stimulate local capital generation and direct it to where it is needed.
4. *Take control of money creation* – We need to restore to the central bank its right and democratic duty to create the money that society needs, instead of leaving this to debt-based creation by commercial banks.
5. *Support local parallel currencies* – We need to regenerate local economies, especially rural enterprise, through developing local currencies and promoting sustainable livelihoods instead of just jobs.
6. *Issue a citizen's basic income* – We need to create a mechanism to raise all people above the poverty line. It is simple to administer and cuts the costs and bureaucracy of means-testing. The multiplier effect of increased local purchasing power will create work for idle hands and skills.

These and other changes will amount to nothing less than shapeshifting, of changing the fundamental nature and dynamics of our global economy, of dethroning the predatory lion-king and embracing the characteristics of a new royal monarch – the sustainability shape of the elephant economy.

Key Associated Topics:

Carbon Tax; Carbon Trading; Consumerism; Ecological Intelligence; Eco Taxes; Environmental Ethics; Environmental Risk & Liability; Population; Poverty; Sustainable Development Section: – Maintaining Profits, or Sustaining People and Planet? – A Corporate Responsibility – The Retail Industry – Banking & Finance – Investing in the JSE SRI Index – Accounting for Natures Services.

Associated Organisations:

Environomics; KPMG; Price WaterhouseCoopers; South African New Economics Foundation; Technikon Pretoria.

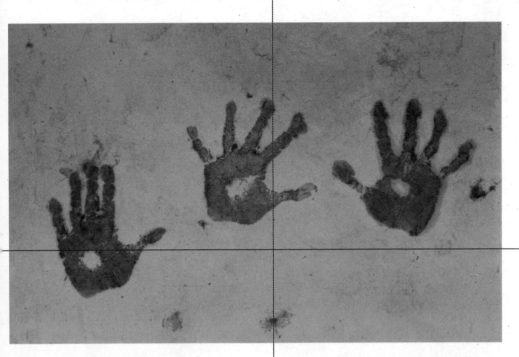

SUSTAINABLE BUSINESS SOLUTIONS

The practice of good corporate citizenship denotes an ethical, socially and environmentally responsible approach to business. Best governance practice increasingly suggests that good corporate citizenship is perceived less as an adjunct to, and more as an integral aspect and requirement of, mainstream corporate strategy and operations.

Our Sustainable Business Solutions practice comprises more than 400 business advisors in over 30 countries. We help clients improve their business performance and create long-term shareholder value by developing sustainability strategy, performance management and reporting solutions.

Corporate governance and ethics
We work with our clients to put in place appropriate governance and ethics frameworks. Our services include:
- Best practice reviews and gap analysis
- Framework design and development
- Policy and code development
- Training and capacity-building
- Design and development of stakeholder engagement programmes

Sustainable management practices
We support our clients in understanding their social and environmental impacts, managing related risks and developing practical, integrated management systems and processes, in areas such as:
- Life cycle assessments
- Climate change
- HIV/AIDS risk management

- Ethical supply chain management
- Transformation
- Corporate social investment
- Environmental, health and safety reviews and due diligence
- ISO 14001 certification

Reporting and non-financial assurance
We provide advice and assurance to our clients relating to sustainability and other forms of non-financial reporting, including:
- Report development
- Report assurance

For details on how we can assist you, please contact Andy Smith:
Tel. no.: +27 (0)11 797 4769
E-mail: andrew.j.smith@za.pwc.com

www.pwc.com/sustainability

ECOSPHERE AND ECOSYSTEMS

The ecosphere (also sometimes called the "biosphere") is that part of the earth's environment in which living organisms are found. The word is normally used to include the atmosphere, hydrosphere and lithosphere (i.e. land, air and water that support living things). The important aspect to remember is that the description refers to the fact that all the organisms live in balance with one another in a system, which is continually evolving and changing, whilst still maintaining the environment, which supports the organisms' survival. This balance is called ecological equilibrium.

The ecosystem is the basic study unit in ecology, which describes a group of interrelated living things (such as plants, animals and insects) and the physical environment (such as a pond, a tree, a dune, etc.) in which they live. There are complex interrelationships and connections between ecosystems and they are only isolated in order to help us understand how each one works. For example, a rotting log is a complex ecosystem but it interrelates with other ecosystems in the soil, water and air. By understanding how ecosystems work, scientists are able to devise methods and processes to help people avoid damaging them through their activities. If humankind could learn the lessons of operations within ecosystems, it could operate its factories and processes in the most optimal manner, which would be the most energy efficient, resource effective and successful manner possible.

Key Associated Topics:

Atmosphere; Biome: Biosphere; GAIA.

Associated Organisations:

Centre for Wildlife Management – University of Pretoria; Cheetah Conservation Fund; Delta Environmental Centre; Eco Systems; EWT; National Botanical Institute; Peace Parks Foundation; SA National Parks; WESSA; WWF.

Some governments levy a tax on dirty industries and refund them some or all of their payment if they implement cleaner technology or reduce their ecological footprint or negative environmental impact. If people and companies ensure that human activities have a limited negative environmental impact, then there would be little need for eco-taxes.

An example of an eco-tax is carbon tax. This is designed to tax those industries that release carbon dioxide into the atmosphere and contribute to climate change. Its purpose is to encourage them to either reduce the amount of carbon dioxide they release or to generate money to help pay for dealing with pollution and climatic change. There are at least three countries in the world that have already introduced a carbon tax. They are the Netherlands (tax of $1.30 per ton released); Finland (tax of $6.10 per ton) and Sweden ($40 per ton). It is, however, very difficult to calculate how much money to tax industries and early, crude calculations suggest that the cost of this pollution may be anywhere between $3 and $107 per ton of carbon. The principle, however, is important and forms part of the "Polluter Pays Principle", which is that the community and the public should not have to "pay" for the problems of pollution caused by industries that do not contribute to cleaning up or preventing the pollution.

Key Associated Topics:

Air Quality; Carbon Tax; Economics; Global Warming & Climate Change; Polluter Pays Principle; Sustainable Development Section: – Maintaining Profits, or Sustaining People and Planet?

Associated Organisations:

Enact International; KPMG; South African New Economics Foundation; SouthSouthNorth.

ECO-TOURISM

The idea behind eco-tourism is to create a tourist industry, which can help the economy by creating jobs but also can help in the wise management of natural resources. Tourists visit areas to see and photograph wildlife and scenic beauty. Their numbers and methods of access must be carefully controlled to retain the spirit and natural status of an area and to avoid unduly disturbing both wildlife and ecosystems. For example, it has been estimated that one lion living to seven years of age in Kenya results in over $515 000 being spent by tourists viewing that lion. However, if the lion were killed for its skin, the skin would fetch only about $1 000. Jobs are created for guides to help the tourists, staff to run the hotels and rangers to manage the reserves. Numbers of wildlife are carefully managed to prevent over-exploitation of the vegetation by too many animals. This is called sustainable management of wildlife resources. The costs of this are subsidised by the income from the tourists. Eco-tourism is seen as one of the fastest and most effective methods of creating new jobs. Studies suggest that every tourist stimulates the creation of up to seven new jobs as a result of the money he spends and the effort and infrastructure that is set up to meet the tourist's needs.

Photo: Conservation Corporation - Londolozi

ECO-TAXES

Eco-taxes are taxes, duties and fees, which are levied by governments to encourage companies and individuals to be more aware of environmental protection requirements and reduce their levels of pollution. A simple example is a compulsory government deposit on a can or bottle, which is refunded when the can or bottle is returned to the store.

SUSTAINABLE ECO TOURISM

The Cederberg boasts one of the greatest natural galleries of San rock art in southern Africa dating back thousands of years - 2,500 rock art sites, together with abundant archaeological and fossil sites, entice you to explore this conservancy.

Based on the principles of sustainable eco-tourism, Oudrif and the Clanwilliam Living Landscape Project are great examples of how to utilise the diversity of the local people, the land and its rich heritage. The area is a real challenge and these ventures work alongside one another in committed synergy to deliver sustainable eco-tourism, which focuses on:

OUDRIF

cederberg, western cape

- Protecting heritage sites and their environs, thereby ensuring continued integrity of the sites;
- Rehabilitating damaged areas by removing aliens and reintroducing indigenous plants which bring with them indigenous insects and animals, thereby rehabilitating the ecology of the area;
- Developing tourism projects that financially benefit the local people and encourage an exchange of knowledge of the environment to everyone's benefit; and
- Ensuring a positive awareness of, and impact on, the environment.

OUDRIF - ON THE DORING RIVER

Oudrif is a haven at which to base yourself while exploring this magnificent and diverse land. Bill Mitchell will indulge you with his culinary skills, passion and in-depth knowledge of the area. Five delightful cottages overlooking the Doring River were built using local materials (straw bales, reeds and alien trees) and labourers, ensuring a distinctly unique, restorative stay while adhering to sustainable eco-tourism principles.

The construction of Oudrif utilised local labour in building the resort, which has resulted in the transfer of practical skills and increased the knowledge of the value of local fauna and flora and heritage sites. Local labour still remains vital for the maintenance and provision of the excellent service at Oudrif.

This is the place to learn to breathe slowly again, to reconnect to your self and to indulge in the rich offerings of this ancient landscape. Whether you hike to San rock art or fossil sites, go river rafting or fishing, or just wallow in the cool waters, you will leave rested and simultaneously revitalised.

Tel. 027 482-2397; or email moondance@49er.co.za; or website www.oudrif.co.za

THE CLANWILLIAM LIVING LANDSCAPE PROJECT (CLLP)

The CLLP is an outreach programme of the Archaeology Department of the University of Cape Town. It focuses on sustainable eco-tourism principles so as to recapture the spirit of the San and bring together the diverse cultural, social and economic elements of the Clanwilliam region. It also champions respect for the environment and encourages local people and visitors to recognise the landscape as a valuable resource for economic growth and social development.

Information transfer is the key. Locals are trained as rock art guides. Their training includes life skills, catering, craft making, office management and nature conservation awareness. The programme works, and scholars gain valuable insight about their own heritage and this living landscape. Tourists also share a unique experience as they interact with the trained guides who share the San rock art sites with them. Unique crafts made by the locals are available as mementoes.

Tel. 027 482-1911, or email chap@lando.co.za

Whilst this may be the case, it is clearly important that the population recognises the value of the natural resources, which attract tourists to the country and subsequently makes sure that those resources are not destroyed or sterilised by inappropriate development, pollution and poor environmental management behaviour.

Key Associated Topics:

Cape Floral Kingdom; Carrying Capacity; Conservation; Cultural Resources; Gondwana Alive Corridors; Hunting; Mountains; National Parks; Poverty Alleviation; Stewardship; Transfrontier Parks; Wildlife; Sustainable Development Section: – South African Tourism.

Associated Organisations:

Africa Birds & Birding; Africa Geographic; Africat Foundation (The); Bio Experience; Birdlife of SA; Cape Enviro Group; Cheetah Conservation Fund; Cheetah Outreach; De Tweedspruit Conservancy; Eco Africa Travel; EcoBound Environmental & Tourism Agency; Engen; Ezemvelo African Wilderness; Ezemvelo KZN Wildlife; Hiking Africa Tours; IUCN; Mondi Wetlands Project; Open Africa; Space for Elephants; Wales Environmental Partnerships; White Elephant; World of Birds Wildlife Sanctuary.

EL NIÑO

El Niño, meaning "the Christ Child" in Spanish, is the name of a warm, Pacific Ocean current which periodically flows southward along the coast of Ecuador, as opposed to the normal, cold, north flowing current. The current begins somewhere between Papua New Guinea and Micronesia (a group of islands in the West Pacific) and flows across the Pacific to Peru. The current was discovered by fishermen working from Spanish ports in the Pacific in the 17th Century who called it "El Niño" because, although it flowed irregularly, its times of flow tended to be around Christmas. When El Niño flows, it causes climatic changes in the Pacific region because water temperatures can reach up to 30 degrees centigrade. El Niño is associated with the Southern Oscillation (a variation in inter-tropical atmospheric circulation), which together are known as an El Niño-Southern Oscillation or an ENSO event. The current can prevent the up-welling of nutrient rich cold waters, which results in the death of large quantities of plankton which, in turn, causes a reduction in population of surface fish that depend on the plankton for food. A serious El Niño event can cause changes in South Africa's climate and in the past has resulted in a number of severe and devastating droughts.

Key Associated Topics:

Atmosphere; Climate Change; Desertification.

Associated Organisations:

SA Association for Marine Biological Research (SAAMBR); Two Oceans Aquarium; Marine Biology Research Institute – University of Cape Town.

ENDANGERED SPECIES

The term 'endangered' has a different meaning to different people. Species such as the Black Eagle are often regarded as endangered but the species is not even listed in the IUCN Red List. The IUCN or World Conservation Union designed the Red List concept as a way of categorising species according to their conservation status. There is a specific set of criteria against which the population and distribution data of a species is tested for listing in different categories. The basic categories are:

Extinct: a species is extinct when there is no reasonable doubt that the last individual has died. South Africa's Quagga and Bluebuck are excellent examples of extinct species.

Extinct in the wild: means that the population is known to only survive in captivity. Many years ago the Egyptian Vulture was extinct in the wild in South Africa but subsequently it has been rediscovered as a vagrant in the north of the country.

Regionally extinct: refers to a species that has disappeared from a particular region.

Critically endangered species: are those that face an extremely high risk of extinction in the wild such as the Wattled Crane and the Riverine Rabbit. Endangered species such as the Cheetah and the African Wild Dog face a very high risk of extinction in the wild.

Vulnerable species: face a high risk of extinction in the wild and this would fit our African Penguin, Cape Griffon Vulture and Bald Ibis.

Near threatened species: are those that have been evaluated but do not fit the abovementioned criteria yet.

South Africa has a host of Red List species with very good data on mammals, plants, amphibians, birds, fish and some of the reptiles. Invertebrates still need a lot data collection before good estimates of the number of Red List species can be produced. The number of invertebrate species is overwhelming and many species still have to undergo taxonomic revision.

It is important to understand the criteria and categories of the IUCN Red List, as this will determine the conservation actions that need to be implemented to improve the status of the species. Focusing on captive breeding is often not even a recommendation of the IUCN process and is thus a waste of funds and time. Most species that are in the Red Lists require concerted efforts to save their habitats and populations through education, awareness, conflict resolution and utilisation management.

Guest Author: Prof Gerhard H Verdoorn ~ Endangered Wildlife Trust

Key Associated Topics:

Biodiversity & The 6th Extinction; Birds; Cape Floral Kingdom; CITES; Extinction; GAIA; Grasslands; Marine Life of Southern Africa; Stewardship; Tropical Rain Forest; Wildlife; Wildlife Management.

Associated Organisations:

Africa Birds & Birding; BirdLife of SA; Cheetah Conservation Fund; Cheetah Outreach; Eco Systems; EWT; Galago Ventures; IFAW; Green Trust; Khama Rhino Sanctuary Trust; National Botanical Institute; SA National Parks; SA Foundation for the Conservation of Coastal Birds; Western Cape Nature Conservation; WESSA; WWF; World of Birds Wildlife Sanctuary.

Endangered Wildlife Trust

Conservation in Action

For information on our projects, corporate and individual membership packages and ways in which you can support the conservation of our natural heritage, please contact the Endangered Wildlife Trust.

www.ewt.org.za
ewt@ewt.org.za
tel: + 27 (0) 11 486 1102
Fax: + 27 (0) 11 486 1506

ENERGY

What are the options to achieve sustainability?

Guest Essay by Dr Eureta Rosenberg

Photo: Bruce Sutherland - City of Cape Town

The biggest 'blackout' in North American history happened not decades ago, but in 2003. On August 14 large parts of Canada and the United States experienced a power failure, which brought business and industry to a standstill and affected 50 million people, trapping many in underground trains or lifts. The cause may have been a single power unit failing and the alarms of the grid monitoring system not working properly.

A month later, on September 28, Italy is without electricity. Espresso machines, televisions, traffic lights – all dead. Affecting almost 57 million consumers, the power failure is said to have been sparked by a tree touching a high-voltage line in Switzerland.

Also in August 2003, France experiences a heat wave and engineers warn that safety at some of its 58 nuclear power reactors is at risk. Due to the weather the reactor casings approach the 500°C danger zone. Attempts to cool them down by spraying cold water from the outside largely fail. Authorities are forced to waver environmental laws and allow the reactors to discharge water at more than 300°C into rivers.

And at the same time, a truck departs from an increasingly bare corner of the Limpopo Province of South Africa, piled high with the remains of indigenous trees, now firewood for sale in the city.

What can we learn from these and other recent incidents?

◆ We are highly dependent on energy. Not just for our daily domestic needs, but our economic development prospects also depend on the availability of energy.

◆ Energy supplies are vulnerable and subject to design, material or human failure – even in the most technologically advanced parts of the world.

◆ We use diverse sources of energy to cook our food, warm or cool buildings, and power engines. These include fuels like oil, coal, gas and uranium. The wealthier among us flick a switch and have electricity. Poorer people use candles and firewood (at least 50% of South Africans). We

have a vision that all South Africans will have equal access to energy sources. But in planning to provide energy to all, there is much to consider:

◆ Energy use and production harm ecosystems. Destroying indigenous trees for firewood reduces biodiversity and the productivity of our land. Coal burned in power stations releases tons of atmospheric pollutants, which contribute to global warming and acidifies farmland and streams. Car exhaust fumes also contribute to global warming and local pollution.

◆ Energy use and production pose risks to human health. Burning coal in many homes leads to air pollution and damage to the lungs. The use of fuels like paraffin is associated with accidents often involving children. Women who repeatedly carry bundles of wood can suffer damage to the spine. Exposure to radioactive nuclear materials can lead to genetic damage, disease and death.

◆ There is clearly room for improvement, and there is choice as to how we produce and use energy.

This overview considers South Africa's energy choices and asks: Are they the most sustainable ways of producing and using energy? Are they, in other words, ecologically sound, economically sound, and socially just? In this article we focus primarily on one form of energy, namely electricity.

What are South Africa's choices for generating electricity?

Mainstream electricity is generated by Eskom, a state monopoly, mainly through its coal-fired power stations on the industrial Highveld. Eskom also invests considerably in generating electricity from nuclear energy, a process which started in the early 80s. To date however it has invested comparatively little in developing the range of renewable energy options available in South Africa, such as the sun, wind or biomass. Let's consider the sustainability of our current energy mix.

1. Coal-fired electricity

Most of our electricity (87%) is generated by power stations situated in Gauteng and Mpumalanga, close to the coal mines which provide their fuel.

Photo: Dr J Ledger

Is coal-fired electricity a sustainable option? Consider this:

◆ We have ample supplies of easily mined coal and can use them to generate electricity quite cheaply, but coal is a non-renewable resource.

◆ Our coal is cheap partly because we exploit abundant supplies of so-called "cheap" labour – an issue of social justice, one of the pillars of sustainable development.

◆ The technology for generating electricity from coal is relatively inefficient and wasteful.

◆ Generating electricity from coal uses up large volumes of water, one of our most limited resources. Vast volumes evaporate daily from cooling towers. (The steam we see from power stations, and may mistake for pollution, is actually wasted water.)

◆ Coal mining contributes to acid run-off, which damages our water quality.

◆ The burning of coal releases ash into the air. When this ash is caught in filters and precipitators and sprayed with water, it results in huge solid waste dumps.

◆ Burning coal contributes to global warming, caused by the build-up of greenhouse gases in the atmosphere. The CO_2, SO_2 and NO_x from power stations concentrated in a 150 km radius around coal mines in Mpumalanga, Gauteng and the Free State, contribute to serious air pollution, and help to place South Africa among the world's top 20 polluters.

◆ The industrial Highveld emits about 94% of the 2 million tons of SO_2 (sulphur dioxide) we produce each year. Eskom can install scrubbers in smokestacks to neutralise SO_2 and NO_x (nitrous oxides), but this reduces the efficiency of combustion, resulting in more CO_2 being released.

◆ Acid rain caused by SO_2 and NO_x from coal power stations can affect 25% of South Africa's surface run-off water, as well as many wetlands, plantations, and large maize-producing areas, where it is already reducing soil fertility and increasing production costs. Downwind of the industrial Highveld, streams in the Drakensberg show signs of acid pollution and a loss of biodiversity.

2. Nuclear energy

Seven percent of South Africa's electricity is currently generated from radioactive uranium at the Koeberg nuclear reactor near Cape Town. Eskom is also developing a high-temperature nuclear reactor called a "pebble-bed modular reactor" (PBMR), which is said to be a safer, cheaper nuclear technology with an increased efficiency. An Eskom-affiliated company PBMR (Pty) Ltd is preparing to erect a demonstration reactor at Koeberg, before taking the technology to scale.

Is nuclear energy a sustainable option? Consider this (and see Nuclear Energy – The Counter Debate):

◆ Nuclear energy is often misleadingly called "clean" because the reactors do not release greenhouse gases; however, the manufacturing of nuclear fuel (enriched uranium) does produce CO_2 pollution.

◆ Releasing the energy of radioactive substances is risky: While radiation is a natural phenomenon, excess radiation is harmful, both at lethal high doses, and at lower doses which can cause genetic abnormalities and cancers. In the

Photo: Guy Stubbs Photographer

process of generating nuclear energy, new 'man-made' radioactive substances are also created. Nuclear power may thus infringe on South Africans' right to an environment which is not detrimental to their health and for this reason, the trade union to which Koeberg workers belong, does not support nuclear energy.

◆ Mining, milling and enriching uranium have environmental impacts, pose risks to workers' health, and are energy-intensive. It takes from 10 to 18 years for a nuclear reactor to produce more energy than what was used to build and fuel it

◆ Nuclear reactors have a relatively short life span of about 40 years, after which they must be decommissioned at great cost and the build-up of radioactive wastes safely disposed.

◆ Nuclear waste takes hundreds of thousands of years to decay; there is no known way to store it safely in the long-term. Some richer countries send their nuclear waste to Africa. Somalia, for example, has three dumpsites where workers handle radioactive wastes, unprotected and unaware of the risk. Trafficking in discarded nuclear fuel is a lucrative business, worth $7 billion in Italy alone.

◆ South Africa has no facility for the long-term storage of nuclear waste. Highly radioactive used fuel from the Koeberg reactor is kept on site; Earthlife Africa estimates that Koeberg has accumulated 1 000 kg of plutonium. Concerned about the accumulation of nuclear waste, the City of Cape Town is officially opposed against the building of the PBMR.

A recently released study of babies' teeth in Florida, USA, indicates much higher levels of radioactive Strontium-90 in children who live near nuclear power plants, and cancer rates three times higher than the national average. Studies in other states show that cancer rates in children decrease within two years after the closure of nuclear reactors. The European Committee on Radiation Risk points out that most risk models fail to explain the growing body of evidence of ill health in populations exposed to nuclear pollution, and believes that they had drastically under-estimated the impact of the nuclear industry on human life.

◆ The accidental release of radiation from nuclear reactors can turn large areas into wasteland and expose thousands of people to a high risk of genetic abnormalities and cancer. After an accident at the Chernobyl reactor in Russia in 1986, some 200 000 people had to be permanently evacuated; an estimated 10 000 people died from exposure to radiation during clean-ups.

◆ Nuclear reactors are expensive to plan, build, fuel and decommission. It will cost an estimated R10 billion to build the proposed demonstration PBMR. The economic feasibility of nuclear power is difficult to prove. The already mature American nuclear industry, for example, still requires large government subsidies.

3. Hydro-electric power

Because South Africa is not water-rich, hydro-electric power (generated, for example, at the Gariep Dam near Colesberg) accounts for less than 2% of the electricity produced here. Eskom imports 4% of its electricity capacity in the form of hydro-electric power from neighbouring countries.

Is hydro-electric power a sustainable option? Consider this:

◆ Hydro-electric power does not pose significant health threats.

◆ It produces few pollutants and is a cleaner form of energy production than most other sources.

◆ Large hydro-electric schemes require big dams, which displace people, destroy their livelihoods, and damage ecosystems. This is something to consider when we buy hydro-power from other countries.

◆ Over time dams silt up with eroded soil; this reduces their capacity to generate power sustainably.

◆ South Africa has few suitable water sources for large-scale hydro-electric power, but micro-hydro power generated from small waterfalls or pumped storage is an option for small-to medium-scale electricity generation.

◆ Micro-hydro power stations (which do not dam whole rivers and divert only part of a stream), have a longer lifespan and a smaller environmental impact (they may affect fish migrations).

4. Renewable energy technology

Eskom has started to explore renewable energy options on a limited scale (as have many energy entrepreneurs). For example, Eskom, working with Shell, supplies solar energy to 50 000 homes in the Eastern Cape and is developing an experimental wind farm in the Western Cape.

Are renewable energy sources sustainable? Consider this:

◆ Renewable energy technologies make use of unlimited natural resources like wind and sun. Biogas taps energy sources that would otherwise be wasted, e.g. the methane given off by rotting materials (biomass) in landfill sites.

◆ Producing power from solar and particularly wind energy results in far less pollution than power from nuclear and fossil (coal, oil, gas) fuels.

◆ 'Renewables' do have environmental impacts, for example pollution from the chemicals used to produce solar batteries, the visual impact of wind generators and the danger of birds flying into them, but these are small compared to the risks associated with nuclear and fossil fuels.

◆ Current generation solar cells may not last long enough to generate significantly more energy than what had been used to produce them.

◆ It takes current generation solar and photovoltaic power nets less than three years, and wind turbines a year, to generate more energy than what had been used to build them.

◆ Renewable energy may be a good economic investment as the international market is currently growing at between 25-45% each year. Shell estimates that by 2050, as much as a third of all energy will be generated from renewable sources.

Photo: Bruce Sutherland - City of Cape Town

The Darling Independent Power Provider wants to start a wind farm in Darling, Western Cape, and in partnership with the City of Cape Town, give consumers the choice to buy 'green electricity'.

◆ The cost of supplying electricity from renewable energy sources was initially higher than from more established technologies, which are often subsidised and have reached their current level of cost efficiency after years of research and development. The cost efficiencies of renewable energy technologies are however improving, in proportion with increased investment to research and develop the technology.

◆ The feasibility of renewable energy options for large scale – and peak demand – use is also improving. Denmark currently produces 15% of its electricity through wind generators. In Germany, private people generate enough electricity at home through photovoltaic cells to sell excess capacity to the main grid.

◆ Unlike 'high-tech' and capital-intensive technologies like nuclear power, 'renewables' allow local ownership and economic empowerment. Per unit energy produced, many more jobs can be created through renewable technologies in general (952 direct jobs) and particularly biogas generation (1341), compared to nuclear energy (80) and liquid petroleum gas (130). (For more information see RENEWABLE ENERGY).

Other energy sources

This article is limited to electricity. Had we reviewed other forms of energy, we would have seen similar trends, and concluded that many current energy sources are not sustainable, but there are alternatives to explore.

Take for example oil, the energy source of choice in most of the world. It has been associated with social injustices (Nigeria being only one case); severe environmental impacts (for example oil spills and exhaust fumes – most petrol is made from oil); harm to human health (oil refineries cause severe air pollution); and vested interests backed by political power (for example the Bush government's agreement to open the Arctic wilderness for oil drilling).

New technologies to generate and store energy include hydrogen fuel cells, which can replace petrol in emission-free cars. Hydrogen can be generated from various sources, including renewables.

Are our current choices sustainable?

South Africa's energy supply is at risk – Eskom's infrastructure is aging and by 2013 most of their power stations and the distribution grid will be reaching the end of their lifespan. How will we provide for our energy needs then? Will the choices we make be sustainable?

Eskom's choice is reflected in the level of investment in the various options included in its "integrated" electricity plan. Nuclear power appears to be highly favoured. Eskom has already spent an estimated R450 million on plans for the demonstration module of a PBMR, which will cost at least R10 billion to build. PBMR (Pty) Ltd's business plan is apparently based on selling 200 reactors, at least 10 of them in South Africa.

Is this a sustainable development plan?

When one evaluates this plan against the criteria for sustainable development, it seems that the impacts of nuclear energy on people and the environment make it a poor choice for social and ecological sustainability. Building 200 – or even just 10 – new reactors will result in the proliferation of nuclear waste – and there is no guarantee that we or the countries to which the technology may be exported, can safeguard it.

The economic sustainability of the nuclear option is also questionable. The market for nuclear energy is growing at only 1% per year. Because nuclear energy is not really 'clean', it has been excluded from Kyoto Protocol subsidies for reducing greenhouse gases. Eskom has not yet found international partners to help cover the cost of the PBMR and ensure markets. Because it is too costly to enrich local uranium, enriched uranium is to be imported and transported between Durban, Pelindaba and Cape Town, adding to the costs and the risks. Eskom's energy plan requires the current generation of South Africans to pay a steep price for an experimental technology and risky investment, and future generations to keep carrying the burden of consequence.

What would be a more sustainable choice? Our review suggests that renewable energy sources should play a far greater role in South Africa's energy mix. This would make social, ecological and economic sense.

The Department of Minerals and Energy believes that renewables have great potential to meet the needs of the developing sector of our society. Others have also noted the commercial benefit of investing in renewables: a R300 million public-private partnership is to supply 50 000 solar home systems in northern Kwazulu-Natal as part of the National Electrification Programme.

But current efforts are still ad hoc and limited, compared to international trends. South Africa should considerably step up and coordinate efforts to develop and use renewable energy sources for sustainable development.

The city of Barcelona plans to replace its main energy source, nuclear power, with solar panels on roofs and terraces, thereby ending the generation of 1 400 kg of radioactive waster per year and moving to a sustainable energy service.

There is yet another and even more critical component to sustainable energy – one involving choices which everyone can make.

Photo: Bruce Sutherland - City of Cape Town

Reducing energy use

Reducing energy consumption is a key challenge. Not even renewable sources would be able to meet ever-increasing demands. Consider this: the fossil fuels (oil, coal, gas) burned globally in one year were derived from 400 years' worth of plant growth! We are therefore using up fossil fuels faster than they can be regenerated. So it is vital to include in our energy strategy – on a much greater scale than is currently the case – steps to reduce energy demand. Ordinary consumers can start by insulating their homes and warm water supplies, and choosing energy-efficient lamps and appliances.

Changing the bulbs in Rio de Janeiro's 325 000 street lights, from mercury vapour to high pressure sodium, was a $2.7 million investment which resulted in saving $8 million per year and enough energy to provide for a small city.

High-rate consumers can make an even greater difference. South African industries squander electricity, making us one of the 15 most energy intensive economies in the world. Yet the technology exists to decrease energy consumption by 20-50% in existing installations, and 50-90% in new ones, while still delivering the same output.

Key Associated Topics:

Air Quality; Carbon Tax; Carbon Trading; Co-Generation; Fossil Fuels; Geothermal Energy; Global Warming & Climate Change; Nuclear Energy – The Counter Debate; Renewable Energy; Solar Heating; Synthetic Fuels; Wave Power; Wind Farming.

Associated Organisations:

African Wind Energy Association; Agama Energy; Divwatt; Energy Efficient Options; Energy Technoloy Unit – Cape Technikon; ESI Africa Magazine; Eskom; Green Clippings; International Solar Energy Society; Iskhus Power; LC Consulting Services; Oelsner Group; Renewable Energy And Energy Efficiency Partnership; Sustainable Energy Association of SA; Solardome SA; SouthSouthNorth; Sustainable Energy Africa.

In the 1980s the United Nations set up the World Commission on Environment and Development. The Commission's work resulted in the 1987 publication 'Our Common Future' (the Bruntland Report). This report is famous for coining the phrase "Sustainable Development" as development which "meets the needs of present generations without compromising the ability of future generations to meet their own needs".

As a strategically based organisation providing an essential service in a developing country, these words are of particular importance to Eskom. Sustainability in Eskom means providing affordable energy and related services by integrating economic development, environmental quality and social equity into all business practices in order to continually improve performance and underpin development.

For this reason Eskom's involvement in sustainable development predates the WSSD with an active participation in global discussions on sustainable development. Recent participation includes involvement in projects initiated by the World Business Council on Sustainable Development, providing a leading role on projects looking at sustainability in the electricity industry and actively participating in the Environment and Energy Chapter of the International Chamber of Commerce.

A key priority for Eskom is Corporate Governance. Corporate Governance practices are continually evolving and recent developments are characterised by a drive towards sustainability. To this end Eskom is in the process of implementing the recommendations of the King Report on Corporate Governance for South Africa 2002, including the code of corporate practices and conduct and the protocol on corporate governance in the public sector 2002. Eskom was converted from a statutory body to a public company in 2002.

The two-tier governance structure of the Electricity Council and the Management Board was replaced by a Board of Directors. The Board Committees exist to assist the Board in discharging its duties. Importantly, a Sustainability Committee was formed

to address all economic, environmental and social issues including safety, health and nuclear issues.

Social Equity

Eskom's signing of the Global Compact demonstrates an appreciation of Human Rights, Equity, Labour Standards and Environmental Issues. This is emphasised by a commitment to public consultation and community involvement, programmes of equity and redress which include employment equity, electrification, black economic and women empowerment and SMME development and training. In addition, the socio-economic benefits of electrification vary from job creation in the construction and consulting industries, to small business development such as hair salons and "spaza" shops. Electrification also has environmental benefits such as improved ambient air quality due to a reduction in domestic coal burning. Overall, these programmes lead to an improved quality of life. Further, Eskom's corporate social investment programmes are overseen by the Eskom Development Foundation.

The Eskom Development Foundation is an independent Section 21 company, which incorporates and integrates Eskom's social responsibility initiatives.

The Development Foundation operates in nine provinces of South Africa in areas that a underdeveloped, especial in rural areas and n urban settlements. It provides grants for soc projects/programmes. The Foundation has six natio programmes. These are the small busine opportunities exhibition, Eduplant, Wome development programme, HIV/AIDS scho development programme, Education programm and the Youth development programm

Environmental Quality

The assessment and measurement of Eskor environmental performance is managed through t operational sustainability index as well as report on additional key environmental indicators and issu to the Sustainability Committee. The environmen component of the operational sustainability ind comprises four equally weighted Key Performar Indicators, namely relative particulate emissions, spec water consumption, customer satisfaction and leg compliance. Other key areas include la management, ambient air quality, gaseous emissic and research and development in these area The Chief Executive bears responsibility for Eskor overall environmental performance.

Economic Development

Eskom's strategic intent is to be the pre-eminent African energy and related services provider, of global stature. Thus, in a bid to bridge Africa's electricity infrastructural gap and in support of the New Partnership for Africa's Development (NEPAD), the African Energy Fund was launched during the WSSD. The Fund, established in partnership with Eskom, the Development Bank of South Africa (DBSA) and the Industrial Development Corporation (IDC), is intended as a vehicle for driving NEPAD's energy and electricity infrastructural projects and developing the electricity interconnection between individual African countries.

Eskom's research and development of clean coal technologies and renewable energy technologies shows commitment to and support for the key energy text of the main WSSD outcome document, the Johannesburg Plan of Implementation, which states: "Diversify energy supply by developing advanced, cleaner, more efficient, affordable and cost-effective energy technologies, including fossil fuel technologies and renewable energy technologies, hydro included, and their transfer to developing countries on concessional terms as mutually agreed". Eskom's renewable energy research culminated in 1998 in the creation of the South African Bulk Renewable Energy Generation (SABRE-Gen) Programme. The SABRE-Gen programme aims to, through research, assessment and demonstration, identify commercially viable renewable energy options for application by Eskom.

Eskom is currently actively researching clean coal technologies, wind energy, biomass, various solar energy technologies and is in the process of gaining approvals for the building of a Pebble Bed Modular Reactor demonstration plant in the Western Cape.

The Solar dish stirling demonstration plant was launched during the WSSD last year on the site of the Development Bank of South Africa in Gauteng. The Eskom demonstration wind farm consisting of three turbines was launched in February 2003 at Klipheuwel, near Bellville in the Western Cape.

This is enhanced by Eskom's Integrated Strategic Electricity Planning (ISEP). ISEP is the way in which Eskom decides by how much the demand for electricity is likely to grow and how best to meet and manage that demand. ISEP's objective is to optimise the supply side and demand side mix to keep the price of electricity to the customers as low as possible and ensure sufficient electricity supply for the future. The overall electricity demand in South Africa varies during the course of a day. The peak demands are usually in the mornings and the evenings when households use most of their electricity. Managing the maximum demand during these periods is the focus of Eskom's demand-side management (DSM) Programme.

An example of the DSM initiatives is the geyser control programme, which can defer the geyser load so that it does not coincide with the system peak. Eskom also co-funds the "Efficient Lighting Initiative" that promotes the introduction of compact fluorescent lights, which use less power during the evening peak. These options reduce the need for expensive generating plant for peak usage, thus reducing the overall electricity production costs.

Eskom has a realistic and realisable view for the future. This includes a diversification of its business and an expansion of its global reach, particularly into Africa. This will be achieved within an agreed framework of business objectives of which sustainability is key.

Eskom

With Energy, Anything is Possible

ENTOMOLOGY

Photo: Bruce Sutherland - City of Cape Town

Entomology is the study of insects. Insects make up by far the largest number of organisms on earth and they are likely to outlive human beings because of their excellent adaptability. However, human activity can have significant effects on their lifestyles and ability to survive. Butterflies, which are also significant pollinators, are being seriously affected by the modifications of habitats and landscapes that affect breeding cycles. One of the most significant examples in recent years was that of the Brenton Blue butterfly. The butterfly's breeding area is a minute section of wood, which was threatened by a developer and it took the intervention of the Minister of Environmental Affairs and Tourism to sort out the problem. It has been stated that insects would probably survive a nuclear holocaust and it could be similarly stated that insects have a greater resistance to chemicals than most other organisms as well as the ability to develop immunity to their effects. The battle to control insects has also touched on chemicals with the use of the previously banned DDT (dichlorodiphenytrichloroethane) arising once again in the control of mosquitoes in northern KwaZulu-Natal. It is clear that the substitute methods have not been able to control the insects and the authorities once again have had to resort to DDT.

Key Associated Topics:

Biodiversity & the Sixth Extinction; Bioremediation; Integrated Pest Control; Pesticide Treadmill; Pesticides.

Associated Organisations:

Albany Museum; Gondwana Alive; University of Orange Free State – Department of Zoology and Entomology.

ENVIRONMENTAL AUDITING

Environmental auditing is a process used mainly in commerce and industry to measure how well environmental policies, programmes and plans are being implemented in practice. The process uses checklists and questionnaires (sometimes called "audit protocols") to understand how suc-

cessfully environmental measures are being implemented and tries to help to improve systems by constructive criticism. The process depends on the co-operation of those people being audited and should not be used merely to police their activities. Simple environmental audits can be carried out in all sorts of different situations such as schools, factories, offices and shops. Although specialists may be needed for more formal and compliance-related environmental audits, simple internal environmental audits can raise a large number of issues, which can be dealt with internally and relatively easily. All that is needed are environmental policies and programmes to audit activities against. A simple checklist should be drawn up which asks performance and action response questions. In the absence of specific environmental policies, use can be made of such environmental policy tools as the International Chamber of Commerce Business Charter for Sustainable Development www.iccwbo.org, or the CERES (Coalition for Environmentally Responsible Economies) Principles www.ceres.org

Key Associated Topics:

Eco Management & Audit Scheme; Environmental Due Diligence; Environmental Liability for Remediation; Environmental Management Systems; Integrated Environmental Management; ISO 14000 Series; Life Cycle Assessment; NEMA.

Associated Organisations:

Africon; Arcus Gibb; Bohlweki Environmental; Centre for Environmental Management; Data Dynamics; Dekra-its Certification Services; Digby Wells and Associates; Eagle Environmental; Environmental Management Centre – Northwest University; Environmental Risk Manage-ment; Groundwater Consulting Services; Independent Quality Services; Kantey and Templer; Ken Smith Environmentalists; Lombard and Associates; Morris Environmental and Groundwater Alliances; NOSA; Peninsula Permits; Price WaterhouseCooper; Stratek; Technikon Pretoria; Wales Environmental Partnerships; WSP Walmsley.

Photo: Bruce Sutherland - City of Cape Town

ENVIRONMENTAL DEGRADATION

Over the past 45 years, just over one tenth of the earth's vegetated soils have become so degraded that their natural functions have been damaged to the point where restoration will be very costly, if not impossible. In the developing world, more than 95% of urban sewage is released untreated into surface waters, where it poses a serious threat to human health. Urbanisation is encroaching on more and more arable land, reducing the available land for farming and forcing existing land to be farmed more intensively. Countries throughout the world are being forced to look at the effects that their activities are having on the natural resources on which they depend. Everyone has a responsibility to prevent environmental degradation, from the individual picking up litter to stop an area looking dirty and neglected, to a country that must stop its waste from poisoning its water supplies. Environmental degradation is a global problem that requires global solutions. Those solutions were started with the initiative of Agenda 21 and Local Agenda 21, which focus on local action to deal with local issues as a start to responding to the wider, regional and national issues. Much of South Africa's new environmental legislation is being driven by the need to respond to the calls within Agenda 21.

Photo: CSIR

Key Associated Topics:

Air Quality; Biodiversity; Deforestation; Desertification; Eco-catastrophe; Environmental Health; Environmental Liability for Remediation; Hazardous Wastes; Integrated Pollution and Waste Management; Land Degradation; Polluter Pays Principle; Pollution; Rivers and Wetlands; Transboundary Pollution; Waste Management.

Associated Organisations:

Apple Orchards Conservancy; Bateleurs – Flying for the Environment in Africa; Gauteng Conservancy Association; Namibia Nature Foundation; Wales Environmental Partnerships; WESSA; WWF.

Photo: Bruce Sutherland - City of Cape Town

ENVIRONMENTAL DUE DILIGENCE

An environmental due diligence exercise or audit is carried out to identify the extent and nature of environmental risk involved when companies are bought and sold. Environmental due diligence audits have developed in response to the numbers of situations where chronic contamination of factory sites and land became a significant factor in determining sales prices, due to the enormous costs of clean up. The issues that tend to arise include illegal or unauthorised toxic or hazardous waste dumps (often consisting of waste chemicals, off-specification products, and hazardous wastes that were deemed too costly to be disposed of by professional waste contractors), workers suffering from occupational illnesses due to poor or non-existent preventative programmes or equipment, unlicensed or unregulated boilers, chimney stacks, waste treatment works, effluent disposal methods or other methods and procedures that impact negatively on the health of the neighbours or the surrounding environment. Most modern sale agreements that involve land or "environmentally unfriendly" processes now include the requirement for environmental due diligence audits as well as clauses that protect the buyer from clean up costs of residual contamination if reported within 12 – 24 months of purchase.

Key Associated Topics:

Contaminated Land; Environmental Auditing; Environmental Law; Environmental Liability for Remediation; Environ-mental Risk & Liability; Hazardous Wastes; Integrated Pollution and Waste Management; Pollution; Polluter Pays Principle.

Associated Organisations:

Barlofco; Eagle Environmental; Envirocover; KPMG; Metago Environmental Engineers; Wales Environmental Partnerships; Winstanley Smith & Cullinan; WSP Walmsley.

DEPARTMENT OF ENVIRONMENTAL AFFAIRS AND TOURISM

Private Bag X447 • PRETORIA 0001

Tel (012) 310 3911 • Fax (012) 322 5890 • www.deat.gov.za

Co-operative Environmental Governance

Sustainable development requires the integration of social, economic and environmental factors in planning, implementation and evaluation of decisions to ensure that development serves present and future generations. Environmental management and natural resource utilisation are key focus areas for sustainable development. Sustainable development implementation is a concurrent functional area of national and provincial legislative competence, therefore all spheres of government and all organs of state must co-operate with, consult and support one another. The South African Constitution (Act 108 of 1996) provides principles for the establishment of structures and institutions to promote and facilitate intergovernmental relations; this mechanism is referred to as co-operative governance. The National Environmental Management Act 107 of 1998 (NEMA) was promulgated within the framework of the Constitution and therefore reinforces the principles of co-operative governance, primarily in the environmental management sector. According to chapter 3 of this Act, every national department listed in schedule 1 and 2 must compile and gazette Environmental Management Plans and Environmental Implementation Plans at least once every four years.

DEAT is responsible for general implementation of NEMA and also for the administration of various aspects of the Act. Other organs of the state, including local government exercise powers and perform duties in accordance with NEMA principles. Officials from all spheres of government apply the NEMA principles in environmental decision-making (e.g. land use applications) and are under duty of care not to cause harm to the environment. This shows that for NEMA administration, there is a separation of powers and delegation of administrative responsibility to other organs of state. For this approach to be effective, there is a need for co-operation amongst the various organs of state.

The Committee for Environmental Co-ordination (CEC) is the key legislative institutional body for co-operative environmental governance. Schedule 1 and 2 departments, in term of NEMA, constitute this Committee. Various subcommittees have been established and report to the CEC on their environmental intergovernmental initiatives. For example, the EIP/EMP Subcommittee reports to the CEC on the implementation of Environmental Implementation and Management Plans. Other co-ordinating institutions include Working Groups. Working groups are project-orientated institution for enhancing (implementation) relations between the DEAT and provinces. Working Groups report to MINTEC. MINTEC provides a platform for aligning the Departmental strategic initiatives and planning process with provinces.

Without cooperation, there can be an exercise of powers in a manner that encroaches on the geographical, functional or institutional integrity of another sphere of government, lack of sharing of information, duplication of functions, poor accountability and lack of coherence amongst government spheres.

The completion of the first round of the implementation of Chapter 3 of NEMA has seen the submissions of the EIPs/EMPs. Stemming from this process the two key fundamental processes that have provided the core instruments towards the improvement of the co-operative governance program have been the Alignment report and the Consolidated Action Plan. Both the EIPs/EMPs provide information on co-operative governance mechanisms with a specific national department or provincial government focus. While useful and important, this approach did not provide a broader understanding of the roles, responsibilities, process, structures and mechanisms to facilitate co-operative governance within impacting and managing sectors across the three spheres of government. The primary purpose of the Alignment Report was to provide information with regard to current roles, responsibilities, process, structures and mechanisms to facilitate environmental co-operative governance within specific sectors.

The second fundamental process was the Consolidated Action Plan process, which provides a summary of all the commitments made by scheduled national departments and provinces in their EIPs and EMPs. Commitments made in EIPs and EMPs have been strengthened with the addition of outputs and timeframes in this action plan. The plan provides the Committee for Environmental Co-ordination with a tool to monitor the implementation of these commitments through the annual reporting system.

Areas that require strengthening relate to the local sphere and currently underway, is a process that interprets the EIPs/EMPs for local government, with particular reference to integration of environmental objectives in the Integrated Development Plans. Although the annual reports to some degree have provided a monitoring mechanism there is still need to improve this. A process, led by the Auditor General's Office (AGO), is the development of a system that will facilitate monitoring of compliance with the EIPs and EMPs.

Monitoring Implementation of commitments made in the EIPs/EMPs: Annual reporting process

Scheduled departments and provinces are required (NEMA section 16 (b)) to report annually within four months of the end of its financial year on the progress made regarding the implementation of its adopted EIP/EMP. The purpose of this is to assist the Committee for Environmental Coordination (CEC) to monitor compliance with the EIPs/EMPs. The role of the CEC (composed of the Director- Generals of the Departments of Environmental Affairs and Tourism, Water Affairs and Forestry, Minerals and Energy, Land Affairs, Constitutional Development, Housing, Agriculture, Health, Labour, Arts, Culture, Science and Technology, provincial heads of departments and local government representation) as coordinating body is to promote the integration and coordination of environmental functions by the relevant organs of state.

Associated benefits and achievements of the co-operative governance programme:

- Municipalities have an available source of information to align their development plans with the provincial priorities, hence reducing the gaps between the provincial priorities and municipal plans. For example, the alignment report. These sources will ensure that municipalities integrate environmental objectives in the Integrated Development Plans.

- A forum or an institution for consultation and information sharing between national departments and provinces in terms of environmental planning exists, that is the CEC subcommittee on EIPs/EMPs.

- EIPs/EMPs assist departments in joint or combined planning on environmental management.

- Minimises conflicting interests and priorities on functions of the environment amongst spheres of government.

LIST OF FIRST EDITION EIPs AND EMPs

Environmental Implementation and Management Plans	Gazette
Department of Environment and Tourism	23232, Notice 354, 28 March 2002
Department of Land Affairs	21562, Vol 423, 2000).
Department of Water Affairs and Forestry	22 929, Vol. 438, 14 December 2001

Environmental Management Plans (EMPs):	Gazette
Department of Minerals and Energy	22080, 23 February 2001
Department of	Labour (Gazette 7350, No. 23395, 27 May 2002
Department of Health	Gazette not available at date of publication

Environmental Implementation Plans (EIPs):	Gazette
Department of Agriculture	23374, Notice 659, 17 May 2002
Department of Defence	22022. Notice 249, 6 February 2002
Department of Housing	23374
Department of Trade and Industry	23254, Notice 404, 28 March 2002
Department of Transport	Gazette 24140, 13 December 2002
Free State Province	89, 13 December 2002
Gauteng province	General Notice No 488 of 2002, Provincial Gazette Extraordinary N0.46 of 22nd February 2002
Limpopo Province	Gazette 755, Notice 186 of 2001
Mpumalanga Province	Gazette 790, Notice 270, 1 November 2001
Western Cape Province	5940, 5 November 2002
KwaZulu-Natal Province	Gazette not available at date of publication
North West Province	Gazette not available at date of publication
Northern Cape Province	Gazette not available at date of publication
Eastern Cape Province	Gazette not available at date of publication

Welcome

ENVIRONMENTAL EDUCATION

Guest Essay by Prof Kader Asmal, MP

Environment and the Curriculum in South African Schools

For a number of decades, environmental learning has taken place in the margins of formal education, both in South Africa and in many other countries around the world. However, since 1997, the curriculum transformation in our country has taken up issues of environment, and the need to enable environmental learning in education has been expressed through legislation and a variety of policy initiatives. South Africa's hosting of the World Summit on Sustainable Development in September 2002 put a spotlight on the country's leaders as well as its citizens to take the lead on issues of sustainability. This socio-ecological imperative was further highlighted in the parallel UNESCO session on Education and Sustainability, where a number of pertinent issues were considered.

As Minister of Education I envisage a future where every teacher in every school in South Africa will be effectively equipped and enthused to use the environment as an essential facet of each learning area. Every child should leave school with a respect for and understanding of the value of a healthy environment, and an appreciation of how they are responsible for managing the world's resources in a responsible manner.

The environment is integral to the Revised National Curriculum Statement from Grade R to 12. Our Constitution calls on us to, "heal the divisions of the past and establish a society based on democratic values, social justice and fundamental human rights and improve the quality of life for all citizens and free the potential of each person." Section 24 in the Bill of Rights enshrines the right of every citizen "to an environment which is not harmful to their health or well-being." It is through education that we will ensure that future generations of South Africans will be able to deliver the society envisaged by the Constitution.

The Education White Paper (1995) emphasised the need for environmental education to be, "a vital element of all levels and programmes of the education and training system, in order to create environmentally literate and active citizens and ensure that all South Africans, present and future, enjoy a decent quality of life through sustainable use of resources." Curriculum 2005 positioned the environment as a phase organiser, which can be integrated into all learning areas. One of the principles of the Revised National Curriculum Statement infuses the principles of human rights, a healthy environment, social justice and inclusivity throughout the curriculum. This is expressed in the Learning Outcomes and Assessment Standards of all the Learning Areas.

This principle is given additional support in the outline of the type of learner that is envisaged by the curriculum. "It seeks to create a lifelong learner who is independent, literate, numerate and multi-skilled, compassionate, with a respect for the environment and the ability to participate in society as a critical and active citizen."

To achieve this grand vision we require a number of things. The first is teachers who are able to educate our children. Fortunately we do have a wealth of committed teachers, school governing bodies, principals and other staff who have embraced the task of delivering the National Curriculum Statement. I encourage them to grasp the opportunities that the curriculum offers and to support the enthusiasm of the learners. They should be setting an example for these learners by being enthusiastic, prepared and ready to deliver the best they can every school day. Environmental learning, although integral to the curriculum, is too often viewed as an additional burden by teachers. Partner organisations that promote environmental education need to gain an in-depth understanding of the curriculum and the changes needed to integrate environmental learning into the curriculum.

Secondly, we need development and support strategies to enable teachers continuously to improve their skills and their commitment to the task.

Delta Environmental Centre

'Environmental Education in Action'

Holcim South Africa support for Delta Enviromental Centre

Delta Environmental Centre (DEC) is a private, independent non-profit organisation (established in 1975) that aims, through innovative education and training programmes and consultation, to enable people to improve the quality of their environment by promoting the management and sustainable use of all resources.

Since 1975, DEC has matured into a leading institution in the field of environmental education, training and consultation. Three factors have contributed to this success:

- The ability to sustain itself financially;

- The uncompromising professionalism of its personnel; and

- Its flexibility in the face of change and development.

Generous sponsors and kind donors have enabled DEC to develop a facility that optimises "ENVIRONMENTAL EDUCATION IN ACTION".

Holcim South Africa has supported DEC's teacher training programme since 1995 as part of their commitment to sound environmental management through capacity building at all levels.

DEC has been at the cutting edge of teacher training in environmental education for the past decade. The professional development and capacity building programmes for educators form the core of DEC's day-to-day activities.

Since the introduction of an outcomes-based approach to education, through Curriculum 2005, "ENVIRONMENT" has been a key component for all grades. DEC is well placed because of sustainable donor support received, to support educators to develop relevant learning programmes that deal with their local environmental issues/concerns.

Learners need to be 'active' participants in these programmes so that they begin to take responsibility for all their actions. The concept of 'ENVIRONMENT' promoted by DEC encompasses the ecological, social, economic and political components of every environmental issue/concern and deals with the interconnectedness of these components.

Approximately 5000 educators, per annum, mainly in Gauteng, North West and Limpopo Provinces, participate in DEC's environmental education training programme. Our vision is to contribute to schools becoming 'centres of environmental action' that develop environmentally aware learners who are committed to having a 'better environment for all'.

Holcim South Africa's support for DEC is a significant contribution to this vision.

Photo: Guy Stubbs Photographer

Through the National Environmental Education Project, The Department of Education is investigating a variety of mechanisms that will facilitate ongoing professional development and activities that will improve on the delivery of environmental learning in our schools. The project also co-ordinates initiatives by other government departments and non-governmental organisations aimed at facilitating environmental learning at schools.

Thirdly, we recognise that although the principle of "a healthy environment" should permeate each learning area, we also need to have resources that enable teachers to deliver. Many environmental resources have been developed, and with some modification would be appropriate for assisting teachers in preparing activities that can help learners to gain the necessary skills and knowledge. New resource materials also need to be developed, and to this end the department has prepared guidelines for resource development, especially for the preparation or modification of environmental education resources.

Teachers and learners also require access to learning support materials. The National Education Portal, through its partnership with Share-Net on Environmental Education Resources, is an exciting new development. For this portal to be a success we need to make sure that teachers have access to Internet services and copiers that will enable them to print resources at low cost. I believe that The Enviropaedia is investigating ways of delivering its services online, and a partnership with the National Education Portal should be considered.

Finally, the Department of Education already has strong partnerships with a range of organisations, including some from other countries, as well as other government departments, particularly Environmental Affairs and Tourism, Water Affairs and Forestry, Health and Transport; environmental non-government organisations, and community-based organisations. These partners have been responsible for supporting a variety of initiatives on the environment in the curriculum. We are grateful for their support, and will continue to foster these partnerships so that they help to deliver the necessary support to teachers and learners.

The need to address issues of sustainability continues to be emphasised to our leaders and citizens. We need to use this awareness to make sure that environmental education is integrated into all our learning from grade R through to degree level. We need to be pathfinders during the 'Decade of Education for Sustainability', that begins in 2005. I believe we have the policy, the infrastructure, the staff and the will to be world leaders in the sphere of environmental learning, and I greatly look forward to this next decade.

Key Associated Topics:

Ecological Intelligence; Environmental Ethics.

Also see the Resource Section: Environmental Websites; Using The Enviropaedia for environmental education; Green Consumer Guide; Eco-Logical Lifestyle guide.

Associated Organisations:

Centre for Environmental Management; Centre for Environmental Studies – University of Port Elizabeth; Cheetah Conservation Fund; Cheetah Outreach; Collect-a-Can; Contact Trust; Crystal Clear Consulting; Delta Environmental Centre; Dept Environmental & Geographical Science – University of Cape Town; Desert Research Foundation of Namibia; Eco-Access; Educo Africa; EWT; Energy Efficient Options; Environmental Education and Resources Unit – University of Western Cape; Environmental Management Centre – Northwest University; Eureta Rosenburg; Green Clippings; Green Gain Consulting; Green Trust; HRH The Prince of Wales's Business & the Environment Programme; Icando; Independent Quality Services; Institute of Waste Management; International Fund for Animal Welfare; Liz Anderson; Minerals & Energy Education & Training Institute; Mountain Club of SA; Museum Park Enviro Centre; Natural Step The; NOSA; Rand Water; SA Wildlife College; Sustainability United (SUN); The Greenhouse People's Environmental Centre; Two Oceans Aquarium; Tyrrell & Associates; Wales Environmental Partnerships; WESSA; Wilderness Action Group; Wilderness Foundation; World of Birds Wildlife Sanctuary; WWF.

Photo: Bruce Sutherland - City of Cape Town

Our influence in the world of fine paper has spread across the globe.

Sappi is a global pulp and paper company that is proud of its South African roots. Today we are the world's largest producer of the coated fine paper used for everything from wine labels to art books to quality magazines as well as the dissolving pulp used to produce viscose fabrics. We manufacture on three continents and export to over a hundred countries. As a focused and growing global enterprise we are in a position to offer our customers better service, more innovative products and a wider choice of the world's leading brands of fine paper wherever they may be.

sappi

The word for fine paper

www.sappi.com

ENVIRONMENTAL ETHICS

Guest Essay by Prof. Johan Hattingh ~ Head: Unit for Environmental Ethics, University of Stellenbosch

Environmental ethics is a sub-division of professional and applied ethics that concerns itself with the responsibilities that we as humans have in our interactions with the environment.

Opinions differ as to how widely or narrowly the term "environment" should be interpreted, but a working consensus seems to have emerged around the notion of "objective encompassing nature", or "the biosphere".

From this broad perspective, the environment not only refers to living nature such as animals and plants, insects and microbes, but also the non-organic basis for life in general, as well as the ecosystemic interactions between all of the above.

Many interpret the environment even wider, to include the built surroundings within which humans live, so that ethical concern for the environment is seen also to include consideration of the aesthetic, cultural, historical and spiritual values that humans may attach to certain aspects of non-human nature.

Environmental ethics thus has to do with the duty of care that we have for the environment in an all-encompassing sense: the earth as a whole, or the whole of the community of life, including the ecosystemic and other processes (for instance the water cycle, the carbon cycle, the nitrogen cycle) that sustain this community of life.

Environmental disputes

Within the circle of environmental ethics a wide range of different positions are taken up on the question of the nature and extent of our duty of care towards the community of life. Views also differ strongly on the reasons we have such a duty, for the sake of whom or what we should care about the environment, what exactly the objects of our concern should be, and how we should discharge our responsibilities, through which actions or policies –

while some skeptics even go further and question whether we should morally care about the environment at all.

Since these different value positions are adhered to by many as ideologies that they live by, and since the enjoyment of certain natural features of our world is strongly identified by growing numbers of people with their health, well-being, quality of life and issues of self-realisation, environmental decision-making is typically fraught with protracted and adversarial disputes. And since many parties to environmental disputes are not easily shifted from their value positions, it seems as if environmental debates often fall into stalemate, going round and round in circles, leading nowhere.

For the purpose of making sense of the different value positions underlying environmental disputes, and with a view to move beyond ideological ping-pong debates about environmental issues, it can be useful to divide the field of environmental ethics into three broad categories, namely instrumental value theories, intrinsic value theories, and radical value theories.

Different value positions

According to instrumental value theories only humans have intrinsic value (i.e. value in and of itself), while everything else only has value in so far as it serves human interests. This human-centered approach at best can lead to the protection of natural areas from consumptive use, while non-consumptive activities aimed at enjoying the recreational, aesthetic, or spiritual value of nature are allowed. At worst, it can lead to the position of those who see nature as nothing but a resource that should be maximally developed for human consumption. Somewhere in between is the position of those who rather would like to see ecologically optimal development with a view to ensure that future generations can also satisfy their needs.

Intrinsic value theories emphasise that human use-value could not be the only consideration in environmental decision-making. Some entities in nature, or nature as a whole, or life in general should rather be respected for the value that it has in its own right, regardless of any use that humans can make of it.

The many variations of radical value theories focus on the root causes of our environmental problems, and make proposals to overcome these causes through a radical transformation of our behavior, mindsets, notions of self and self-realisation, social structures, institutions or decision-making procedures.

Implementing environmental ethics

Regardless of all the distinctions that can be made on a theoretical level amongst different value positions in environmental ethics, there seems to be a growing need in the world in which we function today to articulate a pragmatic environmental ethic that can guide our actions and decisions in individual, social, professional, corporate and public decision-making contexts. We conclude this article with a number of pointers as to how one can go about to develop such an ethic. These should be treated as general guidelines to consider, and not as a recipe.

1. Commit yourself or your organisation to the basic aim of environmental ethics. In general terms the practical aim of environmental ethics could be formulated as the duty to take responsibility for the environmental consequences of our choices and actions. The challenge is to make sure what this means in practical terms of action.

2. Determine who is the subject of this environmental ethics. Who is the agent that will take this responsibility on board? You, yourself, as an individual? You as a citizen, or as a professional, or as an employer, or manager in a business? Or is the agent rather a group, or an organisation, or a decision-making body, or a regulatory authority? The important point is that different subject positions will require different actions to discharge responsibility for environmental impacts.

3. Determine what the impact of this subject is on the environment. The point of departure of any environmental ethics is to know what the environmental impact of our choices and actions are. Different subject positions may have vastly different impacts.

4. Determine where exactly you have influence on the causal links between choices/actions and environmental impacts, and what exactly you can and have to change to minimise negative impacts and maximise benign impacts. As an individual, the effective range of my influence on these causal links is limited to my personal actions in direct interaction with the environment. However, I also have indirect impacts through the mediation of others – which I can influence with more or less success by collective action with others.

5. Determine which steps should be taken to implement these changes. As an individual, one can decide to change one's own behaviour. As a member of civil society, one can form lobby groups to place pressure on corporations, government officials and politicians to take responsibility for their environmental impacts. Professionals and corporations can form associations in which they help one another to minimise the environmental impact of a product or a service throughout its lifecycle. Governing bodies can put in place policies, a regulatory framework and incentives to promote environmentally-considerate choices and actions in business, individual behaviour, and government decisions.

Advantages and obstacles in the way

To develop an ethics of environmental responsibility along these lines has the advantage of engaging with detrimental environmental impacts in the sphere of practical decision-making and action, and has the potential to move beyond endless debates about value differences. We can legitimately differ about the ultimate reasons why we are concerned about negative environmental impacts, but we will find that a working consensus can be found about the measures to take, within a particular context, to address specific negative impacts.

However, it should also be borne in mind that an ethics of environmental responsibility will always be under pressure from different angles. First: there will always be pressure from the institutional frameworks within which we function to cut costs, to take the cheaper option – which is not always the environmentally-optimal option. Second: there will always be pressure from society first to satisfy human needs and then to attend to environmental considerations. Third: we tend to take more seriously impacts that are immediate, direct and unshared, while mediated, indirect, shared impacts are neglected. This last point puts a burden of proof on environmentalists to show why we should be concerned about negative environmental impacts. Environmentalists should take this point to heart when they produce evidence and arguments for their concerns, even if this burden of proof is often placed unfairly upon them, and in many instances should rather be reversed.

The value of environmental ethics

These guidelines may not settle things in environmental debates, but they definitely will help us to follow a line of inquiry on the basis of which we can have fruitful debates about real issues: about the environmental consequences of our actions and policies; about the reasons for our environmental concerns; about which courses of action to follow in order to improve things. They may also help us to ask critical questions about the selves that we are in fact realising through our choices, and if we, or the community of life, can really afford to be realising these selves. They may help us to start asking critical questions about the societal processes that produce the notions of self-realisation with which we so strongly identify.

The value of environmental ethics probably lies more in engaging in these debates from different points of view, than trying to provide final answers from a single point of view. It may be that it has more value to question how we formulate problems and solutions, than to have a monocular vision from which we can see only one kind of problem and one kind of solution. For that matter, the different value positions in environmental ethics together make up a rich mosaic of considerations that can help to open our eyes to the multiple dimensions of environmental problems, and the layered responses that are required to resolve them.

Key Associated Topics:

African Environmental Tradition; Ecological Intelligence; Environmental Justice; GAIA; Global Compact; Green/Brown Debate; Sustainable Development Section: – Corporate Responsibility.

Associated Organisations:

Allganix Holdings; Biowatch SA; Eco Africa Travel; Enchantrix; Environmental Ethics Unit – Stellenbosch University; Ezemvelo African Wilderness; Ubungani Wilderness Experience; Wilderness Foundation.

ENVIRONMENTAL GOVERNANCE

The word "governance" refers to the way in which entities or people are governed. The foundations of good governance are seated in mutual trust and good relations between the government and the people. The primary expectation of any society is that government should fulfil its constitutional, legislative and executive duties to govern in a responsible, transparent and accountable manner.

Environmental governance is founded on the premise that people have environmental rights. These rights include entitlement to an environment inclusive of cultural interests that is not harmful to the health or well being of people as well as a claim to environmental protection for the benefit of present and future generations.

Government is compelled to protect, promote and fulfil the rights of people also in terms of the environment. Such rights are applied accordingly in law to bind the legislature, the executive, the judiciary and all the associated organs of state. More specifically, government is obliged to render the legitimate environmental rights of people a reality by:

◆ Developing and implementing environmental policy

◆ Using its constitutional authority to take decisions and carry out actions

◆ Acting according to the basic values and principles for public administration

◆ Being accountable to the people

◆ Encouraging public participation in environmental governance by providing for mutual exchanges of views and concerns between the government and the people

◆ Monitoring and regulating all actions that impact on the environment

◆ Enforcing environmental regulation.

For government to fulfil its obligations in terms of responsible environmental governance it must co-ordinate all the organs of state at all levels to ensure responsible environmental stewardship. This demands the application of an integrated and holistic approach to environmental governance aimed at achieving a critical balance amongst the aspects of social development, economic development and environmental protection, in this way giving thrust to the global objectives for sustainable development.

Guest Author: Lt Col Etienne F. van Blerk, Staff Officer ~ Environmental Co-ordination, Department of Defence, RSA

Key Associated Topics:

Agenda 21; Environmental Ethics; Environmental Justice; Environmental Risk & Liability; Global Compact; NEMA; World Summit for Sustainable Development; Sustainable Development Section: – Corporate Responsibility.

Associated Organisations:

Centre for Environmental Studies – University of Port Elizabeth; Contact Trust; Data Dynamics; Duard Barnard Associates; Environmental Ethics Unit – Stellenbosch University; Environomics; HRH The Prince of Wales' Business & the Environment Programme; Responsible Container Management; SRK Consulting.

ENVIRONMENTAL HEALTH

Environmental health describes those factors of human health that are determined or affected by environmental influences. Typical environmental health concerns would be the health-related aspects of water and air pollution, for example drinking water quality, sanitation, waste disposal, food quality and the presence of harmful chemicals. Environmental health issues vary in developed and developing countries. For example, in developed or industrialised countries, major concerns might relate to lead in drinking water or pesticide residues in food. However, in underdeveloped countries, the concerns would relate to more basic aspects such as conditions causing the spread of tuberculosis or dysentery, malnutrition, or the lack of a safe drinking water supply. Major determining factors in the maintenance of good health in developed and developing countries are the availability of potable drinking water, safe sanitation systems, clean air and good soil for crops. This is closely followed by the development of sound and accessible primary health care support within easy travelling distance of settlements.

Key Associated Topics:

Air Quality; Contaminated Land; Energy; Environmental Degradation; Environmental Justice; Environmental Liability; Hazardous Wastes; Pollution; Population; Poverty; Urbanisation; Waste Management; Water Quality & Availability.

Associated Organisations:

Art of Living Foundation; Rand Afrikaans University (RAU); Engen; Metrix Software Solutions; NOSA; Southern Health & Ecology Institute.

Photo: Guy Stubbs Photographer

ENVIRONMENTAL IMPACT ASSESSMENT (EIA)

Guest Essay by Dr Merle Sowman,
~ UCT Environmental Evaluation Unit

Sustainable development requires the simultaneous consideration of economic, social and ecological processes, and the optimisation of the trade-offs between and across these three systems. Environmental Impact Assessment or EIA is one of several tools available for improving the way in which decisions are made in order to promote sustainable development outcomes.

Environmental Impact Assessment can be defined as the process of identifying, predicting, evaluating and mitigating the biophysical, social, and other relevant effects of development proposals prior to major decisions being taken and commitments made.

Objectives of Environmental Impact Assessment

◆ To ensure that environmental considerations are explicitly addressed and incorporated into decision-making processes;

◆ To anticipate and avoid, minimise or offset the adverse significant biophysical, social and other relevant effects of development proposals;

◆ To protect the productivity and capacity of natural systems and the ecological processes which maintain their functions; and

◆ To promote development that is sustainable and optimises resource use and management opportunities.

(*Principles for Environmental Assessment Best Practice, International Association for Impact Assessment, in cooperation with the Institute of Environmental Assessment, UK, January 1999. www.iaia.org*)

EIA usually relates to environment assessments of projects, in contrast with Strategic Environmental Assessment or SEA, which aims to integrate environmental considerations into the earliest stages of policy, plan and programme development. Although the National Environmental Management Act (NEMA) of 1998, provides the overarching legislative framework for environmental governance in South Africa, EIAs are still governed by regulations promulgated in terms of the Environmental Conservation Act of 1989. The EIA regulations require that specific procedures are followed, and reports ('Scoping' and/or 'EIA Reports') prepared for those activities that are listed due to their potential to have a "substantial detrimental effect on the environment". The Constitution of South Africa makes environmental management a concurrent competency between national and provincial government; however provincial government is generally the relevant authority for managing the EIA process. The regulations require the appointment of an independent consultant to undertake the EIA. A process for certification of environmental professionals is currently underway and an Interim Certification Board has been established. It is generally accepted that EIA comprises the following key stages: screening, scoping, assessment, evaluation, mitigation and optimisation/enhancement, reporting, reviewing, decision-making, implementation and management/monitoring. Today there exists an emphasis on effective scoping, such that each proposal is examined at a level, and in the detail, appropriate to the activity's potential for environmental change.

To improve its usefulness as a decision-making tool, EIA should:

◆ Provide sufficient, reliable and usable information to decision-makers at every stage of the project planning cycle. EIA thus moves away from taking an 'impacts only' focus (undertaken relatively late in the project cycle), towards approaches that are broadly-based, multi-stage, flexible, problem-solving and start-to-finish in relation to decision-making processes. Furthermore, the process should be customised to the issues and circumstances of the proposal under review, with lessons learned throughout the proposal's life-cycle being incorporated into the assessment process.

◆ Involve the application of best practicable science, using methodologies and techniques appropriate to address the problems being investigated.

◆ Generate outcomes that help to solve problems, are socially acceptable, and able to be implemented by the proponents.

◆ Be transparent and systematic, providing appropriate opportunities to inform and involve the interested and affected parties, whose inputs and concerns should be addressed explicitly in the documentation and decision-making.

◆ Achieve the objectives of EIA within the limits of available information, time, resources and methodologies.

Best practice EIA procedure should have manifold benefits including:

◆ Producing a better project.

◆ Incorporating the principles of sustainable development.

◆ Increasing the level of public acceptability of projects.

◆ Increasing cost effectiveness (through maximum process efficiency).

◆ Eliminating unnecessary duplication in the assessment process.

Early co-ordination between the proponent/consultant and the reviewing authority during the EIA process is considered an essential factor for efficient and comprehensive EIA practice, since early and ongoing dialogue builds flexibility, ensures authority concerns and requirements are integrated into the process, avoids requests for additional information late in the process, reduces cost and time, ensures EIA compliance with policy and law and reduces the authorities' time necessary for review of the EIA report.

Despite many methodological and administrative advances in EIA, experience from many countries indicates that considerable scope still exists for strengthening the EIA process both locally and abroad.

In the South African context, various issues of concern have been raised regarding EIA processes and practices. In particular, concerns regarding public involvement in the EIA process and environmental decision-making exist. Many environmental organisations and members of civil society groupings argue that the EIA process has become a rubber-stamping exercise and that concerns raised by the public during the public process are seldom taken seriously.

Key issues include: limited access to information in a format that can be understood; the selective inclusion of issues raised by the public; exclusion of certain sectors of society, particularly historically-disadvantaged groups, in the process due to inappropriate participatory methods being employed; environmental experts rather than public participation practitioners managing the public involvement process. Other key questions revolve around who holds the decision-making power with respect to which issues and impacts are addressed in the investigation, the process of assigning significance ratings to these impacts, and the trade-offs made amongst viable alternatives. Concerns regarding political interference in the decision-making process and even the appeal process have come to the fore in recent years and have undermined the integrity of the EIA process.

In order for EIA to fulfill its key objective, namely to inform decision-making, it is imperative that the inadequacies identified above be addressed. The recent IEM information series, published by the Department of Environmental Affairs and Tourism, should contribute to improving EIA practice and addressing some of the problems identified above.

Key Associated Topics:

Development of Land; Ecological Footprints; Environmental Management Planning; Interested and Affected Parties; Public Participation; Strategic Environmental Assessment.

Associated Organisations:

Acer; Africon; Agency for Cultural Resource Management; Aldo Leipold Institute; Aqua Catch; Bapela Cave Klapwijk; BKS; Bohlweki Environmental; Brousse James & Associates; Cape Town Heritage; Cebo Environmental Consultants; Centre for Environmental Management; Centre for Environmental Studies – University of Port Elizabeth; Centre for Wildlife Management – University of Pretoria; Coastal & Environmental Services; Crowther Campbell; Data Dynamics; Desert Research Foundation of Namibia; Dial Environmental Services; Digby Wells; Doug Jeffery Environmental Consultant; EcoBound Environmental &Tourism Agency; EcoSense; Enviroleg; Environmental Evaluation Unit – University of Cape Town; Environmental Management Centre – Northwest University; Environmental Risk Management; Envirowin; Eskom; Galago Ventures; Green Inc; Groundwater Consulting Services; Hilland & Associates; Holgate & Associates; Icando; Ilitha Riscom; Institute of Waste Management of Southern Africa; Ishecon Kantey & Templer; Ken Smith Environmentalist; Landscape Architects - Uys & White; Liz Anderson & Associates; Metago Environmental Engineers; Museum Park Enviro Centre; National Air Pollution Services; Newtown Landscape Architects; Ninham Shand Consulting Services; Oryx Environmental; Peninsula Permits; South African Association for Marine Biological Research (SAAMBR); SRK Consulting; Strategic Environmental Focus; Terratest; Tribal Consultants; Wales Environmental Partnerships; Web Environmental Consultancy; Wilbrink & Associates; Winstanley Smith & Cullinan; WSP Walmsley.

---

delivering **solutions**

WSP Walmsley

WSP Walmsley, which is part of the WSP Group plc, is one of the leading inter-disciplinary environmental consultancies, offering a broad range of services in the natural and built environment.

Our aim is to help our clients outperform their competitors with enhanced environmental management and commitment to sustainability. The provision of sound environmental management is becoming an essential ingredient for business success. Stronger regulatory control, market pressures, stakeholder awareness and global concerns, have caused businesses to adopt an innovative, proactive approach to the evaluation of environmental issues. Our vision is to provide an independent, innovative and professional service whereby we strive to achieve a balance between environmental protection, social desirability and economic development. The working expression of this philosophy is found in our integrated approach to environmental management. We offer the following services:

- Environmental project management
- Integrated environmental management (EIAs, Scoping, EMPs, EMPRs, Closure Plans)
- Environmental management systems
- Environmental auditing
- Environmental training
- Due diligence, compliance and liability assessments
- Rehabilitation and revegetation programmes
- Environmental and financial risk and scenario assessment
- Air quality management services
- Public participation
- Ecotourism and wildlife management services
- Environmental legal advice
- Contaminated land and environmental remediation

For more information contact:

Brent Ridgard
WSP Walmsley (Pty) Ltd
PO Box 5384
Rivonia
2128
South Africa
Tel: +27 11 233 7800
Fax: +27 11 807 1362
Email: br@wspgroup.co.za

Vicky King
WSP Walmsley (Pty) Ltd
PO Box 1442
Westville
Durban
3630
Tel: +27 31 240 8860
Fax: +27 31 240 8861
Email: wspwd@wspgroup.co.za

WSP Group plc is an international business providing management and consultancy services to the property, land and construction sectors. The company employs 5,000 people worldwide. WSP is an equal opportunity employer.

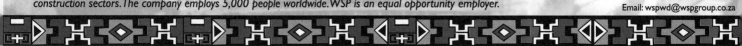

ENVIRONMENTAL JUSTICE

Photo: Dr J Ledger

Environmental justice tries to right the wrongs of past practices, which put the poor and the disadvantaged next to dirty and polluting industries or on poor and degraded land. It tries to ensure that the wrongs are not repeated. Those past wrongs denied the people the political or economic abilities to resist or change their situation. Past apartheid policies often "dumped" black communities near to industrial areas to ensure that a labour supply was close at hand, not thinking that those people could have their health seriously affected by the smoke and effluent emissions from the factories that provided the people with work. One of the principles of the National Environmental Management Act 107 of 1998 states, "…Environmental justice must be pursued so that adverse environmental impacts shall not be distributed in such a manner as to unfairly discriminate against any person, particularly vulnerable and disadvantaged persons …"

Guest Author: Terry Winstanley ~ Winstanley Smith & Cullinan Inc. Environmental Law Specialists

Key Associated Topics:

Environmental Ethics; Environmental Law; Environmental Risk & Liability; Interested and Affected Parties; Land Use Planning; NEMA; Stakeholders.

Associated Organisations:

Biowatch SA; Earthlife Africa; Enact International; Environmental Justice Networking Forum; Environmental Monitoring Group; Icando; Imbewu Enviro; Mark Dittke; Western Cape Environmental Ethics Forum.

ENVIRONMENTAL LAW

Although environmental law is a relatively new discipline in South Africa, a considerable number of new laws have been passed in the last five years. These, together with the environmental rights contained in the Constitution of the Republic of South Africa Act, 108 of 1996, result in a very comprehensive framework of environmental laws in the country. Unfortunately, to date, enforcement of these laws has been poor, although the Department of Environmental Affairs and Tourism has indicated that it will, in the near future, focus its efforts on better enforcement.

This enforcement by public agencies is supported by very liberal rules relating to legal standing, thereby enabling the public to enforce environmental laws, too. This has resulted in litigation, some of which has been reported, and which gives flesh to the environmental legal framework.

The introduction of mandatory environmental impact assessments in 1997 marked a significant shift towards attaining sustainable development. In addition, the duty of care provisions contained in both the National Water Act, 36 of 1998, and the National Environmental Management Act, 107 of 1998, create a general duty not to pollute and to remediate where pollution has been caused. In addition, these provisions create retrospective liability.

The next parliamentary session (in 2004) will see the repeal of the Atmospheric Pollution Prevention Act and the coming into force of the National Environmental Management: Air Quality Act, as well as the enactment of the Biodiversity Act and the Protected Areas Act.

Guest Author: Terry Winstanley ~ Winstanley Smith & Cullinan Inc. Environmental Law Specialists

Key Associated Topics:

CONNEP (P); Environmental Due Diligence; Environmental Liability for Remediation; Environmental Risk & Liability; NEMA; Polluter Pays Principle.

Associated Organisations:

Cameron Cross Incorporated; Centre for Environmental Management; Centre for Environmental Studies – University of Port Elizabeth; Dittke Mark; Enact International; Environmental Management Centre – Northwest University; Environomics; Green Gain Consulting; Imbewu Enviro; Technikon Pretoria; Winstanley Smith & Cullinan.

ENVIRONMENTAL LIABILITY FOR REMEDIATION

Guest Essay by Terry Winstanley ~ Winstanley Smith & Cullinan Inc. Environmental Law Specialists

HOW CLEAN IS CLEAN?

Photo: Bruce Sutherland – City of Cape Town

Introduction

An important aspect of environmental law is the liability for clean-up following pollution or degradation of the environment. The basis of this requirement lies in the Constitution of the Republic of South Africa Act, 108 of 1996, but rehabilitation obligations also arise under the National Environmental Management Act 107, of 1998 (NEMA) and the National Water Act, 36 of 1998 (NWA), insofar as the remediation of polluted water is concerned. What is not entirely clear is the extent of the remediation, which is required in law.

The Constitution

The Constitution guarantees the right to an environment that is "not harmful to human health and well-being" and to have the environment protected from, among other things, pollution and ecological degradation (section 24(b)). The duty to take reasonable measures to uphold this right is binding on the state and natural and juristic persons (section 8(2)).

The National Environmental Management Act

The Constitutional obligation, referred to above, is fleshed out in NEMA. It imposes a "duty of care" in relation to the environment, which requires that:

Every person who causes, has caused or may cause significant pollution or degradation of the environment must take reasonable measures to prevent such pollution or degradation from occurring, continuing or recurring, or, insofar as such harm to the environment is authorised by law or cannot reasonably be avoided or stopped, to minimise and rectify such pollution or degradation of the environment. (Section 28(1))

The categories of people upon whom the duty to take reasonable measures is imposed is very wide and includes an owner of, a person in control of, or a person who has a right to use land or premises where significant pollution or degradation of the environment is caused or is likely to be caused. Section 28(3) gives examples of the type of "reasonable measures" that may be required to comply with the duty under section 28(1). They include, but are not limited to, obligations to contain or prevent the movement of pollutants or the cause of degradation; eliminate any source of the pollution or degradation; or remedy the effects of the pollution or degradation.

The National Water Act

The NWA imposes a duty of care similar to that contained in section 28 of NEMA in relation to the pollution of water resources (Section 19). It also specifies the reasonable measures required as a result of this duty of care, which include requirements to contain or prevent the movement of pollutants; eliminate any source of the pollution; and remedy the effects of the pollution.

The extent of remediation required

There are essentially two opposing views on the extent of remediation required. Some authorities (e.g. the Department of Water Affairs – DWAF) apply the so-called "fitness for purpose" standard. This involves determining the purpose for which, for example, a water resource is likely to be used and then requiring that rehabilitation or "clean-up" measures be taken until the water quality is sufficient for that purpose. The "fitness for purpose" standard is based on the assumption that pollution or environmental degradation is acceptable provided that it does not exceed the level at which it would no longer be possible for humans to continue to use the environment for current (or reasonably foreseeable) purposes.

The use of the "fitness for purpose" standard in relation to the clean up of contaminated groundwater in South Africa appears to have originated from the application of the now repealed Water Act, 54 of 1956 (the old Water Act). It provided that it was an offence for any person to wilfully or negligently do anything which:

could pollute public or private water, including underground water or sea water, in such a way as to render it less fit – for the purpose for which it is or could be ordinarily used by other persons …; for the propagation of fish or other aquatic life; or for recreational or other legitimate purposes … (Section 23)

The old Water Act required anyone who polluted water to remediate it in accordance with prescribed measures (Section 22). Since no measures were prescribed, DWAF adopted the "fitness for purpose" test when authorising remediation measures. Importantly, the wording of the repealed

Water Act is significantly widened by the NWA which, although it defines pollution as an act that renders a water resource less fit for its purpose, includes in the definition any alteration of a water resource that could potentially cause harm to the resource quality (section 1(xv)).

The restorative approach under NEMA and the NWA

The other view regarding remediation is the "restorative approach", which seeks to restore environmental integrity and health by repairing damage to the environment caused by humans. This latter approach is based on the understanding that the environment is an integrated whole and that, given our limited understanding of its functioning and of the interests of future generations in inheriting a healthy environment, environmental damage should be avoided where possible and, in other cases, minimised and, as far as possible, rectified.

The preamble to NEMA makes it clear that one of the purposes of the Act is to establish a framework that will promote the progressive fulfilment of section 24 the Constitution. Furthermore, the interpretation and implementation of NEMA must be guided by the National Environmental Management Principles set out in NEMA (Section 2). These principles emphasise the importance of avoiding pollution and, where it cannot be avoided, minimising and remedying it. This, read with the very wide definition of pollution and the duty of care, impose, in our view, a wide-ranging duty to remediate pollution. This obligation is limited by the qualification that only "reasonable" measures need be taken and if such measures are taken, then the authorities may not recover so-called "clean-up costs" from the person concerned.

Accordingly, in relation to rehabilitation of contaminated land and water, in our view, the question is not, "what standard must be applied to determine what level of environmental degradation is acceptable?" Instead one must ask, "were all reasonable measures taken to rectify the pollution and degradation of the environment?" In other words, if by taking reasonable measures the site could have been restored virtually to its natural state, then those measures should, by law, have been taken even though lesser measures would have sufficed to render the site fit for the purpose that was intended.

Key Associated Topics:

Contaminated Land; Eco-catastrophe; Environmental Due Diligence; Environmental Law; Environmental Risk & Liability; Exxon Valdez Disaster; NEMA; Polluter Pays Principle.

Associated Organisations:

Cameron Cross Incorporated; Dittke Mark; Enact International; Edward Nathan & Friendland; Environomic; Envirocover; Green Gain Consulting; Imbewu Enviro; Winstanley Smith & Cullinan.

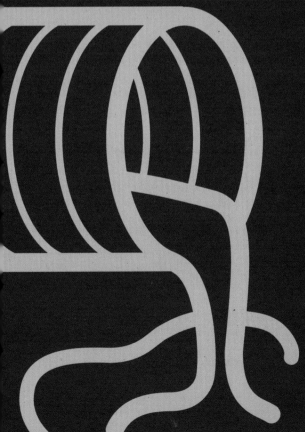

ENVIRONMENTAL MANAGEMENT PLANNING

Environmental management is a deliberate, multi-disciplinary process, which requires careful preparation and planning. Information on natural and human activities, processes and systems must be gathered, plans and procedures decided upon, documents drawn up and systems implemented. The results add environmental perspectives to existing planning processes. The process will result in the identification of new issues and concerns, but in most cases, the matters identified are well known in concept, if not in environmental jargon. Although formal environmental management systems (such as ISO 14001) are desirable, for most smaller businesses a simple environmental management system is all that is needed and, in most cases, it can be linked to the existing operational management systems. Planning for the implementation of environmental management needs to be undertaken and it should be carried out in a participative manner to ensure that all stakeholders and interested and affected parties who will be subject to it, will understand its purpose and the manner and form in which it is to be applied.

Key Associated Topics:

Environmental Management and Audit Scheme; Environmental Management Systems; Integrated Environmental Management; Integrated Pollution and Waste Management; ISO 14000 Series; Safety Health & Environment (SHE); Waste Management.

Associated Organisations:

Africon; Aldo Leipold Institute; Bapela Cave Klapwijk; BKS; Bohlweki Environmental (Pty) Ltd; Cebo Environmental Consultants; Centre for Environmental Management; Coastal & Environmental Services; Conservation Support Services; Crystal Clear Consulting; Dekra-its Certification Services; Department Environmental & Geographical Science – University of Cape Town; Doug Jeffery Environmental Consultant; EcoBound Environmental &Tourism Agency; EcoSense; Enviroleg; Environmental Evaluation Unit – University of Cape Town; Environmental Management Centre – Northwest University; Enviro-nomics; Envirowin; GeoPrecision Services; Green Gain Consulting; Green Inc; Groundwater Consulting Services; Landscape Architects Uys & White; Morris Environmental & Groundwater Alliances; Newtown Landscape Architects; Ninham Shand; Peninsula Permits; Settlement Planning Services; Strategic Environmental Focus; Terratest; Tribal Consultants; Wales Environmental Partnerships; Web Environmental Consultancy; Wilbrink & Associates.

ENVIRONMENTAL MANAGEMENT SYSTEMS (EMS)

Care for the environment includes concern about how industry manufactures products and the impact that such products and manufacturing processes could have on the environment. Controlling this requires a documented environmental management system (EMS) so that everyone understands how they must work to protect the environment from pollution and degradation. There are many different types of environmental management systems, which can be used by companies of differing sizes. In 1996, the South African Bureau of Standards (SABS) published a code of practice, SABS ISO 14001 that provides companies with guidelines and requirements for adopting an environmental policy and implementing an environmental management system. This code helps companies to set their own environmental policies and objectives. The code is an international standard and provides the basis for a uniform EMS, which will conform to wider international standards and requirements.

Key Associated Topics:

Environmental Management and Audit Scheme; Environmental Management Planning; Integrated Environmental Management; Integrated Pollution and Waste Management; ISO 14000 Series; Safety Health & Environment; Waste Management.

Associated Organisations:

Acer; Advantage ACT; Aldo Leipold Institute; Bohlweki Environmental; Cebo Environmental Consultants; Centre for Environmental Studies – University of Port Elizabeth; Clothing Textile Environmental Linkage Centre; Crystal Clear Consulting; Data Dynamics; De Beers; De Beers Marine; Dekra-its Certification Services; Duard Barnard Associates; Environmental Risk Management; Envirowin; Eskom; Green Gain Consulting; Independent Quality Services; Kantey & Templer; Liz Anderson & Associates; Metago Environmental Engineers; Metrix Software Solutions; Mondi Wetlands Project; Ninham Shand; Oryx Environmental; Softchem; Wales Environmental Partnerships; WSP Walmsley.

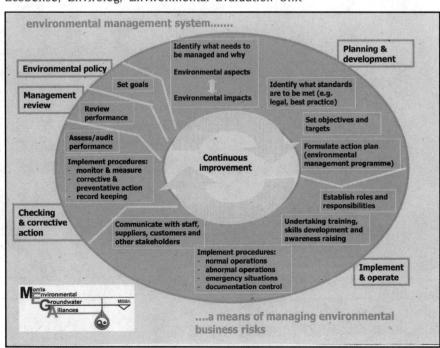

environmental management system.......

Identify what needs to be managed and why
Environmental aspects
Environmental impacts

Environmental policy

Management review

Set goals

Review performance

Assess/audit performance

Implement procedures:
- monitor & measure
- corrective & preventative action
- record keeping

Continuous improvement

Checking & corrective action

Communicate with staff, suppliers, customers and other stakeholders

Implement procedures:
- normal operations
- abnormal operations
- emergency situations
- documentation control

Planning & development

Identify what standards are to be met (e.g. legal, best practice)

Set objectives and targets

Formulate action plan (environmental management programme)

Establish roles and responsibilities

Undertaking training, skills development and awareness raising

Implement & operate

Morris Environmental Groundwater Alliances

....a means of managing environmental business risks

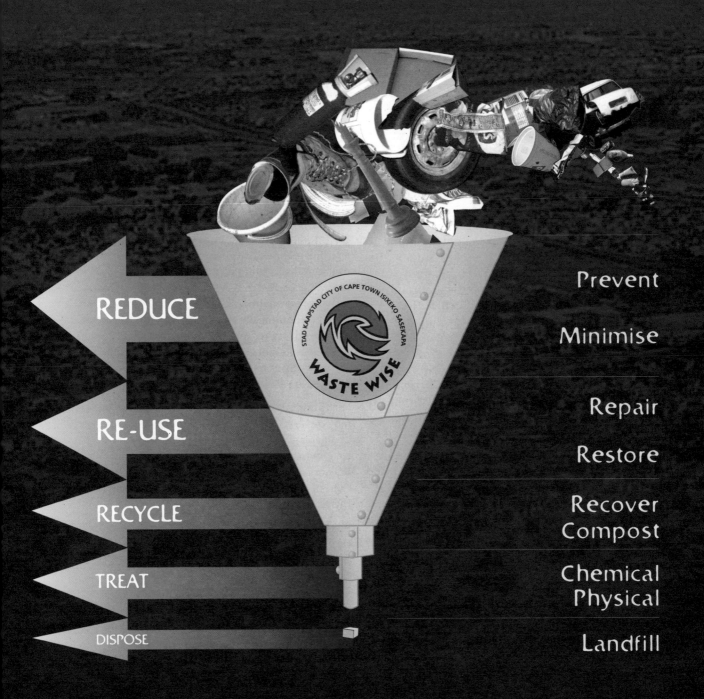

ENVIRONMENTAL REFUGEES

The most compelling feature of so-called environmental refugees is indeed the obscure distinction of an estimated 20-25 million people worldwide forced to abandon their lands through an intricate host of causes involving flooding, drought, soil loss, deforestation, earthquakes, nuclear accidents, toxic spills and the like. The United Nations High Commission for Refugees (UNHCR) was established in 1951 to act on behalf of one specific group of displaced people – refugees. Refugees are legally defined by the UNHCR in classical terms as persons forced to flee across an international border as a result of a well-founded fear of persecution based on race, religion, nationality, political opinion or membership of a particular social group.

Photo: Earthyear Magazine

Socio-political changes in the past few decades have spurred a contemporary debate around this definition. It is felt that millions of environmental migrants, some internally displaced within their own countries and others displaced across international borders, should also be categorised as refugees to receive the kind of legal and material assistance from the international community they would otherwise be denied.

In the late 90s the International Federation of the Red Cross and Red Crescent Societies revealed that humanitarian assistance provided to victims of major floods, drought and earthquakes had risen from 500 000 to 5.5 million in only six years. More people were suddenly being affected or displaced for environmental reasons than by the turmoil of armed conflict that was once the norm.

Researchers rank global climate change as an emerging key imperative for a burgeoning 150 million refugees by 2050 as the planet's warming atmosphere, rising seas and expanding deserts steadily maroon multitudes.

Another key imperative is given as the world's soaring population growth. Almost 10 400 newcomers arrive every hour, of whom more than 90% will be born into poverty, disease, malnutrition or thirst in the developing world. Here countries now suffer acute land shortage as people urbanise around over-saturated cities in their droves to flee deprivation in remote rural areas. At the same time, 40 of the world's fastest growing cities – being located within earthquake zones – are under constant threat of natural disaster.

Technological disaster constitutes a further imperative for instigating longer-term mass exodus in the wake of arable lands contaminated by landmines, industrial or nuclear toxins and waste or the four million people per year who are displaced by public works projects such as large new dams.

As this problem deepens, the leaders of the world are just beginning to redefine the state's responsibility toward environmental refugees with the ultimatum to either stem key migration imperatives or find homes for a further 125 million people in decades to come.

Guest Author: Lt Col E.F. van Blerk ~ Staff Officer Environmental Co-ordination, Department of Defence

Key Associated Topics:

Chernobyl; Contaminated Land; Desertification; Eco-catastrophe; Land Degradation; Poverty Alleviation.

Associated Organisations:

Environmental Justice Networking Forum; Groundwork.

ENVIRONMENTAL RISK AND LIABILITY

Environmental risk is the way of describing the possibility (the technical word is the "probability") of an environmental accident (such as a spill of wastes or chemicals polluting a river) occurring. Environmental liability is the threat of an environmental accident or disaster that may exist in a certain situation. For example, a factory that has no procedures to deal with spills will have a greater environmental liability in the eyes of insurance companies and the authorities than a factory that has taken precautions and made plans.

"Environmental risk and liability" is a general term that may be used to describe the readiness and preparedness of organisations and individuals to deal with, and respond to, environmental disasters, problems and accidents. Risk must be evaluated in order to quantify the level and type of remedial actions required to reduce or eliminate the risks.

It is most important that environmental risks and liabilities be identified so that precautionary measures and plans can be prepared, beforehand, to deal with the eventualities, should they occur.

Key Associated Topics:

Air Quality; Contaminated Land; Eco-catastrophe; Environmental Due Diligence; Environmental Law; Exxon Valdez Disaster; Hazardous Wastes; NEMA; Polluter Pays Principle; Pollution; Safety Health & Environment.

Associated Organisations:

Advantage ACT; Africabio; Arcus Gibb; Biowatch SA; Ceatec; Cameron Cross; Chemical Emergency Co-ordinating Services; Crystal Clear Consulting; Denys Reitz; Dittke Mark; Envirocover; Ilitha Riscom; Ishecon; Edward Nathan & Friedland; GeoPrecision Services; Groundwater Consulting Services; Imbewu Enviro; KPMG; NOSA; Responsible Container Management; SRK Consulting.

ENVIRONMENTAL SKEPTICISM

Guest Essay by Maarten de Wit
~ Department of Geological Sciences, UCT

The Skeptical Environmentalist ~ Bjorn Lomborg.

The Skeptical Environmentalist is a popular book, aimed at a broad audience of laypeople, according to its author, Bjorn Lomborg. This 500-page book, published by Cambridge University Press in 2001, argues that over the last 3 decades and more, most environmental scientists have grossly misled society with "doom and gloom" stories that spell out limits to present rates of consumption and plundering of Nature's store. Lomborg claims scientific models, which imply that the state of the present global natural environments and ecosystems are in disarray, and in need of better management and more careful conservation, are wrong. The subtitle of the book "Measuring the real state of the World", reflects the results of Lomborg's own statistical evaluation of the scientific data, and from which he concludes that the state of the world is in fact improving and getting better all the time: there is no need to increase environmental protection; scientists are wrong about forecasts of global warming over the next century; the aerial extent of natural forests are increasing; and estimates of the rates of species extinctions are flawed and highly exaggerated; pollution of the hydrosphere and atmosphere is not critical nor worthy of further action; and the Kyoto agreement is a waste of economic effort. Not surprisingly, the book has met with stern and alarming criticism from a number of sectors of society, including many NGOs (www.anti-lomborg.com). The harshest and most damning comments have come from the scientific communities, particularly from several leading scientists whose work is disputed by Lomborg. A series of rebuttals, written by respected scientists such as EO Wilson, S Pimm, P Gleick, J Mahlman and others, through the Union of Concerned Scientists based in Washington, were published in Scientific American in 2002 (www.acsusa.org/environmental/lomborg).

A review by S Pimm and J Harvey states that the book "is a mass of poorly digested material, deeply flawed in its selection of examples and analysis". In addition, the World Resources Institute and World Wildlife Fund have published critiques of the controversial book informing journalist and environmental educators of Lomborg's misuse of data and lack of scientific knowledge. They advise readers to approach the book with caution (www.wri.org/press/mk_lomborg-10_thingsl).

Lomborg concedes that he has no scientific qualifications: he is a right-wing political analyst, with elementary skills in statistics; an associate Professor and Director of the Institute of Environmental Assessment of Aarhus University in Denmark, from which he is on sabbatical leave for 4 years until mid 2004 (www.lomborg.com).

In 2002, a number of complaints about the book were analyzed by the Danish Committees of Scientific Dishonesty (DCSD). Their verdict (06 Jan 2003) is that Lomborg misused scientific data to support his arguments, and "objectively speaking, [The Skeptical Environmentalist] fall(s) within the concept of scientific dishonesty"; the committee adds, however, that Lomborg is not guilty of intentional gross negligence (www.forsk.dk/uvvu/nyt/udtaldebat/bl_decision), because the book is judged not to be a scientific work: it was not peer reviewed, but published, rather, as a socio-political monograph in which Lomborg disagrees with the prevailing broad scientific consensus and, like other influential dissidents (G Bush on climate change and pollution; T Mbeki on HIV/Aids) implies that he basically does not trust scientists, nor their research and communication endeavors, to provide society with true viewpoints.

Key Associated Topics:

Biodiversity & the Sixth Extinction; Ecological Intelligence; Global Warming & Climate Change.

Associated Organisations:

Gondwana Alive; WESSA; WWF.

EUTROPHICATION

Eutrophication is a process whereby high levels of nutrients such as phosphates and nitrates result in excessive and abnormal growth in plants. The additional growth not only consumes nutrients but also large quantities of oxygen, which affects the water quality. The rapidly growing plants have a high BOD and fairly soon exceed the DOC of the water body. This results in a deterioration of water quality and the characteristic stink of stagnant water. Eutrophication is a natural process, which occurs in lakes or inland water bodies. It is also an inevitable consequence of human activities. The nutrient loads that cause man-made eutrophication result from runoff from farmlands, wastewater treatment plants, garden runoff and septic tanks seepage, or untreated human and animal waste. Eutrophication can be lessened by better human management of nutrient cycles but it can never be eliminated.

Key Associated Topics:

Biological Oxygen Demand; Dissolved Oxygen Content; Freshwater Ecology; Rivers and Wetlands.

Associated Organisations:

Aqua Catch; Unilever Centre for Environmental Water Quality; Freshwater Research Unit – University of Cape Town.

Photo: SAPPI (Earthyear Magazine)

EXTINCTION

The biodiversity found on earth is the result of billions of years of evolution. A natural part of the evolutionary process is extinction, where species disappear owing to changes in their living conditions, which they are unable to survive. In recent times, however, the rate of extinction has increased dramatically as a result of the impact of human activities. An extinct species is one for which there is a historical record, but which no longer exists in the area under review.

The variety of life on earth forms a huge gene pool, which is a resource of crucial importance to humanity for food, fuel, shelter and to maintain health. While modern technology has given people greatly increased power over nature, it has done little to reduce our dependence on biodiversity. As species loss increases, the question now being posed is: "How does loss of species impact on the quality of human life?"

Potential impacts of loss of diversity include:

◆ Loss of potential cures for diseases
◆ Increasing global climate change
◆ Decreasing genetic diversity within food crops
◆ Loss of habitats.

What can be done? We all need to reduce our ecological footprint and live 'more lightly' on the earth. Demands for goods and services place pressure on the environment – the less we use, the less our pressure on natural resources.

Guest Author: Alison Kelly ~ WESSA

Key Associated Topics:

Biodiversity; Biodiversity & The Sixth Extinction; Birds; CITES; Endangered Species; GAIA; Habitat; Marine life of Southern Africa; Stewardship; Wildlife.

Associated Organisations:

Africat; Centre for Environmental Studies – University of Pretoria; Centre for Rehabilita-tion of Wildlife; Cheetah Outreach; EWT; Gondwana Alive Society; Khama Rhino Sanctuary Trust; Southern African Foundation for the Conservation of Coastal Birds; WESSA; WWF.

EXXON VALDEZ DISASTER

On 24th March 1989, the oil tanker *Exxon Valdez* went off course in a 16 km wide channel in Alaska's Prince William Sound and hit submerged rocks on a reef. The hull of the tanker was gashed in several places and 42 million litres of crude oil (22% of its cargo) leaked out, causing one of the worst oil spills and environmental disasters in US waters. By the end of the clean up (which cost over $4 billion), the oil spill was known to have killed 580 000 birds (including 144 rare bald eagles), 5 500 sea otters, 30 seals, 22 whales and countless fish. It also oiled over 5 100 km of shoreline. The American oil company, Exxon, eventually paid approximately $400 million in fines. The disaster has become a significant and valuable case study to illustrate a number of different environmental points. For example, oil tankers should have double skinned hulls to reduce the risk of pollution from damage or disaster; countries with sea lanes close to their coastlines should prepare emergency plans to cope with pollution disasters; and insurance companies need to give more thought to environmental liability and practical means of reducing it. As a direct result of the disaster, a new NGO, CERES (Coalition for Environmentally Responsible Economies) was established, which set out to try and prevent the occurrence of another *Exxon Valdez* disaster. The group has produced what has become known as the CERES Principles, which can be used to audit a company's environmental performance against simple but clear environmental standards.

Key Associated Topics:

Eco-catastrophe; Environmental Degradation; Environmental Liability for Remediation; Environmental Risk & Liability; Marine Environment; Pollution; Polluter Pays Principle.

Associated Organisations:

Coastal and Environmental Services; Crystal Clear Consulting; Department of Environmental Affairs & Tourism.

 Printed on Sappi recycled paper

FIRE

Fire is a natural process that has large consequences for natural and urban environments. Fire affects vegetation composition, structure and numerous ecological processes. The effects however, differ depending on the frequency, season, intensity (heat energy emitted) and the severity (damage caused) of fire.

Fires in the natural environment

Most vegetation cover in South Africa burns periodically. The maintenance of many southern African vegetation types, particularly grasslands, savannas and fynbos, rely on fire. Fire-prone vegetation species have evolved with fire and many depend on fire for regeneration and the completion of their life cycles.

Fire intensity is dependent on the amount of fuel available to burn (the more fuel, the more intense the fire) as well as the chemical composition of the fuel. Some plants contain substances that stimulate fire and are described as being "born to burn". Intensity also increases with higher temperature, lower humidity and strong winds.

South Africa does not have the same fuel loads as in the forested areas of countries like Australia and the USA and so does not experience the consequent bush fire problems. However, fuel loads are increased significantly when our natural vegetation becomes infested with invading alien plants. These invading alien plants, such as exotic acacias, hakea and myrtle, out-compete our indigenous plants and have higher growth rates and biomass.

Landowners are obliged by law to clear their properties of invading alien plants, and can be held responsible if a fire starts on their land and causes damage to neighbouring property. The higher fuel loads produced by these invaders, lead to hotter, more intense fires that pose a significantly higher risk to people and property, increasing the cost of fighting fires for authorities, and causing greater ecological damage by destroying seed banks, damaging soil and exacerbating erosion. Every house that was burned down in the devastating fires of January 2000 on the Cape Peninsula, was surrounded by invading alien plants.

Fires in the urban environment

In the urban environment, fire poses a risk to people and property, particularly those living adjacent to open fire-prone vegetation on the interface between natural and urban areas. This poses serious challenges to managing these areas. Because of the inherent risk involved, managers are often reluctant to undertake deliberate burning to reduce fuel loads where this may endanger adjacent property owners. But by not undertaking controlled burning, fuel accumulates over time to become an even greater hazard because when it does eventually burn, it is usually under uncontrolled wildfire conditions that can have devastating consequences.

The large quantities of smoke emitted from veld fires can negatively affect air quality including visibility, and can also have health consequences with the fine particulate matter in smoke affecting sufferers of asthma and other respiratory diseases.

In the urban environment, the highest number of fires occur in informal settlements. The combination of the use of flammable building materials, close proximity of dwellings and reliance on gas, paraffin or wood fuel for energy, increases the risk from and frequency of fire, often with the most devastating of consequences. For example, on 26 November 2000 when nearly 1 000 homes were destroyed in the Joe Slovo informal settlement in Langa, Cape Town, a national disaster was declared by President Mbeki. In response, an innovative approach was pioneered by the Ukuvuka campaign in close collaboration with the City of Cape Town, to rebuild dwellings in groups of 60-100 surrounded by tracks which served both as firebreaks as well as access for fire tenders. This pilot dramatically decreased the number of dwellings lost to fire by over 90% in the 2 years since the test. This approach has now been replicated in the whole of the informal settlement and is being planned for other informal settlements.

Photo: Santam/Cape Argus Ukuvuka

People rather than nature cause the vast majority of wildfires. A fire education programme is, therefore, of huge importance and should be an ongoing quest with the aim to create a fire-conscious nation. Ukuvuka is currently piloting an innovative behaviour change programme in another informal settlement. Results of the pilot should be available in the next edition of *The Enviropaedia*.

Guest Author: Sandra Fowkes and team members ~ Santam/Cape Argus Ukuvuka: Operation Firestop Campaign

Key Associated Topics:

Alien Invasive Plants; Atmospheric Pollution; Biomass; Cape Floral Kingdom; Deforestation; Eco-catastrophe; Grasslands.

Associated Organisations:

Department of Water Affairs and Forestry; Ilitha Riscom; Santam/Cape Argus Ukuvuka.

FISHERIES
~ THE ECOSYSTEM APPROACH

Guest Essay by Deon Nel ~ WWF-South Africa and Lynne Shannon - Marine and Coastal Management

The evolution of ecosystem-based fisheries management

Ecosystem-based fisheries management is a relatively recent concept that has slowly taken formal shape over the past three decades. In the past fisheries management relied almost exclusively on a single-species approach. This entailed calculating the Maximum Sustainable Yield (MSY) for an individual fish stock. The MSY is greatest harvest that can be taken from a self-regenerating stock of animals year after year, while maintaining the average size of the stock. Although it was probably always appreciated to some degree, scientists

began to express their concern about the need to take into account the knock-on effects that the harvesting of a commercial species may have on the functioning and integrity of the ecosystem within which the harvested species occurred. In 1980 the Convention for the Conservation of Ant-arctic Marine Living Resources (CCAMLR) became the first Regional Fishery Management Organisation (RFMO) to implement an ecosystem approach to marine resource management.

Under CCAMLR provisions harvesting activities need to take account of three principles:

♦ Prevention of decrease in size of the harvested population.

♦ Maintenance of ecological relationships between harvested, dependent and related living resources

♦ Prevention (or minimisation of the risk) of changes to the ecosystem.

Concurrent to these developments, was a growing need for management systems to be integrative and take into account the equitable needs of all stakeholders in the marine ecosystem (e.g. the different fisheries, tourism, recreational users, scientific and medicinal uses, etc), while maintaining biodiversity and ecosystem integrity and catering for needs of species of special concern (e.g. species threatened with extinction, or having geographically restricted ranges). Although this approach is often cited as being progressive and innovative in terms of fisheries management, in reality it is based on one of the earliest principles of common law; that is the law of nuisance i.e. the activities of one stakeholder (or stakeholder group) should not have an undue negative impact on the activities of other legitimate stakeholders. Lastly, the ecosystem-based fisheries management concept was influenced by the "precautionary principle", which requires that in the case of lack of full scientific understanding, management should err on the side of caution and the conservation

of the ecosystem. These principles have shaped what we today interpret broadly as ecosystem-based management. The concept is widely accepted and is included in many prominent international fisheries agreements, including the 1982 UN Law of the Sea Convention, the 1995 UN Straddling Fish Stocks Agreement and the 2001 Reykjavik Declaration. However, its effective implementation in national fisheries management remains sporadic and only a few countries, including Canada, Australia, New Zealand, and the USA have attempted to integrate these principles in their fisheries management regimes. South Africa has very recently embarked on the process towards an ecosystem approach to fisheries management.

What is an ecosystem approach to fisheries?

Ecosystem-based fisheries management makes ecological sustainability the primary goal of management and recognises the dependence of human well-being on ecological health. The ecosystem approach to fisheries therefore takes account of the interactions of fishing with the ecosystem and is best achieved through an integrated management approach involving all stakeholders. Stakeholders need to collectively set a vision for the marine environment that meets their needs equitably, but also recognizes the ecological constraints of the system.

In a study by the World Wildlife Fund (WWF), it was recognized that ecosystem-based management requires a hierarchical approach that is based on five basic principles. **(Table 1)** Nested under these principles are the key elements and the operational components of the system. From this framework specific objectives, enabling processes and performance indicators are developed that meet the needs of the key elements.

Examples of enabling processes include:

1. Harvesting levels for ecologically important species are set at such a level to allow "escapement" of sufficient prey items to support other predators in the marine ecosystem (e.g. whales, dolphins, seabirds, seals and other predatory fish).

2. Fishing gear modifications may be needed to reduce the bycatch of non-target species. For example, Turtle Excluder Devices (TEDs) allow turtles to escape from shrimp trawl nets, and bird-scaring lines deter seabirds from becoming unintentionally hooked and drowned on longline fishing gear. Modifying times at which fishing gear is deployed may further minimise harmful interactions.

3. Marine Protected Areas (MPAs) may be needed to preserve the natural functioning on the ecosystem from both the effects of harvesting and habitat destruction. For instance, the use of bottom trawl nets not only reduces the prevalence of the target species, but also disturbs and modifies habitats on the sea bottom. It may therefore be necessary to declare 'no trawl' areas that can remain undisturbed. Furthermore, the use of 'no take' MPAs is increasingly recognised as an effective fisheries management tool that bolsters fish stocks and increases overall catch rates for the fishery.

4. Areas that are exceptionally sensitive to disturbance by fishing and other operations (e.g. fish spawning grounds) need to be identified and appropriately protected. For instance, many seamounts of the high seas are home to highly sensitive and geographically restricted ecosystems.

Printed on Sappi recycled paper

Many of these ecosystems are being destroyed by trawl nets even before the species that inhabit them have been described.

Ecosystem approach to South African fisheries

The concept of taking ecological considerations into account in environmental resource management is embedded in the Constitution of South Africa. Section 24 of the Constitution, which expresses general environmental rights, requires protection of the environment that (amongst other) "secures ecologically sustainable development and use of natural resources, while promoting justifiable economic and social development."

In South African law, legislation pertaining to the harvesting of marine resources is governed by the Marine Living Resources Act (1998). The need for an ecosystem-based fishery management is explicitly stated in the objectives of this Act, and requires the Minister to have due regard for the following (amongst other objectives):

◆ The need to achieve optimum utilisation and ecologically sustainable development of marine living resources;

◆ The need to protect the ecosystem as a whole, including species which are not targeted for exploitation.

Furthermore, the Act endorses the precautionary principle as a fishery management principle. It is therefore clear that this Act gives a clear mandate for South African fisheries to be managed according to ecosystem-based principles.

South Africa is a signatory to various legally binding international agreements (e.g. the 1982 UN Law of the Sea Convention) that uphold ecosystem-based management principles, as well as other "soft law" international instruments – most notably the Reykjavik Declaration on Responsible Fisheries in the Marine Ecosystem. Additionally, the World Summit for Sustainable Development, held in Johannesburg in 2002, encouraged the application of the ecosystem approach to sustainable development of the oceans by 2010.

This political and legislative commitment by South Africa has recently been crystallized by the creation of a special task group within the Directorate of Marine and Coastal

Photo: Bruce Sutherland - City of Cape Town

Management to progress the "Ecosystem Approaches to Fisheries Management". It has been decided to use the pelagic fishery to pilot the concept, and a preliminary stakeholder meeting was held during the latter part of 2003. In early 2004, with the collaboration of a larger group of stakeholders and affected parties, a policy document will be drafted for the ecosystem approach to fisheries in South Africa.

Key Associated Topics:

Aquaculture; Carrying Capacity; Ecosphere & Ecosystems; Marine Environment; Marine Life of Southern Africa.

Associated Organisations:

I&J; Global Ocean; Two Oceans Aquarium.

Table 1. The components of an effective ecosystem-based management system framework. Taken from *Ward et al.* 2002.

Principles of the Framework	Key elements of the framework	Operational components
1. The central focus is maintaining the natural structure and function of ecosystems, including the biodiversity and productivity of natural systems and identified important species.	1. Management operates within a policy framework designed to facilitate and enable effective implementation of all the principles of ecosystem-based management.	1. Develop out-come oriented objectives for management activities, i.e. clearly express what the resource management is trying to achieve.
2. Human use and values of ecosystems are central to establishing objectives for use and management of natural resources.	2. Recognition of economic, social and cultural interests as factors that may affect resource management objectives, targets and strategies and activities.	2. Delineate boundaries for the management system, including ecologically defined spatial boundaries, and all ecologically and socio-economic factors influencing the productivity of the resource and the integrity of the ecosystem.
3. Ecosystems are dynamic; their attributes and boundaries are constantly changing and consequently interactions with human uses are also dynamic	3. Ecological values are recognized and incorporated into the management system through developing agreed objectives, targets, strategies and activities that reduce the risk of impacts of resource exploitation.	3. Involve stakeholders in all aspects of the management system leading to shared and agreed individual and collective aspirations for the resource and associated ecosystems.
4. Natural resources are best managed within a management system based on a shared vision and set of objectives developed amongst stakeholders.	4. Information on the utilized species is adequate to ensure that there is a low risk of over-harvesting and population and genetic diversity is maintained.	4. Have a functional information system, including monitoring activities for the objectives and targets, and research activities for the key uncertainties.
5. Successful management is adaptive and based on scientific knowledge, continual learning and embedded monitoring process.	5. The resource management system is comprehensive and inclusive and uses an adaptive approach.	
	6. Environmental externalities that may affect the resource or the ecosystem are properly included in the resource management system.	

FORESTRY

Forestry can be split into two different types of management. The first type relates to indigenous forestry and involves the management and utilisation of natural indigenous forests. The second refers to plantation forestry, which requires a different approach. Plantation forestry is a large consumer of water and plantation forests can pose problems to downstream water users. Indigenous forests are vitally important because they are a source of biodiversity and habitat for a wide range of species. Plantation forests are less biologically diverse but provide a product, pulp and wood, which is essential to the needs of society. It is clear that compromise will always have to be the case when examining the balance between the two different types of forestry management. The 1992 Earth Summit produced a statement of 15 forest principles and Agenda 21 includes a chapter on forests and deforestation. The growing demand for forestry products and the pressures on indigenous forests pose serious challenges for sustainable forestry management. A number of efforts have emerged to certify forestry developments as being sustainable timber production. One example is the Forest Stewardship Council (FSC), a voluntary effort involving NGOs, industry and government participants, which has negotiated a consensus on criteria for credible certification programmes for sustainable timber production.

Key Associated Topics:

Afforestation; Agriculture; Agroforestry; Biodiversity; Deforestation; Monoculture; Tropical Rain Forest.

Associated Organisations:

Bainbridge Resource Management; Conservation Support Services; Department of Water Affairs and Forestry; Ken Smith Environmentalist; Sappi.

Photo: Guy Stubbs - Photographer

FOSSIL FUELS

Fossil fuels are non-renewable (limited) natural resources, which means that one day they will run out and alternative energy sources will have to be found for them. They include crude oil, coal, gas or heavy oils, which are made up of partially or completely decomposed plants and animals. These plants and animals died millions of years ago and, over long periods of time, they became a part of the earth's crust and were exposed to heat and pressure which, through carbon chemistry, turned them into fossil fuels and sources of energy for people. Fossil fuels, when burnt, release gases and particles, which can cause pollution if not managed correctly. Carbon dioxide, one of the gases released from burning fossil fuels is contributing to Global Warming. South Africa's energy supplies are based mainly on power stations, which burn low-grade coal and emit vast quantities of sulphur dioxide, carbon dioxide and other pollutants. Over the years, studies have revealed that the fossil fuel emission from Eskom's Highveld power stations have had a marked impact upon the vegetation of Mpumalanga. There is no doubt that at this point in time, most countries are not in a position to completely replace their fossil fuel burning power stations with cleaner options such as wind, wave, hydro-electric and solar. However, many countries have begun serious research and development work into alternative power options. For example, many remote microwave or cellular towers are powered by solar power. This is an easier, cheaper and safer option than extending power lines to often remote areas just for a power supply.

Key Associated Topics:

Air Quality; Atmospheric Pollution; Carcinogen; Catalytic Converter; Carbon; Carbon Monoxide; Energy; Global Warming & Climate Change; Natural Gas; Industrial Ecology; Non-renewable Resources; Particulates; Petrochemicals; Photochemical Smog.

Associated Organisations:

Engen; Eskom; Rose Foundation; Toyota; Wales Environmental Partnerships.

FRESHWATER ECOLOGY

Freshwater ecology is the discipline for measuring, managing and monitoring the health of freshwater systems. It looks at water quality from a chemical perspective and then examines the organisms that are living in the water and their relationship to the water chemistry. There are many organisms that can act as indicators of the state of health of a particular water system. The South African Scoring System (SASS) is a method of bio-assessment, which was developed as a rapid method of assessing water quality in rivers using invertebrate species, allocating scores for the presence of particular species in a body of water. Tracking these organisms over time can help to build up a profile of the manner in which the system functions and the kind of pressures that they are placed under through pollution, over-utilisation, etc.

Photo: Dr J Ledger

Freshwater ecologists can make predictions and model the behaviour patterns of freshwater systems. This is valuable when planning water use and trying to predict carrying capacity and yields.

Key Associated Topics:

Carrying Capacity; Dilution; Dissolved Oxygen Content; Eutrophication; Groundwater; Rivers and Wetlands; Suspended Solids; Water Quality & Availability.

Associated Organisations:

Afrodev Associates; Aqua Catch; Freshwater Ecology Unit – University of Cape Town.

Photo: Dr R Parry-Davies

GABIONS

Gabion is a derivative of the Italian word *Gabbione*, which means a large cage.

Gabions are made of various types of material, the earliest dating back 2 000 years to the time of the ancient Egyptians, and were made from willows, filled with small stones and used to protect the banks of the River Nile.

Today gabions are traditionally made from steel wire with various forms of coating, which will influence the life span of the product. Typical examples of coating include galvanizing and extruded PVC coating, which have life spans of at least 40 and 120 years respectively. The steel wire, with its protective coating, is then woven into a double twisted hexagonal mesh arrangement and made into a basket known as a gabion. Gabions, which come in various shapes and sizes, are then filled with suitable rock whose diameter is slightly larger than the diameter of the mesh configuration. Adjacent units are joined to one another in a special way and used to form structures that are used primarily in the civil engineering and soil conservation industries for erosion control and embankment stabilization. The rock filled gabions form flexible, permeable, monolithic structures such as retaining walls, channel linings and weirs. Gabion construction is labour intensive and can be conducted using unskilled workers. Gabion works are made primarily of a natural resource: rock, and can be easily vegetated to blend into the natural environment. This makes them a very good solution for conserving South Africa's very precious, non-renewable resource: soil.

Guest Author: Ronel Suthers ~ Environmental Manager, Maccaferri (African Gabions)

Key Associated Topics:

Green Architecture; Landscape Architecture; Rehabilitation of Land; River Catchment Management; Rivers and Wetlands; Urban Greening.

Associated Organisations:

ERA Architects; SA Gabions.

GAIA

Guest Essay by Glenn Ashton
~ Ekogaia Foundation

Could Lovelock's legacy be our Salvation?

James Lovelock is the grand old man of environmental science whose expansive ideas carry considerable weight. His Gaia hypothesis has been described as "the second Copernican revolution". Gaia is an idea whose time has come.

In the late 1960s Lovelock was working as an atmospheric scientist with the North American Space Administration (NASA). He evolved his hypothesis about our planet acting much like an organism when he was considering systems and procedures to discern the presence of life on Mars for the NASA Viking project.

Lovelock is a brilliant practical and theoretical scientist who was responsible for inventing the electron capture detector, which assisted Rachel Carson in measuring global pesticides for her seminal work "Silent Spring", often cited as the start of the global environmental movement. His inventions also assisted accurate measurement of CFCs above the South Pole, which gave credence to theories of ozone depletion.

In spite of his practical excellence, it is for the Gaia Hypothesis that he will be remembered. Together with Lynn Margulis and others Lovelock laid the foundation of his hypothesis in his 1979 book *Gaia; a New Look at Life on Earth*.

Ge, Ga or Gaia was the Greek Goddess of the Earth; the root of the word can be traced in the names of earth sciences such as geography and geology.

The beauty of Gaia is its capacity to include a broad range of interpretations of life on earth, which are restricted only by the infinite possibilities of Planetary/Gaian interaction with the rest of "Life, the Universe and Everything". The Hypothesis proposes that life does not act in isolation but in concert with all natural events. Our world is compared to a single organism, in which life is intrinsic to the operation of global systems, to the extent that it modifies and is modified by the global environment, in order to maximise successful interactions. Lovelock defines Gaia "as a complex entity involving the Earth's biosphere, atmosphere, oceans, and soil; the totality constituting a feedback or cybernetic system which seeks an optimal physical and chemical environment for life on this planet." The Gaia Hypothesis is not simply an extension of Darwinian evolutionary theory but a quantum leap into interdependent complexity. It has been described as a "systemic wisdom", and as a multidimensional bio-geographical planetary matrix.

In the Gaia Hypothesis, the planet is not viewed simply as a dead rock with individual life forms tenuously hanging on, but as a repository of raw materials that has been altered by all the varieties of life and environment over the last 3.5 billion years. Gaia is a mix of rocks; gases; radiation; liquids; magnetism; movement and life that is bound together in its complexity, forming an interconnected web. When any aspect is altered, adjustments are made in the system in order to move in the direction of homeostasis or toward beneficial change.

Thus the Gaian matrix is sustained and constantly evolves in myriad ways through its integral self-regulatory systems.

The wisdom of the philosophy behind the Gaia theory naturally attracts many people. Similar beliefs have long been held by numerous global cultures but Gaia is the first Western scientific model to describe the planet and the biosphere as an integrated, self-regulating organism. Gaia has predictably drawn a strong following from environmentalists, feminists, new-agers, environmental theologians, communists, deep-ecologists and many other so-called fringe groups that compliment its growing mainstream support. The irony is that so many fringe groups have claimed and adapted the concept that it has gained broad acceptance by default.

Photo: *Andrew Parker* Photographer

The entire Gaia Hypothesis cannot be defined or claimed by a single branch of science. Even evolutionary biology, a broad discipline that examines aspects of Gaian thought, has limitations in its ability to analyse all the relevant factors that may play a role in moderating the system. Narrow and rigid scientific disciplines have been especially slow to accept the hypothesis but a perceptible shift is becoming apparent. However, the concept remains more easily accepted by systems thinkers such as biologists and physiologists, than by reductionists, such as microbiologists, working at the coalface of detail.

Nevertheless, sceptics continue to question some of the core concepts of the Gaia Hypothesis. The antithesis of Lovelock's theory is typified by the ideas contained in Steven Dawkins' acclaimed books the *Blind Watchmaker* and *The Selfish Gene*. Dawkins' central tenet is that evolution as explained by classical scientific theory is responsible for our diversity. Dawkins' books broaden evolution by systematically working through the steps of evolutionary reductionism, where complex issues are analysed through their individual parts. He suggests that the processes that govern life on earth are predictable and consequently are able to be controlled.

There are valid points in both the Selfish Gene and the Gaia theories. However the Gaia Hypothesis can readily assimilate Dawkins' theory while Dawkins cannot comfortably accommodate Gaia. This is mainly because of the prevalent system of scientific reductionism. Reductionism is very good at examining details but often misses (or purposefully ignores) the larger, often more relevant picture of the infinite variability of the Gaian matrix. Because of this "Mechanistic" and "Reductionist" thinking, the Gaia Hypothesis lies largely outside the boundaries of the present, dominant scientific viewpoint.

Gaia can readily accept both the Selfish Gene and the Darwinian evolutionary theories, both of which it asserts are incomplete. Instead of a single theory, Gaia is a group of kindred theories that complement and inform one another. Gaia is certainly a lot more humanly satisfying than the detached and dangerous scientific objectivity that results in our present mechanistic myopia of looking at details whilst ignoring the view. It can be argued that scientific reductionism is a major factor in our perception and acceptance of present 'progress' that consists of an extractive and destructive cycle of exploitative consumption.

The Gaia Hypothesis is far more complex than most other explanations of life on earth. It would seem to be closer to life than either Darwin's or Dawkins' model. Interestingly, Dawkins' mentor, the eminent evolutionary biologist William Hamilton (who created the term "the selfish gene" and opposed the Gaia hypothesis) has recently called Lovelock "a Copernicus awaiting his Newton". After overcoming his significant scepticism of the Gaia Hypothesis, Hamilton even went so far as to examine a case study that extends classic Darwinian analysis to more closely fit the Gaian model, perhaps in a quest to become that Newton!

Hamilton's example is a wonderfully elegant illustration of how algae create and sustain their own clouds to assist their survival. They do this by creating their own sulphur-based aerosol, around which water vapour readily condenses. The algae can thus regulate their local climate, thereby showing how life and natural systems work toward an interdependent equilibrium. After all, clouds are critical in modifying climate (without clouds the earth would be 20 degrees warmer). Hamilton's example neatly shows how local effects are induced in a feedback system that has global implications, as suggested by Lovelock.

In fact we owe a lot to algae and Lovelock is convinced that the most serious danger to life on earth is posed by global warming. This has profound ramifications for algal and planktonic life, which are critical links in the chain of life. His insists that most other environmental concerns are secondary.

Lovelock's perception views the world as more than a rock pointlessly orbiting the sun, populated by humans pitted against nature in a struggle of survival. Acceptance of the Gaia Hypothesis indicates that we have moved beyond this inbred insecurity and our acceptance of the Gaia hypothesis offers a hope of redemption from our destructive and exploitative cycle. By acceptance, we are confronted by our role in planetary metabolic systems.

Another common objection to the Gaia Hypothesis is its teleology (teleology being the explanation of phenomena by the purpose they serve rather than by postulated causes). Teleology implies that a divinity or "godness" is inherent in the system. This implication creates a further barrier for mechanistic, reductionist analysis. However, it can easily be argued that teleology lies within in the perception of the beholder and is essentially dependent on personal belief systems. Even so, surely there can be little moral objection to seeing the hand of God evident in our support systems? After all, many scientists today are able to safely embrace religious beliefs without undue conflict!

Religious beliefs need be neither integral to nor separate from the Gaia Hypothesis. They may remain simply as components and form subsets of the Gaian matrix. Nevertheless, Lovelock was so concerned about this criticism that he modelled a theoretical planet called "Daisyworld" to counter it. With a mathematical world populated by white and black daisies he constructed a system that was both adaptable and self-regulating. This model is complemented by Hamilton's algae and cloud relationship. Lovelock has sketched the evolution of our planet from the primeval soup to the present time, looking mainly at the interconnectedness of living organisms with the natural resources available at that time. Thus the patterns that emerge provide glimpses of an infinite matrix.

The beauty of the hypothesis is the manner in which existing, evolving and new information may be incorporated. Each detail can be shown or assumed to form an aspect or part of Gaia. We are constantly introduced to fresh and powerful ways of perceiving our role amongst all that surrounds and supports us. Recognition of Gaia offers a way to end our present environmental deadlock by recognising and admitting our dependence on the Gaian matrix. This, in turn, emphasises our obligation to demonstrate evolutionary intelligence by recognition of this interdependence. More importantly, Gaia allows a feasible transition between a more satisfactory integration of the theoretical and the practical, the scientific and the divine. It comprises the strongest single thrust toward an integrated theory of life.

Human intervention in the planetary balance is accelerating the tempo of extinction. We are unravelling the very fabric of our support system, by causing the extinction of thousands of species. Humanity faces two choices; either to indirectly cause our own extinction by the destruction of our support system, or to recognise our integral role in Gaian systems and reverse the impacts that we have on them. A more widespread acceptance of the Gaia Hypothesis will improve the prospects for the collective health of life on this small, blue planet.

Key Associated Topics:

Biodiversity & the Sixth Extinction; Ecological Intelligence; Ecosphere & Ecosystems; Sustainable Development.

Associated Organisations:

Ekogaia; Gondwana Alive; Wilderness Foundation.

This is a place to read.
It's where you build your son's first house.
A place to shelter from the sun.
It's a place to hang a swing from.

This is the seed of the tree Combretum Erythrophyllum.

You probably know it as the River Bush Willow,

a tree indigenous to our beautiful country.

But what does the future hold for the Bush Willow?

Will it flourish as it does today in a world fifty years

from now?

Inaugurated in 1994, the **Audi Terra Nova Awards**

recognise individuals who conserve and protect our

environment for future generations. Air, land, water

and all living creatures, including man himself.

Every year we nominate four individuals and assist

them in their environmental projects.

If you would like to find out more about the Audi

Terra Nova Foundation, the nominees and the

Audi Terra Nova Awards, please visit www.audi.co.za

or www.auditerranova.co.za.

It's just one of the many ways we strive to conserve

the incredible beauty and wonder of our world.

Because Vorsprung isn't just a way to build cars.

It's a way to build life.

Vorsprung durch Technik www.audi.co.za

GAME FARMING

Game farming is a method of farming that uses wildlife species as livestock instead of domesticated animals. The wildlife species are better adapted to the veld and other conditions, cause less damage and are less affected by disease and illness. Species used include eland, crocodile, guinea fowl, zebra and ostrich. There are an increasing number of game farms in Mpumalanga and the Cape and, in many cases, they are more profitable than domestic stock farms. By ensuring that the farms are run according to the carrying capacity of the total land area, it is possible to run successful farms and combine these with a tourist function such as game viewing or hunting. There is a need for strict control of these operations to avoid the occurrence of "canned lion" or other cruel or unethical animal hunting. Experience has shown that game farms do not have serious problems with grazing shortages and, provided that the animals are kept separate from foot-and-mouth infected animals, financial stability can be sustained.

Photo: Copyright South African Tourism.

Key Associated Topics:

Agriculture; Biodiversity; Carrying Capacity; Eco-tourism; Hunting; Wildlife; Wildlife Management.

Associated Organisations:

Brousse-James and Associates; Centre for Wildlife Management – University of Pretoria; Conservation Management Services; Eco Systems.

GENETIC MODIFICATION ~ THE "ANTI" POSITION

**Guest Essay by Glenn Ashton ~ SAFEAGE
(South African Freeze Alliance on Genetic Engineering)**

"Genetic Modification of food crops is not the solution for feeding the world."

Promoters of genetically modified (GM) food insist that the process of genetic engineering is simply an extension of traditional breeding that will enable us to feed a growing global population. Claims of higher yields, lower chemical use and improved nutrition from GM crops are trumpeted. However, these claims have little factual basis.

In conventional breeding 50% of the genetic information of each sex is transferred between closely related species. With GM, genetic information from completely unrelated species can be transferred between organisms. The most widely grown GM crops presently contain fragments of viral, antibiotic, bacterial, as well as synthetic genes.

The promoters of GM crops are abusing public trust in science by misrepresenting basic facts about the performance, testing, safety and regulation of GM crops and food.

Better, healthier, less chemicals?

Promoters of GM crops claim they use fewer chemicals and increase yield. These claims are dispelled by independent analysis that shows more pesticides are used on GM crops, while industry analysis shows a marginal reduction. Who is to be believed; industry or independent sources? It has also been shown that the most widely grown GM crop – Soya – yields lower than conventional Soya. GM maize yields are similar to those of conventional maize while cotton is slightly higher in the short term but the advantage is reduced by higher seed costs. In several cases, including India and the southern USA, GM cotton has failed spectacularly and developers of GM seed have been sued by farmers and farm organisations for poor product performance.

Most tested?

GM crops have never been tested on humans; only test tube simulations have taken place. Limited testing has taken place on animals and these raise some profound concerns.
For more information on these go to
http://www.actionbioscience.org/biotech/pusztai.html

Most GM safety evaluations have not used the actual GM crop to test for safety but have used the original – but by not genetically identical – bacterial toxin in test tube studies. A remarkable amount of genetic rearrangement has been shown to take place in engineered crops. For more on this, go to http://www.i-sis.org.uk/UTLI.php Similarly to the testing for food safety, environmental testing has relied on testing the related bacterial toxins that are not part of the GM crop. This testing ignores the effects of the different 'cassettes' that are used to introduce the genetic constructions into the plants. Both Canada and US/EU studies have emphasised shortcomings in such studies. http://europa.eu.int and http://www.rsc.ca give further background to this and the journal "Science" quoted US government scientists saying that: "key experiments on both the environmental risks and benefits are lacking."

GM products are deemed to be substantially equivalent to conventional crops, unless proven otherwise.

If crops are substantially equivalent, surely there can be no nutritional improvement in GM crops, as so zealously claimed?

Secrecy and control

Scientists from some lead agencies for the testing of GM crops like the United States Environmental Protection Agency (EPA) and the US Food and Drug Administration (FDA) have voiced concerns about their dependence on GM crop developers for background data. Developers have a vested interest in finding that the food components of crops do not differ substantially from their conventional counterparts, thus avoiding any further testing. This shortcoming is compounded by the veil of secrecy that operates in the name of commercial confidentiality with these crops. In South Africa there are currently two legal actions against the government, which are attempting to obtain basic information about GM crops. Predictably, both the commercial developers and the state oppose these. We are told that nobody has become ill from eating GM food, yet. The problem is, nobody is looking. And just where do you begin to look if the developers of GM crops refuse to disclose fully tests or label food containing GM?

More food for the starving

Using third world hunger to justify the introduction of GM crops is perhaps one of the most misleadingly cynical marketing ploys in recent history. This ignores the fact that enough food is produced globally to supply a balanced diet for every human on earth. It is far more important to address the excessive agricultural subsidisation within developed nations while also improving soil health and transport and storage infrastructure in developing nations. Africa should retain as wide a variety of plant food genetic material as possible and should resist the apparent "temptations" of using GM technology as this would inevitably lead to a narrowing of the gene pool, and create a reliance on high input hybrid seeds supplied by trans-national corporations. The global south does not need a technical quick fix but should rather address its systemic shortcomings. GM technology fails to address core differences in food production between developing and developed nations and the reality is that GMO's actually stand to hinder resolution of food security and sovereignty in the south in both their present and proposed configurations.

Photo: Earthyear Magazine

Key Associated Topics:

Activism; Agriculture; Biodiversity; Eco-labelling; Genetically Modified Organisms – The Pro Position; Monoculture; Organic Farming; Organic Foods; Precautionary Principle.

Associated Organisations:

Africabio; Biowatch SA; Green Clippings; Heinrich Boell Foundation; SA Freeze Alliance on Genetic Engineering; Southern Health & Ecology Institute; Winstanley Smith & Cullinan.

GENETIC MODIFICATION ~ THE "PRO" POSITION

Guest Essay by Prof Jennifer A Thomson ~ Department of Molecular and Cell Biology, UCT

Genetically modified (GM) crops are those in which genes from another species have been inserted to express a particular trait, which the original crop lacked. Genetic modification differs from traditional breeding in that GM crops may contain genes from a completely different organism. The major GM crops currently in commercial use worldwide are herbicide tolerant soybean, insect resistant maize and cotton, with smaller amounts of cotton and maize carrying both herbicide tolerance and insect resistance. Those in the development and field-testing stages include virus resistant fruits and vegetables, insect resistant potatoes and vitamin A enriched rice. The main countries growing GM crops are the USA, Argentina, Canada and China. However, increasing numbers of farmers in developing countries are adopting the technologies. More than three-quarters of the farmers that benefited from GM crops in 2001 were resource-poor farmers planting insect resistant cotton, mainly in China but also in South Africa. Most of the smallholder cotton farmers in South Africa are in the Makhathini Flats region of KwaZulu-Natal. In 1998/99 there were 75 adopters, growing less than 200 hectares of GM cotton. By 2000/01 this number had risen to 1 184 with about 1 900 ha. Thus, in only three years, 60% of the producers in this region had adopted the new technology. They did this due the increases in yield, the improved quality of the cotton and the decrease in the use, and therefore the cost, of insecticides. It is estimated that these farmers have reduced pesticide sprays by 60 to 70%. Some of the "knock on" advantages of GM crops are with both herbicide-tolerant and insect-resistant plants. As spraying of herbicide can now be at the discretion of the farmer and does not have to occur before the crop is planted, there has been a marked decrease in the use of tilling. This "low till" farming practice results in a decrease in the loss of topsoil and resultant soil erosion. In addition, because insect resistant maize does not suffer from insect damage, it is far more resistant to post-harvest fungal infection. Many of these fungi produce highly toxic compounds called mycotoxins that can result in toxic hepatitis and oesophageal cancer. Concerns have been raised that herbicide tolerant crops could result in the formation of superweeds and that insects could develop resistance to the insect resistant crops. Although to date no evidence of this has been found, it is still important to use sound agricultural practices to mitigate against this happening. For instance, farmers must plant a certain percentage of non-GM crops among their insect resistant varieties to decrease the chances that the insects will develop resistance. Recent reports from the USA and China indicate that these measures are working and insects in the field are not developing resistance to the GM toxin. Concerns have also been raised about the safety of foods derived from GM crops. No evidence of lack of food safety has been found despite years of rigorous testing. In fact, conventional foods are not tested for food safety; only GM foods are. Thus we know more about the safety of foods derived from GM crops than about any other food we eat. Another concern is that the use of GM crops will result in a decrease in biodiversity. However, the opposite may well be true. It is much easier for plant breeders to introduce a single gene, giving a desired trait, into many different crop varieties than it is with traits resulting from conventional breeding. All countries wishing to test or commercialise GM crops are required to have in place biosafety regulations to ensure the correct use of such crops. In South Africa the GMO (Genetically Modified Organisms) Act was passed in 1997. It is administered by the National Department of Agriculture and has on its executive committee members of all the government departments involved e.g. Health, Environment, Trade and Industry, etc. All applications for imports, field trials and commercial releases have to be subjected to scrutiny by panels of experts before permission is given or withheld. Other countries in Africa either have regulations in place or are developing them. In terms of the development of regulations, the affected countries are being assisted by the United Nations Environmental Programme and the Global Environment Facility, which aim to assist over 100 developing countries worldwide in the next few years. Despite opposition to GM crops, most notably in Europe, developing countries are adopting the technology as part of the answer to sustainable food production. It will be important to monitor the use of GM crops on a case-by-case basis, but in most resource-poor countries the benefits are likely to outweigh any possible risks.

Key Associated Topics:

Activism; Agriculture; Biodiversity; Eco-labelling; Genetically Modified Organisms – The Anti Position; Monoculture; Organic Farming; Organic Foods; Precautionary Principle.

Associated Organisations:

Africabio; Department of Molecular and Cell Biology – University of Cape Town.

Photo: Guy Stubbs Photographer

GEOGRAPHICAL INFORMATION SYSTEMS (GIS)

GIS is essentially a set of computer-based software systems (database and mapping techniques combined) that allow managers, decision-makers and scientists to look at spatial data (information that often appears on maps) in new ways. It can be used to manage utilities such as water and sewage mains, roads, power lines, or help to understand relationships between different issues "on the ground" (e.g. areas of high population and the location of amenities such as schools and clinics).

Different "sets" (or types) of data and information can be "mixed" on the computer to see what types of overlaps or parallels there may be. It is also possible to identify problems. For example, one set of data may map the flows of smoke from a factory chimney and, superimposed over a land use map, it would be possible to identify what residential areas would be affected by the smoke from the chimney. Using the mapping capabilities of GIS information, it is possible to represent facts and circumstances in real life and make predictions about the likely outcome of changing events or different circumstances.

Key Associated Topics:

Ecological Footprints; Environmental Auditing; Environmental Due Diligence; Environmental Impact Assessment; Environmental Management Planning; Environmental Management Systems; Land Use Planning; Strategic Environmental Assessment.

Associated Organisations:

Bapela Cave Klapwijk; Brousse James & Associates; Conservation Support Services; Imagis; Namibia Nature Foundation; Open Africa; South African Association for Marine Biological Research (SAAMBR)

GEOTHERMAL ENERGY

Geothermal energy is energy obtained from heat sources contained within the earth. Water is heated to steam on hot rock located below the surface of the ground and the steam is captured in order to drive steam turbines, which generate electricity. For every 30 metre drop below the surface, the temperature increases by 1° C. Iceland and New Zealand both make extensive use of geothermal energy for heating and electricity generating purposes and 65% of all homes in Iceland are heated by geothermal energy. The areas with the greatest potential as sources of geothermal energy are the Americas, western Siberia and the Pacific Rim. Internationally, 20 countries are using geothermal energy sources. Geothermal energy is regarded as "clean" energy, which does not produce significant pollution in the form of particulates or noxious gases. It does, however, sometimes produce small quantities of hydrogen sulphide ("rotten egg" smell). South Africa does have some sources of hot springs but none are used for significant production of geothermal energy. Geothermal energy is an alternative energy source, which conforms to many of the principles of sustainable development.

Key Associated Topics:

Energy; Eco-efficiency; Industrial Ecology; Renewable Energy.

Associated Organisations:

Energy and Development Group; Eskom; Society of South African Geographers.

GLOBAL COMPACT (THE)

The United Nations – Global Compact is a voluntary corporate citizenship initiative with two objectives: mainstreaming the nine principles in business activities around the world and to catalyse actions in support of UN goals. In order to achieve these objectives, the Global Compact offers facilitation and engagement through several mechanisms: Policy Dialogues, Learning, Local Structures and Projects.

United Nations Secretary-General Kofi Annan first proposed the Global Compact in an address to The World Economic Forum on 31 January 1999. The Secretary-General challenged business leaders to join an international initiative – the Global Compact – that would bring companies together with UN agencies, labour and civil society to support nine principles in the areas of human rights, labour and the environment. These principles are:

Human Rights: Business should:

◆ Support and respect the protection of internationally proclaimed human rights.

◆ Ensure that they are not complicit in human rights abuses.

Labour: Business should uphold and support:

◆ The freedom of association and the effective recognition of the rights to collective bargaining.

◆ The elimination of all forms of forced and compulsory labour.

◆ The effective abolition of child labour.

◆ The elimination and discrimination in respect of employment and occupation.

Environment: Business should:

◆ Support a precautionary approach to environmental challenges.

◆ Undertake initiatives to promote greater environmental responsibility.

◆ Encourage the development and diffusion of environmentally friendly technologies.

The Global Compact is a network. At its core are the Global Compact Office and five UN agencies: the Office of the High Commissioner for Human Rights; the United Nations Environment Programme; the International Labour Organisation; the United Nations Development Programme; and the United Nations Industrial Development Organisation.

Today, hundreds of companies from all regions of the world including South Africa, international labour and civil society organisations are engaged in the Global Compact.

South African Global Compact Network

On the 24th April 2003 the South African Global Compact Network was launched and this led to the formation of a Taskforce. The Taskforce was mandated by the conference to work towards establishing a South African Global Compact Network.

The overall aims of the Network would be inter alia to promote, support and entrench the United Nations Global Compact principles and corporate social responsibility through policy dialogue workshops, seminars, and a national learning forum. Corporations in the taskforce have committed to developing case studies to showcase the value and benefits of participation and share this information amongst members and other participants in the South African Global Compact Network.

In the southern African region there are opportunities for linking up different national initiatives into a regional network. The Malawi Global Compact network was launched on the 10th December 2003. The Mozambique Global Compact was launched in 2003 as well and preparations for launches in Botswana and Lesotho are under way for 2004.

For more information contact: Nkosithabile Ndlovu Tel: 011 643 6604, Fax: 011 643 6918 and email: nkosi@aiccafrica.org or Mr. Mokhethi Moshoeshoe Tel: 011 643 6604, Fax: 011 643 6918 and email: mokhethi@aiccafrica.org and for the AICC Malawi office contact: Mr Sean de Cleene Tel: +265 1 772404 and email: sean@aiccafrica.org

Guest Author: Mokhethi Moshoeshoe ~ African Institute of Corporate Citizenship

Key Associated Topics:

Ecological Intelligence; Environmental Governance; Environmental Justice; Responsible Care; Poverty Alleviation; Sustainable Development.

Associated Organisations:

African Institute of Corporate Citizenship; Eskom.

GLOBAL WARMING

Global warming describes the gradual increase of the air temperature in the Earth's lower atmosphere. Why is global warming called the greenhouse effect? Greenhouses are not common in Africa, so don't be surprised if you have never seen one! They are used mainly in the cooler Northern Hemisphere to grow vegetables and flowers. A greenhouse is made entirely of glass. When sunlight (shortwave radiation) strikes the glass, most of it passes through and warms up the plants, soil and air inside the greenhouse. As these objects warm up they give off heat, but these heat waves have a much longer wavelength than the incoming rays from the sun. This longwave radiation cannot easily pass through glass, and thus causes everything in the greenhouse to heat up.

The natural greenhouse effect

The term greenhouse effect is used to describe the warming effect that certain gases have on the temperature of the Earth's atmosphere under normal conditions. Sunlight (shortwave radiation) passes easily through the Earth's atmosphere. Once it strikes and warms the Earth's surface, longwave radiation is given off and goes back into the atmosphere. While some of this longwave radiation, or heat, escapes into space, most of it is absorbed or held by carbon dioxide and other gases that exist in small quantities in the atmosphere. Thus these gases keep the Earth at an average of 33°C warmer than it would be if this greenhouse effect did not occur. Without these gases the whole planet would be an icy wasteland with an average temperature of 16°C below freezing!

How have people made the atmosphere warmer?

Human population growth and related industrial expansion, have led to greater air pollution and a change in the composition of the Earth's atmosphere. Some pollutants enhance global warming, resulting in increased global atmospheric temperatures.

What gases are responsible for global warming?

Water vapour is the main greenhouse gas. Human activities are not known to have had a significant influence on the atmosphere concentration of water vapour.

Carbon dioxide (CO_2) is the pollutant most responsible for increased global warming. It is released into the atmosphere mainly through burning of fossil fuels (e.g. coal, petrol, diesel).

In addition, widespread destruction of natural vegetation, particularly forests, has contributed to increased atmospheric CO_2 levels (see Enviro Facts "Deforestation"). This has occurred for two reasons. First, plants take up CO_2 through the process of photosynthesis. The destruction of vegetation, as occurs in deforestation, reduces the amount of CO_2 that is removed from the atmosphere. Second, when forests are cleared, and burnt or left to rot CO_2 is released.

Methane (CH_4) doubled in concentration, mainly as a result of agricultural activities, between 1750 and 1990. Nitrous oxide (N_2O), also a product of burning fossil fuels, increased by 8% over the same period. Chlorofluorocarbons (CFCs), in addition to damaging the ozone layer, are potent greenhouse gases. Their concentrations in the atmosphere are increasing by about 4% every year. (See Enviro Facts "Ozone").

Signs that global warming has begun

◆ The average global temperature is about 0,5°C warmer than it was 100 years ago.

◆ Snow and ice-cover have decreased this century, deep ocean temperatures have increased, and cloud cover over North America has also increased over this period. The latter indicates increased atmospheric water vapour.

◆ Over the last century, global sea levels have risen by between 100mm and 200mm.

Further effects of global warming

If current pollution trends continue, some scientists estimate that the Earth could probably be about 1°C warmer by 2025 and 3°C warmer by 2100. This rapid temperature rise could have several effects:

◆ These changes in global temperature, although apparently small, could cause very large changes in climate. For example, the last Ice Age, which ended approximately 15 000 years ago, was only 5°C colder than current temperatures, but the resulting climate changes were massive: most of North America was covered in a layer of ice about 1,5 km thick and sea levels in the Cape were about 120m lower than at present. In those days, if you had wanted to go for a swim at Cape Agulhas you would first have had to walk about 150km to reach the sea.

◆ A rapid extinction of species.

◆ Rising sea levels – water expands as it warms and glaciers melt, adding water to the oceans, thus we can expect widespread flooding of coastal areas as sea levels rise.

◆ Greater frequency and scale of extreme weather conditions, e.g. drought and flood.

◆ Changes in the distribution of disease-bearing organisms so that people, domestic animals and crops might be exposed to diseases previously absent from the area.

Did you know?

Many scientists believe that global warming is not a threat. They argue that the processes affecting atmospheric temperature are not fully understood, and that it is not possible to come to any conclusion regarding the warming of the earth.

What can we do?

Find out about the disagreements between scientists on the issue of global warming. Reduction of pollutants responsible for global warming requires greater energy efficiency. Industrialists and governments have a key role to play here. As individuals we can: Reduce electricity consumption; use lift clubs, public transport, bicycles or our feet for transport; reduce, reuse, recycle, and save energy – the manufacture of all products requires energy.

This topic has been copied directly from Enviro Facts which are information leaflets sponsored by Pick 'n Pay through the World Wide Fund for Nature – SA, published by Share-Net and supported by many environmental organisations including: Wildlife and Environments Society of SA; KZN Conservation Service; Endangered Wildlife Trust; Sea World Education Centre; National Botanical Institute; National Water Conservation Campaign and the Department of Environmental Affairs and Tourism.

Key Associated Topics:

Air Quality; Carbon Tax; Carbon Trading; Climate Change; Kyoto Agreement.

Associated Organisations:

Earthlife Africa; Renewable Energy And Energy Efficiency Partnership (Reeep); SouthSouthNorth; South African Climate Action Network; National Botanical Institute; WWF.

GLOBAL WARMING AND CLIMATE CHANGE
THE WORLD RESPONSE

Guest Essay by Lester Malgas ~ Environmental Writer, SouthSouthNorth (SSN)

Global warming is a natural phenomenon. It occurs when solar radiation bouncing off the Earth's surface is absorbed by carbon dioxide and other so-called greenhouse gases (GHGs). This causes what has been termed the greenhouse effect and is responsible for heating up the atmosphere. Though global warming was identified as far back as the late 1800s, evaluating the effects of human activities on the global climate, only gained ascendancy in the late 1960s. Trade, the flow of materials after World War II, the prominence of global powers and space exploration, gave currency to the idea of the Earth as a closed system with finite resources. This perspective, and the determination that human activities damage the Earth's natural resources, informed the agenda of the 1972 United Nations Conference on the Human Environment (UNCHE) in Stockholm and inspired the conference's catchphrase: "Only One Earth".

Climate change occurs as a result of man's contribution to global warming, and has been identified as a global threat since the early eighties. The role of humans in contributing to global warming is two-fold. Firstly, industrial activities starting with the Industrial Revolution have caused an increase in the concentration of CO_2 and other GHGs, exacerbating the normal degree of global warming of the Earth's atmosphere to dangerous levels. The second concern is the denuding of the Earth's forests, reducing the Earth's potential to absorb the increase in CO_2.

Scientific studies suggest increasingly grim scenarios about the future effects of climate change. Temperature increases due to global warming may cause ice caps to melt, bringing about a rise in sea levels, more frequent storms, droughts in some areas and floods in others. And as fragile ecosystems collapse on account of temperature fluctuations, desertification and water shortages, higher incidences of vector-borne diseases, large scale population shifts and consequent losses in agriculture and fishing, threaten not only damage to infrastructure, but also to health and food security, with low-lying coastal regions and developing countries found to be the most vulnerable to the resultant economic instability and loss of livelihoods.

In response to this threat, and amid various environmental conferences, the World Meteorological Organisation (WMO) and the United Nations Environment Programme (UNEP) established the Intergovernmental Panel on Climate Change (IPCC) in 1988. The IPCC's task is to assess existing scientific knowledge on climate change, its environmental, economic and social impacts; and to provide a list of response options. The IPCC is the authoritative body on climate change and its establishment occurred in recognition of the need for a vital reassessment of global economic development.

The IPCC's First Assessment Report (FAR), released August 1990, stated that a 60%–80% emissions cut would be required to stabilise the concentration of GHGs in the atmosphere. This report spurred the UN General Assembly to establish an Intergovernmental Negotiating Committee (INC) which drafted the Framework Convention on Climate Change (UNFCCC), which was adopted in May 1992 at the UN headquarters in New York, and opened for signature a month later at the Earth Summit in Rio de Janeiro, Brazil.

The United Nations Framework Convention on Climate Change (UNFCCC)

The UNFCCC entered into force in March 1994. Its stated objective was the "stabilization of greenhouse gas concentrations in the atmosphere at a level that would prevent dangerous anthropogenic (man made) interference with the climate system." The convention required that "such a level should be achieved within a time-frame sufficient for ecosystems to adapt naturally to climate change, to ensure that food production is not threatened and to enable economic development to proceed in a sustainable manner."

The Convention recognised that certain countries are vulnerable not only in adapting to climate change, but also that others will experience "special difficulties" as a consequence of global mitigation measures. It also recognised that industrialised developed countries were historically responsible for the majority of GHG emissions, and it provided for these countries (listed as Annex I countries in the Convention) to voluntarily reduce their GHG emissions to 1990 levels by the year 2000.

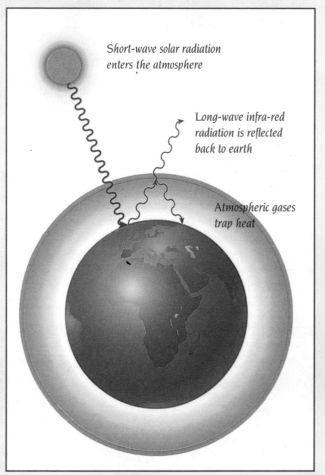

Short-wave solar radiation enters the atmosphere

Long-wave infra-red radiation is reflected back to earth

Atmospheric gases trap heat

It also provided that the richest of the Annex I countries (listed in Annex II) were to make available "new and additional financial resources" and "promote, facilitate and finance, as appropriate, the transfer of, or access to, environmentally sound technologies and know-how to other Parties." Developing (non-Annex I), countries received no targets. Though the convention was seen to have taken a soft approach, as opposed to adopting legally binding targets, it did establish obligations, objectives and principles, as well as an international participatory basis for multilateral negotiation and co-operation.

Photo: Inspire!

The Conference of Parties (COP) of the UNFCCC has met every year since the Convention came into effect. At each COP the implementation of the Convention, its objectives and new scientific findings are reviewed. It was clear at COP 1 that voluntary commitments set out by the Convention were inadequate. Accordingly, the IPCC's Second Assessment Report called for even more rigorous policies. It was clear that legally binding targets would be required and the Berlin Mandate was adopted. The Ad Hoc Group on the Berlin Mandate (AGBM) was established in April 1995 to toughen Annex I Party commitments for after 2000. The Group met eight times, and spurred on by the political momentum of COP 2, which called for a legally binding protocol with emissions reduction targets within a specific period, conducted the negotiations that eventually led to the adoption of the Kyoto Protocol.

The Kyoto Protocol: Creative solutions for global multilateral cooperation

The Kyoto Protocol is named for the city in Japan where it was adopted on the 11th December 1997 at COP 3. The Protocol was opened for signature at UN headquarters in March 1998 and requires to enter into force, signatures from "not less than 55 Parties to the Convention, incorporating Parties included in Annex I which accounted in total for at least 55 per cent of the total carbon dioxide emissions for

1990". The ratification of the Protocol became an uncertainty when the United States, accounting for 36% of global emissions, withdrew from it claiming that it was "fatally flawed" and would cost too much to implement. Furthermore, the United States was reluctant to take on targets unless developing countries made a "meaningful contribution". U.S. objections led to considerable weaknesses in the negotiated text. There has, however been widespread support and affirmation from other nations and the Protocol is likely to be ratified without the U.S.

The Kyoto Protocol sets individual emission limitations and reduction targets on 6 greenhouse gases for Annex I Parties. These gases are Carbon dioxide (CO_2), Methane (CH_4), Nitrous oxide (N_2O), Hydrofluorocarbons (HFCs), Perfluorocarbons (PFCs) and Sulphur hexafluoride (SF_6). Achieving these targets will contribute to an average global emissions reduction of 5% from 1990 levels, and must be achieved within a commitment period, which has been set for 2008 – 2012. Emissions reduction targets range from -8% for the EU, to +10% for Iceland. Russia's target is 0% as GHG emissions have been lower since 1990 due to the economic and political upheaval of the post-Cold War period. The Protocol requires that by 2005, Parties with targets show "demonstrable progress" in the achievement of their emissions reductions.

The Kyoto Protocol with its legally binding emissions reduction targets set the stage for the challenges of implementing climate change mitigation through financial mechanisms, technology transfers and sustainable development principles. The protocol also represented a milestone in the climate change process. A significant development with regard to international environmental policy, its effects will engage mankind well into the next century.

The Flexible Mechanisms: Market instruments for Climate Change Mitigation

In order to facilitate financial efficiency in emissions reductions, and satisfied that such reductions would help mitigate climate change regardless of where they occur, the Kyoto Protocol provides for three flexible mechanisms. These are: Joint Implementation (JI); Emissions Trading (ET); and the Clean Development Mechanism (CDM).

Under JI, developed (Annex I) countries may finance emissions reduction projects in another developed country and earn from this activity emissions reduction units (ERUs) which it may count as part of its emissions allowance.

Emissions trading allows for developed countries to trade their emissions allowances with each other for the purposes of achieving emissions reductions where it is cheaper to do so.

Because developed countries are historically responsible for the majority of GHG emissions, developing (Non Annex I) countries were not allocated emissions reduction targets. The CDM allows for Annex I countries to invest in emissions reduction activities in developing countries and earn from these activities Certified Emissions Reduction Units (CERs).

The idea of using market-based mechanisms to help reduce mitigation costs became a dramatic solution, which had both supporters and objectors. Supporters were satisfied that these mechanisms would ensure wide and realistic participation within the Protocol. Opponents pointed to the difficulties of implementation and the potential problems for ensuring honesty and integrity. The flexible mechanisms have been of particular controversy, as they have been seen as loopholes through which developed countries may escape their emissions reduction commitments. The Protocol does state, however, that the use of the flexible mechanisms by developed nations must be "supplemental" to domestic action.

The Clean Development Mechanism

The Clean Development Mechanism, Like JI, is project-based. Article 12 of the Kyoto Protocol states that the purpose of the Clean Development Mechanism is to assist developing countries "in achieving sustainable development", and assist developed countries "in achieving compliance with their quantified emission limitation and reduction commitments."

The rules of the CDM are that projects should have "voluntary participation", that project activities "result in real, measurable and long term benefits to the mitigation of climate change", and that the said benefits "be additional to any that would have occurred in the absence of the project activity". The Kyoto Protocol provided only a framework for the CDM, and the rules for it were fleshed out in the Marrakesh Accords adopted at COP 7 in Morocco in 2001. This mechanism is unique for several reasons. It is the first attempt to lever internationally an opportunity for developing countries to attract investment while achieving the ends of reducing global emissions, and it emphasises Sustainable Development as a criterion for approval.

CDM: Low Hanging Fruit and Sustainable Development

The Clean Development Mechanism is of particular importance to developing countries. Not only does it provide for developing countries to participate in the alleviation of climate change, it also provides for resource flow from developed nations in the form of finance and clean technologies to ensure that developing nations progress along a lower carbon path. There are, however, possible future commitments, as well as other risks, such as capability to adapt to climate change, which developing nations face as the climate negotiation continues beyond the end of the first commitment period (2012), extending its ambit to non-Annex I countries.

As Annex I Parties may be more likely to invest in high-yielding, low-cost CER projects, developing countries run the risk of selling off their easily achieved emissions reductions or "low hanging fruit" and consequently, experience difficulties in meeting their own possible future commitments.

It seems to be the case that projects that contribute significantly to sustainable development seldom yield many, or low-cost, CERs. There is thus also the danger that some developing countries will maintain relaxed sustainable development requirements for the projects they host in order to attract investors, a phenomenon referred to as "the race to the bottom" for investment. This strategy may well result in an influx of foreign direct investment (FDI) but generally leads to small translatable benefit for the majority of the country's population.

Developing countries participating in CDM activities should therefore enforce stringent sustainable development criteria when approving projects, taking into consideration: economic issues such as employment, technological self-reliance, and balance of payments; environmental issues such as pollution, biodiversity as well as soil, air and water quality; and social issues such as access to energy and other essential services, empowerment, education, race and gender equity and a telling indicator for developing countries, poverty alleviation. These are the indicators that contribute to a stable economic, social and political environment, which is key to attracting investment. A highly refined set of criteria devised for assessing sustainable development has been developed by SouthSouthNorth (SSN) and has subsequently been adopted by the World Wildlife Fund (WWF). Details can be found on www.southsouthnorth.org

CDM provides a means for developing nations to play a substantial role in climate change mitigation, as well as benefit from technological leapfrogging and investment. More importantly, however, CDM contributes to the shaping of a global collective consciousness endorsing social, economic and environmental sustainability for present and future generations.

....

Key Associated Topics:

Air Quality; Carbon Tax; Carbon Trading; Climate Change; Global Warming; Kyoto Agreement.

....

Associated Organisations:

Earthlife Africa; National Botanical Institute; Renewable Energy And Energy Efficiency Partnership (Reeep); SouthSouthNorth; South African Climate Action Network; WWF.

Photo: Copyright South African Tourism

GONDWANA ALIVE CORRIDORS

Guest Essay by Dr John M Anderson ~ Gondwana Alive

In these earliest years of the Third Millennium, we surely have a unique opportunity. If we are to stem the Sixth Extinction there is no alternative but to unite – now – as the one genetically-close human family that we have only in the last few years discovered we truly are. We have a cause far greater than warfare or religious crusades to energise us. The pervasive, divisive conflicts that have employed and drained our energy in the past could, in view of our new scientific awareness, be relegated to history. All six billion of us should now have a clear and common goal. We must commit to it or stand condemned in the courts of our grandchildren.

The Gondwana Alive (GA) Society has a book – in the final stages of preparation – entitled "A set of 100 strategies towards stemming the Sixth Extinction". Of these 100 strategies, we find the "Gondwana Alive Corridors" the most holistically compelling. The strategy holds the certain potential to persuade each one of us to adopt a new worldview: a view acknowledging, profoundly, that 'business as usual' is untenable.

Historical background

The GA Corridors concept was launched simultaneously, in August-September 2002, at two world events at opposite sides of the globe: the 11th International Gondwana Symposium (Christchurch, New Zealand) by Maarten de Wit; and the World Summit on Sustainable Development (Johannesburg, South Africa) by myself. Continuing the global emphasis, the first two papers on the Corridors were published by Maarten and I in Japanese (mid 2003) and Indian (late 2003) publications respectively. A book on a selection of 40 Corridors across Gondwana is planned for early 2005.

Defining the concept

Two essential elements define the concept of the GA Corridors – "Unique core themes" and "Holistic management".

Unique core themes: Each Corridor is a defined geographic strip of land that is unique or unsurpassed – geologically, biologically or culturally – in telling a part of the story of our world. Each illustrates and represents a chapter in the extraordinary 4,6 billion-year story of this most diverse and beautiful of planets. Together these corridors weave an autobiographical tapestry of the planet.

Holistic management: In view of each Corridor being a priceless part of our global heritage, like a superb art gallery or a towering body of music, they lend themselves ideally to holistic management. Those persons living within a corridor or adjacent to it will feel a sense of pride, of belonging, of joint curatorship. They will cherish it and feel impelled to help preserve it and enhance its value, and to spread awareness of its particular uniqueness.

Network of Corridors across Gondwana

We visualise within 10 years a lattice of corridors spreading across the southern continents and southern oceans. The corridors may be further characterised along the following lines:

Geographic extent: They are meandering belts of territory around 1 000 km in length and 50 km in breadth.

Critical nodes: Each can be likened to a string of pearls, incorporating UNESCO World Heritage sites or Biosphere Reserves; national, provincial or private parks; geological, biological or cultural hotspots.

Autonomous yet integrated: They are independently governed, yet are part of a federated international network.

All persons: Every individual living along a corridor is encouraged to take pride of curatorship and proprietorship and are partners in the holistic management thereof.

All plant and animal species: All other species, and all biological communities, living along the corridor are respected as having an equal right to optimal existence.

Uniqueness of the five southern African Corridors

The set of five corridors thus far defined for southern Africa are individually and collectively unmatched. They embrace our planetary heritage – geologically, biologically and culturally – from the earliest signs of life to the present day.

Cradle of Life to Cradle of Humanity Corridor: Celebrates 3.5 billion years of life on Earth, from the earliest-known bacteria in the Barberton Mountains to our hominid origins at the "Cradle of Humanity World Heritage Site" in Gauteng. The CCC tracks a walk through more than two thirds of Earth history far more comprehensively than is possible elsewhere.

Great Karoo Corridor: For 150 million years (325 to 175 Ma), when the mammals and dinosaurs evolved from the early reptiles and the stem-angiosperms (flowering plants) from the gymnosperms (cone plants), the Karoo basin filled with a great thickness of sediments and abundance of fossils. It lay deep within the heartland of Gondwana. The GKC, which spans the apocalyptic end-Permian Extinction, is the undisputed Oscar winner for preserving this critical phase in the story of life.

Cape Coast-Fynbos Corridor: Here we witness one of the supreme ironies. Two themes intertwine: our Homo sapiens story and prodigious biodiversity. The impact of the one on the other, the Sixth Extinction, is more poignantly expressed here than in any other corner of our world.

Kalahari-Khoisan Corridor: All six billion persons alive today can trace their ancestry back directly to a single woman, Mitochondrial Eve, who lived somewhere in eastern or southern Africa between 150 000 and 140 000 BP. Of all extant clans worldwide, the Khoisan peoples have the deepest roots within our human family. They are the richest bearers of our genetic and linguistic history.

Great Zimbabwe Corridor: The kingdoms of Mapungubwe and Great Zimbabwe flourished from ca 1000 to 1600 AD, coinciding essentially with the Romanesque, Gothic and Renaissance periods in Europe. They were built up largely around gold, copper and iron crafts, with trade links as far afield as Persia and China, and are quite unique in human history.

Eco-cultural tourism and education

Whilst the corridors offer endless innovative opportunities in all sectors, we will comment here just on eco-cultural tourism and lifelong learning. As vehicles for both, the value of the corridors is perfectly evident.

Eco-cultural tourism: In focusing on unique chapters of our Earth's story, an entirely new genre of interest and engagement for the tourist is opened. All traditional attractions remain undiminished, of course, from 'the big five' to Zulu dancing, yet now there is a broader picture, the global context. Regular international travelers might find themselves 'reading' different chapters of our story on an annual basis.

Education through involvement: As one example, every scholar, in a new era of hands-on, outcomes-based education, could find themselves participants in building up a comprehensive inventory of life, from soil bacteria to elephants, along that corridor in which they live.

GIS and the Corridors

Geographic Information Systems, combining satellite imagery, electronic data basing, and holistic thinking, make inventorising the Corridors and managing them sustainably – taking all factors into consideration – clearly feasible. All layers of information are fungible.

Financing the Corridors

To convert the concept of the Corridors to the process of the Corridors will cost large sums of money. However, everyone in global and national finance is fully aware that there are indeed hardly comprehensible sums of money potentially at hand. The issue is how best to spend that money and how to manage the spending of it.

Wealth of nations to wealth of Nature: With the great majority of the world's financial wealth concentrated in the northern (Laurasian) countries and the overwhelming majority of the world's natural resources preserved in the southern (Gondwana) countries, an ideal situation prevails for forging a new balance. Substantially increasing the standard of living of all persons everywhere, whilst preserving biodiversity and wilderness, can be clearly envisaged and negotiated.

Adopt a Corridor and adopt a School: The idea is simple. Billionaires, top corporations and foundations from the north are approached to adopt particular Corridors – with their identity appropriately advertised. Millionaires and middle-ranking corporations adopt individual schools and projects within the corridors.

Joint authorship

By taking co-authorship in writing the story of our past, we can fashion an altogether different future. Why would we not take this choice? Look back from a century hence and we undoubtedly would take this choice!

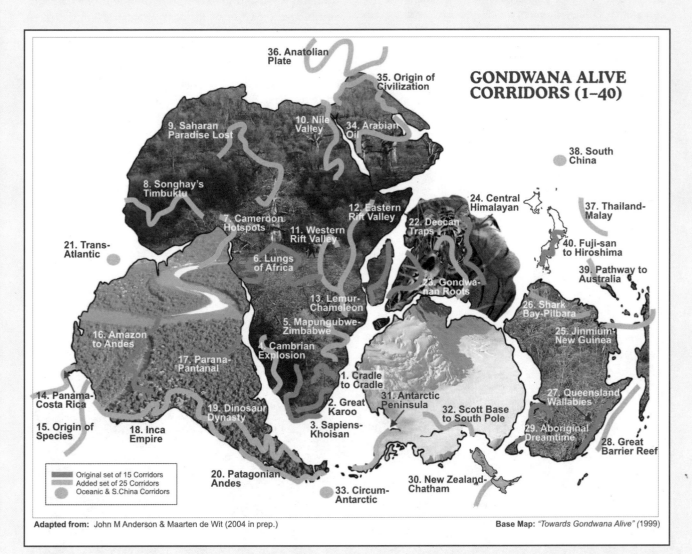

GONDWANA ALIVE CORRIDORS (1–40)

36. Anatolian Plate
35. Origin of Civilization
9. Saharan Paradise Lost
10. Nile Valley
34. Arabian Oil
38. South China
8. Songhay's Timbuktu
24. Central Himalayan
37. Thailand-Malay
12. Eastern Rift Valley
22. Deccan Traps
40. Fuji-san to Hiroshima
7. Cameroon Hotspots
11. Western Rift Valley
21. Trans-Atlantic
6. Lungs of Africa
39. Pathway to Australia
23. Gondwanan Roots
13. Lemur-Chameleon
26. Shark Bay-Pilbara
5. Mapungubwe-Zimbabwe
25. Jinmium-New Guinea
16. Amazon to Andes
4. Cambrian Explosion
17. Parana-Pantanal
1. Cradle to Cradle
14. Panama-Costa Rica
31. Antarctic Peninsula
27. Queensland Wallabies
2. Great Karoo
32. Scott Base to South Pole
15. Origin of Species
18. Inca Empire
19. Dinosaur Dynasty
3. Sapiens-Khoisan
29. Aboriginal Dreamtime
28. Great Barrier Reef
20. Patagonian Andes
30. New Zealand-Chatham
33. Circum-Antarctic

Original set of 15 Corridors
Added set of 25 Corridors
Oceanic & S.China Corridors

Adapted from: John M Anderson & Maarten de Wit (2004 in prep.)　　　**Base Map:** *"Towards Gondwana Alive"* (1999)

GRASSLANDS

Humanity owes much of its development as a civilization to the world's grasslands. Grasslands have provided our ancestors with many of the domesticated food crops that we now consume, for example: wheat, sorghum and sugarcane. Grasslands have also provided the grazing resources that allowed humans to domesticate vast herds of livestock and ultimately to create large settlements and cities. In modern South Africa, our economic powerhouses occur in our grasslands. Coal and gold mines, agriculture and major urban complexes both depend on and threaten these areas.

Grasslands are very old, complex and slowly evolved systems of diverse plant communities. Most plant reproduction takes place vegetatively with the implication that many grass clumps are genetically identical to the grass clumps present in the area as much as 2 000 years ago, emphasising the incredible complexity of this often "taken for granted" ecoregion. Only one in six plants in a grassland community are in fact grasses, the remainder are bulbous plants (such as orchids, arum lilies, red hot pokers, watsonias) that include many rare and valuable medicinal plants. Once a grassland is destroyed, through ploughing or mining, its diversity is lost forever and cannot be regained.

In South Africa, the ecoregion provides a habitat rich in plant and animal species, some of which are found nowhere else. Our grasslands contain high numbers of often-threatened, species of birds, mammals, invertebrates, reptiles and amphibians. They are probably most well known for the special bird species such as blue swallows, cranes, larks and pipits that occur in them and also provide huge ecotourism potential.

Tourism is one possible economic benefit to be gained from grasslands, but grasslands on the whole already contribute an enormous amount of benefit to the South African economy. As an agricultural resource the remaining grasslands of Mpumalanga and KwaZulu-Natal alone provide natural grazing to sheep producing more than eight million kilograms of wool annually, with a value exceeding R52 million (1995 figures). Production of maize, wheat and timber also occurs extensively in this biome. In the informal economy, the value of medicinal plants harvested from the grasslands is considerable.

Less evident, but no less important to the economy are the critical ecosystem services which pristine grasslands provide, the most essential being that of collecting and purifying water. In a water-stressed country such as South Africa, their role as catchments and their ability to hold the water in the form of ground water or wetlands and then slowly release it throughout the year through seepage zones is crucial. Many of the power stations and petrochemical plants that occur in the highveld grasslands depend on the water services of this biome. In addition, grasslands play a role in flood attenuation, soil creation and the prevention of soil erosion as well as providing essential pollination services. Yet, despite the vital role they play, up to 60% of South Africa's grasslands have been irreversibly transformed and less than 2% are currently formally conserved.

The threats facing grasslands are enormous and it is ironic that, because they were the very habitats that enabled humanity's economic advancement, that they are now so threatened. Threats include voracious urban development, agricultural expansion and pollution, not to mention climatic change. There is some effort to address these problems however. 1,6 million hectares of high altitude, moist grasslands spanning the corners of three provinces, Mpumalanga, KwaZulu-Natal and the Free State are the project area for an initiative known as the Ekangala Grassland Project to conserve these relatively intact grasslands. A National Grasslands Initiative is also in development to try to establish a coordinated approach for the protection of our remaining grassland fragments. It comes none too soon to protect this, one of our very crucial life support systems.

Guest Author: Thérèse Brinkcate ~ WWF SA

Key Associated Topics:

Agriculture; Alien Invasive Plants; Biodiversity; Conservation; Desertification; Ecosphere & Ecosystems; Land Degradation; Monoculture; Rehabilitation of Land.

Associated Organisations:

BirdLife of SA; Grassland Society of SA; Green Trust; WESSA; WWF.

Photo: Nigel J Dennis Photographer – www.africaimagery.com

GREEN ARCHITECTURE

It has been estimated that 50% of the world's energy is used in the production and servicing of buildings. Through carbon dioxide and CFC emissions, buildings may be responsible for up to 32% of global warming. Enormous quantities of non-renewable resources and water are consumed in the production of modern buildings. Most architecture is oblivious to these facts and their consequences, and only attempts to address the issues of cost, function, aesthetics, and construction. Green architecture takes a much broader view, and attempts to address the issues of resource depletion, environmental degradation, pollution and social imbalance. It encompasses the three pillars of sustainable development: Social, Economic and Environmental sustainability.

Srawbale house at Oudrif – Cederberg

Green architecture has intrinsic in its philosophy the notion of a caring social ideology in the design process. The site is considered more than a place to build the project, but rather its geology, geography and ecology are regarded as a resource with which the architecture synthesizes to produce built form. Climate is most often an opportunity for synergy to produce comfortable internal and external environments with much less energy expenditure than in traditional architecture. This is particularly true in areas such as southern Africa with its temperate climate. It is recognised that any development that is not economically sustainable is an even greater waste of precious resources than traditional architecture, and therefore green architecture has to work within the current financial systems, while still producing less environmental impact than its traditional counterpart.

In the language of the building industry, the building is broken down into separate elements: wall, roof, floor, etc. There is little mention of the interaction between these elements, and how they perform as a whole. Green architecture represents a holistic approach, where the building is considered as a whole; an organism or machine, in which every component has an effect on the other components and on the building as a whole. Working in this way, it is possible to have single elements performing several functions, producing more resource efficient buildings.

Guest Author: Ken Stucke - ERA Architects

Key Associated Topics:

Aesthetics; Ecological Intelligence; Eco-efficiency; Greywater; Renewable Energy; Solar Heating; Sustainable Development Section: – The Built Environment.

Associated Organisations:

Canstone; ERA Architects; Eco-City; Greenhaus Architecture; Greenhouse Project; Institute of Landscape Architects of SA; LC Consulting Services; The Greenhouse People's Environmental Centre.

THE GREEN / BROWN DEBATE

The so-called "Green/Brown Debate" only became prominent in the early to mid 1990s when community environmental Non-Governmental Organisations (NGOs) entered the wildlife/environmental debate. "Green" means a more wildlife focus while "Brown" refers to a more people's approach. Concerns were raised that the environment is seen as something that is all about saving species of plants and animals – mainly ignoring the urban environment and people's basic right to a clean and healthy environment.

The "Green" environmental NGOs were initiated because of real concern for the dwindling number of species and loss of habitats. Until the late 1890s, subsistence hunting and the eradication of predators that posed a threat to livestock by the earlier European settlers and then later the indiscriminate slaughter of game by so-called "sport hunters", caused an alarming drop in wildlife numbers. It was this fear that there may not be wildlife left to hunt that was the catalyst for our first "green" NGO, the Wildlife and Environment Society of SA. A more genuine interest in the conservation of bio-diversity, locally and internationally, is the driving force behind many of the more modern "Green" movements. With slogans such as "Save the Whale", "Save the Rhino", etc. these organisations raise enormous amounts of funding to preserve species and wilderness for prosperity. At the turn of the last century another dimension was added when "animal rights" groups joined the fray – some of them demanding that animals be accorded rights on an equal footing with humans.

The "Brown" environmental NGOs have their roots in the political struggle against apartheid. The Apartheid laws and most specifically the Group Areas Act forced most South African citizens into overcrowded residential areas. Lack of clean water, inadequate sanitation, insufficient housing and poverty created ghettos in many places with appalling environmental standards. Understandably people living in these conditions would view the environment from a completely different perspective. To make matters worse some rural communities were also uprooted in certain provinces to make room for wildlife parks.

Wildlife at the expense of people did nothing to foster a "Green" feeling amongst those who were negatively affected by the system. A host of community-based organisations sprung up to improve the lot of people and to tackle air and water pollution, sanitation, illegal dumping, toxic waste, food security, recycling, etc. The name of the most prominent of these organisations, The Environment Justice Networking Forum, perhaps describes the aim of this movement the best.

Of course in reality "the environment" includes both people and wildlife and it is very unfortunate that our history has politicised the debate to the extent of polarizing organisations into two different camps.

We need to heal this division and work toward mutually beneficial "Sustainable Development" goals. Our Constitution protects every citizen's right to a clean and healthy environment and because of the legacy of apartheid, much effort is needed to fix what is wrong in some residential areas. On the other hand, South African is also a signatory to the International Bio-diversity Treaty, which means we have to protect species and habitats critical to their survival – the world is still losing species at an alarming rate.

Some organisations have broadened their base to become more inclusive. The Wildlife Society of SA changed their name to the Wildlife and Environment Society of SA (WESSA). Some of the rural organisations have made wildlife their main activity – forfeiting ancestral land to maintain wilderness areas.

The environment's worst enemies are poverty and greed. South Africa needs economic growth – which could however easily lead to the further exploitation of our natural resources if we do not have these "environmental" watchdogs to keep the greedy in check. Whilst our potential to manufacture and export products to create wealth is severely limited, our priceless natural heritage holds the key to a better future as "eco- tourism" provides a wonderful and sustainable activity to can alleviate poverty – for the present and future generations.

Guest Author: Danie Van der Walt ~ Executive Producer 50/50 Programme, SABC 2

Key Associated Topics:

Conservation; Environmental Justice; Poverty Alleviation; Sustainable Development.

Associated Organisations:

Environmental Justice Networking Forum; SA New Economics Forum.

GREYWATER

Greywater is any water that has already been used in homes, offices or factories and which can be directly reused for other purposes without any further purification. Greywater excludes water from toilets as well as industrial effluent that contains problematic chemicals.

Generally speaking, in a household greywater comprises some 50% to 80% of what is normally considered waste water. This water can be reused in a variety of ways. An immediate use is to irrigate the garden or factory lawns. This usage of greywater has a number of benefits, for example, a lower fresh

water usage, groundwater recharge and the reclamation of otherwise wasted nutrients in the water. Using greywater in this way also reduces the load on septic tanks and water treatment plants.

More complex greywater systems can be engineered, for example, routing greywater to flush toilets, or to use in factories for washing machinery, or for use in cooling towers and other industrial activities that do not need high-grade water.

Guest Author: Dr Kelvin Kemm

Key Associated Topics:

Eco-efficiency; Green Architecture; Recycling; Water Quality and Availability; Urban Agriculture.

Associated Organisations:

Unilever Centre For Environmental Water Quality; Water Institute of Southern Africa.

GROUNDWATER

Groundwater is the name given to water that is stored within the pores (air spaces) of soil and in rock formations. The upper limit of the ground water is the water table and the lower limit is any layer of impermeable material. Groundwater is the underground "link" in the water cycle. Water from wells is supplied from groundwater sources and this is particularly important where there are surface water shortages. People are tending to place stress on groundwater sources by over-pumping them, which dries out some areas (causing subsidence) and pollutes others through seepage of agricultural chemicals and industrial wastes. For example, the city of Shanghai has sunk by 2.63m and some areas of the Californian San Joaquin valley have sunk by as much as 9m. Once groundwater sources are polluted, they are almost impossible to cleanse because of the relatively slow water movement, lack of micro-organisms to breakdown bacteria, and limited dilution opportunities. South Africa, as a water deficient country, is at particular risk because the pollution of groundwater supplies will seriously affect the availability of potable water supplies to the growing population.

Photo: Earthyear Magazine

Key Associated Topics:

Agriculture; Desertification; Hazardous Wastes; Leachate; Pollution; Population; River Catchment Management; Rivers & Wetlands; Water Quality.

Associated Organisations:

Barlofco; Biolytics Southern Africa; Divwatt; Enviro Options; Groundwater Association of SA; Groundwater Consulting Services; Unilever Centre For Environmental Water Quality; Wales Environmental Partnerships.

HAZARDOUS WASTES

In order to provide for the increasing demand for consumable products in today's society, industries often produce hazardous and dangerous by-products, which cannot be used and must be disposed of by being dumped in landfill sites. Responsible companies use properly designed, managed and licensed landfill sites, which minimise the risk of pollution to the environment. Responsible companies also try to reduce their production of hazardous wastes by looking at alternative materials, substances and chemicals that are less hazardous. Hazardous wastes that are dumped into the natural environment not only affect natural ecosystems but are also a human health hazard. It is an ethical and legal responsibility of industries and the government to inform communities of the hazardous substances to which they may be exposed. Unfortunately, South Africa currently has a shortage of properly contained, High Hazard (H: H) waste disposal sites, which is placing considerable pressure on hazardous waste generators. Hazardous waste disposal is controlled in terms of the Environment Conservation Act but the Department of Water Affairs and Forestry implements and polices the regulations. Changes will take place over the next 12 months as the new *White Paper on Integrated Pollution and Waste Management's* recommendations are implemented into law.

Key Associated Topics:

Air Quality; Basel Convention; Carcinogen; Environmental Due Diligence; Environmental Liability for Remediation; Environmental Risk & Liability; Environmental Law; Incineration; Integrated Pollution & Waste Management; Landfill; Leachate; Marine Environment; NEMA; Nuclear Energy – The Counter Debate; Pollution; Polluter Pays Principle; Radioactive Waste; Used Oil; Waste Management.

Associated Organisations:

Chemical Emergency Co-ordinating Services; Environmental Risk Management; Enviroserv Waste Management; Groundwork; Icando; Institute of Waste Management; Lombard and Associates; Millennium Waste Management; Tiger Chemical Treatment; Wales Environmental Partnerships.

HERITAGE

Heritage is that which we inherit: wildlife and scenic parks, sites of scientific or historic importance, national monuments, historic buildings, works of art, literature and music, oral traditions and museum collections, together with their documentation. Heritage is closely linked to culture in the sense that different cultures have differing priorities on what particular type of heritage they want to see protected and preserved. The fusing of different cultures in the new South Africa is likely to produce an interesting mix of heritage items for preservation.

The preamble of the South African National Heritage Resources Act states that our heritage "…helps us define our cultural identity and therefore lies at the heart of our spiritual well-being and has the power to build our nation. It has the potential to affirm our diverse cultures, and in so doing, shape our national character. Our heritage celebrates our achievements, and contributes to redressing past inequities…"

Key Associated Topics:

Cultural Resources; Environmental Ethics; Gondwana Alive Corridors; Indigenous Peoples; Indigenous Plants; Stakeholders; Stewardship; Wilderness Area.

Associated Organisations:

Agency for Cultural Resource Management; Cape Town Heritage Trust; Centre for Conservation Education; Hout Bay & Llandudno Heritage Trust; Open Africa; Wales Environmental Partnerships.

HORMESIS

Hormesis is an effect in which a biological entity actually benefits from a low dose of chemical toxin or a low dose of nuclear radiation.

Chemical hormesis is actually the basis of medical homeopathy. The principle is that a small dose of toxin causes the biological entity to respond and to defend itself, thereby stimulating its natural defences. Chemical hormesis has been found in a large number of toxicological systems. The general finding is that toxins often have a positive effect on the health of organisms, at low doses.

The beneficial effect of a chemical toxin usually rises to about 150% of the control entity in the experiment, and extends over a ten-fold dose before falling to the NOAEL (No Observed Adverse Effect Level). This principle of hormesis is in direct contrast to the concept of 'linear extrapolation to zero' that is so often used in the development of legislation in many advanced countries. The word 'linear' means 'in a straight line' and this principle in effect says: if a large dose of toxin is poisonous to a person or organism, then a low dose must be bad as well, just not as bad as a high dose. The adverse effect is then assumed to lessen in a linear fashion (straight line), all the way down to zero dosage. In other words, this theory says that there is no safe limit or 'no threshold' at which some small quantity of a toxin can be assumed to be safe.

This 'no threshold' concept then results in legislation, which develops targets of zero pollution or zero chemical emissions and so on. The 'no threshold theory' therefore has serious cost implications. It can be very costly for industry to really try to remove the very last minute fraction of a chemical labelled as 'toxic'.

Hormesis has also been found to occur in the case of nuclear radiation, which similar to chemical hormesis, is also in contrast to the 'no threshold theory.' It has been found that low levels of radiation (from the gas radon for example) actually decreases the death rate from lung cancer by a significant amount.

It would appear that slight amounts of damage to biological systems from low doses of radiation allows the biological system to repair itself, and to also simultaneously get rid of damaged material such as cells already invaded by cancer.

Guest Author: Dr Kelvin Kemm

Key Associated Topics:

Carcinogen; Dilution; Dioxins; Environmental Health; Medicine & Health; Pops and Pops Convention; Precautionary Principle; Radiation.

Associated Organisations:

Stratek.

HUNTING

Guest Essay by Peter Borchert
~ Africa Geographic Magazine

Photo: WESSA

Few subjects in the world of conservation and wildlife and, indeed, the world at large, work up more of a froth of emotion and opinion than the word 'hunting'. The image that immediately comes to mind is that of a person, usually a man, armed with a high-powered rifle, having a go at some hapless beast. The aficionados would call it sport and those at the other end of the spectrum a shameful exhibition of power over an animal with no say in the matter.

There are of course many forms, and purposes, of hunting. Take, for example, a poor man in a rural environment who clambers over or through a game fence and lays a trap to catch a small antelope in a neighbouring reserve or farm. Most would say that he was guilty of poaching; stealing something that doesn't belong to him. At one level yes, but at another he is simply catching food for his family as rural, pastoral and nomadic folk have done since the dawn of mankind. Then there is the hunter who could, and mostly does, pop down to his local supermarket to buy the meat of his choice, but who occasionally enjoys going out into the bush to track and shoot an animal for the pot. In reality, is he any different from a recreational fisherman bent on landing a juicy trout or bream? And, before non hunters such as me become anything close to smug, are we not just hunters by proxy who, by virtue of our urban lifestyles, simply turn a blind eye to the horrors of abattoirs where the killing is so conveniently done for us, and supermarkets who, even more conveniently, package protein in a way that it no longer remotely resembles the creatures it comes from? But that's another story.

Then there is the trophy hunter for whom the thrill is killing the best, biggest, most magnificent specimen of a species he can possibly find. It is the size of the tusks, the colour and luxuriousness of the mane, the length of the horns, the height, the weight, that is the goal. And even within this body of
so-called sport hunters there are divisions: those for whom the tracking on foot and sense of outwitting an opponent are an integral part of the process; and those who are willing to be taken with the least physical effort as close to their quarry as possible – and if that involves shooting an already trapped or drugged animal from a vehicle or through a fence, then so be it.

It is this latter form of hunting that causes the greatest outcry and makes the pages, airwaves and TV screens of the media. And quite justifiably so, for in anyone's book, including the majority of passionate hunters, this 'canned hunting' is a travesty that has nothing to do with sport, conservation or anything for that matter other than shame, greed and cruelty.

Hunting is big business by any standard. Ian Michler, writing for Africa Geographic in June 2002, estimated that in South Africa alone some R16 billion is invested by private landowners in game farms, breeding centres and hunting ranches spanning more than 11 million hectares. He further estimated that the turnover from these and other supportive and associated activities exceeds R1 billion a year and that hunting is one of the fastest expanding sectors in the economy. It also has to be acknowledged that many of these landowners have played an important role in rehabilitating land, restoring it to near pristine condition and in some instances reintroducing indigenous animal species that had long been absent from the landscape.

Hunting can, therefore, be defended as a legitimate use of a renewable wildlife resource, as long as it is practised at levels that are sustainable and that the practitioners embrace a code of ethics that shows accountability for the way in which they conduct their business and shows respect for animals. This is strongly reinforced by Gerhard Damm, the president of the Safari Club International African Chapter: 'The culture of trophy hunting requires the hunter to be conscious of animal welfare and the well-being of the biosphere. Animal welfare includes not only effective and humane killing but, more importantly, a commitment to the maintenance of healthy animal populations within vibrant ecosystems.' (Africa Geographic February 2003).

Regrettably, as mentioned earlier, there is an element of the hunting fraternity that does not subscribe to this notion and where, as Ian Michler contests, 'operators and groups have corrupted these principles of utilisation and conservation by using wild animals as mere commodities in order to generate substantial financial returns. Animals are being bred, captured, relocated, auctioned, hunted, killed, imported and exported in an ongoing process and in a manner that not only displays an alarming lack of concern for their welfare, but in many instances has become nothing less than cruel exploitation.'

The 'ethical' hunters of course roundly resent the publicity given to the reprehensible canned hunting fringe, saying that they are unjustly tarred with the same brush. Well, that might be so, but is it any different from bad doctors and lawyers giving their kind a bad name? At least in those professions, the law often takes its course and irresponsible or criminal practitioners are 'struck off'. Perhaps it behoves the associations of hunters to be more vigorous in bringing the bad elements in their profession to book?

It is unlikely that hunters, even those who follow their own code of ethics to the letter, will ever be able to convince non-consumptive use protagonists that their sport is legitimate or morally defensible. And likewise, it is unlikely that non-hunters will ever be able to sway the hearts and minds of those who hold the power of life and death in their trigger fingers. As pointed out earlier, hunting is big business and, unless the laws of the land are changed, it is here to stay. Perhaps, therefore, if the worst elements of hunting were to be eliminated through censure and prosecution, hunters and non-hunters could find a more comfortable middle ground where they could focus on their common love of wild, open spaces and use their combined energy and passion in its defence.

Key Associated Topics:

Biodiversity; CITES; Conservation; Ecological Intelligence; Eco-tourism; Environmental Ethics; Extinction; Game Farming; Stewardship; Wildlife Management.

Associated Organisations:

SA Gamebirds Research; SA Wildlife College; Centre for Wildlife Management – University of Pretoria. WESSA; WWF.

Don't Dump - Recycle!

Environmental responsibility begins with the individual.

Sappi Waste Paper's recycling programme is an initiative

aimed at leveraging individual responsibility. The programme

empowers the community to take care of the environment,

earn a living and contribute to a cleaner future.

Sappi Waste Paper

sappi

www.sappi.com

INCINERATION OF WASTE

The incineration (high temperature burning, using sophisticated high technology equipment) of wastes, particularly non-recyclable municipal waste and a variety of industrial and hazardous waste, is practised in many countries, predominantly in Western Europe. In most cases incineration plants include cogeneration technology, whereby the heat of combustion is also used for electricity generation and/or heat recovery. One of the other advantages of incineration is that this process typically reduces the amount of waste volume by 80 or 90 percent and destroys a large variety of hazardous chemicals including persistent organic pollutants (POPs). The alternative to incineration is normally land filling, which has a number of disadvantages, including a large space demand. Typically leachate is also produced in landfills, which, depending on the source of the waste, can be very hazardous to humans and nature. Often special treatment of the leachate is required. Persistent pollutants can also be retained for a very long time period in landfill sites and can be released at any time in the future if development of the landfill site should ever take place. Land filling is preferred in countries where the cost of both land and of waste disposal by landfill is low. In many countries, including South Africa, public opinion is generally against incineration of waste. Many environmental NGOs share this opinion. There are several reasons for this, ranging from idealistic to more pragmatic. In the first category, falls the thinking that wastes and specifically hazardous wastes should not be produced at all. In the second category, however, it is recognised that in the past some incinerators were built cheaply, operated poorly and caused unacceptable hazardous emissions. The public and NGO reason for opposition to incineration is that in their opinion it is not possible to guarantee that hazardous emissions would not be emitted in the future. It is my personal opinion that for certain types of wastes including waste solvents, waste oils, POPs and infectious waste there is no good alternative to incineration. However, it is necessary to ensure that incinerators are designed correctly for adequate destruction and removal efficiency and are equipped with gas cleanup systems that remove potentially hazardous emissions to a level that can pose no risk to nature or humans. Furthermore, the operating staff must be well educated and must keep records of crucial operating parameters. Lastly, appropriate continuous stack monitoring is required to ensure that the plant is operated according to specifications.

Guest Author: Dr Sibbele Hietkamp ~ Senior scientist, CSIR

Key Associated Topics:

Air Quality; Atmospheric Pollution; Cogeneration; Dilution; Hazardous Wastes; Integrated Pollution & Waste Management; Landfill; Scrubber; Waste Management.

Associated Organisations:

Enviroserv; Groundwork; CSIR – Cleaner Production Centre; Groundwork; WESSA.

INDIGENOUS PEOPLES

Indigenous peoples are people who have lived in their traditional way of life for many centuries and who still retain close ties with the land. The San (Bushmen) in South Africa are an example of indigenous peoples. Agenda 21 says that governments should allow indigenous peoples to take an active part in all political decisions affecting them and their land. In the past, indigenous peoples were denied these rights and the land taken from them. Indigenous peoples like the San have successfully managed to look after the land they used for many centuries in a sustainable manner. Their knowledge of natural processes is irreplaceable and has a crucial contribution to future management of our environment.

Key Associated Topics:

African Environmental Tradition; Communities; Cultural Resources; Ecological Intelligence; Environmental Justice; Green/Brown Debate; Heritage; Stakeholders.

Associated Organisations:

Biowatch SA; Africa Geographic; Eco Africa Travel.

INDIGENOUS PLANTS

Indigenous plants are plants that are native to a particular area. South Africa has many indigenous species, which are found only in South Africa. "Fynbos" is one example. The fynbos region is the seventh floral Kingdom of the world and has thousands of indigenous plants, particularly in the protea family, which is only found in the Southern Cape region. It is important to protect indigenous species because they have often evolved to cope with particular conditions, circumstances or situations. Fynbos, for example, has evolved to cope with veld fires and, in many cases, requires the smoke and heat of the fire to germinate its seeds. There are many indigenous species, which have been found to contain unique properties, which have been used to develop specialised drugs, which have saved thousands of lives. Once indigenous plants become extinct, they are lost forever.

Key Associated Topics:

Alien Invasive Plants; Biodiversity & the Sixth Extinction; Bio-prospecting; Cape Floral Kingdom; Medicinal Herbs; Pioneer Species.

Associated Organisations:

Albany Museum; Apple Orchards Conservancy; Botanical Society of SA; Cebo Environmental Consultants; Environ-mental Education and Resources Unit – University of Western Cape; Gauteng Conservancy Association; Green Inc; National Botanical Institute; Thorntree Conservancy; Ubungani Wilderness Experience; Viridus Technologies; Vula Environmental Services; Zandvlei Trust.

Harry Gazendam, Executive Vice-President: Human Resources and Public Affairs.

Toyota South Africa is currently involved in various social development projects in various parts of the country, mainly in Gauteng and KwaZulu-Natal, where the company draws its major workforce, says Harry Gazendam, the company's Executive Vice-President: Human Resources and Public Affairs.

"Toyota has focused on education as its major empowerment thrust.

"The company does not invest in empowerment for humanitarian or political reasons. We believe that a business cannot run successfully in a dysfunctional society. Toyota's long-term vision is to be a successful business and the best at what we do. A prerequisite for that is to have an orderly society into which we can market our products.

"The social need in South Africa is vast, and Toyota assists where appropriate, but, you can spend millions of rands on humanitarian projects and you may alleviate isolated problems in the short term. However, for the long term, we believe education is the key and that is why we have focused on education as our main empowerment tool.

> "Knowledge gives people the power to participate successfully in an economic system."

"Knowledge gives people the power to participate successfully in an economic system."

He says this philosophy is in line with Toyota South Africa's major shareholder, Toyota Motor Corporation (TMC) of Japan.

"Toyota's South African operation, with the powerful TMC behind it, is also set for rapid expansion in the next five years. "We expect to double production in that time, meaning we will need more people with added skills," says Mr Gazendam.

He says Toyota SA previously exported cars only into the rest of Africa. However, now Toyota SA is set to become a major exporter, initially to Australia, starting from this year, and ultimately into European markets.

"To do this, we have to meet international quality and production standards. We have to compare with the best in the world, and to do that, we need a pipeline of skills into our business.

"The company began the Toyota Teach Primary School Project (TTPSP) programme in 1991. More than 120 primary schools, 50 000 pupils and 1 500 teachers have benefited from the programme since it was introduced in KwaZulu-Natal.

"The aim of the programme, which costs Toyota SA about R2-million a year, is to improve the skills of teachers, particularly in mathematics, science and English, thereby giving pupils a better grounding, knowledge and understanding of these subjects. Through the programme, teachers gain a Further Diploma in Education (FDE), a qualification that is accredited with the national education authority. More than 250 primary school teachers have already completed the programme successfully."

"And," says Mr Gazendam, "any day now, we will be able to start recruiting the first pupils who

benefited from this programme."

"A second phase of the programme, an FDE in School Management, was started five years ago to enable teachers to participate in the management and running of schools.

Mr Gazendam says the company has launched a corporate university. The Toyota Academy of Learning for Africa will assume responsibility for all training and development activities within Toyota SA and amongst dealers and distributors within the entire Africa region.

"The company will use its existing training infrastructure, consolidate it, and restructure it along more academic lines. We will link up with mainstream universities and technikons, meaning that employees who successfully complete the training will end up with a qualification that is nationally accredited.

"Initially there will be three campuses: one in Johannesburg and two in KwaZulu-Natal. The Johannesburg campus will focus on dealer development; one of the KwaZulu-Natal campuses will focus mainly on employee development and the other on supplier development. This will help to ensure that our quality and production standards will compete with the best in the world," says Mr Gazendam.

"The establishment of the Toyota Academy of Learning is part of Toyota Motor Corporation's global strategy of improving quality and service standards throughout the Toyota world. Toyota SA has been a leader in training within the motor industry for many years and is the first company in the sector to announce a project of this nature. Over the past 10 years Toyota SA has funded training to the value of approximately 3% of annual payroll and was awarded the 2001/2002 State President's Award as the top corporate in the area of human resource development.

"Following on the success of our early programmes, TMC, has requested that we expand our training initiatives further. The proposal now being implemented is for an all embracing training establishment modelled on the 'Corporate Universities' that have become both popular and successful in the United States.

"The Toyota Academy of Learning will build on the already comprehensive training activities within Toyota SA and unify the various training departments within one body. The Academy will be based on established academic organisational principles and will feature four faculties or centres for development. A unique feature will be the extension of training opportunities to second and third tier suppliers in association with the Department of Trade and Industry and the Department of Labour.

"Aside from the approximately 7 500 direct Toyota employees, the Academy has the potential to impact on as many as 28 000 employees within the total Toyota franchise network throughout Africa, and a further pool of in excess of 35 000 employees within the dealer and supplier networks in South Africa.

"The Toyota Academy of Learning will have as one of its key points of focus the development of

black economic empowerment and promote South Africa's position of economic leadership in Africa. It is anticipated that the Academy will have a full time staff of between 60 and 70 people. Once fully established, the Academy will have the potential to provide training in specific areas of expertise for as many as 15 000 people per annum.

"In this way, we will be taking empowerment into Africa, which links up well with President Thabo Mbeki's New Partnership for Africa's Development (Nepad) initiative," added Mr Gazendam.

Toyota, he says, is not only determined to keep going right. It's just getting better and better.

> "Toyota is not only determined to keep going right. It's just getting better and better."

In another social development project, Toyota SA intends increasing the number of students joining the company and its dealer network as automotive technicians after passing Grade 12 (Matric).

John Hawkins, Toyota SA's Senior Manager: Technical and Service Training says the company has embarked on an intensive drive to recruit as many students as possible to join the company and its dealer network as automotive technicians after completing their Matric, hence the company has launched a total of 14 Toyota Technical Education Programme (T-TEP) schools in various parts of the country. The T-TEP project is a joint venture between Toyota South Africa and Toyota Motor Corporation (TMC), of Japan, and was first introduced in the country in 1991.

Celebrating the launch of a new Toyota Technical Education Programme (T-TEP) school at Klerksdorp Technical High School, North West, recently are (from left): the school's headboy, Charl Drake; the headmaster, Koos Viljoen; Toyota Motor Corporation's Director, Hironobu Ono; and Toyota SA's Vice-President of Customer Care and Technical Services, Neil Smith.

The main purpose of these T-TEP institutions is to introduce students into the motor manufacturing industry and to support Toyota dealers' recruitment activities.

A total of 1 153 students have gone through T-TEP since the project was launched in South Africa in 1991. Of these, 411 have been employed by Toyota SA and its dealer network and 289 of these students are still employed currently by Toyota SA and its dealer network. A total of 14 of these students currently hold senior positions within the Toyota dealer network; including one dealer principal, one technical training manager, five service managers, two master technicians and five master service managers.

The T-TEP project has been introduced in 355 educational institutions in 48 countries worldwide, and more than 20 000 students, have graduated from these institutions. More than 13 000 T-TEP graduates, worldwide, have joined the Toyota dealer network.

TOYOTA TECHNICAL EDUCATION PROGRAM

Looking for a better future

TOYOTA MOTOR CORPORATION (TMC), OF JAPAN, AS PART OF ITS CONTINUING EFFORTS TO SET THE TONE FOR THE 21ST CENTURY, HAS MADE COMPREHENSIVE REVISIONS TO THE TOYOTA EARTH CHARTER, WHICH OUTLINES TMC'S BASIC POLICIES, ACTION GUIDELINES AND ORGANISATIONAL STRUCTURE FOR ENVIRONMENTAL PRESERVATION. TO BACK ITS WORDS UP WITH ACTIONS, TMC HAS ADOPTED THE THIRD ENVIRONMENTAL ACTION PLAN FOR THE YEARS 2001 TO 2005.

The Third Environmental Action Plan will be carried out to help develop solutions to the world's increasingly serious and complex environmental problems and to contribute to the sustainable development of a prosperous society in the new era. TMC realises that to achieve these goals, a determined effort - one greater than ever - must be made.

The revision of the Toyota Earth Charter is the first since the charter was adopted eight years ago, in January 1992, on the basis of TMC's Guiding Principles. The revisions incorporate new perspectives including a "quest for zero emissions" and participation in the "creation of a recycling society" to respond to rapidly growing environmental issues.

Earth Charter

1. **Contribution toward a prosperous 21st century society**
   In order to contribute toward a prosperous 21st century society, aim for growth that is in harmony with the environment, and work to achieve zero emissions throughout all areas of business activities.

2. **Pursuit of environmental technologies.**
   Pursue all possibilities in environmental technologies and work on developing and establishing new technologies in order to enable the environment and economy to coexist.

3. **Voluntary actions**
   Develop a voluntary improvement plan that is based on complete prevention and compliance to laws and that also addresses environmental issues on the global, national, and regional scales, and promote continuous implementation.

4. **Working in co-operation with society**
   Build close and cooperative relationships with governments, local municipalities, and a wide spectrum of people who are involved in environmental preservation, in addition to cooperating with related companies and related industries.

Environmental Policy: Toyota SA Manufacturing

Toyota SA Manufacturing accepts its responsibility to the environment and in the manufacturing of motor vehicles and parts is committed to:

1. Strive to comply with all applicable environmental policies, legislation, regulations, and other requirements to which Toyota SA subscribes.
2. Liaise closely with the relevant authorities and local environmental interest groups and maintain a policy of openness and cooperation with these bodies.
3. Prevent pollution through effective control measures and, where pollution may occur, take the necessary steps to prevent it from recurring.
4. Consider possible environmental impacts of its processes, activities or materials before implementation and to take effective measures to minimise these environmental impacts.
5. Reduce or prevent environmental impacts by setting and regularly reviewing environmental objectives and targets, which are supported by sustainable environmental management programmes in all relevant operations.
6. Promote environmental awareness amongst our employees and provide training to those employees involved in activities or processes, which may impact on the environment.
7. Keep the public, interested parties and our employees informed of our environmental performance, which is focused on continual improvement.

Environmental purchasing guidelines

Toyota South Africa has launched its Environmental Purchasing Guidelines for suppliers to its manufacturing and assembly operations in Durban. According to the guidelines, Toyota SA requires all suppliers of components as well as suppliers of direct and indirect raw materials to be ISO 14001 certified by December 2005. Toyota also identified and listed more than 460 chemical substances it regards as harmful to the environment and any chemical product currently in use, containing any of these substances, will be either replaced or phased out. The target date set for this phasing out of present products containing any of these environmentally harmful chemical substances, is also December 2005.

Toyota will also, with immediate effect, stop purchasing any new chemical product containing a chemical substance listed.

According to the action guidelines supporting Toyota's global environmental policy, the Toyota Earth Charter, Toyota must strengthen its green purchasing programme and enlist the support of its suppliers, says Henry Pretorius, Toyota SA's Senior Vice-President: Product Development and Procurement.

He continued by saying: "Toyota has already issued these guidelines to its operations in Japan, USA and Europe, and is now rolling it out to its other global operations like South Africa, Thailand and Turkey. It is part of what Toyota calls global environmental management to ensure all its operations adhere to the same stringent environmental standards.

Suppliers will be monitored by Toyota to ensure they adhere to these guidelines, says Riaan Olivier, Toyota SA's General Manager: Safety,

Health and Environment. "We have already phased out chemical products containing listed substances and we are adamant to have them all phased out by December 2005. Toyota is not only serious about addressing its own environmental impacts, but also wants its suppliers to co-operate and share this responsibility towards the environment.

"These environmental purchasing guidelines have been sent to all our suppliers and Toyota's requirements are very detailed. The response from our suppliers, so far, has been very promising. Not only will these guidelines assist with the protection of the environment but also with the protection of the health and safety of our employees. By phasing out the use of these listed substances we also ensure that our employees are not exposed to substances which can cause serious harm to them."

Toyota's "Green building"

Meanwhile, the largest "green" building complex in the United States officially opened its doors for business early last year as part of Earth Day celebrations at the new headquarters of Toyota Motor Sales U.S.A., Inc. The project received Gold Level Certification from the U.S. Green Building Council's Leadership in Energy and Environmental Design (LEED(TM)) Green Building rating system.

The complex, the largest ever to receive a LEED(TM) Gold rating, has 624 000 m" of office space and will initially house about 2 000 Toyota associates.

"Office buildings have a significant impact on the environment, using about one-third percent of the electricity and 12% of the drinking water in the U.S.," said Christine Ervin, USGBC President and CEO. "Fortunately, there are ways to reduce the environmental impact buildings have, while also enhancing the overall work environment. This complex demonstrates what can be accomplished when concern for the environment plays a role in every aspect of the design and building process."

Toyota elected to pursue LEED(TM) certification for the new complex as part of Toyota Motor Corporation's Earth Charter guidelines established in 1992, calling on the company to reduce its impact on the environment in every aspect of its business. Building a green complex, however, also had to be based on smart business practices.

Toyota FINE-S (Fuel Cell Innovation Emotion - Sport) concept car.

TOYOTA MOTOR CORPORATION HAS LONG BEEN A MAJOR PLAYER IN SEARCHING FOR WAYS OF ENSURING HARMONY BETWEEN MOTOR VEHICLES, SOCIETY AND THE ENVIRONMENT. THE COMPANY CONTINUALLY PURSUES MANY AVENUES IN ITS BID TO PRODUCE ENVIRONMENTALLY-FRIENDLY CARS, TRUCKS AND BUSES THAT ARE EFFICIENT IN THEIR USE OF RESOURCES AND LOW ON UNWANTED EMISSIONS.

The company is involved in developments on many fronts as it seeks the ultimate eco-car. There are projects involved with alternative fuels, such as natural gas, liquid petroleum gas and hydrogen, as well as major advances in technology for improving diesel and petrol engines, research on electric cars, petrol-electric hybrid cars and fuel-cell electric vehicles.

These are not dreams. Many of these projects have already reached production, including the electric RAV4, petrol-electric Toyota Prius, Toyota Highlander, Toyota Estima and Lexus RX400h as well as the FCHV4 wagon that uses liquid hydrogen as a fuel. Truck company Hino, a Toyota subsidiary, is already testing a fuel cell bus, while Daihatsu, another affiliate, also has a wide range of eco-friendly vehicles under development.

Toyota's multi-faceted eco-research programme

aims firstly to reduce CO^2 and other exhaust emissions, secondly to make better use of dwindling fossil fuel reserves, thirdly to develop alternative power sources and fourthly to devise more efficient traffic control systems to help alleviate environmental problems in urban areas.

Achievement of these goals would mean the realisation of the corporate philosophy of harmonious growth. Toyota believes this is a mandate from Planet Earth.

The company believes that if it can develop the ideal means of mobility - something that ensures future sustainability - and if this technology can be transferred to other countries, then the motor vehicle will be more important and more valuable in the current century than it was in the previous one.

Recycling vehicles when they have completed their useful life is another priority with Toyota's research teams.

The company has many technical agreements with a number of other motor manufacturers to ensure high productivity in searching for eco solutions on a basis of co-operation instead of "going it alone." One of the recent agreements will see Nissan using Toyota's petrol-electric hybrid systems.

Major programmes are also underway at Toyota factories and dealerships around the world to make significant reductions these businesses have on the environment.

A great deal of progress has been made already

and this environmental thrust is an ongoing priority with Toyota in its position as a trendsetter among the major motor manufacturers of the world. They have to succeed and they will.

The Toyota FCHV-4 wagon.

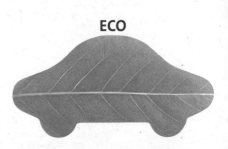

ECO

The Hino fuel cell hybrid bus: FCHV-Bus2.

More accolades for Toyota's new Prius petrol/electric hybrid

Toyota's new Prius petrol/electric hybrid car continues to make a major impact on the international motoring scene. The latest accolade to come the way of this innovative hatchback is the North American Car of the Year award.

The award was announced at the recent Detroit Motor Show and this is the first Toyota product to be honoured in this annual contest. A jury of 50 leading automotive writers

and broadcasters voted Prius No. 1 by a wide margin over two other cars considered as finalists out of a "short list" of 17 new cars on the American market. Prius received 298 points, compared to 167 for the Mazda RX8 and 113 for the Cadillac XLR. The award, which has been made since 1994, honours vehicles that establish new benchmarks in areas including design, innovation, safety and handling. It is the only independent, international vehicle award in North America.

This latest accolade follows Prius being named 2004 Car of the Year, by Motor Trend, a leading American motoring magazine, and one of the Top 10 cars of 2004 in the prestigious ratings of

> "A jury of 50 leading automotive writers and broadcasters voted Prius No. 1 by a wide margin over two other cars considered as finalists out of a "short list" of 17 new cars on the American market."

Car and Driver, another US motoring publication. It is the first hybrid vehicle to achieve these honours. The latest Prius, introduced last year, initially in Japan and then in the United States, was also adjudged winner of the Technology section for 2003 among 10 category winners in the annual automotive awards made by Autocar, the leading British motoring magazine.

Demand for the new Prius is outstripping supply and Toyota is instituting plans for overtime working and increased production capacity, with normal production going up from the original 7 500 units a month to 10 000 a month, starting in the second quarter of 2004.

The very economical and low emission Prius hybrid has had the fastest sales start of any new Toyota model in the United States besides strong demand in the Japanese domestic market. Already TMC has increased its Prius sales plan for the United States for 2004 from an original 36 000 to 47 000 units; 10 000 Prius orders had been placed in the US before the car was even launched.

The new hybrid petrol/electric Prius sedan.

INDUSTRIAL ECOLOGY

A considerable amount of international debate about the meaning of sustainable development and how it can be applied continues to dominate many development decisions. Much of this debate has begun to centre on minimum resource use, efficiency requirements, or sustainable production. With a rising global population and ever-increasing levels of consumption among developed and developing countries, pressure on the Earth's many ecosystems and the biosphere as a whole promises to continue to mount well into the foreseeable future.

Increasing levels of consumption at the limits of the biosphere's capacity to provide resources will require dramatic improvements in resource use efficiency. At the limits to material throughput, sustainability requires that the growth in the consumption of goods and services be accompanied by a proportional decline in the energy and material intensity of that consumption. Industrial ecology research is focused on various strategies aimed at reducing the energy and material intensity of the economy. Immediate attention should also be given to the use of policy tools such as ecological tax reform in order to motivate such reductions. An example of what industrial ecology aims to achieve in practical terms is, if you currently drive a car that achieves a fuel efficiency of 100 kilometres per litre, by the year 2040, your new, super light car will achieve 100 kilometres from one tenth of a litre. Reductions of material and energy throughput by factors of 9 and 10 over the next quarter century are not as farfetched as they may at first seem. Japan, for example, reduced the energy and material inputs for its manufacturing sector by 40 percent between 1973 and 1984, at a time of considerable economic expansion.

'Industrial ecology' refers to the exchange of materials between different industrial sectors where the 'waste' output of one industry becomes the 'feedstock' of another. Industrial ecology represents a relatively new and leading edge paradigm for business. It emphasises the establishment of public policies, technologies and managerial systems, which facilitate and promote production in a more co-operative manner. Implementing industrial ecology involves such things as life cycle analysis, closed loop processing, reusing and recycling, design for environment and waste exchange. Hypothetically, in a completely efficient economy functioning in harmony with ecosystems, there would be no waste.

The diagram below illustrates the changing nature of industrialisation culminating in full industrial ecology, whereby "all process systems and equipment, and plant and factor design, will eventually be fully compatible with existing industrial ecosystems as a matter of course."

Photo: Sasol (Earthyear Magazine)

(Arthur D. Little, "Industrial Ecology: An Environmental Agenda For Industry," Industrial Ecology Workshop: Making Business More Competitive. Toronto: Ministry of Environment and Energy, February 1994.)

The Emergence of an Eco-Industrial Infrastructure

Industrial ecology and the development of technologies which eliminate waste and maximise efficiency will be critical to achieving the required reductions in material and energy throughput in order to maintain a basic quality of life into the twenty-first century. The extent to which these measures make economic as well as ecological sense – i.e. are 'eco-efficient' – will determine whether market forces play a role as an important driver of change.

The successful establishment of industrial ecology linkages requires continuing implementation of projects that identify industrial ecology opportunities. Work is needed to clearly identify the regulatory and other policy barriers in order that they be removed.

Governments, industry, academics – and other organisations which focus on establishing the right institutional, fiscal and policy environment for the practical implementation of pollution prevention technologies, sustainable technologies and industrial ecology – can help to ensure prosperity for their citizens and secure an important role for their countries in global efforts to achieve sustainable development.

Guest Author: Prof. P. Marjanovic ~ SASOL Chair of Environmental Engineering, School of Civil and Environmental Engineering, University of the Witwatersrand

Key Associated Topics:

Cleaner Production; Cogeneration; Eco-efficiency; Environmental Management Planning.

Associated Organisations:

Icando; Natural Step; Rose Foundation; Richards Bay Minerals; Sasol Centre for Innovative Environmental Management; Wales Environmental Partnerships.

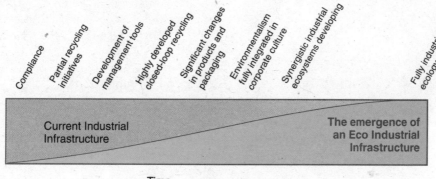

Compliance

Partial recycling initiatives

Development of management tools

Highly developed closed-loop recycling

Significant changes in products and packaging

Environmentalism fully integrated in corporate culture

Synergistic industrial ecosystems developing

Fully industrial ecology

Current Industrial Infrastructure

The emergence of an Eco Industrial Infrastructure

Time

INTEGRATED ENVIRONMENTAL MANAGEMENT (IEM)

Integrated environmental management is a philosophy that is concerned with finding the right balance (sometimes called the "golden mean") between development and the environment. It provides a framework of published guidelines (available from the Department of Environmental Affairs and Tourism) to ensure that environmental considerations are taken into account at every stage of the life of a project, process or policy. In other words, IEM is not only used in looking at, for example, the environmental impact of converting grazing land to the growing of wheat (project), but it also looks at the environmental impact of increasing the wheat price (policy), which may encourage more marginal land to be put under wheat. It also looks at the impact of assistance programmes, for example, the implications of encouraging new wheat farms (programmes). IEM considers the environmental elements in a "cradle to grave" concept (i.e. from the inception of the idea right through to the decommissioning or the end of the project). The difference between IEM and EIA (Environmental Impact Assessment) is that IEM is a whole philosophy whereas EIA is just one tool or technique used to gather and analyse environmental information that is a part of the IEM process. IEM forms a part of the *White Paper on Environmental Management Policy in South Africa* and has been written into the new National Environmental Management Act (No. 107 of 1998) and is thus formally and legally recognised.

Key Associated Topics:

Agenda 21; Cleaner Production; Eco-efficiency; Environmental Impact Assessment; Environmental Management Planning; Land Use Planning; Life Cycle Assessment; Social Impact Assessment; Sustainable Development; Waste Management.

Associated Organisations:

Acer; Africon; Afrodev Associates; Aldo Leipold Institute; Arcus Gibb; BKS; Bohlweki Environmental; Digby Wells; Doug Jeffery Environmental Consultant; Duard Barnard Associates; EcoBound Environmental & Tourism Agency; EcoSense; Environmental Evaluation Unit – University of Cape Town; Fairest Cape Association; Greenhaus Architects; Hilland & Associates; KPMG; Landscape Architects Uys & White; Namibia Nature Foundation; National Cleaner Production Centre; Newtown Landscape Architects-Limpopo; Ninham Shand; Oryx Environmental; Softchem; Solar Engineering Services; Technikon Pretoria; Tribal Consultants; Umgeni Water Services; Wales Environmental Partnerships; Web Environmental Consultancy; WSP Walmsley.

INTEGRATED PEST CONTROL (IPC)

Integrated pest control is the modern approach to managing all kinds of pests, be they insects, plants, birds or animals. The theory is to use methods that control the pests, whilst minimising the possible damage to the environment. For example, instead of using pesticides, farmers can introduce creatures that are the natural enemies of the pests. This is less dangerous and polluting and makes use of Nature's normal population control systems. Other methods would be to remove the nesting areas of the pests or to disturb their life styles in order to encourage them to break up their flocks or move elsewhere. In the case of plants, for example, it would involve using insects, which would attack or eat the seeds of the plants or trees. The IPM approach accepts that there may be circumstances where chemical methods must be used but this will only be done after all other methods have been investigated or tried and failed.

Key Associated Topics:

Accumulation; Agriculture; Alien Invasive Plants; Aquatic Weed Control; Biological Pest Control; Entomology; Genetically Modified Organisms – the Pro Position; Pesticide Treadmill; Pesticides.

Associated Organisations:

Plaaskem; Agricultural Research Council; EWT – The Poison Working Group; New Africa Skills Development.

INTEGRATED POLLUTION AND WASTE MANAGEMENT (IPM&WM)

Integrated pollution and waste management (IPM&WM) is an integrated approach that has been adopted by the government to deal with the current problems relating to waste management and pollution. In the past, South Africa has had extensive environmental, pollution and waste management legislation but the responsibility for implementation and enforcement was scattered over a number of different departments and institutions. This resulted in inconsistent implementation and enforcement of laws and regulations throughout the country. The White Paper on IP&WM was published in May 2000 and the legislative programme that is included will result in new pollution and waste legislation. A new framework of legislation, regulations, policies, and information will simplify pollution control and ensure better management and protection of the environment. The policies include greater transparency and more availability of pollution and environmental information to the public. The White Paper can be found on the government website at www.gov.za/documents

Key Associated Topics:

Air Quality; Atmospheric Pollution; Contaminated Land; Environmental Law; Hazardous Wastes; Incineration; Landfill; NEMA; Pollution; Polluter Pays Principle; Recycling; Waste Management.

Associated Organisations:

Beco Institute for Sustainable Development; Cape Town – Mess Action Campaign (MAC); Jarrod Ball & Associates; Softchem; Solar Engineering Services; Technikon Pretoria; Wales Environmental Partnerships; Water Institute of SA; Wilbrink & Associates.

INTERESTED AND AFFECTED PARTIES

This is a term used to describe those people who have a concern about a development, project, policy or action and who need to be consulted during the process of decision making. For example, if a factory was planned to be built near a school, the school board, teachers and parents would be interested and affected parties because they would all be concerned about the possible effects of the factory on the school children. Usually, when an Environment Impact Assessment is carried out for a project, one of the tasks is to identify all the interested and affected parties so that they can have an opportunity to comment, once the details of the project are known. This is part of the public consultation process. Interested and affected parties are those directly interested or affected whilst the public consultation process can involve people from a much wider area and with less specific interest. The broad principles embodied in the new Constitution and the new framework of government includes the twin aspects of transparency and consultation. Recognition of interested and affected parties and consideration of their views and opinions is a valuable insight into acceptance of new plans and activities or local concerns and issues, which may need to be taken into account.

Key Associated Topics:

Activism; Communities; Development of Land; Environmental Impact Assessment; Environmental Justice; Environmental Law; Indigenous Peoples; NIMBY; Public Participation; Social Impact Assessment; Stakeholders.

Associated Organisations:

Eagle Bulletin; Groundwork; Icando; Ninham Shand; Sustainability United (SUN); WESSA.

ISO 14000 SERIES

ISO stands for "International Standards Organisation". This is an international organisation that designs and writes codes of practice (called standards) to ensure that actions and processes are carried out in a correct and uniform manner. This is especially important where operations require that equipment must be made to fit different countries' systems and the tolerance or margin for error is very small. Similarly, there may be products that are manufactured in different countries that need to be standardised or specific processes or procedures included. ISO 14000 is a series of international standards for environmental management, which sets guidelines on how to manage environmental matters in different companies in different countries. (Download the document www.iso.ch/9000e/iso14000.pdf for more detailed information and background on the ISO 14000 series.) There are a number of different detailed standards in the ISO 14000 series, for example, ISO 14001- Environmental Management Systems, ISO 14004 - Environmental Management Systems – General guidelines on principles, systems and supporting techniques (recommended), ISO 14010 - Guidelines for Environmental Auditing – General principles, ISO 14011 - Guidelines for Environmental Auditing - Audit procedures – Auditing of environmental management systems, ISO 14012 - Guidelines for Environmental Auditing – Qualification criteria for environmental auditors, and ISO 14040 - Life Cycle Assessment.

Key Associated Topics:

Cleaner Production; Environmental Auditing; Environmental Management Planning; Environmental Management Systems.

Associated Organisations:

Advantage ACT; Crowther Campbell & Associates; Data Dynamics; Dekra-its Certification Services; Independent Quality Services; KPMG; Metago Environmental Engineers; Green Gain Consulting; NOSA.

KYOTO AGREEMENT

The Kyoto Protocol is an internationally negotiated framework for regulating greenhouse gas emissions of signatory countries, which aims to limit the potential negative impacts of global warming. The Protocol serves the aims of the Framework Convention on Climate Change adopted at the United Nations in May 1992, and opened for signature at Rio de Janeiro in June, namely to "achieve stabilization of the greenhouse gas concentrations in the atmosphere at a level that would prevent dangerous anthropogenic interference with the climate system." Greenhouse gases specified by the Protocol are carbon dioxide, methane, nitrous oxide, hydrofluorocarbons, perfluorocarbons, and sulphur hexafluoride. The Protocol originated in December 1997, when representatives of more than 160 countries met in Kyoto, Japan, to negotiate binding emission limits for developed nations. The resulting Kyoto Protocol established emissions targets for each of the participating developed countries – so-called Annex I (developed) countries – in relation to their emissions in 1990. Targets range from 8% reduction for EU states, 7% reduction for USA, to a 10% increase for Iceland. Non-Annex I countries have no targets set by the Protocol, but all parties are encouraged to develop and implement climate change mitigation and adaptation programmes. The Protocol allows for flexibility in meeting targets, such as providing credits for carbon storage (sequestration in "sinks") through land afforestation or reforestation, allowing international trade in carbon, and allowing countries to meet their targets jointly. In the original form of the Protocol, emissions targets for Annex I countries would have had to be achieved on average over the commitment period 2008 to 2012. However, in March 2001, the U.S.A. withdrew from the Kyoto Protocol, plunging its future into doubt. But in July 2001, the EU, Japan, Canada, Russia, Australia, and 170 other nations reaffirmed their commitment to the Protocol. In order to secure the support of industrialised nations, emissions targets set by the Protocol were substantially weakened, being reduced by two-thirds from the original goals.

Guest Author: G. Midgley ~ National Botanical Institute of South Africa

Key Associated Topics:

Air Quality; Atmospheric Pollution; Carbon Tax; Carbon Trading; Energy; Global Warming & Climate Change – The World Response.

Associated Organisations:

Department of Foreign Affairs; Department of Trade and Industry; Department of Minerals and Energy; Renewable Energy And Energy Efficiency Partnership (Reeep); SA Climate Action Network (SACAN); SouthSouthNorth.

HP South Africa's Take Back Programme.

South Africa's First Large-scale Corporate Return and Recycling Programme for HP LaserJet and Inkjet Print Cartridges.

Recycling and reuse are fundamental principles of nature, not human inventions and therefore recycling doesn't need to be complicated. It can be simple.

Worldwide HP is committed to providing customers with inventive, high quality products and services that are environmentally sound throughout their life cycles and to conduct its operations in an environmentally responsible manner. This commitment continues to be one of the guiding principles at the heart of HP's values.

Core to this commitment is the consideration given to the environmental impact of a product throughout its entire lifecycle, from design and manufacturing as well as use, return and finally recycling. HP designs its supply material with recycling in mind. Already, tens of millions of HP LaserJet and inkjet cartridges have been recycled, diverting a huge volume of waste from landfills.

Returned original HP LaserJet print cartridges are disassembled at a recycling plant in France, where more than 95 percent of the material can be recovered and reclaimed. In turn original HP inkjet printer cartridges are shipped to a recycling facility in Germany where, depending on the model, up to 70 percent by weight of each cartridge can be separated into its individual components and subsequently recycled. The remaining parts that cannot be used are disposed of in an environmentally friendly manner.

In 2003 HP introduced its Take-Back return and recycling programme to South Africa, an initiative to collect HP inkjet and LaserJet cartridges from corporate customers for recycling purposes, building on the company's commitment to doing business in an environmentally responsible manner. This initiative will be expanded in 2004 to also include the return and recycling of inkjet and LaserJet cartridges by consumers and Small and Medium Businesses.

As a recognised leader in environmental stewardship, HP internationally began providing end-of-life solutions for its original print cartridges in 1991, starting with a LaserJet print cartridge return and recycling programme. Since its inception, this programme has recycled and recovered almost 68 million kilograms of HP LaserJet print cartridge materials worldwide. Coupled with these efforts, HP's Design for the Environment programme integrates efforts in reducing the ecological footprint of a product in all stages of its design, manufacture, use and end of life.

HP has been recognised for its ecologically sound product design, manufacturing and recycling processes with various certifications and awards among them the Ecohitech Product Award, the German environmental label Blue Angel, the Energy Star and the Green Cross Millennium Award.

LAND DEGRADATION

Guest Essay by Dr Eureta Rosenberg

WHAT IS IT?

The experts disagree on how to define land degradation and associated processes such as desertification, but as an issue it is not difficult to understand. Land degradation occurs when the economic and biological productivity of land is lost, primarily through human activities. This can happen, for example, when:

- Fertile soils erode away,
- Indigenous trees are removed,
- Alien plants invade an area,
- Farm land is used for housing,
- Soils become salty through poor irrigation, or
- Soils are degraded by acid pollution and heavy metal contamination.

The loss of productive land obviously affects farming and rural communities. As the land degrades, more fertiliser, machinery and supplementary feeds are needed and the cost of production increases. Small-scale, subsistence farmers are often unable to meet extra costs and even large-scale, commercial farmers can find that farming becomes impossible. As a result farm workers and others may be forced to move to towns and cities, only to face unemployment and poverty.

So land degradation also affects urban areas, through spreading informal settlements and rising food prices. Water also becomes more expensive as soil erosion makes rivers muddy and causes dams to fill up with silt, adding to the costs of water purification and storage.

HOW BIG IS THE PROBLEM?

After looking at the vast array of food on display in a South African supermarket it may be hard to imagine that food security could be an issue. However, many households struggle to feed themselves and in parts of the country, notably the Eastern Cape, children suffer from malnutrition. The Earth Policy Institute predicts that, globally, insecure food supplies will become an even more a widespread source of conflict and hardship in the future.

Only 13,5% of South Africa's land surface area is considered arable, or suitable for food production. Every year an estimated 34 000 hectares of farmland is converted for other purposes to other uses such as urban expansion. At this rate, by the year 2050 the experts predict that there will be just 0,2 hectares per person available on which to produce food in South Africa. This is considerably less than international norms. Food imports are expensive and, like other environmental problems, they hit the poorest people hardest.

While it is difficult to estimate the extent of the problem of land degradation, there is no doubt that people around the world are suffering from its effects. The term "environmental refugees" has been coined to refer to people who have had to abandon their homes because, for various reasons, the land has lost its capacity to sustain them.

Globally, a staggering 70% of all drylands (or non-tropical regions) are already classified as degraded. This represents 14% of the Earth's land surface area. Africa may be the worst affected continent, as 73% of its agricultural drylands are thought to be degraded. The number of people affected is vast, for it is estimated that more than 70% of Africa's 500 million people depend directly on the environment for their livelihoods.

The United Nations Conference on Environment and Development drew up a Convention to Combat Desertification to address the problem of land degradation. As a signatory, South Africa is required to develop a National Action Programme to Combat Desertification and commissioned a survey of the extent and causes of land degradation in the country. The survey results are summarised in a user-friendly book called Nature Divided: Land Degradation in South Africa by Hoffman and Ashwell. Of the many aspects of land degradation it covers, we can note only a few.

WHAT DOES IT INVOLVE?

Desertification is one aspect of land degradation. It is related to veld degradation (when rangelands lose their vegetation cover through over-grazing or inappropriate fires), deforestation and soil erosion.

Deforestation refers to the loss of trees. South Africa's trees are heavily used for construction, herbal medicine and fuel. Up to 99% of rural households use firewood for energy and, despite electrification projects, 38% of township residents also use firewood. Shortages are already being experienced and it is estimated that, if unchecked, and at the current rate of harvesting, trees will have disappeared from communal areas within 20 years. While the number of people in villages plays a role in deforestation, urban demand for wood and traditional medicine also contributes to the problem. Deforestation makes life harder for rural people, destroys the habitats of numerous creatures and contributes to soil depletion and erosion.

Soil erosion causes the loss of the fertile layer of topsoil in which food crops can grow. Through factors such as reduced plant cover, topsoil is removed by wind and water. The eroded soil can cause eutrophication of dams and rivers and can harm the marine environment. The environmental issues that face us today are usually the result of complex chain reactions! They always have an economic impact. In this case, the farmer has to enrich his depleted soils with fertilizers – costing agriculture an estimated R1,5 billion per year.

Loss of soil quality is another contributing factor to land degradation, through pollution with heavy metals and acid from mines and power stations. Again, the loss of soil productivity is expensive; for example, it costs about R25 million to neutralise the effects of acid rain in Mpumalanga.

Invading alien plants are plants that are not indigenous to South Africa but which grow so prolifically that they threaten indigenous plants and decrease the land's biological productivity. Of the 161 classified invaders, syringa, black wattle, eucalyptus, lantana, rooikrans and Port Jackson are the most common, covering an area about the size of Gauteng. They spread or "invade" quickly, partly because the local environment has few means of keeping them in check. As they do so, they push out indigenous plants, reducing biodiversity. In addition, invading alien plants contribute to soil erosion, reduce grazing areas and reduce the capacity of indigenous plants to reproduce. Some alien plants burn more easily and intensely than the indigenous vegetation, thus increasing the risk level and damage caused by fire. Some also use far more water than indigenous plants; for instance, woody aliens use about 3 300 cubic litres per year. In the dry interior, mesquite (suidwes-doring) threatens precious groundwater supplies. Alien trees on the Western Cape coastal mountains and lowlands may reduce the mean annual run-off by one third. The need to build new dams in ecologically sensitive areas could be substantially reduced if thirsty alien species were removed.

It is interesting to note that many plants now classified as invasive were originally brought to South Africa as solutions to problems! They were imported to provide fodder, to stabilise driftsands, or to supply the leather and mining industries.

Another example of a technical solution causing unforeseen problems is the use of inappropriate irrigation systems on some commercial farms. Not only is inappropriate irrigation wasteful it can also cause waterlogging and salinisation. Many agricultural lands along the Fish River have been rendered useless by poor irrigation.

Black people remaining in what became the Republic of South Africa were eventually confined to only 13% of the land. "Over-crowding" in the homelands was blamed for land degradation. Ironically, "under-farming" may have played a bigger role. The migrant labour system resulted in few decision-makers and able-bodied people being available to do labour-intensive farm work. Furthermore, there was very little money to invest in farming. Farming around the world requires subsidies and, unlike their counterparts in the Republic, homeland farmers had no government support or access to credit.

As a result of the social and environmental engineering of apartheid, people were reluctant to respond to government schemes for improving the land. An impression developed that black people did not care for the land. Yet, even today, farmers in Venda spend hours and months, with little more than their own labour to draw on, building terraces to combat erosion on the fertile slopes of the Soutpansberg mountains.

Photo: Earthyear Magazine

Potential "quick-fixes" should be considered very carefully, as they can create further problems in the long-term. We should bear this in mind when coming up with strategies to tackle land degradation.

WHAT CAN BE DONE?

Solutions must take into account the history of land degradation. Political practices of moving people from where they were settled into reserves and homelands contributed to the pattern of land degradation in South Africa. In former homeland areas such as Ciskei and Transkei the productivity of the land has been reduced to such an extent that most residents are unable to feed their families, let alone make a small profit from the land.

Black South Africans were not always impoverished as farmers. In the 1800s Basotho farmers out-competed European settlers, producing more grain more cheaply. Not only did they feed themselves, they also exported sheep, horses, grain and wool to the mines and the Cape Colony. But the government, being concerned about competition with white farmers, restricted the Basotho's access to markets and land, and forced them into the barren mountains of Lesotho.

Solutions to land degradation problems should take account of the conservation methods that some land users already practise. VaVenda farmers' traditional soil conservation methods include contour ploughing, stone walls and grass strips. Large-scale commercial farmers in the Karoo achieve considerable success with "holistic land management" methods.

These are just two of the initiatives that we can draw on to develop flexible strategies for improving the quality of the land and encouraging sustainable land use throughout this country of diverse environments and cultures.

Key Associated Topics:

Afforestation; Agriculture; Alien Invasive Plants; Contaminated Land; Deforestation; Desertification; Environmental Refugees; Grasslands; Landfill; Population; Poverty; Rehabilitation of Land; Soil; Tropical Rain Forest.

Associated Organisations:

Barlofco; Denys Reitz; Dept Environmental & Geographical Science – University of Cape Town; Wales Environmental Partnerships; WESSA; WWF.

LANDFILL

A landfill site is a formally registered dumpsite for mainly domestic or industrial waste. The site will be classified as being general, low hazardous waste or high hazardous waste. Most municipal sites are, at best, classified as low hazardous waste. New sites are carefully planned and are usually lined with a double liner to help detect and prevent leaks. Many older municipal sites are not registered and were not really planned to prevent leachate from going down through the dump and out into nearby streams or watercourses. Thus they are potentially hazardous from a pollution point of view, and the authorities are trying to close them down as quickly as they can. Most low-level industrial waste is sent to low hazard sites but managers should strive to send as little waste to landfills as possible. The law says that the waste generator and the disposer (contractor) are jointly responsible for the load. This means that the generator never relinquishes responsibility.

Key Associated Topics:

Hazardous Wastes; Incineration; Integrated Pollution and Waste Management; Land Degradation; Leachate; Waste Management.

Associated Organisations:

Enviroserv; Jarrod Ball; Millennium Waste Management; Vula Environmental Services; Wales Environmental Partnerships.

LANDSCAPE ARCHITECTURE

Landscape architecture is a diversified and creative design profession that integrates ecological, social, functional and aesthetic aspects of everyday life into designing places to improve people's quality of life.

The increasing demand for the professional services of landscape architects reflects the public's desire for better housing, recreational and commercial facilities, and its expanded concern for effective environmental management.

Clear differences exist between landscape architecture and other design professions. Architects primarily design buildings and structures with specific uses, such as homes, offices, schools and factories. Civil engineers apply scientific principles to the design of city infrastructure such as roads, bridges, and public utilities. Urban planners develop a broad overview of development for entire cities and regions.

While having a working knowledge of architecture, civil engineering and urban planning, the task of landscape architects is to integrate elements from each of these fields to design aesthetic and practical relationships with the land.

A diverse profession

Landscape architects are therefore involved in designing the built environment while also guiding the management of the natural environment A variety of interwoven specialisation exist within the profession, including:

Landscape Design, the historical core of the profession, is concerned with detailed outdoor space design for residential, commercial, industrial, institutional and public spaces. It involves the treatment of a site as art, the balance of hard and soft materials in outdoor and indoor spaces, the selection of construction and plant materials, infrastructure such as irrigation, and the preparation of detailed construction plans and documents.

Site Planning focuses on the physical design and arrangement of built and natural elements of a land parcel. More specifically, site design involves the orderly, efficient, aesthetic and ecologically sensitive integration of man-made objects with a site's natural features including topography, vegetation, drainage, water, wildlife and climate. This may also include the field of urban and town planning that deals with designing and planning cities and towns and the places within them.

Regional Landscape Planning merges landscape architecture with environmental planning. Landscape architects working in this area require a knowledge of real estate economics and development regulation processes, as well as an understanding of the physical and ecological opportunities of developing and working with the land. The challenge is to integrate economic factors with good design and thus create high quality environments.

Ecological Planning and Design studies the interaction between people and the natural environment. This specialisation includes, but is not limited to, analytical evaluations of the land and focuses on the suitability of a site for development. It requires specific knowledge of environmental and development regulations. It includes preparation of environmental impact statements, visual analysis, landscape reclamation and coastal zone management. Landscape architects also develop plans for extensive natural areas as part of national parks, forests, or wildlife refuge systems.

Historic Preservation and Reclamation of Sites such as farms, parks, gardens, grounds, waterfronts, and wetlands involves increasing numbers of landscape architects as growing populations lead to additional development. This field may involve preservation or maintenance of a site in relatively static condition, conservation of a site as part of a larger area of historic importance, restoration of a site to a given date or quality, and renovation of a site for ongoing or new use.

The profession of the future

The years ahead promise many new developments and challenges to the ever-broadening profession. With environmental concerns becoming increasingly important, landscape architects are increasingly being called upon to help solve complex problems relating to farmland preservation, small town revitalisation, landscape preservation, and energy resource development and conservation. Advances in computer technology have opened the field of computerised design and geographic information systems, which has become a significant tool in support of well-informed and complex environmental decision-making.

Guest Author: Dr. Gwen Breedlove ~ Department of Architecture and Landscape Architecture, University of Pretoria; Chairperson Institute of Landscape Architects of South Africa (ILASA)

Key Associated Topics:

Aesthetics; Environmental Impact Assessment; Green Architecture; Urban Greening; Sustainable Development Section – In the Built Environment; Urbanisation.

Associated Organisations:

Bapela Cave Klapwijk; Biolytics Southern Africa; ERA Architects; Green Inc; Institute of Landscape Architects; Landscape Architects Uys & White; Newtown Landscape Architects Gauteng; Newtown Landscape Architects Limpopo; South African Landscapers' Institute; Wales Environmental Partnerships.

LAND USE PLANNING

Land use planning makes use of environmental base data. During the course of the evolution of different planning strategies, various environmental management tools are utilised. South Africa's land use planning legislation and regulations at local, provincial and national level are geared up to make use of the environmental impact assessment (EIA) and the strategic environmental impact assessment. Broad-based land use planning needs to consider environmental issues at all levels, particularly because there are many instances where individual projects might not have an impact but cumulatively, a number of projects together will make a significant difference. Similarly, when considering cumulative water and air impacts, it is cumulative impacts that make the difference. The Integrated Environmental Management (IEM) process is complementary to most of the planning procedures and is designed to dovetail with the public participation initiatives that form part of most land use planning exercises.

Key Associated Topics:

Carrying Capacity; Coastal Zone Management; Development of Land; Environmental Impact Assessment; Environmental Justice; Habitat; Interested and Affected Parties; Land Degradation; Population; Poverty; Public Participation; Strategic Environmental Assessment; Sustainable Development Section – The Built Environment – Land Reform.

Associated Organisations:

Bateleurs Flying for the Environment; Cebo Environmental Consultants; Doug Jeffery Environmental Consultant; Master Farmer Programme; Newtown Landscape Architects; Settlement Planning Services; Web Environmental Consultancy; Wales Environmental Partnerships.

LEACHATE

Leachate is the liquid that is formed and drains out of a landfill site. It is derived from three sources:

◆ Liquid waste disposed on site

◆ The natural liquid product of waste decomposition

◆ Rainfall and storm water catchment entering the site.

Liquids percolate down through the waste body and in the process dissolve and carry organic and inorganic compounds in the waste creating a highly contaminated and potentially toxic liquid (landfill leachate). This liquid is collected in the leachate collection system of the landfill cells and stored in ponds.

Leachate is often the source of undesirable odours emanating from landfill sites. Aerators can be installed into leachate dams in order to minimise odour problems, enhance biological degradation and encourage evaporation. Special aspirating sub-surface aerators, which result in air being drawn down into the leachate, rather than the more familiar aerators used at sewage treatment plants, result in considerably lower odour formation.

Non-toxic leachate can be used on site to control dust by spraying access roads and working areas.

Normally, leachate would be treated chemically and released into the sewer system once approved quality standards have been achieved.

Guest Author: Jeff le Roux ~ Enviroserv

Key Associated Topics:

Bio-remediation; Contaminated Land; Environmental Liability for Remediation; Groundwater; Hazardous Wastes; Landfill; Pollution; Waste Management.

Associated Organisations:

Enviroserv; Groundwater Association of KZN; Groundwater Consulting; IBA Environnement.

Photo: Enviroserv

LIFE CYCLE ASSESSMENT (LCA)

Life cycle assessment (LCA) looks at the entire life cycle of a product, process or activity "from cradle to grave" or from the first moment it was thought of until its final disposal or destruction. The purpose of LCA is threefold: 1) to provide as complete a picture as possible of the product, process or activity and its relationship to the environment; 2) to add to humankind's understanding of the interdependence of human activities and the environment; and 3) to provide decision-makers with information that clearly defines the direct and indirect, positive and negative environmental impacts of a product, process or activity. LCA forms a part of the philosophy that encourages holistic thinking and integrated environmental management approaches. There are more details in SABS ISO 14040 "Environmental management – Life cycle assessment – principles and framework". The custodians of the ISO standards in South Africa are the South African Bureau of Standards. There is a useful Australian network for LCA, which can be contacted through http://auslcanet.rmit.edu.au/

Key Associated Topics:

Eco-efficiency; Environmental Management Planning; Environmental Management Systems; Industrial Ecology; ISO 14000 Series; Recycling; Waste Management.

Associated Organisations:

National Cleaner Production Centre; Price WaterhouseCooper; Responsible Container Management.

LOCAL AGENDA 21

The Agenda 21 document says that local government (local authorities and municipalities) should draw up their own "Local Agenda 21" to reshape the policies, laws and regulations of their districts to fit in with sustainable ideals. This includes the participation and co-operation of local communities and authorities to address the problems and solutions in terms of the local environmental, economic and social circumstances. Local Agenda 21 is the name given to the initiatives aimed at the local authorities. A number of metropolitan areas, Cape Town, Durban, Johannesburg and Pretoria have already prepared State of the Environment Reports in compliance with the directions given in Agenda 21. These reports, along with the national State of the Environment Report, have been made available on a website, where they will be supplemented by updated information, as it becomes available. The reports can be obtained by contacting the municipalities and the Department of Environmental Affairs and Tourism, direct at: www.environment.gov.za/soer/index

Key Associated Topics:

Agenda 21 is a central tenet of Sustainable Development – and therefore relates to practically every other Topic within The Enviropaedia, but in particular; Local Agenda 21; Sustainable Development; World Summit on Sustainable Development.

Associated Organisations:

Cape Town Environmental Management; Strategic Environmental Focus; Sustainable Energy Africa; The Greenhouse Project.

MARINE ENVIRONMENT

Photo: Bruce Sutherland - City of Cape town

The oceans cover about 71% of the earth's surface and provide habitats for over 250 000 species of marine animals and plants. They also function as a climate regulator, absorbing solar heat and re-distributing it via ocean currents. The heat, in turn, causes evaporation of ocean waters, which forms a part of the global hydrological cycle, which redistributes freshwater, which is vital to the survival of many terrestrial organisms, including humans. (If 100 litres were to represent all the water on earth, 3 litres would represent all the freshwater, most of which is tied up in icecaps, glaciers and groundwater, and 0.003 of a litre or half a teaspoon, would represent the readily available freshwater on earth.)

The oceans also act as a reservoir for carbon dioxide which, in turn, helps to regulate the temperatures in the lower atmosphere, called the troposphere. Carbon dioxide is dissolved into the oceans or released from them, depending on water temperatures. In this way, the oceans act as a heat sink, absorbing excess heat and releasing it. Heat absorption and release is brought about through ocean currents moving around the globe, winds pushing waters back and forth, and differing densities of cold and warm water rising and sinking at various parts of the earth's surface. It has been estimated that 29% of human output of carbon dioxide is absorbed into the oceans.

The oceans also serve as a transport mechanism for nutrients, gases, organisms, animals and people. In this way, the elements vital to the survival of organisms (such as iron, oxygen, carbon, nitrogen, phosphates, magnesium, sodium and chlorides) are recycled. The oceans also perform the function of absorbing and diluting many of the less harmful wastes of humans and re-distributing them more widely round the globe.

The oceans support vast quantities of life. The two main areas are the shallow coastal zones and the open sea. The shallow, relatively warm, nutrient-rich, coastal zones on the edge of the continental shelves are important breeding and living habitats for over 90% of all marine species.

Also linked to these areas are the coastal estuaries (which act as nurseries for many organisms) and coastal wetlands (which provide food and space for many marine and freshwater species).

THREATS

Over fishing

Over fishing occurs when the annual harvest of fish does not leave enough breeding stock to renew the species for the next harvest. Estimating the size of stocks is difficult so over fishing can be "tolerated" by the species if not all the stocks are affected or if counts have underestimated the size or extent of stocks. It has been estimated that 15 of the earth's 17 major fishing zones have either reached or exceeded their estimated maximum sustainable yield for commercially valuable fish species. The need to find cheap animal feed has driven many nations to range far and wide in the oceans to gather up as much fish as possible. This "vacuuming" of the seas is having a significant impact on biodiversity and is rapidly turning what were renewable resources into non-renewable resources, which are fast becoming extinct resources as a result.

Ocean dumping

Although limited by the London Dumping Convention, many nations still use the oceans as a convenient dumping ground for their solid and liquid wastes. Until quite recently, the city of New York dumped tens of thousands of tonnes of its domestic garbage into a submarine trench just offshore. Some countries are still persisting in dumping their nuclear wastes into the oceans.

Pollution

Large quantities of wastes from streams and rivers drain into the oceans and a concerted effort is needed to reduce these flows of pollution. Strategies such as waste reduction and minimisation, control of agricultural chemical releases and the sound management of sewage and storm water are crucial in this respect. Air pollution is also a major culprit adding to ocean pollution, and it has been estimated that 33% of all pollutants entering the oceans worldwide come from air emissions from land-based sources.

Key Associated Topics:

Aquaculture; Fisheries – The Ecosystem Approach; Marine Life of Southern Africa; Wave Power.

Associated Organisations:

De Beers Marine; Global Ocean; Irvin & Johnson; Two Oceans Aquarium; Wales Environmental Partnerships.

MARINE LIFE OF SOUTHERN AFRICA

Guest Essay by Prof. Rudy van der Elst ~ Oceanographic Research Institute, Durban

South Africa is endowed with an exceptionally rich marine environment. Besides the huge diversity of ecosystems, at least 11 130 species of marine animals and numerous species of marine plants and seaweeds have been identified. These many species represent an enormously valuable resource to South Africa. Aside from the great industrial value of several species, there are additional less obvious benefits attributable to many of the other organisms. While traditionally much attention is devoted to the management of harvesting resources, it is equally important to protect the overall biodiversity of the South African marine environment. These two activities need not be divergent.

Biodiversity

It is known that there are some 2 200 species of fish in our seas, equivalent to about 15% of the total number of marine fish species worldwide. There are 270 families of fishes represented in South Africa, equivalent to 83% of all marine fish families known. Strikingly, up to 13% are endemic, ranking amongst the highest anywhere. There are several families considered to be typically South African. These are the klipfishes (38 species), the gobies (28 species), the seabreams (25 species), the cat-sharks (11 species) and the toadfishes (7 species). Even more impressive are the invertebrates with an estimated 36% considered endemic. Notable families are the sea cucumbers (122 species), squids (195 species), jellyfishes (469 species) and pelagic copepods (354 species).

This marine species richness is largely attributable to the diversity of habitat and the fact that South Africa is located at the confluence of three great oceans: the Indian, Atlantic and Southern Oceans. However, the distribution of these species is not uniform along the 2881 km coastline, clearly an implication for their conservation. Accordingly, several attempts have been made to group all these species into zoo-geographic provinces. One of these defines three or four regions as follows:

Region	From	To
Cool temperate south-west coast	**Luderitz**	**Cape Point**
Warm temperate south coast	**Cape Point**	**East London**
Subtropical east coast (south)	**East London**	**Durban**
Subtropical east coast (north)	**Durban**	**Maputo**

In a more detailed analysis, the coast was divided into 52 sections of 50 km each. Using distributional fish records, it was shown that there were huge differences along the coast. Analysis of the data reveals a progressive increase in diversity from west to east and, significantly, the proportion of endemic species was highest along the south coast region. Protecting these different species is clearly a challenge, but the analysis revealed that protecting 25% (650km) of the coastline would secure the conservation of all inshore fish species. In order to conserve only the 227 species of coastal endemics, 21% (550km) of coastline required protection.

Conservation status of marine fishes

The conservation status of marine resources can be determined in different ways, for example:

1. Quantitative stock assessment including the regenerative capacity of fish stocks.
2. The Red Listing of species according to IUCN categories.

Both methods are of inherent value, the former especially for the benefit of fisheries development and the latter for protection of biodiversity. Indeed, the two methods are inextricably linked and should be viewed simultaneously if any real progress towards conservation of marine resources is to be achieved.

Global status of marine fisheries

Each year the Food and Agricultural Organization of the UN (FAO) collects the catch and effort statistics from fishing nations around the world. For statistical purposes, these data are allocated to 16 oceanic regions, before being processed on an individual species and family basis. The results of this process are summarised into a global assessment of the status of marine (and freshwater) fishes. Obviously, only exploited species are given consideration. However, as fishing represents one of the main pressures on the survival of fish species, this represents an important indicator of species conservation status.

The first global assessment was made in 1971 and since then there have been regular updates. Several general conclusions can be drawn from these results.

◆ Despite substantial increases in effort and technology, the historic annual growth in total catch has ceased and the world's harvest of marine fish has reached a plateau.

◆ Considering individual species-area complexes, 200 of the world's top fish species have been classified into one of four categories: undeveloped, developing, mature and senescent stocks. These results show that more than 35% of these top species are in serious decline with a further 25% having reached maximum exploitation levels. In other words, 60% of the world's top marine fish species are at, or beyond, their levels of maximum sustainable yield and hence their regenerative capacity.

◆ Considering the above information on a regional basis, 9 out of the 16 regions are overfished. South Africa falls in two of the regions, one of which (the SE Atlantic) is considered overfished, while the other (SW Indian Ocean) is still indicating an annual growth in landings.

Photo: Bruce Sutherland - City of Cape Town

◆ One of the more significant deductions from the global fishery analysis is the progressive change in species composition of landings. In the 1950s, the bulk of the global catch was made up of some 50 species of fish. This progressively changed with more species being added as historically important species became depleted – referred to as serial overfishing. At present, more than 600 species constitute the world's catch – and so the number of potential new species will eventually run out and only depleted species will remain.

◆ The level of wasted by-catch in the world's fisheries is staggering. A total of 27 million tons of marine life is dumped each year, equivalent to almost a third of all fish harvested.

South African fish and fisheries

The big industrial marine fisheries of South Africa have endured large-scale fluctuations over the years. Much of this is attributable to environmental variability and interactions between species. However, in the 1960s and 1970s, the landings of several species were allowed to escalate beyond levels of sustainability, posing a threat to the biodiversity of the region. Fortunately, this situation has since stabilised, largely due to improved assessment and management strategies implemented by Marine and Coastal Management (MCM). This is especially true for the pelagic and demersal fisheries, but less so for some of the inshore resources such as abalone. Indeed, several serious issues remain, despite growth in landings. Serial overfishing is not only a global issue and South African fisheries have also shown a progressive increase in the number of species harvested over the years. In 1964 only 17 species of marine fish were reported in national landings. Today there are at least 51 species involved, with additions such as swordfish, orange-roughy and toothfish. The concern over by-catch is also a local problem. For every ton of prawns caught on the Tugela Banks, up to four tons of fish are dumped, including endemic and rare species.

The situation is even less optimistic with demersal linefish species. Comprised predominantly of endemic species of the seabream family (Sparidae), these fishes have been severely depleted. Most famous amongst these are the seventyfour, red steenbras, red stumpnose and similar species.

Following concerted efforts by South African linefish researchers, the various species were assessed and allocated to different levels of conservation status. This was a quantitative assessment based on the species' regenerative capacity termed the "spawner biomass per recruit" (SB/R). Based on extensive data from many different species and studies, it was deduced that if SB/R fell below 25% of virgin stock levels, its regenerative capacity would be severely damaged. The only conclusion that can be drawn is that demersal, and especially endemic, linefish species are seriously depleted, many to levels below regenerative capacity. It can be argued that these depletions, some of more than 90%, ranks amongst the most serious impacts on biodiversity and species richness in the region.

While human pressures through fishing have impacted severely on many species, the decline in the quality and extent of suitable habitat also plays a part. Degraded estuaries and increased silt loads all contribute to increasing pressure on some fish species.

South Africa red data of fishes

The original approach to redlisting involved the allocation of species to four categories: threatened, endangered, vulnerable and rare. This was later expanded to include data deficient and recovered species. While the system proved useful for the protection of terrestrial plants and animals, it was rather simplistic and quite inadequate for marine organisms, especially fishes. In fact, only 4,2 % of the global redlisted species are marine. This shortcoming was realised by WWF and IUCN, who arranged a joint expert workshop in 1996 in London to identify specific shortcomings in redlisting criteria for marine fishes and thence to develop guidelines for improving the classification.

The first attempt at classification of fishes in red data categories was done by Skelton in 1977, when he recorded 28

Table of Red Data Categories

	TOTAL SPECIES LISTED	RED LISTED	% LISTED	EXTINCT	CRITICAL	ENDANGERED	VULNERABLE	LOW RISK	DATA DEFICIENT
Global	25000	148	0.6	0	10	18	90	3	27
Australia	4200	114	2.7	0	3	6	8	44	53
Southern Africa	2200	53	2.4	0	5	2	11	21	6

species of fishes, four (14%) of them marine linked. A decade later, Skelton listed 50 species, 14 (27%) of them marine related – all estuarine. Now, 14 years later, IUCN and the Department of Environmental Affairs and Tourism has released a new South African red data list for marine animals as part of a CITES species management plan. This time the list has 53 marine fish species, a 13-fold increase since 1997, and with 33 species exclusively marine.

How does this list compare to international trends? Data are available for the global listing of marine fishes as well as for Australia. This comparison in numbers listed is reflected in the table on the previous page.

The table of Red Data Categories indicates that southern Africa and Australia have reasonably similar levels of redlisting, while the proportion allocated to "data deficient" in Australia is much higher than that in southern Africa. In contrast, South Africa has listed a greater proportion in the critical and low risk categories.

Regional challenges

South Africa cannot protect its marine resources in isolation. Many of our species are shared with other nations and many others migrate across national boundaries. This calls for concerted regional collaboration in resource management and conservation. About 30% of the world's population live in countries surrounding the Indian Ocean, but this ocean generates only 5% of the global fish catch. It is estimated that there are more than half a million people in this region who are engaged in a daily routine of harvesting marine resources. These range from subsistence harvesting along intertidal shorelines to small scale fishing ventures at the individual or community level, ranging to the nation's territorial limits. However, the capacity to manage these resources, as well as the data required to monitor stocks is often inadequate. For this reason a number of regional research and management programmes have been introduced. South Africa actively participates in several of these.

Conclusion

There is no doubt that the challenges facing marine conservation in the southern African region are enormous. Added to this has been a steady decline in scientific expertise, as many scientists have been lured overseas. But this also provides new opportunities that need to be grasped and developed. Notwithstanding several serious conservation problems, South Africa has an excellent marine conservation record. Technical knowledge remains good, but this must also be matched with political will to implement conservation strategies based on good science.

Key Associated Topics:

Aquaculture; Fisheries – The Ecosystem Approach; Marine Environment; Wave Power.

Associated Organisations:

Global Ocean; Irvin & Johnston; Two Oceans Aquarium; Knysna Estuarine Aquarium; SA Association of Marine Biological Research; Global Ocean.

MEDICINAL HERBS

Traditional medicine and healing has used herbs as medicine for thousands of years and over 65% of all formal medicines are based on herbal or plant-based structures or formulas. The CSIR has commenced a programme of identifying several hundred medicinal herbs with a view to developing them as registered drugs that can be sold to drug companies. There have already been a number of breakthroughs and the CSIR team are confident that they can produce some special medicines. In the past, some who have recognised the commercial opportunity to sell traditional herbs to drugs companies and makers of "traditional remedies", have carelessly hacked trees and plants down and caused enormous damage. Partnerships have therefore been formed between traditional healers and nature conservation officers to enable the healers to harvest their needs without destroying the sources. The nature conservation officers are also promoting bushplanting programmes to ensure that the specimens are available in the future.

Key Associated Topics:

African Environmental Tradition; Biodiversity; Bioprospecting; Cultural Resources; Indigenous Plants; Medicine and Health.

Associated Organisations:

National Botanical Institute; Weleda.

Photo: Earthyear Magazine

MEDICINE AND HEALTH

Guest Essay by Chamilla Sanua ~ Weleda Pharmacy

Do we need a change from our current perspective?

Never before and never in the future will the human being be able to improve on God's magnificent creation of the human body. The liver, alone, performs more than 500 functions per second! We are a perfect creation. We can digest food, eliminate toxins, concentrate and grow a baby all at the same time without being conscious of these processes.

The human body is a highly effective and efficient system that has been designed to heal itself. When one is presented with an illness, one should learn to respect the natural process of healing and support the body's endeavour to heal itself through the use of well-known safe and effective natural products, which are designed to aid the body's healing mechanisms rather than suppressing them. Symptoms of an illness are the natural expression of a disharmony within the person. A headache may be the result of short-sightedness; a neck or back problem may be the result of constipation; an infection may be the result of hormone imbalance or overindulgence! Taking away the pain (symptom) is only a short-term solution. It is important to treat the cause and not to simply suppress the symptoms with (suppressive) drugs. Conventional medicine expresses itself through the words it uses: *Anti*biotics, *anti*-hypertensive, *anti*-depressants, *anti*-histamines, *anti*-diuretics, *anti*- etc.

How the body works – the "bath analogy"

If one compares the body with a bath, one can see that they both function in a similar way. In the bath the water is eliminated through the plug. If the rate at which the water enters the bath exceeds the rate of elimination, then the water fills up the bath until it reaches the secondary plug. This may be because of a blocked drain or because the rate at which water comes into the bath exceeds the rate at which it can leave the bath. In our bodies, the primary elimination process takes place between the navel and the knees i.e.: menses, urine, and faeces. If the rate at which toxins come into the body exceeds the rate of elimination then there is a build up of toxins in the body, which will be eliminated through the eyes; nose; throat; tonsils; prostate; vagina etc. The conventional approach of dealing with uncomfortable symptoms is to block up this elimination with agents such as cortisone, anti-histamines etc. These agents effectively block this elimination and add toxins into the pool. If one has blocked the elimination channel, added toxins and done nothing else, the 'water' in the bath must rise until it flows over the edge! This is far more serious than water passing through the outlet channel! One has then reached the tertiary elimination channel, which, in the body, represents mucous membranes inside the body, the lung system (asthma, bronchitis, pneumonia, etc.), the joints (arthritis) and the brain (depression).

Consuming chemical substances

The whole issue relating to whether food should be Genetically Modified (GM) should be considered carefully and debated vigorously. Will GM foods play havoc with the human genetic pool? It could be a weapon or a tool depending on whose hands it is in!

The human digestion is a powerful system able to separate and delegate nutrients/toxins to where they are required in the body. All natural substances are easily recognised and dispatched to the correct 'department'. As soon as a substance is synthetic or chemically produced, it becomes a "foreign" substance in the body and it therefore becomes difficult for the body to recognise and eliminate, which often results in nasty side effects. This may occur with GM foods and often occurs with almost all chemically produced drugs used today. Sadly, 50% of all drugs approved by the Food and Drug Administration (FDA a very powerful and stringent organisation in America) are banned after 10 years as they are later found to be too toxic for human consumption. That means that one out of two drugs on the market today, proclaimed to be safe, efficacious, bigger and better, may be off the shelves in ten years' time and you might have been the Guinea Pig! Claims made such as "Estrogen or HRT do not cause breast cancer" are often later proved false.

It may be worthwhile considering the fact that the pharmaceutical industry will not financially benefit from a healthy population!

An important and motivational element of the "health industry" today is the issue of patenting laws. Only synthetic, man-made substances can be patented. Naturally occurring substances such as Vitamin C or E cannot be patented as they are considered "God given" and are thus for everyone. Many doctors are influenced and "educated" by the pharmaceutical industry, which strongly promotes its so-called new "wonder drugs". The pharmaceutical industry may however keep quiet about the side effects or lack of efficacy of their drugs, due to economic pressure to perform on the stock exchange. For example, Calcium carbonate, found in certain well-known effervescent brands of Calcium tablets, was found to increase the bone density of the wrist but have no effect on the hips and spine. As most elderly people do not break their wrists but their hips, this makes this product of almost no use to most of the population who use it. Most doctors and patients still do not know this fact and the product continues to be sold! Premarin, a well know synthetic Estrogen, was known to cause cancer if not prescribed together with a progestin (synthetic progesterone) 25 years ago; yet some doctors are still prescribing it without also prescribing progesterone to avoid cancer!

The pharmaceutical industry will not undertake the huge trials necessary to prove the safe, effective approach of natural substances, as it is not in their interests to do so. This means that few trials are done on natural substances such as vitamins, minerals, herbs and homeopathic medicine.

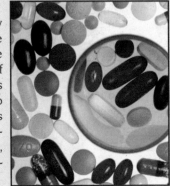

We should also be aware that vast quantities of unused or outdated chemical medicines are thrown away by households instead of being properly destroyed. These chemical drugs or their residues may end up entering our rivers and groundwater supplies – with immeasurable possible ecological effects in the future.

The immune system and supplementation

Insufficient information is available about the important need to supplement our bodies with minerals and vitamins, despite the fact that a large proportion of the food we eat is deficient in nutrients.

Many of the problems that we face today are largely due to a deficient immune system. These problems include AIDS, leukaemia and cancer as well as deficiency illnesses such as osteoporosis and arthritis.

Today, more than ever before, we need to strengthen the immune system so that we will be able to defend ourselves against a new generation of 'super bugs' that have evolved as a result of the large amount of antibiotics in the food we eat and due to excessive antibiotic use by the well-meaning medical profession. Immune system boosting can be achieved through constant good nutrition, good hygiene and by allowing the body to heal itself naturally. Skin rashes, colds and flu, etc., should be welcomed, as this is what helps to keep the system clean and well.

Supplementation with vitamins is also very necessary today because of the vitamin and mineral deficiencies in the foods that we eat. Hydroponically grown food is only given the barest minimum in nutrients' such as potassium, sodium, nitrogen and phosphorous in order to grow the plant. What has happened to the 78 other minerals found in the body, let alone essential fatty acids, essential amino acids and vitamins required by the body to function adequately? Added to the deficiency in our food is the excessive processing of our food and the lack of awareness as to how best avoid mineral and vitamin loss during the actual cooking and preparation of food. We also regularly pollute ourselves with what we eat. Well known 'problem' foods include: dairy products; sugar; excess tea and coffee; alcohol; colourants and flavourants. These products should be kept to a minimum especially during acute illness.

The human body does have coping mechanisms, but why should we force our body to go into emergency mode when this is not necessary?

Treating the cause instead of suppressing symptoms

There is a growing realisation that physical illness is often a reflection of or a result of our inner mental state. Could a heart attack result from a deficiency in the 'heart' realm – from a stifling of emotional feeling, or could asthma result from a conscious or subconscious 'grief'? A leading thinker in this field – Deepak Chopra, firmly believes that this is the case, and there seems to be substantial empirical evidence to back up his point of view.

The conventional medical approach of suppressing uncomfortable symptoms is very likely to push the illness to deeper levels and more chronic illness. Today, when a dermatologist treats a baby with cortisone for eczema, he does not realise that he may be the cause of the asthma that the child develops later because the illness was pushed from the skin to the lungs. Which is more serious?

The natural approach to dealing with the same symptoms is to unblock the elimination channel and decrease the toxins coming into the body. The liver governs the colon and the kidneys govern the urine. By detoxing the liver with simple herbs and stimulating the kidneys, much can be achieved.

What should we do to have a healthy lifestyle?

◆ **Reduce or avoid "polluting foods"**, including artificial sweeteners; colourants and flavourants; preservatives; saturated fats (animal fats except fish oils); wheat (which, due to modification, now has up to 25 different chemicals in it); dairy products.

◆ **Increase** your consumption of fruit and vegetables, especially those that are organically grown.

◆ **Decrease** your exposure to all plastics and all petrochemical products such as clingwrap, Tupperware, aqueous creams and other petroleum-based cosmetics. Weleda and Wala make plant based cosmetics such as aqueous creams that do not have the estrogenic side effects of petrochemical products.

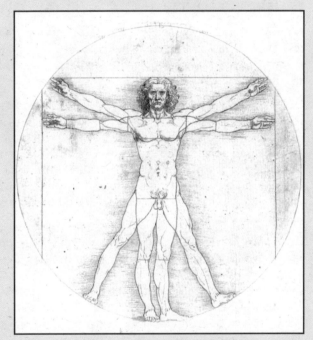

◆ **Avoid** using pesticides and herbicides. These are often highly toxic.

◆ **Dry bristle brush** your body with a natural bristle brush daily.

◆ **Commit yourself** to a moderate amount of daily exercise.

◆ **Connect to yourself** or your Creator daily in meditation.

◆ **Practise** a daily appreciation of nature such as watching the sun rise or observing the beauty of nature.

◆ **Last but not least remember to thank your Creator and help someone each day.**

If we can respect nature inside and outside ourselves we will go a long way towards making ourselves and the world around us more healthy and balanced – mentally, emotionally and physically.

METHANE

Methane is an odourless and colourless hydrocarbon gas produced either by natural or artificial anaerobic decomposition of organic material. The gas burns with a pale flame to produce water and carbon dioxide and releases no hazardous air pollutants. The gas is a by-product of landfill sites and, with careful planning, it is possible to tap the gas from the site, pressurise and dry it and use it to power machinery or run trucks on site. The gas is also emitted from intensive animal husbandry operations (cattle, pigs, or other stall fed animals) in such large quantities that it has been cited as a contributor to increases in carbon in the atmosphere.

Key Associated Topics:

Air Quality; Atmospheric Pollution; Composting; Landfill; Pollution.

Associated Organisations:

EnviroServ; Environmental and Chemical Consultants.

MINERAL RESOURCES

Minerals refers to commodities such as gold and silver, but also to coal, oil and iron ore. When one refers to *mineral resources* it implies the amounts of these various minerals that are spread around the world.

Usually in referring to mineral resources it implies those resources that are still in the ground and not yet mined, but occasionally, mined stockpiles of ore that are above ground but not yet beneficiated, can be included. The term 'resources' is reasonably difficult to quantify because a country's proven mineral reserves refers to minerals still in the ground, but which can be recovered and beneficiated at a particular price.

If a country has a large quantity of some mineral resource then they will not exploit deposits of low grade. But a country that has little of that mineral resource will exploit a low-grade ore that another country may regard as sub-economic. So the true amount of an exploitable resource is largely linked to political and economic decisions as to what is worthwhile.

Therefore the total amount of exploitable mineral resources at any time can change from year to year, either because new deposits are discovered or because previously sub-economic deposits became economically viable as political and economic circumstances changed.

Guest Author: Dr Kelvin Kemm

Key Associated Topics:

Energy; Natural Gas; Natural Resources; Non-Renewable Resources; Opencast Mining; Petrochemicals; Sustainable Development Section – The Mining Industry.

Associated Organisations:

Anglo American; Chamber of Mines; Contact Trust; De Beers; De Beers Marine; Department of Minerals & Energy; Kumba Resources; Minerals & Energy Education & Training Institute; Mining Review Africa; Richards Bay Minerals; Strategic Environmental Focus; Terratest.

MONOCULTURE

Monoculture describes the repeated growing of a single crop on a given piece of land over and over again. This usually leads to exhaustion of nitrogen in the soil and the build up of pests and diseases, which would normally be wiped out by crop rotation. An example of this would be the American wheat prairies where there are tens of thousands of acres under wheat every year. One new pest or disease could wipe out the entire crop. The lesson from nature is that there is a need for diversity and variance to ensure that there is sufficient flexibility to deal with problems that may occur.

Key Associated Topics:

Afforestation; Agriculture; Biodiversity; Carrying Capacity; Crop Rotation; Integrated Pest Control; Land Degradation; Organic Farming; Organic Food; Permaculture; Pesticides; Pesticide Treadmill; Soil.

Associated Organisations:

Agricultural Research Council; Elsenburg College of Agriculture; SA Sugar.

MONTREAL PROTOCOL

The Montreal Protocol was the original international instrument drawn up to ensure that measures were taken to protect the ozone layer. South Africa ratified the London amendments to the Protocol in May 1992 designed to restrict the use of CFCs (Chlorofluorocarbons) and halons. For a more detailed history, go to the following website: www.ec.gc.ca/ozone/en/index.cfm. CFC producing industries in South Africa phased out their production of CFCs two years before their specified deadlines. Until recently, South Africa was regarded as a developed country. At the ninth meeting of the members of the protocol in September 1997, South Africa was granted its request to be downgraded to developing country status to avoid early phase-out dates imposed on the developing countries and also to avoid the annual R3 million payments to the Protocol's Multilateral Fund for poorer countries who needed assistance with compliance. Subsequent meetings of the signatories to the Protocol have made further concessions to withdraw or reduce their consumption of the ozone damaging products.

Key Associated Topics:

Air Quality; Atmosphere; Atmospheric Pollution; Ecosphere and Ecosystems; Ozone; Ultra-violet Light; VOCs.

Associated Organisations:

Department of Trade and Industry; Department of Minerals and Energy; National Business Council.

... A world of resources

The earth is a treasure house of natural resources
When wisely unlocked they continue to enrich mankind.

Which is why Anglo American remains committed
To resourcing our future
And to spearhead the global transition
Towards securing a sustainable future for all.

By exploring a world of resources
We'll continue to make a world of difference.

MOUNTAINS OF SOUTH AFRICA ~ A NATIONAL TREASURE

Guest Essay by Dr W.R. Bainbridge
~ Environmental Consultant and Vice-Chairman of The Wilderness Action Group

High mountains are internationally recognised by organisations such as IUCN (the World Conservation Union) as amongst our most important global assets. It is recognised that not only do high mountains represent some of the most extreme and remote environments on the planet, but that they are also of great importance: as sources of water, for their unique biodiversity resources and for a range of natural and agricultural products. Though they may physically tower above their surrounds, which are usually more densely populated lowlands, they are intimately associated with them through ecological, social and economic links. While mountains may present the appearance of permanence and splendid isolation, in reality, they are fragile environments, inextricably linked to and affected by their surrounding landscapes. By virtue of their fragility and sensitive nature, their survival is therefore dependent on how well they are managed. The United Nations declared 2002 as the International Year of the Mountains, in order to increase awareness and knowledge of mountain ecosystems in terms of their importance for the provision of strategic services and products; improvements to the well-being of rural people; promotion and defence of the cultural heritage of mountain communities; as well as promotion of conservation and sustainable development issues, including sympathetic tourism. 2002 was also declared the International Year of Eco-tourism.

The mountains of South Africa are mainly, but not exclusively associated with the Great Escarpment, which stretches the entire breadth of the southern regions of the country. Whilst they constitute only about one tenth of our land surface, they form some of the most prominent physical features and beautiful landscapes of our country and also feature prominently in our history, mythology and spiritual life. Their high scenic value also makes them prime tourism destinations. Most significantly, however, they form the principal catchments for our water supplies which are a crucial (and sometimes limiting) factor for our industrial and agricultural economies as the high yield of water that flows from our mountains is disproportionate to their limited extent. Our mountains are also characterised by a great diversity of landscapes, which support globally significant biological diversity, rich in species and with high levels of endemism. Two of the seven floristic regions, which occur in Africa south of the Sahara (the Afro-alpine and Afromontane Regions), are entirely associated with mountain systems. Many of our biodiversity hotspots (areas where high levels of diversity and endemism are under threat), are also associated with our mountain systems.

Despite the importance of our mountains and the strong conservation messages that emerged during the Year of the Mountains celebrations, which called for the protection of global mountain systems, there is cause for concern about their state of conservation in this country. Only about 15% of the total mountain catchments of the country are under

Photo: John Hone - Art Publishers (Pty) Ltd

State protection (and this includes most of our proclaimed wilderness areas). The majority of our mountains lie in private or communal ownership, and are subject to a range of significant threats. Of particular concern is the fact that mountains are not protected by a single specific statute, and no individual official agency or department appears to take responsibility for their protection. Yet, a range of international conventions such as that on Biodiversity Conservation, which bind the State, are applicable to our mountain systems. Twelve years after the Rio Conference, only lip service appears to be paid by our government to the provisions of Agenda 21 relating to mountain conservation. A White Paper defining official policy for conservation of mountainous areas based on this was produced in 1991, but its provisions appear to have been largely ignored. Of particular concern is that the new Municipal Demarcation Act and Municipal Structures Act decree that all land, including mountains, fall within municipal areas, and are therefore rateable. There is serious concern that landowners will be forced to initiate non-sustainable or inappropriate land-uses, in order to generate revenue to meet these new taxes.

Many therefore believe that a new and comprehensive approach to conservation of our mountain systems is urgently needed. Government, private and communal landowners, environmental NGOs, user groups and other stakeholders should collaborate to reach consensus and to formulate effective policies and guidelines for sustainable land-uses by revision and updating of the White Paper, implementation of existing legislation and where appropriate, possible promulgation of new statutes dedicated to mountain conservation. Two major projects currently underway (the Cape Action Plan for People and the Environment, and the Maloti-Drakensberg Transfrontier Conservation and Development Project) while not specifically directed exclusively at mountain conservation, are expected to make important contributions in this direction.

However, the apparent lack of commitment by official organisations to effectively conserve our mountain environments, is surely a matter of the greatest concern to those of us who love mountains, and are convinced of the vital role they play in the national economy, in bringing the many benefits outlined above, and for their intrinsic value.

Key Associated Topics:

Biodiversity; Biome; Birds; Birding; Conservation; Ecosphere & Ecosystems; Eco-tourism; GAIA; River Catchment Management; Stewardship; Water Quality & Availability; Wildlife; Wildlife Management.

Associated Organisations:

Bainbridge Resource Management; Hiking Africa Safaris & Tours; Imagis; Mountain Club of SA.

NATIONAL PARKS IN SOUTH AFRICA

Guest Essay by Lauren Gardner ~ WESSA

A New Strategy of Commercialisation and Concessions

It was in 1926, amidst the excitement of the introduction of the new National Parks Act, that the Minister of Lands outlined to the House of Parliament his plans for the control and administration of national parks. He explained a very important safeguard he had included in the Act: "As long as the [park] boundary is in the hands of the government, the government will always be exposed to being pressed by supporters to alter the boundary. Politics must be kept out of it…Therefore I propose a Board of Control, representing the interests of the whole people." South African National Parks (SANParks) has evolved from these early beginnings and today is the authority responsible for managing the parks on behalf of the people of South Africa. In terms of the National Parks Act (Act 53 of 1976), this mandate encompasses the dual obligations of protecting the National Parks in their natural state while allowing managed access by the public. We must now ask the question: "Are Government and the SANParks management honouring their mandate and commitment to the people of South Africa?"

In response to the increasing scarcity of migrations of healthy herds of purple buffalo (the monetary kind) from government coffers, SANParks has been obliged to seek other sources of revenue and embarked upon a Commercialisation Programme (CP) in 2000 as a means of making up for the shortfall in government funding. This Programme, termed 'Commercialisation as a Conservation Strategy', according to SANParks, is about a focus on their core business, namely the establishment, expansion and management of National Parks. The non-conservation functions related to the management of the Parks, most notably retail and hospitality aspects are to be increasingly outsourced to the private sector.

One controversial component of the Programme is the allocation of concessions in National Parks to private operators to build and operate tourism facilities on a long-term basis. According to the 2001 Government Yearbook, this is quite a radical departure from past policies where SANParks has traditionally both provided and managed the accommodation. In terms of the concession contracts, the rights over a defined area of land are granted exclusively to the concessionaire until the termination or expiry of the 20-year contract. In return for this privilege, SANParks is guaranteed a total minimum income from the profits generated by each of the concessions for the 20-year period. At the higher end of the tariff scale, most of these concessions will only be accessible to international tourists and very wealthy South Africans.

The first concessions were awarded in the Kruger National Park in 2002. As a consequence, just less than 5% of the Park has already been allocated to private interests for exclusive use. An extra 570 kilometres of roads have been added to the Park to support these concessions, substantially enlarging the human footprint on the ecology. Concessions have also

been awarded in many of the other National Parks, and more concessions are probably on the cards.

Provincial authorities are now also following suit. For example, according to the Wildlife and Environment Society of South Africa (WESSA), the Greater St Lucia Wetland Park Authority, in association with Ezemvelo KZN Wildlife, intends to squeeze 7 200 permanent beds into concessions in the Greater St Lucia Wetlands World Heritage Site. Many of these beds are in development nodes that are in ecologically sensitive areas. Seven thousand two hundred beds is a lot, considering St Lucia is only one tenth the size of the Kruger National Park, which has 4 200 permanent beds.

What's the fuss about?

One of the widely held concerns about the CP is that the concessions fly in the face of the spirit of the National Parks Act and the National Environmental Management Act (Act 107 of 1998). National Parks belong to the people of South Africa and by allocating portions of the Parks to private concessionaires many believe that SANParks is violating the tenet that the land is held in trust for the public. Even SANParks' advisor to the CP, the International Finance Corporation, states that "a slight ambiguity in the

Act's wording is noted with regard to accommodation" but in the end concludes that the interpretation that the accommodation can be contracted out, is defensible.

The controversy doesn't stop there. Are the special allowances that have been made for the concessions potentially detrimental to the environmental integrity of the Parks? What damage will the off-road driving that is permitted in the concession areas inflict? And what about the noise pollution that is likely to result from the private airstrip that has been allowed at one of the concessions? And will it worsen? It is probably only a matter of time before other concessions raise their hands and ask for similar aviation privileges.

Whilst concessions are an internationally recognised means to an end, how they are implemented is a critical variable. Interest groups, such as WESSA and the National Parks Support Group Trust, have voiced their concerns in this regard. One common theme is that there are too many procedural shortcomings related to the evaluation of the sustainability of the concessions.

Missing from the equation in the calculation of sustainability are parameters to define what, where, when, how and critically, how much. Many of the concessions are being

THE SOUTH AFRICAN NATIONAL PARKS

awarded in Parks that have no up-to-date environmental management plans to guide the extent, location and implementation of concessions. Furthermore, there is a lack of commitment from SANParks to a formalised assessment of the larger scale and cumulative impacts of the concessions on the Parks' ecosystems. This is despite the fact that the desirability of such an assessment is stated in environmental legislation. Another concern is that SANParks is reluctant to define what the Programme encompasses. Will Parks in the future be pockmarked with concessions, changing the once 'virgin' ecosystems into overcrowded pleasure resorts? Will all accommodation be run by the private operators, pricing South Africans out of the market? Where will it end? How much is enough?

According to SANParks, the Commercialisation Programme is ongoing. To many, this is most worrying. Where money is the driving factor, as it is in this instance, and without the benefit of formalised strictures to guide implementation, it is quite possible that concessions become to SANParks what the ant is to the fluke: an easy source of ongoing sustenance that ultimately kills the host.

Whither to, Conservation?

At the very heart of the matter lies the uncomfortable philosophy of 'if it pays, it stays'. That seems to be what the Minister of Environmental Affairs and Tourism was intimating when he said in an interview: "Parks can pay for themselves ..." Many therefore feel that our government is reneging on its financial responsibility to National Parks by placing pressure on the Parks to become viable financial entities. But if it is necessary to look at things this way, then perhaps a more comprehensive way of calculating 'viability' should be considered. Has anyone calculated the economic value that the Parks proffer by way of supplying clean air and clean water? Or the value provided by the wetlands that assist in flood attenuation? Or the fact that the protected areas act as a genetic seed banks? Have the total macro-economic spin-offs from our tourism industry, which relies so heavily on the pull of wildlife and wilderness, been accurately calculated? Until the real value is determined, many continue to share the opinion of independent environmental policy analyst Rachel Wynberg, expressed in her article in African Wildlife: "The rapid commercialisation of our protected areas is a major concern, as is the Government's readiness to absolve itself of financial responsibility for the management of our natural heritage."

Key Associated Topics:

Biodiversity; Birds; Birding; CITES; Conservation; Ecotourism; Green/Brown Debate; Heritage; Hunting; Public Participation; Stewardship; Stakeholders; Transfrontier Parks; Wilderness Area; Wildlife; Wildlife Management; World Parks Congress.

Associated Organisations:

Centre for Wildlife Management – University of Pretoria; Ezemvelo KZN Wildlife; SA National Parks; SA Wildlife College; WESSA; EWT; WWF.

NATURAL RESOURCES

Photo: Mossgas (*Earthyear Magazine*)

"Natural resources" is the term used to describe the basic materials and resources that are produced through the earth's own inherent natural processes and systems. They include the planet's air, water, and land; nutrients, minerals and resources in the soil; wild and domesticated plants and animals and the entire range of nature's natural systems for waste disposal, purification, dilution, recycling abilities and pest control. Humankind must be sensitive to the use and management of renewable and non-renewable natural resources. If renewable resources such as fish stocks are not used in a sustainable manner (i.e. they are over-fished and extracted or utilised beyond their capacity to recover from the harvest), they are doomed to become non-renewable resources, resulting ultimately in extinction. Non-renewable resources have a basic point where their reserves will run out. Non-renewable resources must be used wisely and economically to try and stretch their life span a little longer. For example, it is currently estimated that there are approximately 200 years of reserves of oil on earth at present. If those reserves are made to last longer, then not only will they be available longer but there may also be a chance that a suitable alternative may be discovered.

Key Associated Topics:

Agenda 21; Air Quality; Atmosphere; Birds; Biodiversity; Ecological Intelligence; Ecosphere and Ecosystems; Energy; Fisheries; Forestry; Fossil Fuels; GAIA; Geothermal Energy; Groundwater; Land Degradation; Marine Environment; Marine Life of Southern Africa; Mineral Resources; Mountains; Non-renewable Resources; Renewable Resources; Rivers & Wetlands; Soil; Stewardship; Tropical Rain Forest; Water; Wildlife; Sustainable Development – Accounting for Nature's Services.

Associated Organisations:

Anglo American; Delta Environmental Centre; Department of Minerals & Energy; Energy Efficient Options; Environmental Education and Resources Unit – University of Western Cape; ESI Africa Magazine; Galago Ventures; Imagis; Iskhus Power; Mining Review Africa; Umgeni Water Services.

NEMA
(The National Environmental Management Act 107 of 1998)
Guest Essay by Cathy Kay ~ WESSA

In South Africa the National Environmental Management Act 107 of 1998 serves as a framework for environmental legislation. It was planned to be supported by sectoral-specific subordinate legislation. Hence, the omissions from NEMA of some sectoral-specific issues such as waste management. No policy is, however defined in NEMA and neither does it provide for the formulation of environmental policy, as was the case in the Environment Conservation Act 73 of 1989. Flexibility is intended to be achieved by means of an extended list of generic principles.

Photo: Guy Stubbs Photographer

NEMA compares well with the sustainability principles and rights as identified at WSSD, however, as a framework, it has many shortcomings, including the following:

◆ Two significant omissions are:
  i. The principle of rehabilitation of existing disturbed ecosystems.
  ii. The duty to develop norms and standards that define environmental quality.

◆ A significant inadequacy is the fact that roles, responsibilities and authorities of different environmentally related line functions as well as between the various spheres of Government are not clearly defined. For example both NEMA and the National Water Act 36 of 1998 make provision for the control of emergency incidents. This can give rise to conflict between the different Departments if one or the other Department or a private person decides to investigate the incident through a specific Department in order to negate or reduce the other Departments authority.

◆ Another problem that needs to be addressed is formalizing procedural arrangements between Government Departments that have joint jurisdiction over an activity. NEMA is not explicit on how to pro-actively make arrangements to avoid conflict.

◆ NEMA does not provide for compliance auditing by the national and provincial Governments. Provision should also be made for the appointment of environmental inspectors and their duties and powers should be spelled out in the legislation. NEMA does not refer to environmental inspectors.

◆ NEMA fails to establish an integrated platform or process where applications for environmental authorizations are lodged, decisions are made and authorizations issued. NEMA compels the Committee for Environmental Co-ordination only to investigate and recommend the establishment of mechanisms for a single point in the province where applications for authorizations, licenses and other permissions could be considered in a co-ordinated manner.

◆ NEMA fails to provide for statutory reporting on environmental performance by the private sector; this is highlighted by the notable exemption of the reporting of pollution incidents.

◆ NEMA is deficient in that it does not provide for the establishment of a standard generating and enforcement review committee nor does it provide an interface or integration among the various role-players such as the Department of Water Affairs and Forestry, the Department of Environmental Affairs and Tourism, the Department of Mineral and Energy, the Department of Health, the Department of Agriculture and the Department of Labour.

This fragmentation is exacerbated by the absence of a uniform standard setting and enforcement procedure. The Committee for Environmental Co-ordination is tasked only to make recommendations on securing compliance with national norms and standards.

◆ NEMA is totally silent on the deployment of economic instruments as a modern strategy for environmental governance and management.

◆ NEMA is silent when it comes the appeal process. The Act states that once a record of decision is issued, the public has 30 days to appeal that decision. However it does not state that the developer may not proceed with the activity during the 30-day appeal period.

◆ Similarly, it remains silent on whether, after the 30-day appeal process and during the period when the Minister assesses the appeal, the developer may continue with the development or not. In short the Record of Decision is an instruction and the developer is entitled to proceed irrespective of the appeals and the review period.

◆ NEMA has its main emphasis on people first. The triple bottom line principle often only includes the social and economic perspectives and more often than not leaves the ecological and environmental issues in the shadows.

In conclusion

NEMA was drafted by way of broad-based representation and input. One of the main features of NEMA is the attempt to achieve co-operative governance between different line functions in the international context as well as between same and different spheres of government. NEMA makes extensive arrangements for management by outsiders by way of empowering civil society and providing for co-operation agreements.

Despite its shortcomings, NEMA is a definite departure from the command and control practices of the past. It also allows for the integration of multiple environmental management tools and governance instruments.

Key Associated Topics:

CONNEP (P); Environmental Law; Environmental Liability for Remediation; Environmental Risk & Liability; Sustainable Development Section: – Is it achievable under our current Legal framework?

Associated Organisations:

Contact Trust; Enviroleg; WESSA; Department of Environmental Affairs & Tourism; Winstanley Smith & Cullinan.

NIMBY (NOT IN MY BACK YARD)

NIMBY describes the attitude of some communities and interested and affected parties who do not oppose necessary but unpleasant or undesirable projects and developments, such as waste dumps and industrial projects, until they find out they are to be located in their own neighbourhoods. (Not in my back yard) Those same people will accept benefits from these projects (e.g. being able to get rid of their wastes, or the products made by an industry) but will not accept that the waste dump or industry must be located somewhere and that this may be near them. With increased urbanisation and the fact that waste disposal and treatment facilities need to be located as close to the generators as possible to save transport costs, residential developments are being located closer and closer to landfill sites and treatment facilities. The other problem relates to the fact that, although initially these types of undesirable facilities are located as far away from development as possible, urban creep tends to bring neighbours closer over time.

Key Associated Topics:

Communities; Environmental Impact Assessment; Interested and Affected Parties; Land Use Planning; Public Participation; Stakeholders.

Associated Organisations:

IBA Environnement; Jarrod Ball and Associates; Ken Smith Environmentalists.

NITROGEN

Nitrogen is a colourless, odourless gas that comprises 78% of the atmosphere. Nitrogen is cycled between the air and the soil in the nitrogen cycle, which includes a process known as nitrogen fixation. Nitrogen fixation can be carried out by industry (to make fertilisers) or naturally by micro-organisms producing combined nitrogen as ammonia and nitrates, which are absorbed by plants from the soil and used to make proteins for nutrition. Nitrogen Oxide (N_2O) is a colourless gas with a sweet odour, which is used as a weak anaesthetic ("laughing gas"). Nitrogen Dioxide (NO_2) is a red-brown toxic gas, which is one of the products of vehicle combustion and is an active part of smog.

Key Associated Topics:

Air Quality; Fossil Fuels; Petrochemicals; Photochemical Smog; Pollution.

Associated Organisations:

Chemical and Allied Industries' Association.

NOISE POLLUTION

There are different levels of noise all around us all of the time. When it appears to be very quiet there it still a background noise, even if it is crickets chirping or a gentle breeze blowing. All of the different day-to-day noises are collectively referred to as the ambient noise, or background noise. From one location to another the background or ambient noise is different.

Sometimes a different or unusual noise can intrude and become irritating, such as a construction worker using a jackhammer next door. This noise intrusion is referred to as noise pollution. Noise pollution need not be loud to be irritating; for example, a diesel generator in the distance at night can produce a rhythmic sound that just feels wrong for that circumstance. Another type of noise pollution is an out-of-place sound even if it is not loud or rhythmic. For example, if you are sitting on the beach at night listening to the roaring of the surf breaking on the beach, this is pleasant and relaxing, but if someone nearby starts playing a radio, even if it is not too loud, this can become an intrusive irritating background noise, which is 'noise pollution' under those circumstances.

At times noise pollution is not only irritating, but can become dangerous; for example, when trains pass near a building and the constant vibration leads to cracks and damage in the structure of the building. This vibration is passed through the ground as sound waves, but also through the air as low frequency sound waves which batter buildings.

Generally noise pollution is rather subjective, being the opinions of people rather than a sound intensity measurement, in decibels, measured using audio-measuring instruments.

Guest Author: Dr Kelvin Kemm

Key Associated Topics:

Acoustics; Environmental Health; Environmental Impact Assessment; Environmental Law; Interested and Affected Parties; Nuisance; Stakeholders.

Associated Organisations:

Ceatec; National Air Pollution Services; Wales Environmental Partnerships.

NON-RENEWABLE RESOURCES

These are environmental resources that cannot be replenished. Once they have been used up, there will be no more. Most non-renewable resources are minerals, which are mined, for example, gold, iron ore, titanium. Coal and oil are known as fossil fuels and are also non-renewable. Besides being significant sources of pollution, these fuels will not last for ever (some scientists have suggested that the earth has only 200 years of exploitable oil reserves left at current consumption levels) and alternative energy sources will have to be found. Solar energy poses potential solutions to many problems relating to the use of non-renewable resources. Photo-voltaic cells technology has advanced substantially and conversion of the sun's rays to electricity is commonplace. The principles behind sustainable development seek to stretch the availability of non-renewable resources by finding sustainable replacements, and reducing existing usage.

Key Associated Topics:

Consumerism; Fossil Fuels; Mineral Resources.

Associated Organisations:

ESI Africa; Institute of Natural Resources; Renewable Energy And Energy Efficiency Partnerships (Reeep).

NUCLEAR ENERGY
~ THE COUNTER DEBATE

Guest Essay by Muna Lakhani ~ Earthlife Africa

Eskom wish to build a version of a nuclear reactor that has failed overseas, called the Pebble Bed Modular Reactor (PBMR). They hope to build, use and sell over 200 reactors, and manufacture the fuel for them all locally.

10 reasons why Nuclear Power is not a solution for our country:

1) There is no such thing as a "safe" dose of radiation. Indeed, there is a growing body of evidence that low doses may actually be more dangerous, as they may mutate cells more easily than high doses, which can kill the cell. There is no debate as to whether radiation kills; maims; causes mutations; is cumulative; causes leukemia (mainly in children), cancers, respiratory illnesses and attacks the immune system (with children, pregnant women and the elderly the most vulnerable). The only disagreement is about what is legally considered an allowable dose. There is clear evidence that the incidence of leukemia in children dropped in areas surrounding 3 nuclear power stations that closed in the USA.

2) No responsible way to "dispose" of radioactive waste – there is not a single repository worldwide for high level radioactive waste, and safe management remains as elusive as it was 50 years ago. The problem lies with the longevity of the radiation hazards (many thousands of years), and it is impossible to guarantee safety for that long a period. The best solution is not to produce any in the first place. This also implies that future generations will be carrying both the health and financial cost of this waste, an issue that goes against our Constitution, regarding future generations.

3) Nuclear power is expensive electricity. All States in the USA with nuclear power charge on average, 25% more for their electricity. The economics are speculative and already escalating, as has been the history of the nuclear industry regarding cost over-runs. Eskom claim a cost for the demonstration module as "about R1 billion". Their information document of January 2001 shows the costs of establishing and fueling the pilot plant at a total between R2. 86 billion and R3.24 billion. (If we factor in the accuracy range of "between 10% and 30%", the highest cost already anticipated by the developer is R4.21 billion). Their USA partner, Exelon, is on record as stating that the cost of the demonstration unit would be in the region of US$300 million – approx. R2.7 billion at Oct 2001 exchange rate. Even getting all 10 initial nuclear reactors proposed for South Africa right the first time, will cost (according to US academics) about R35 billion over the proposed 40 year life, for which amount we could install more than the same amount of generating capacity using safe and clean renewable energy sources.

4) The entire fuel chain (mining, transport, enrichment, manufacture, etc.) is extremely energy intensive and dirty. Some debated research even shows that nuclear power generates less energy than it uses in the entire fuel chain over its lifetime.

Photo: Bruce Southerland - City of Cape Town

supply. The proponents state that only a 10 reactor PBMR site is viable, making this a barrier to community ownership of basic resources. RE can be developed to include community ownership and responsibility, resulting in local economic empowerment. This would boost the quality of life for communities, both from an energy security point of view, as well as enabling them to become Independent Power Producers.

9) The rationale for the PBMR programme requires high volumes of export yet there is little prospect of an export market. Internationally respected analysts have shown that the worldwide market for nuclear power grew at less than 1% per annum over the last 10 years. The market for RE is growing in leaps and bounds between 25% and 45% per annum. The export benefits are obvious. Benefits claimed by PBMR Ltd are also based on the assumption that other companies will relocate their operations to South Africa, a highly unlikely scenario.

10) Any extension of the nuclear industry increases risks of nuclear weapons proliferation, particularly when involving international movement of materials. Spent fuel will contain weapons grade uranium, which is simple to extract. The documentation provided by the proponents shows that all that would be needed would be a crushing device, and boiling it in nitric acid to retrieve the uranium. Enriched uranium is already being traded illegally.

5) Transport of materials is also problematic. Just for the proposed 10 PBMRs (as opposed to the planned production of 216 units), we would see a vehicle carrying radioactive materials about every second day, and about 7 carrying chemicals every working day, for 40 years. This could grow to 9 radioactive, and 145 chemical trucks, every day at full production. The container carrying the enriched uranium would cause a catastrophic radioactive incident if it fell more than 9 meters into water. The impact will last for generations.

6) Renewable Energy (RE) provides more jobs, on average at least 4 times more than nuclear, and requires less imported expertise. Currently, the only local content of the proposed reactor would be some steel tanks, and some construction work. RE, such as wind, is already at about 60% local content, and increasing. In the USA, wind is already cheaper than coal, especially when the health impacts are included. In addition to wind there are many other renewable energy options. These include wave (a few kilometers of coastline could supply Cape Town), photovoltaic, solar thermal, bio-mass, micro-hydro, etc. A mix of these technologies can easily provide all of the energy requirements for South Africa.

7) Energy Efficiency (EE) provides/ releases more capacity for serving new customers while cutting overall costs. Estimates are that we could easily save up to 30% of current capacity (about 12 900 MW, about 100 PBMR's) and utilise it for development, be it for communities or for business. This will also allow us time to grow our local RE manufacturing capabilities.

8) Nuclear power is capital intensive (see above), expensive to site (according to the proponents, siting the proposed nuclear reactor at currently non-licensed sites would render the project unviable) and uneconomic in sizes suitable for community and small business energy

Furthermore Nuclear power stations and spent fuel facilities are vulnerable to terrorist attack. Eskom's plans for the PBMR do not include secondary containment, i.e. a safe and solid building to minimise the possibility of, for example, aircraft impact. It must also be remembered that the fuel is graphite based, which is a particular fire risk. The massive radiation impacts from Chernobyl were spread mainly due to a graphite fire, which is difficult to extinguish.

Earthlife Africa's "Nuclear Energy Costs the Earth" Campaign against a nuclear development path for South Africa seeks to support safe, clean and sustainable alternatives

Contact: Muna Lakhani – Earthlife Africa
Tel: 082-416-9160
email: muna@iafrica.com
or visit www.earthlife.org.za

Our actions today mould the future....

The "Bottom Line" is that there will be no future for future generations unless we act together.

Holcim South Africa is committed to promoting the principles of Sustainable Development.

The very nature of our industry forces us to limit adverse effects on the environment and our track record demonstrates our ongoing focus on improved environmental performance.

With an established rehabilitation programme at all mining operations and a recent R320 million investment at the Dudfield cement plant, our manufacturing facilities are being transformed into the most environmentally friendly in Southern Africa. Our recently introduced low CO_2 cement products further emphasise our commitment to sustainable development.

Holcim

Customer Service: 0860 141 14
www.holcim.co.z
the building sit

NUISANCE

In environmental terms, a nuisance is something that causes irritation and distress (e.g. a bad smell, or a loud or piercing noise, or a continuously barking dog) but not necessarily to the degree that it contravenes laws or regulations. There are municipal by-laws which control nuisances such as burning rubbish in back gardens, playing portable radios loudly in public places, and barking dogs but these are often quite difficult to enforce in practice. Once again, what is a nuisance to one person may be perfectly acceptable to another. It is difficult to quantify nuisance. The best way to manage nuisance is for people to discuss actions that cause each other irritation and to try resolving differences of opinion to everyone's satisfaction. This is not easy and not everyone will be satisfied as different people have different levels of nuisance tolerance. However, communities must try and reach a consensus where most people are reasonably satisfied.

Key Associated Topics:

Air Quality; Alien Invasive Plants; Environmental Degradation; Landfill; Methane; Noise Pollution; Photochemical Smog; Pollution.

Associated Organisations:

Imbewu Enviro; Ken Smith Environmentalists; Icando; WESSA–Friends Groups; WESSA.

OPENCAST MINING

Opencast mining is where, instead of digging a hole and going underground to mine, the topsoil is stripped away from the surface and the materials are mined from an open pit or trench in a layered fashion. Legislation requires that opencast mines must be rehabilitated after mining has been completed so that the land can be re-used responsibly once the minerals have been extracted. Usually, when open cast mining is undertaken, top soil is safely stored and protected so that when it is replaced, it still has its stock of seed to help replace lost vegetation for stabilisation and aesthetic reasons. A number of mining companies in South Africa have so successfully rehabilitated their opencast mines that the land can once again be used for high grade farming purposes. If mining plans are co-ordinated with rehabilitation proposals it is possible to plan for contoured land, which fits in with the surroundings. Landscape architects who are able to analyse landscape and form and guide mining plans for the long term often carry out the work of rehabilitation planning and aesthetic design.

Key Associated Topics:

Mineral Resources; Non-renewable Resources; Rehabilitation of Land; Sustainable Development Section: – The Mining Industry.

Associated Organisations:

Dept Minerals & Energy; GeoPrecision Services; Holcim(South Africa); Kumba Resources; Mining Review Africa; PPC Cement; Chamber of Mines; Department of Minerals & Energy; Digby Wells; Wales Environmental Partnerships.

ORGANIC FARMING

Photo: Abalimi Bezekhaya

Organic farming, understood most easily when compared to chemical farming, began during World Wars I and II, as soil fertility declined and chemical by-products became plant food. The Organic Farming Movement, driven by civil society, understands all "life bits" as the holistic result of a wise created whole. The Chemical Farming Movement, centrally controlled, is highly profitable, typically supporting huge high-tech single-crop farms. Nature is skilfully manipulated as an input-output Bio-Machine, deconstructed into chemicals, DNA and genetic codes. Genetically Engineered (GE) Super Plants, bearing unknown side effects, are now common. Super-Food derived without soil, sun, seed or mating, produced directly from cloned bio-matter and minerals are its future. Organic farmers on the other hand, prefer to gradually enhance, not replace, what already works naturally. Commonly, small organic farms grow a variety of slowly adapted crops and animals, using many hands and no chemicals, for modest profit. The recognised benefits of organic farming are sustainable soil fertility and renewable biodiversity. Emerging additional benefits include: the use of organic food as medicine – with remarkable immune-enhancing properties; organic land-work – as an effective wide-range therapy, and as a catalyst for community building economies where none are hungry; organic food is also found to be nutritionally superior – less feeds more. It is important to recognise two new emerging adaptive movements, which may offer fresh initiatives and benefits:

◆ Pure-Business Organic Farming: grows mega-crops organically, supplying the well-off.
◆ Adaptive Chemical Farming: profits by combining organic and chemical approaches.

The choices we make determine our collective future.

Guest Author: Robert Small ~ Abalimi Bezekhaya and The Cape Flats Tree Project

Key Associated Topics:

Agriculture; Biodiversity; Composting; Eco-efficiency; Eco-labelling; Ecological Intelligence; GAIA; Medicine and Health; Organic Foods; Permaculture; Soil; Stewardship; Trench Gardening; Urban Agriculture.

Associated Organisations:

Abalimi Bezekhaya; Allganix Holdings; Ecolink; Food Gardens for Africa; Food and Trees for Africa; Master Farmer Programme; Permacore.

ORGANIC FOOD
Guest Essay by Thys Strydom ~ Allganix Holdings

Probably the most significant choice any individual can make to enhance both their own as well as the environment's health is to eat certified organic food. There are two main reasons for this, namely the vanishing nutrients and the toxic trail within non-organic foods.

The widespread use and developed 'dependency' on chemicals in agriculture since the 1940s has increased so much that as many as 16 different chemicals might be applied to a single crop during one growing season! It has been shown in a number of government studies in the U.S. that the residues of these chemicals are still present on food when they reach your plate. As many as 500 different synthetic chemicals can be found in the average North American's blood and they will eat a total of 2 pounds of toxic chemicals in the form of trace residues found on food each and every year. Children are even more at risk due to their small bodies and fast metabolisms. Today babies are born with pesticide residues in their bodies and are being fed on breast milk that contains pesticide residues. This has led to the dramatic increase of previously rare diseases such as cancer, liver disease, kidney disease, Parkinson's disease, birth defects and behavioral disorders that all have their roots in chemical exposures. The Environmental Protection Agency (EPA) concluded, after taking 29 environmental problems under their wing, that only worker exposure to chemicals and indoor radon exposure posed greater risks for cancer than the pesticide residues found on our food. They also rate pesticide exposure as a "high risk" for non-cancer disorders and diseases. Claims regarding the safety of any individual chemical will not usually take into account the so-called "chemical cocktail" effect whereby the accumulative effect of many different ingested chemicals together may have a very different effect (on plants, animals and humans) than that originally intended or than can be possibly envisaged.

Unfortunately chemical pesticides are not partial to whom and what they kill. Beneficial insects that eat crop destroying insects, insect eating birds and fish as well as beneficial bacteria, insects and worms that live in the soil are also being killed. This leads to a depleted soil and elimination of a living vibrant medium that allows strong healthy plants to grow and flourish. Plants that grow in such a depleted soil have a low resistance to pests and diseases. The 1992 Earth Summit stated that the U.S. has the worst soil on the planet. 85% of the soil has been depleted to the extent that it can no longer nourish healthy plants, which has led to the problem of vanishing nutrients within plants (and animals) and thus a marked reduction in the nutrient content of our food.

Organic farming, on the other hand, focuses on revitalizing and enriching the soil. This results in healthy plants that are disease and pest resistant as well as rich in the vital nutrients necessary for our health. The SOEL Survey, Feb 2003, shows that South Africa has 45 000 hectares of organic certified land compared to the U.S.'s 950 000 hectares and the U.K.'s 679 631 hectares.

Research at Thurman State University, Florida, found that organically grown oranges contained up to 30% more vitamin C than conventional chemically grown oranges. (*Science Daily Magazine*, June 2002) The Doctor's Data Study (*Journal of Applied Nutrition*, 1993) suggests further significant differences between organic and conventionally grown food. It found that organic pears, apples, potatoes and wheat had on average over 90% more mineral content than their respective conventional chemically treated counterparts. It was also found that organic crops contain substantially higher levels of several nutrients e.g. 27% more vitamin C, 21.5% more iron, 29.3% more magnesium and 13.6% more phosphorous.

By choosing to eat and use certified organic products we help to establish a sustainable chain of supply that does not only benefit our health and well being immensely, but also that of the environment. We help to restore farms and their surrounding environment to thriving ecosystems. We help to ensure the quality of groundwater, streams, rivers, estuaries and coastal areas. We consume foods that truly nourish and sustain us.

Photo: Copyright South African Tourism

Key Associated Topics:

Eco-labelling; Ecological Intelligence; Food Gardens for Africa; GAIA; Medicine and Health; Organic Farming; Permaculture; Steward-ship; Trench Gardening; Urban Agriculture.

Associated Organisations:

Abalimi Bezekhaya; Allganix Holdings; Ecolink; Food Gardens Foundation; Food and Trees for Africa; Master Farmer Programme; Permacore.

OXYGEN

Oxygen is a colourless, odourless, tasteless gas that is essential to life on earth. Its most important compound is water (H_2O). The name "oxygen" comes from the Greek oxys "sharp acid" and -genes "born" meaning "acid former". The occurrence, by weight, of oxygen in the atmosphere is 23%, in seawater 85.5%, and in the earth's crust 46.6%. It is the most abundant of all elements in the earth's crust, occurring in rock, water and air. In the lower atmosphere, oxygen is found as O_2 but in the upper atmosphere it appears as ozone (O_3). During respiration, animals and lower plants take oxygen from the atmosphere and return carbon dioxide (CO_2) whereas using photosynthesis, green plants assimilate carbon dioxide in the presence of sunlight and produce oxygen. Almost all free oxygen present in the atmosphere is the result of photosynthesis.

Key Associated Topics:

Air Quality; Atmosphere; Dissolved Oxygen Content; Eutrophication.

Associated Organisations:

South African Institute of Ecologists and Environmental Scientists.

OZONE

The ozone layer occurs in the upper atmosphere (stratosphere) surrounding the earth. It protects life on earth by absorbing most of the harmful ultra-violet rays (such as UV-B radiation) coming from the sun before they reach the earth. This layer is therefore very important in protecting all living things from UV-B radiation. This radiation can be very damaging to plants and animals as it suppresses the growth and functioning of the immune system. UV-B radiation is also a major cause of skin cancer and so any destruction of the ozone layer increases the health risk for humans. The layer is destroyed by ozone depleting substances (ODS), which are chemicals such as CFCs (chlorofluorocarbons) and halons, used in fridges, aerosols, fire extinguishers and released through other industrial processes. Whilst ozone in the stratosphere performs a useful function, ozone in the lower atmosphere (troposphere) is harmful to living things at high concentrations and is thus a pollutant. Tropospheric ozone pollution is caused by the reaction of volatile organic compounds (VOCs) with sunlight, causing a photochemical reaction. Ozone is an unstable form of oxygen (the molecule contains 3 oxygen atoms) and is found in the atmosphere. The ozone in the stratosphere is more stable than tropospheric ozone and forms the so-called "ozone layer" between roughly 15 and 50 kms above the earth. The chemistry of the atmosphere is still not fully understood but it is known that the stratospheric ozone layer protects life on earth from the harmful effects of short wave (UV) radiation.

Key Associated Topics:

Aerosol; Air Quality; Atmospheric Pollution; Biosphere; Montreal Protocol; Ultra-violet Light.

Associated Organisations:

Aerosol Manufacturers' Association; Amphibion Technologies.

PARTICULATES

Particulates is the term used to describe either particles of solid matter or droplets of liquid that are small or light enough to remain suspended in the atmosphere for short periods of time. The solid particles would be made up of dust, soil, soot, ash, asbestos, lead nitrate and sulphate salts and the liquid drops would include sulphuric acid, PCBs, dioxins and pesticides. Those particulates that are smaller than 10 microns (1 micron = one millionth of a metre) are small enough to reach the lower parts of the human lungs and can contribute to lung and respiratory disease. Control of particulates is an important aspect of any air quality management programme and it is important that particulates are measured and controlled to prevent them causing health problems to communities. Coal is still a major source of cooking fuel and heating in South Africa, and, combined with the fact that much of the coal burnt is low quality with a high ash content, means that particulates are a specific problem in air quality management.

Key Associated Topics:

Aerosol; Air Quality; Atmospheric Pollution; Climate Change; Environmental Health; Fossil Fuels; Pesticides; Incineration; Petrochemicals; Photochemical Smog; Pollution; Sulphur Dioxide; Transboundary or Transfrontier Pollution.

Associated Organisations:

Dial Environmental Services; Institute of Waste Management; South African Institute of Mining and Metallurgy; National Clean Air Association.

PERMACULTURE

The concept of permaculture was developed in the 1970s by two Australians, Bill Mollison and David Holmgrem. They were concerned about soil, water and air pollution from agricultural and industrial systems, loss of plant and animal species, reduction of natural non-renewable resources and an insensitive and destructive economic system. By thinking about the consequences of humankind's actions, combining this with old wisdom, skills and knowledge of plant, animal and social systems and adding some new ideas, they came up with permaculture.

One of the earliest definitions of permaculture comes from Mollison and Slay's *Introduction to Permaculture*, "...Permaculture is about designing sustainable human settlements. It is a philosophy and an approach to land use which weaves together microclimate, annual and perennial plants, animals, soils, water management, and human needs into intricately connected, productive communities..."

The main features of permaculture can be summarised as:
◆ It is a system for creating sustainable human settlements by integrating ecology and design;

◆ It uses natural systems as a basic model and works with nature to design sustainable environments, which provide for basic human needs and the social and economic infrastructure that support them;

◆ It synthesises modern science, and traditional knowledge, both of which are applicable to rural and urban situations;

◆ It encourages the idea that people must become a conscious part of the solutions to the many problems at local and global level.

Key Associated Topics:

Agriculture; Biodiversity; Composting; Eco-efficiency; Ecological Intelligence; GAIA; Organic Farming; Organic Foods; Soil; Stewardship; Urban Agriculture.

Associated Organisations:

Abalimi Bezekhaya; Ecolink; Food Gardens Foundation; Food and Trees for Africa; Permacore.

PESTICIDE TREADMILL

The pesticide treadmill is a term indicating a situation in which it becomes necessary for a farmer to continue using pesticides regularly because they have become an indispensable part of an agricultural cycle.

This can occur if pesticides are used on crops or animal herds such that other natural remedies are no longer effective. The farmer then has no option but to use the pesticides year after year in each agricultural cycle.

A more severe aspect of the pesticide treadmill is when it escalates. It can happen that the effective elimination of one target insect pest allows other insect pests to thrive, resulting in the farmer having to use other insecticides to eliminate the new pest problem. This can lead to more pests and the need for yet more insecticide types to deal with the additional problems. In starting to use any agricultural insecticide, the farmer should be very aware of the danger of unintentionally ending up in a treadmill situation, which becomes self-perpetuating.

Guest Author: Dr Kelvin Kemm

Key Associated Topics:

Agriculture; Alien Invasive Plants; Biological Pest Control; Carcinogen; Entomology; Groundwater; Integrated Pest Control; Land Degradation; Organic Farming; Pesticides; POPS & POPS Convention; Soil; Water Quality and Availability.

Associated Organisations:

Agricultural Research Council; EWT – The Poison Working Group; New Africa Skills Development; Plaaskem.

Illustration: Lennie Sak - Earthyear Magazine

PESTICIDES

Pesticides can be categorised as a diverse group of chemicals that kill insects. Domestic pesticides are used indoors to kill small clusters or individual insects, whereas agricultural pesticides are used on a large scale on crops, or livestock.

Domestic pesticides have been an aspect of life for centuries in the form of 'old wives remedies.' For example, in South Africa it has been known for many years that putting Khakibos on carpets drives out fleas, and also that filling a thatch house with smoke rids the house and thatch of many insects.

However, in more modern times a variety of aerosols, powders, liquids and smoke pellets have been available in supermarkets to attack a wide spectrum of insect pests found in and around the home. Some of these chemicals are deceptively benign, and cases are known of children – who have used the spray in large quantities in a confined space – who have suffered headaches, swollen eyes, constricted throats or even quite serious poisoning.

In the case of agricultural pesticides, they have been of great benefit to agriculture in that they have enabled crops to be grown in areas that would normally have been impossible due to resident problematic insects. They have controlled transient insect plagues before whole crops are destroyed, and they have enabled the authorities to tackle large-scale problem pests such as the major locust swarms of earlier eras that wiped out hundreds of kilometres of crops. Insecticides are used on animal herds to tackle diseases and, as a preventative measure, to prevent parasites from infesting animals, which could lead to sickness or reduced market value of the herd. However, agricultural pesticides need to be used with skill and foresight, which has not always been the case. Incorrect use or overuse can eliminate beneficial insects as well as the target insects. Furthermore, pesticides can migrate by wind or water to areas that they were not intended to reach, thereby causing unintended damage to insect ecological systems which, in turn, can have a detrimental effect on insect pollination and other essential agricultural systems.

If pesticides are used extensively and regularly there is a danger of falling into a pesticide treadmill situation in which continued regular use becomes a necessity. As agricultural circumstances change, research and development continues to develop more advanced insecticides that have refined properties such as being more target specific.

Guest Author: Dr Kelvin Kemm

Key Associated Topics:

Agriculture; Alien Invasive Plants; Biological Pest Control; Carcinogen; Entomology; Groundwater; Integrated Pest Control; Land Degradation; Organic Farming; Pesticide Treadmill; POPS & POPS Convention; Soil; Water Quality and Availability.

Associated Organisations:

Agricultural Research Council; EWT – The Poison Working Group; New Africa Skills Development; Plaaskem.

PETROCHEMICALS

The petrochemicals industry is broadly defined as that industrial activity which uses petroleum or natural gas as a source of raw materials, and whose products are neither fuels nor fertiliser.

The petrochemicals industry begins with oil refineries or extracting plants built to remove ethane and higher hydrocarbons from gas streams. In the case of South Africa, SASOL is a world leader in extracting a range of petrochemicals from coal. The original coal gasification process was pioneered by SASOL because one of the few raw materials that South Africa does not possess is oil.

The petrochemicals have a range of complex names for groups such as Alkenes, Olefins and Aromatics. Individual chemicals form a long list and have names such as ethylbenzene, formaldehyde, dichloroethane, toluene and many more.

Many of the petrochemicals are then further processed to manufacture the everyday items that we see around us such as plastic, lipstick, polystyrene and nail varnish remover.

Guest Author: Dr Kelvin Kemm

Key Associated Topics:

Air Quality; Carbon; Carbon Monoxide; Catalytic Converter; Cleaner Production; Dioxins; Eco-Taxes; Energy; Fossil Fuels; Non-Renewable Resources; Photochemical Smog; Plastics Volatile Organic Compounds.

Associated Organisations:

Chemical and Allied Industries' Association; Plastics Federation of SA.

PH SCALE

The pH scale is a measure of acidity or alkalinity. It is a logarithmically derived scale, which is presented as a scale from 1 to 14. The number 1 represents highly acidic while the number 14 represents highly alkaline. The number 7 is neutrality. A domestic swimming pool is typically maintained at a pH of about 7,3.

Guest Author: Dr Kelvin Kemm

Key Associated Topics:

Air Quality; Atmospheric Pollution; Contaminated Land; Freshwater Ecology; Groundwater; Water Quality & Availability.

Associated Organisations:

Chemical and Allied Industries' Association.

PHOTOCHEMICAL SMOG

Photochemical smog is composed of a number of chemicals, mainly ozone, aldehydes, peroxyacetylnitrates (PANs) and nitric acid. These substances are formed through the action of strong sunlight upon a mixture of hydrocarbons and nitrogen oxides released into the air from vehicles and industrial processes. The severity of the smog is linked to the atmospheric concentrations of ozone at ground level. In the past, the city of Los Angeles in the USA has experienced serious problems with photochemical smog. To combat this, the city introduced programmes to restrict the number of vehicles travelling on freeways when climatic conditions develop that encourage the formation of photochemical smog. South Africa has not generally seen the development of severe photochemical smog though it does, in certain areas such as Cape Town, experience the effects of smog caused by a combination of inversions and excessive pollution during the winter.

Key Associated Topics:

Air Quality; Atmospheric Pollution; Carbon Monoxide; Catalytic Converter; Environmental Health; Fossil Fuels; Nitrogen; Ozone; Petrochemicals; Pollution; Sulphur Dioxide; VOCs.

Associated Organisations:

Cape Town – Environmental Management Department; Ceatec; Groundwork; Holgate; National Air Pollution Services; SI Analytics.

PIONEER SPECIES

Pioneer species are the first species to colonise a new area (for example, a new island that has emerged from the sea or an area that has been disturbed by large-scale activity such as a volcanic eruption or man's clearance of a piece of land for development purposes). These species include wind-dispersed microbes, mosses and lichens, which grow close to the ground and establish quickly over large areas. Their basic characteristics include rapid growth, the production of copious, small, easily dispersed seed and the ability to germinate and establish themselves quickly and easily on open sites. These species then trigger the process of soil formation on bare rock by trapping wind-blown soil particles and by secreting acids, which eventually break down rock. This chemical breakdown is supplemented by physical weathering. When the plants die, they decay and their remains help to form new soil. Plants that are commonly recognised as weeds are often a form of pioneer species establishing on disturbed soils or newly cleared areas.

Key Associated Topics:

Indigenous Plants; Land Degradation; Soil; Rehabilitation of Land.

Associated Organisations:

Botanical Society of South Africa; Coastec; National Botanical Institute; School of Botany & Zoology – University of Natal; Department of Botany – Stellenbosch University.

PLASTICS

Plastics are synthetic substances, which can be moulded and formed when soft and then set. There are two types of plastics: 1) Thermoplastics , which can be softened by heating and will harden again on cooling (this softening and hardening can be repeated as often as required). 2) Thermosets, which are plastic materials that, once set, cannot be softened again.

Photo: *Earthyear Magazine*

Education for a cleaner South Africa

An industry commitment to eradicating littering

In 1997 the Plastics Federation of South Africa embarked on a campaign to help address the problem of littering in our country, through the Plastics Environmental Initiative.

Education programmes/Environmental exhibitions

Since the root of the problem is not waste itself, but the attitude towards the disposal of waste, the emphasis has been on changing the mindset of the population towards one of environmental care and consideration.

Funds and efforts have been concentrated on informing the public through clean-up campaigns, education programmes, environmental exhibitions and workshops and the production of educational material for schools. By working in tandem with various like-minded bodies the efforts of the Plastics Environmental Initiative have been maximised.

Clean-up Campaigns

We are committed to continuing our efforts to educate all South Africans to contribute towards a cleaner South Africa.

Recycling, reusing, recovering, reducing, removing

Plastics Federation of South Africa
18 Gazelle Avenue, Corporate Park,
Old Pretoria Road, Midrand
Private Bag X68, Halfway House, 1685
Tel (011) 314-4021
Fax (011) 314-3764
e-mail enquiries@plasfed.co.za

A Plastics e nvironmental Initiative

Photo: Plastics Federation

President Thabo Mbeki with Ministers Ronnie Kasrils, Valli Moosa and Geraldine Fraser led the 2001 National Water Week clean-up campaign supported by the Plastics Enviromark team.

Plastics are usually synthesised from petrochemicals and natural gas, although in South Africa most plastics originate from coal. Plastics are recyclable but tend to attract negative attention because of their presence in litter – particularly in the form of plastic bags. Plastics are identified by a recycle triangle with a number in the centre to help classify them for recycling. The seven main groupings are as follows:

1 Polyethylene terephthalate or PET
   (used for cool drink bottles);

2 High Density Polyethylene or PE-HD
   (buckets, crates, motor oil containers);

3 Polyvinyl chloride or PVC
   (clear trays for foods, pipes and gutters);

4 Low Density polyethylene or PE-LD
   (frozen vegetable bags, garbage bags);

5 Polypropylene or PP (bottle caps, car battery cases);

6 Polystyrene or PS (yoghurt containers, takeaway food tubs, disposable cups)

7 Other
   (usually covering specialised engineering plastics).

Plastics are very useful in everyday life because they are light, easy to clean, durable and relatively inexpensive. If used wisely, recycled and disposed of correctly, they can therefore make a valuable contribution to society. However, recent research has suggested that there may be chemicals such as Phthalates, which leach out of the plastic and have an adverse effect on health.

Key Associated Topics:

Petrochemicals; Pollution; Recycling; Responsible Care; Waste Management.

Associated Organisations:

Chemical and Allied Industries' Association; Plastics Federation of SA; Recycling Forum (The National).

PLUTONIUM

Plutonium is a heavy metal not found in nature because it undergoes radioactive decay over a period of time, and millions of years ago all the Plutonium in the Earth's crust had already disappeared due to the steady radioactive decay.

It is however possible to make Plutonium by bombarding Uranium with neutrons in a nuclear reactor. Of commercial importance is the fact that Plutonium is produced as a natural consequence of Uranium being used as fuel in a conventional nuclear reactor, of which there are hundreds in the World. Plutonium and Uranium have a significant property in common where, when hit just the right way by a neutron, they undergo splitting or fission. This means that Plutonium can be used as a nuclear reactor fuel. As a Uranium-fuelled nuclear reactor burns up its fuel, it produces some Plutonium as a natural consequence of the nuclear chain reaction. This Plutonium can be extracted and used as fuel in a Plutonium-reactor. A major technological problem is that the Plutonium produced is all mixed in with unused Uranium, and also with all the radioactive by-products of the nuclear process, so it is extremely difficult to separate it from the highly radioactive spent fuel. Only a very few countries possess the technology to do this.

Another issue that has made Plutonium rather notorious is that it is also toxic like any other heavy metal, including lead (this has nothing to do with its radioactivity). As far as the radioactivity is concerned, it is not very dangerous, but if Plutonium dust becomes inhaled into one's lungs it can induce cancer, because a dust particle can sit for some time against the lung tissue emitting short-range radiation. It is therefore a potential carcinogenic hazard for Plutonium workers if it is not handled correctly. Plutonium has been incorrectly labelled as the 'most toxic substance known to man'; this is far from the truth.

Another issue that has brought Plutonium to public attention is the Plutonium shipments from France to Japan that pass the Cape. This Plutonium is taken to Japan to fuel Japanese Plutonium-reactors. The cycle is that Japan sends some of its spent Uranium fuel to France where France extracts the Japanese Plutonium and then sends it back to Japan. In many quarters, considerable public fear was created relating to the possibility of the ship having an accident and so releasing some of the Plutonium.

Plutonium and Uranium are the only two atoms that will fission for the purposes of the production of commercial nuclear power.

Guest Author: Dr Kelvin Kemm

Key Associated Topics:

Basel Convention; Carcinogen; Chernobyl; Energy; Hazardous Wastes; Nuclear Energy – the Counter-Debate; Precautionary Principle; Radiation; Radioactive Waste.

Associated Organisations:

Eskom; Green & Gold Forum; Koeberg Nuclear Power Station; Stratek.

POLLUTER PAYS PRINCIPLE

This is a basic economic principle (currently being adopted in legislation by many countries) that requires the producers or generators of pollution to pay for all of the costs of avoiding pollution or of cleaning up or remedying its effects. Thus any pollution generated by a process must be paid for from within the cost structure of production. In broad terms, this means that companies must either stop producing polluting emissions and effluents or ultimately stop production. The new National Water Act is tightening up controls on water pricing and effluent disposal. The proposals for the re-writing of the current air pollution control legislation will have a similar air quality focus, meaning that the Polluter Pays Principle will have much more practical application in future. The concept has only been introduced recently, and many industries are not in a position to make the radical changes necessary. This is because of the lax pollution controls of the past, which allowed equipment that had limited pollution control mechanisms to be designed and installed. In South Africa, the principle is being applied through the *White Paper on Environmental Management Policy for South Africa* and through the principles embodied in the National Environmental Management Act. An interesting case study on the Polluter Pays Principle can be found at: www.arava.org

Key Associated Topics:

Air Quality; Atmospheric Pollution; Carbon Tax; Contaminated Land; Eco-catastrophe; Eco-taxes; Environmental Law; Environmental Liability for Remediation; Environmental Risk and Liability; Exxon Valdez Disaster; Hazardous Wastes; Integrated Pollution and Waste Management; NEMA; Precautionary Principle; Transboundary or Transfrontier Pollution.

Associated Organisations:

Edward Nathan and Friedland; Envirocover; Green Gain Consulting; Imbewu Enviro; Jarrod Ball Associates; Mark Dittke; Winstanley Smith & Cullinan.

POLLUTION

Pollution comes in many forms and types. In its broadest sense, pollution can be described as the result of the release to air, water or soil from any process or of any substance, which is capable of causing harm to man or other living organisms supported by the environment. Long lasting pollution can also affect the physical environment by changing its characteristics through speeding up or slowing down natural processes. One other way of describing pollution is that it is waste that has been inadequately managed or controlled and which results in unnecessary damage.

An example of "air pollution" is acid rain or "acid deposition", which occurs when polluting sulphur dioxide and nitrogen oxides released into the atmosphere from industries, towns and cities are transported by prevailing winds to form droplets or particles that fall back to earth. The droplets or particles fall in either a wet or dry form. The wet form falls as acidic rain, snow, fog or cloud vapour and the dry form as acidic particles. Rain is normally slightly acidic (around pH 5.0 – 5.6) because it contains carbonic acid from chemical reactions in the atmosphere.

However, industrial activity can push the acidity up to pH 3 (same pH as vinegar) and even up as high as 2.3 (same pH as lemon juice), which can be about 1 000 times higher than normal rainfall. Excessive acidic rain can cause substantial damage to plants and soil.

Photo: Bruce Sutherland - City of Cape Town

An example of water and land pollution is the toxic solid or liquid waste that is dumped on land, which can filter into the soil, percolate down and contaminate groundwater. Pollution from radioactive waste (e.g. Chernobyl) can pollute vast areas in a different way, making them uninhabitable to man and beast because of the high levels of radiation, which ultimately cause death. As a basic principle, pollution should be minimised as far as possible.

Key Associated Topics:

Air Quality; Atmospheric Pollution; Carbon Dioxide; Carbon Monoxide; Carbon Tax; Carcinogen; Chernobyl; Contaminated Land; Dilution; Dioxins; Eco-catastrophe; Eco-taxes; Environmental Degradation; Environmental Health; Environmental Law; Environmental Liability for Remediation; Environmental Risk and Liability; Exxon Valdez Disaster; Global Warming; Groundwater; Hazardous Wastes; Integrated Pollution and Waste Management; Land Degradation; Leachate; Montreal Protocol; NEMA; Noise Pollution; Nuclear Energy – The Counter Debate; Particulates; Pesticides; Photochemical Smog; Plutonium; Polluter Pays Principle; POPS & POPS Convention; Precautionary Principle; Radiation; Radioactive Waste; Scrubber; Solvents; Sulphur Dioxide; Transboundary or Transfrontier Pollution; VOCs; Waste Management; Water Quality.

Associated Organisations:

Bateleurs – Flying for the Environment in Africa; City of Cape Town Solid Waste; Enviro Options; Environmental Risk Management; Enviroleg; Enviroserv Waste Management; Fairest Cape Association; Gauteng Conservancy Association; IBA Environnement SA; Winstanley Smith & Cullinan.

POPS AND POPS CONVENTION

POPs stands for Persistent Organic Pollutants, which is a group of chemicals that have been targeted for elimination. These chemicals are all products of the industrial age and were initially used by the first world in industrial processes.

POPs are organic chemicals (this means carbon-based) and there are quite a number of them. The POPs chemicals consist of complex molecules and they generally take a long time to degrade; hence the term 'Persistent.'

In recent years there has been a move to have the use of 12 particular POPs banned. These 12 are composed of; eight pesticides: Aldrin, Endrin, Chlordane, Heptachlor, DDT, Mirex, Dieldrin and Toxaphere; two industrial chemicals: HCBs and PCBs; and two unintended by-product chemical families: Dioxins and Furans.

A POPs Convention was developed by the United Nations to have these 12 POPs banned. A problem, however, is that these chemicals are no longer used by the first world, but some are still used by the developing world. The UN is seeking a world ban. For example, a particular problem case is DDT. Malaria is a disease that has been eliminated from Europe and North America, but it still kills one child every minute in Africa. DDT has been used to control malaria mosquitoes but the POPs Convention seeks to ban it. At the POPs Convention meeting in Johannesburg in December 2000, some relief was allowed in the case of DDT for some countries to use it for malaria control. South Africa has resumed DDT use after malaria figures rose dramatically in recent years.

Wildlife and people are generally exposed to POPs through their food supplies, but workers and residents of communities near POPs sources can also be exposed through inhalation or skin contact. There is still considerable debate as to the health effects of POPs. It is claimed that they are carcinogenic, but this is disputed. It is also claimed that POPs can cause genetic defects. This has led the protagonists of the POPs Convention to argue that prevention is better than cure, so the POPs should be banned.

As a counter argument, developing world countries say that the first world used the chemicals to grow rich and become healthy and now they do not want the developing world to follow in their footsteps.

Another fear voiced about the POPs Convention is that if it is adopted, it will set a precedent and in future it will be easy for the first world just to ban more chemicals that may be beneficial to the developing world.

The POPs convention has resulted in significant confrontation, which seems set to continue for some time.

Guest Author: Dr Kelvin Kemm

Key Associated Topics:

Carcinogen; Dioxins; Environmental Health; Pesticides; Pollution; Transboundary or Transfrontier Pollution; Volatile Organic Compounds (VOCs).

Associated Organisations:

Chemical and Allied Industries' Association; Environmental Evaluation Unit – University of Cape Town.

POPULATION GROWTH AND SUSTAINABLE DEVELOPMENT

Guest Essay by Dr Eureta Rosenberg

ERA	TIME PERIOD	POPULATION
Beginning of Neolithic	–	10 million
	3000 years	20 million
5000 – 3000 b.c.e.	2000 years	50 million
3000 – 1400 b.c.e.	1600 years	100 million
1400 – 0 b.c.e.	1400 years	200 million
0 – 1200	1200 years	400 million
1200 – 1700	500 years	800 million
1700 – 1900	200 years	1.5 billion
1900 – 1960	60 years	3 billion
1960 – 2000	40 years	6.1 billion
2000 – 2028	28 years later	8 billion (UN *projections*)

In the year 2000 the human inhabitants of the Earth reached six billion. Now, in 2003, there are 6.1 billion of us. And not only is the world population increasing, the rate at which it is growing has also been increasing. The above table illustrates this, including the tremendous acceleration over the past 200 years. It forces us to ask the sustainability question: What will happen if the human population continues on its current growth path?

Impacts

Growing populations are faced with the harsh reality of limited natural resources. The issue of water supply is a good example to demonstrate that unrestrained population growth is not sustainable. Consider this:

1. Water, like other natural resources, is not evenly distributed around the globe. The countries described as 'developed' or 'industrialised' have in general more abundant sources of water, or the technology to use water more efficiently.

2. The supply of fresh water is essentially fixed. While technical means are being explored to increase the supply of fresh water (such as Desalination) their impact is likely to be limited.

3. We are already consuming close to the planet's limits. Worldwide, 54% of the annual available fresh water is already being used. This may seem to leave a lot to spare, but scientists have demonstrated that we need to leave a certain volume of water in rivers and other wetlands as an ecological 'reserve', in order to maintain their functional viability. When we use up this reserve, we destroy these ecosystems and reduce the overall available volume of water.

4. This level of use (54%) is based on unequal consumption: Around the world, some 1.1 billion people do not have access to fresh water, or consume less than the basic daily requirement of 50 litres.

Now consider what will happen with the projected population growth. It is estimated that by the year 2025 the increased number of people on earth would use up 70% of the available fresh water. This is not taking into account the needs of those people who currently do not get enough. If consumption increased everywhere to current developed country levels, we will be using 90% of all available fresh water. The effects on ecosystems would be devastating. Springs, rivers and other wetlands as well as underground water sources would run dry; lakes and estuaries where fish stocks breed would be irreversibly damaged; the list of impacts is long.

> A child born today in an industrialised country will consume more and pollute more in his or her lifetime than 30 to 50 children born in a developing country.

5. Population growth will also result in greater volumes of pollution. In developing countries, 90-95% of sewage and 70% of industrial waste are dumped into surface waters thus polluting the water supply. Water quality is also affected by chemical run-off from pesticides and fertilisers and acid rain from air pollution, requiring expensive, energy-intensive processes to clean it for human use. In a recent case in Brits, residents could not use council water because of pollution in the Hartebeespoort Dam. Pollution clearly decreases the volume (and increases the cost) of available water.

In 2000, 508 million people lived in 31 water-stressed or water-scarce countries. By 2050, this will have rocketed to 4.2 billion people living in countries that cannot meet the minimum requirement of 50 litres of water a day.

Water is not only a basic human need, without which we die. It is also the basis of health, food security and economic development. For individual families, lack of access to clean water is associated with unhygienic living conditions, already one of the biggest causes of deaths among infants. On a national and regional level, cash crops and other industries depend on water supplies. As water becomes scarcer, we see not only a decrease in the quality of life, but an increase in social conflict.

The same scenario will play out (and already does) for land and other non-renewable natural resources. These resources limit the number of people the earth can bear sustainably. This is why the rate at which the world population is growing, is such a serious ecological and social threat.

Demographics and trends

Just as the world's natural resources are unequally distributed, the world population is also unequally distributed. High population numbers are associated with those regions where natural resources are generally more limited. Here the population increase is also the fastest, the consumption per person the lowest, and the negative impacts of growth most acutely felt.

Photo: Guy Stubbs Photographer

At the current global population growth rate of 1.3%, there are 77 million more people living on this planet per year. Six countries alone are responsible for half of this growth: India (for 21%), China, Pakistan, Nigeria, Bangladesh and Indonesia.

There has been a general decline in fertility in countries described as 'developing' (to an average of just under 3 children per woman – half of the 1969 figures), and the figure is expected to decrease further to 2.17 by 2045-2050.

But despite this trend, most of the projected growth in the world population will take place in developing countries. By 2050, 85% of the world population will be living in developing countries. (The comparative figure for industrialised countries is 1.6 children per woman.) The 49 'least developed' countries will almost triple in size. This level of growth will almost certainly have devastating effects for their environment and inhabitants, with rippling impacts on their neighbours and other countries to which people may migrate.

One of the effects of population growth can be seen in cities. As rural environments become less able to sustain people, an estimated 160 000 rural dwellers move to cities every day. This results in sprawling, densely populated urban areas under great social, economic and environmental stress.

The effects on people range from social friction (such as crime and xenophobia) to health impacts. Problems like traffic congestion and pollution are common: Air pollution levels in many fast-growing cities far exceed World Health Organisation guidelines, and is said to cause ill health and death for millions of people each year.

City surroundings are depleted through the concentrated extraction of resources ranging from water to firewood; the conversion of farmland or wetlands for housing, roads and shopping centres; and the spill-over of pollution which is often worsened by local governments failing to provide the necessary facilities for the swelling numbers.

The local situation

Do we have a problem in South Africa? Our population growth rate is slowing down – from 2.1% in 1975-2000, to a predicted 0.2% for 2000-2015. This figure takes deaths due to HIV/AIDS into account. While HIV/AIDS slows the growth rate (it would probably have been 1.4% in 2010 without HIV/AIDS), the epidemic will probably not result in negative population growth (a smaller population). In 2000, our population was 43.3 million.

Is this a problem? Are there too many of us? For the answers, we need to look towards the environmental resources, which must sustain us. In each locality, we need to consider whether that particular environment is able to support the people living in it. And critically, we need to ask whether current needs are being met adequately.

Our apartheid history has much to do with how the population has been distributed, and associated environmental degradation. In former Bantustan areas like KwaMhlanga and Transkei, too many people were forced to live in resource-poor areas with limited capacity to support their artificially high numbers. The effects on both the environment and people's ability to sustain a livelihood were devastating. Looking for alternative livelihoods, many rural South Africans moved to cities, where the majority of us (56.9%) now live. Here informal settlements are expanding, often into sensitive or high-risk areas such as flood plains or waste dumps. Lacking resources, they contribute to environmental and health issues, e.g. pollution of ground water through untreated sewage. Local governments struggle to meet the demand for housing, water, sanitation, waste removal, transport and health services. Biodiversity is lost as remnants of indigenous vegetation are destroyed. Considering that the economic and ecological demands of our existing population are already straining the Earth's capacity, and that many basic human needs are not yet met, can we afford to continue growing in numbers?

Solutions

If we want to achieve a sustainable relationship between natural resources, development and human numbers, we need to consider the fact that many people still do not get a big enough slice of the cake, as well as the reality that the Earth's cake is of a limited size. As we saw from the water example, natural resources are essentially fixed, and taking strain under the demands of consumption and growing populations.

Yes, we can produce more food and we should distribute resources more fairly and efficiently around the globe. This, along with reducing over-consumption and discarding discriminatory economics, can alleviate a great deal of hunger and hardship. (See ECONOMICS.) Technological advances towards energy-efficient and resource-light production can reduce resource use and pollution, but these steps will not reverse the impact of the population explosion. Something must be done to slow down population growth – but what?

The answer is not as simple as "family planning" or reducing the number of pregnancies. We need to understand what will make family planning and fertility control possible and likely, from a social point of view, and how these factors can be addressed.

The greatest threat to sustainable development is consumption; consumption is linked to unsustainable production and over-consumption among the consumer classes in both the North and South, and to population growth, particularly in the South. Different camps fight about who has the greater environmental impact – the rich over-consumers or the poor population-growers. The fight may just defer responsibility, however, because BOTH over-consumption and over-population must be addressed, if we are to achieve ecological sustainability and social justice – and we cannot deal with them as separate issues.

Most agencies involved in population development advocate a multifaceted and integrated approach. They point out that high population growth rates are associated with poverty, environmental degradation, limited opportunities and unequal power relations. High fertility is still a feature of rural life in many areas, even when the rationale for having large families (such as needing many hands for harvesting) are no longer valid.

In these situations women often do not have the power to choose the size and spacing of their family. Addressing these vicious cycles requires an integrated approach, in which family planning goes hand in hand with increasing women's rights and their ability to exercise them, reducing poverty, protecting the environment and increasing livelihood options. These goals are all interrelated and cannot be separated from economic models and over-consumption. (See POVERTY ALLEVIATION).

Social development that empowers women and girls will support family planning. This would include:

◆ Cultural, legal, political and economic changes which acknowledge women's rights as equal citizens; such changes are frequently resisted in overt or subtle ways, but while the resistance is often said to be based on 'culture', it may reflect short-term interests rather than fundamental values.

◆ Legal and community protection against sexual harassment and violence.

◆ Education (in many parts of the world it is still unusual for girls to be educated).

◆ Establishing women's right to own land (many women farm without the ability to make decisions about the management of the land).

◆ Reproductive health care (UN agencies lack funds to provide fertility control to all the people requiring it).

◆ Adequate general health and social support, so that each child born can be treated with the necessary care and respect.

◆ Reducing social strife and conflict so that life is no longer treated as 'cheap' anywhere on Earth.

Key Associated Topics:

Biodiversity & the Sixth Extinction; Carrying Capacity; Communities; Indigenous Peoples; Poverty Alleviation; Urbanisation; Women and the Environment.

Associated Organisations:

Afrodev Associates; Art of Living; Sapler Population Trust.

Photo: Guy Stubbs Photographer

ADVANCING AFRIKATOURISM

Engen has committed itself to the development of tourism in this beautiful country and across the African continent. Over the years, we have lent our support to various projects that further the aims and spirit of nation building, with specific emphasis on job creation.

In particular, we have channelled our funds and energy into supporting Open Africa, an organisation established in 1995 with Nelson Mandela as its patron. Its aim is to take advantage of the natural synergy that exists between tourism, job creation

and conservation in Africa.

What has emerged from this process is Afrikatourism, a branding strategy entirely unique to Africa. It highlights

the diversity of the continent, together with its cultural and environmental splendour in order to promote tourism, under the premise that these experiences are exclusive to Africa. The manifestation of this vision is to link new and existing tourist attractions in a continuous network of Afrikatourism routes, from the Cape to Cairo.

To achieve this vision, a systemised method of establishing routes has been developed and is represented in its entirety on a special website. A first for Africa, the website allows potential visitors to peruse maps of the various routes online (www.africandream.org). This project already has 35 routes, covering 11623 kilometres in five countries and involving 83 towns and 813 establishments.

Overall, this accounts for 5988 full time jobs in season and 2429 part-time positions. The corporate, institutional, professional, individual and partner members who subscribe to the accomplishment of this vision, are known as Team Africa. The team presently has 2474 members and is growing rapidly on a daily basis.

The challenge from Engen's side is to integrate its involvement in tourism with its extensive retail network, by providing a variety of new services intended to benefit travellers and tour operators alike, whilst drawing attention to the sensitivity, attractiveness and value of our numerous cultural and environmental assets. In this way, tourism will be linked to hundreds of Engen sites, thereby adding value to the already outstanding service that accounts for our position of market leadership in South Africa.

Besides sponsoring the development of routes, Engen also publishes a series of tabloids called **Discovering SA**. In addition to informing travellers about what to see and do, they also provide a platform for service providers in the surrounding areas to advertise and display their products. In keeping with the fundamental principles of Afrikatourism, these publications provide a medium for communities to interact and engage with the tourist market. The series is available free of charge at all Engen 1-Stops, covering almost all the provinces. A helpline also exists for travellers to obtain information on tourist attractions, make holiday accommodation bookings and to summon emergency help in breakdown situations. The number to dial is (083) 123 2345.

Engen and Open Africa have also developed a series of **Fun Routes**

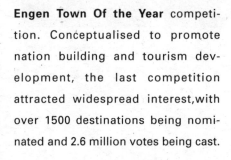

Guides, which are on sale at Engen 1-Stops around the country. The brochures are designed to show people, particularly children, the country's natural and cultural history as well as interesting sites on the main highways in South Africa, thereby promoting Afrikatourism across the country. Interesting information, puzzles, quizzes and landmarks along the road are included in each of the 24 Fun Routes. The routes run for 200 kilometres either side of the 1-Stop service stations, and are a practical example of how the Afrikatourism concept can be channelled into educational activities for consumers.

Another practical commitment to tourism development is the annual

Engen Town Of the Year competition. Conceptualised to promote nation building and tourism development, the last competition attracted widespread interest, with over 1500 destinations being nominated and 2.6 million votes being cast.

Engen also publishes an annual **Tide Table**, containing useful information about our coastline and its numerous marine reserves. This, together with the well known **Engen Road Atlas** – which itself has reached bestseller status – complements

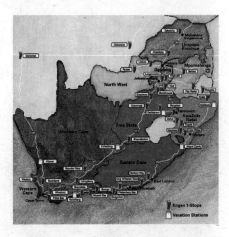

the afore-mentioned host of activities aimed at conservation-based tourism in South Africa.

International experience has conclusively shown that an investment in tourism holds many positive implications for job creation. Boosting the tourism sector in South Africa therefore has a significant multiplier effect for the country's economy. This is why Engen's investment in tourism will always remain a priority.

With us you are Number One | **ENGEN**

POVERTY ALLEVIATION THROUGH SUSTAINABLE DEVELOPMENT

Guest Essay by Dr Eureta Rosenberg

What is the most important thing we can do with land – set it aside for nature conservation, or invite poor people to build homes and grow food?

What is more important – reducing plastic bag pollution or protecting jobs in the plastics industry? Saving whales or saving people?

I hope you find these questions difficult, for they are non-choices, really. There is no need to choose between human suffering and the environment. And yet there is a common belief that we need to address poverty first, before we can think about the environment. Actually, looking after the environment and looking after destitute people go hand in hand – a point which the idea of sustainable development is meant to convey.

What will help the poor?

Not presuming to speak on behalf of poor people, I will restrict this article to the links between poverty alleviation and environmental care. My aim is to illustrate the folly of pitting these two concerns against each other.

> Approximately 18 million South Africans (45% of the population) live in poor households, which earn less than R352 per adult per month. Power relations in poor households are characterised by violence, fear, food insecurity, exploitative work and other pathologies. Three of every five South African children grow up in poor households. (Human Development Report for SA, UNDP, 2000).

1. Alleviate poverty through environmental protection

The poorer we are, the harder environmental issues hit us. Desperate for employment, poor people suffer the most unhealthy work environments. Without much choice as to where to live, they are more exposed to disasters like floods and fire, and to toxic dumps, polluted air and water.

As their quality of life declines and their health deteriorates, these environmental issues render them even less able to make a living.

Land degradation and the loss of biodiversity affect poor people most, as they often depend directly on natural resources (for firewood, food, building material). When environmental degradation destroys opportunities for development in one area, those with the financial means to do so can still move on to greener pastures. Being poor means having nowhere better to go.

It is therefore in poor people's interest that laws protecting environmental quality and safety are enforced, and that they have access to the law when their right to a healthy environment is violated. The common natural resources on which rural people depend should be protected by laws for sustainable use. Far from being 'anti-development', environmental policies can be used to protect the health and livelihoods of poor people, and increase their political and economic power.

But look – someone disagrees. There he is, collecting bait in the Knysna lagoon. His spade has once again been confiscated by the authorities, and he received another fine he cannot pay. He disagrees that environmental laws work for the poor. Hey lady, he says, this is my only means of survival, and Conservation is taking it away!

But he and millions of others in the same boat should not be in this position in the first place – having too few livelihood options. Relaxing the environmental laws will not solve their problem. It will only provide a short reprieve – and soon afterwards a resource may be destroyed forever, thus affecting the livelihoods of many others in the future.

2. Strengthening livelihood strategies and increasing options

Those with few livelihood options often feel forced to exhaust even the few resources to which they have access. Examples include overgrazing the land, hunting out the wildlife around ever-growing settlements, collecting muti plants to sell until none are left. This creates vicious cycles. When people can no longer survive on depleted land, they go to 'squat' in towns and cities, where the chances of a better life are also bleak.

Caring for natural resources so that they can be used indefinitely (sustainably) is an important way of increasing people's livelihood options. Care means both protection and restoration. If we protect the soil and increase its fertility, remove alien invasive plants and re-establish indigenous vegetation, we help people to make a living off the land – through farming, crafts, tourism – for generations to come. By protecting estuaries, where many marine species breed, we protect people's ability to make a living from the sea – sustainably. Renewing urban landscapes can help people grow food at home and work and relax in safe and pleasant surroundings – indefinitely.

A livelihoods-centred perspective informed the ANC's Reconstruction and Development Programme. The RDP puts people first, not profits. This is not the case with export-led and growth-oriented economic strategies, especially when these fail to take care of the ecosystems on which livelihoods are based. Current economics identify growth per capita (GNP) as development, ignoring the way in which growth often depletes natural and social capital in order to produce money capital.

For the poor, boosting livelihood strategies and options makes a greater difference than the profits of economic growth, which by and large fail to trickle down to them from the consumer class. Some growth strategies can actually harm poor people. The livelihoods of millions around the world are threatened by exploitative 'developments' that benefit those who are already powerful. Logging, the extraction of oil and minerals, big dams – all displace people from where they had been able to derive a livelihood, onto barren lands and into city slums and menial labour. India's Green Revolution was famous for increasing the profits of a minority of richer farmers while devastating those with more limited means or social status.

3. Basic incomes

An income does not always improve quality of life, but it can help. Economic growth in South Africa has failed to address the vast inequality in incomes, and Government recently responded by increasing grants to the poorest 20% of households. Social grants are thus supplementing (or substituting for) the trickling down of wealth through economic growth. A Basic Income Grant has been recommended, to alleviate hardship. This could also relieve the over-harvesting of natural resources in some areas.

Conservation of the natural environment (as opposed to using up its resources) is increasingly recognised as a source of income for both Government and communities. Nature-based tourism is now the fastest growing industry in the world travel business. Conserving nature on the Cape Peninsula yielded an income for South African National Parks of more than R15 million in visitor fees in 2001-2002. Communities in the North West and Limpopo Provinces who set aside land for conservation now generate income from eco-tourism; if managed well, this form of land-use can provide a sustainable income for many generations.

Two initiatives, which provide incomes through environmental restoration are Working for Water (Department of Water Affairs and Forestry) and Land Care (National Department of Agriculture). Working for Water removes alien invasive trees, which harm the environment. Land Care aims to prevent and reverse rural land degradation. Several recent conservation activities in National Parks also make use of Poverty Relief Funds. These initiatives all employ destitute people on a temporary basis. However, as a solution to poverty they unfortunately tend to lack sustainability, often encouraging people to give up existing livelihood initiatives.

More sustainable efforts to improve the land help people to secure their own livelihoods while generating an income. In examples from Uganda, low-input agriculture helps owners of small plots to farm with nature, using environmentally-friendly pesticides and fertilizers, which cost little. Not only do these farmers have food security, they can export their produce to new markets for organic produce, fetching a 20% higher price.

Securing subsistence is an important form of sustainable development for the poor – a point to consider in the current trend to advocate a move away from subsistence to commercial activities.

4. Improving basic services

Providing basic services is a huge task, because of the apartheid backlog as well as a rapidly growing population. Struggling to deliver, Government seeks to hand over the responsibility through various forms of privatisation.

But privatising common natural resources such as water and energy has proven to be to the disadvantage of the poor. Around the world, privatisation has pushed up costs of services, despite promises that it won't. The privatisation of Telkom saw increases in the price of local phone calls, while international calls became cheaper, to the advantage of wealthier customers.

However, providing services to the millions of still disadvantaged South Africans can present Government and service providers with a great opportunity. We could become a world leader in sustainable development, if we built new infrastructures in a resource-light, pollution-light manner and invested in public transport; grey-water recycling; water-less sewage; low-energy housing … to name a few. Such innovations also create new job opportunities.

5. Meaningful jobs

There was a time when economic growth meant more jobs – when growth was built on mining and other labour-intensive industries. Increasingly, economic growth is generated by industries, which are NOT labour intensive, either because they are mechanised or because they require small numbers of highly-skilled people to run them. Creating jobs by investing in industry is a capital-intensive process and industrial jobs grow at a slow rate – slower than populations and the numbers of unemployed. And, in a free-trade regime businesses need to be competitive, and industrial developments cannot be competitive and job-intensive at the same time.

For these reasons, the world economy is now characterised by jobless growth, and South Africa's strategy of seeking world market integration has not supported job creation. There have been 800 000 job losses between 1995 and 2000, and the number of unemployed people increased from 1.9 million in 1995 to 4.2 million in 2002. Better-than-predicted growth in the recent past had little impact on unemployment.

Government recommends self-employment in the informal sector. In the process, we also look towards the environment for employment opportunities.

Indeed, conserving nature and places of natural and cultural significance does create jobs, either directly in conservation, or indirectly but more extensively, through eco-tourism. This is why we increasingly understand conservation to be a form of development. Jobs can also be created by 'greener' lifestyles (e.g. making durable shopping bags), greener production (e.g. producing renewable energy technology), better waste management (e.g. jobs in the recycling industry) and environmental restoration (e.g. alien clearing contractors). These kinds of activities require both high-skill and low-skill work.

Protecting the environment can also lead to temporary job losses, for example when a polluting business is closed down. However, if that business starts a more sustainable enterprise,

jobs will be re-created. Failing to look after the environment, on the other hand, leads to long-term job losses – for example when a fishing industry collapses.

6. Economic reforms

Efforts to alleviate poverty will not be successful without a complete overhaul of the economic systems, which have maintained poverty for so long. We are using outdated economic models that lead to the redundancy of people, and poverty thrives. (See ECONOMICS).

Campaigns to alleviate poverty often call for international assistance. However, the track record of donor aid suggests that it may benefit governments more than it benefits the poor. Donors can use aid to reinforce their assumed supremacy and create positions and markets for their own citizens; recipient governments have been known to expand their consumer class and secure their own power base, under the banner of poverty eradication.

A more equal relationship between nations would be fostered by facilitating fair trade. This means scrapping the trade barriers, which rich countries erect against imports from poorer countries, while they ask for the liberalisation of poorer countries' markets. Economists estimate that reducing unfair trade barriers could allow southern countries to generate $130 billion a year, roughly three times the sum total of the official development aid assistance.

7. Alleviating affluence

Poor people need to consume more environmental resources. What are the implications for the Earth?

Not good, if they were to consume at the same rate as the 'consumer class'. The environmental footprint of a wealthy person is far greater than that of a poor person. Even when we multiply the number of poor, we still find that a mere 20% of the population – the wealthy – is responsible for the bulk of the world's consumption and pollution.

The global environmental resources are finite and unequally divided. To obtain more resource rights for the low-consumers, the over-consumption of the affluent must be reduced. The idea is to have both convergence – of the consumption of the poor, towards a more sustaining livelihood – and contraction – of the consumption of the rich, towards more sustainable lifestyles.

In practice, this means choosing a development path that is both pro-environmental and pro-poor, and de-linking economic growth from an increase in resource use, and social progress from economic growth. For individuals, it means that as our incomes grow, we should find ways other than excessive consumption to fulfil our personal ambitions and social needs. (See CONSUMERISM).

It is a matter of promoting both ecology and equity – proving the point with which we started out – that poverty alleviation and environmental care go hand in hand.

8. Acknowledging rights

Poverty is about a lack of power. Poverty alleviation should therefore also address situations in which some people have few opportunities to exercise their rights. Rather than thinking of the poor as needy persons awaiting hand-outs, recognise their basic rights to common resources.

> "Existing macroeconomic policies [in SA] leave existing power relations intact and reinforce the subordinate position of women and poor people" (HDR for SA, UNDP, 2000)

South Africa's democracy brought equal rights, but not yet equal access. Historical inequalities remain largely untouched by current economic policy, including black empowerment strategies. Unemployment figures are higher among women, and women-headed households are more likely to suffer from poverty.

9. Slowing down population growth

Most measures to alleviate poverty will be easier if the exponential growth in human numbers slows down.
(See POPULATION).

10. Education and capacity-building

Education and training must help the unemployed, under-employed and youth at risk, to move from being unskilled or redundant in one kind of economy, to contribute productively to a new economy based on social justice and ecological sustainability.

Equally important is capacity-building for government departments, which must deliver on poverty reduction and development. Current policies such as the Integrated Rural Development Programme are well-intentioned, but lack substance, particularly on sustainable development. Rather than recognising the many links between environment and poverty alleviation illustrated here, 'environment' is either ignored or tacked on to development policies and initiatives – when in fact it is at the heart of sustainable measures to alleviate poverty.

WHAT YOU CAN DO

Three years ago, at the Millennium Summit, 189 countries pledged to pursue an ambitious global poverty fighting agenda embodied in a set of eight Millennium Development Goals. The first Goal is to half extreme poverty and hunger by the year 2015.

The Goals have been broken down into a list of targets that are specific, practical and realistic. They are technically feasible and financially affordable. But without social mobilisation we will not progress from concept to achievement. A movement is needed to create awareness, trigger policy reforms, mobilise resources, and motivate actions to meet the goals, both globally and locally.

The world cannot remain neutral ... Doing nothing is not an option. Each of us can make a difference.

From an address by The UNDP Administrator on The International Day for the Eradication of Poverty, 17 October 2003.

Key Associated Topics:

Environmental Justice; Population; Permaculture; Trench Gardening; Urbanisation; Women and the Environment; Sustainable Development Section: – Maintaining Profits, or Sustaining People and Planet? – Eco-Tourism – Land Reform.

Associated Organisations:

Afrodev Associates; Eco-City; Master Farmer Programme; Art of Living; Eco-City; Environmental Justice Forum.

PRECAUTIONARY PRINCIPLE

This principle argues that if the environmental consequences of a particular project, proposal or course of action are uncertain, then the project, proposal or course of action should not be undertaken. It is sometimes possible in these circumstances to use predictive tools such as risk assessments, to make value judgements in the absence of full information. In cases where there is poor communication between project developers and interested and affected parties, the precautionary principle is often well advised. If, however, there is trust between the various stakeholders, then it is often possible to make decisions without the fullest of information being available and based upon the professional judgements and opinions of the experts involved. Calculated risk is the basis of advances in science and technology, as an over-cautious policy could stifle any advances to the detriment of society as a whole. South African environmental policy and legislation now have the precautionary principle and the principle of polluter pays firmly entrenched within their structures. This means that should the officials have any doubt regarding the environmental merits or demerits of a proposal, they can apply the precautionary principle and delay development or formal legislative approval, pending further investigations or evidence. For a position paper on the Precautionary Principle, go to www.cefic.org

Key Associated Topics:

Accumulation; Atmosphere; Biodiversity; Biological Pest Control; Bio-Prospecting; Carcinogen; Carrying Capacity; Climate Change; Dilution; Dioxins; Eco-catastrophe; Eco-sphere & Ecosystems; Environmental Health; Environmental Management Planning; Environmental Risk & Liability; Genetically Modified Organisms; Global Warming; Hazardous Wastes; Non-renewable Resources; Pollution; Nuclear Energy – The Counter Debate; Radiation; Radioactive Waste; Waste Management.

Associated Organisations:

Biowatch SA; Earthlife Africa; Jarrod Ball; Responsible Container Management; Association of SA; SA Freeze Alliance on Genetic Engineering; WESSA.

PUBLIC PARTICIPATION (CONSULTATION)

There is a growing trend throughout the world that requires projects and developments to be discussed with the public, especially local communities and interested and affected parties. The purpose behind this is to get people's opinions and to learn from the knowledge and experience that they possess through living in a particular area. This does not mean that projects and developments are stopped by public consultation but rather that decisions and plans can be made, which are best for the largest number of people or communities. There will always be people who are disadvantaged, inconvenienced or upset by decisions because it is very rare that everyone can be fully satisfied. Public consultation needs people to be responsible both in participating and also in accepting the final decision reached through negotiation and consensus.

Photo: Bruce Sutherland - City of Cape Town

An important part of public consultation is that people should contact their local political representatives and voice their concerns and feelings. If political representatives do not listen to the opinions and feelings of the public and communities, then they can be voted out of office at the next elections. This is the basic operating principle behind the sound working of a democracy, and also contributes towards the growing awareness and sensitivity to the importance of environmental justice.

Key Associated Topics:

Activism; Communities; Development of Land; Eco-Logical Lifestyle Guide; Environmental Education; Environmental Impact Assessment; Environmental Justice; Green Consumer Guide; Indigenous Peoples; Interested and Affected Parties; NIMBY; Scoping; Social Impact Assessment; Stakeholders; Stewardship; World Summit on Sustainable Development; Women and the Environment.

Associated Organisations:

Acer; BKS; Crowther Campbell; Groundwater Consulting Services; Ken Smith Environmentalist; Liz Anderson; SRK Consulting; Strategic Environmental Focus; Terratest.

DON'T THROW IT AWAY, THROW IT OUR WAY.

To us, there's no such thing as waste paper. As one of the largest paper recycling companies in Southern Africa, our mills process huge volumes of recovered fibre into useful products like tissue paper and corrugated board. Here's how you can help – and even benefit:

FORMAL SECTOR

Deal with us direct and we will:
- design a tailor-made paper collection system to cater for your needs
- provide recycling aids & equipment
- train your staff
- shred confidential documents
- provide input on total recycling programmes

ENTREPRENEURS

Contact us and we will:
- help set you up as a paper collector
- provide recycling aids & equipment
- train you
- give ongoing advice & assistance
- assess viability of business plans

Nampak Paper Recycling

Call us on +27 11 799 7111 or e-mail us at recycling@nampak.co.za

RADIATION

Radiation can be split into two main types: ionising and non-ionising radiation. Ionising radiation is high-energy short wave radiation, which has enough energy to knock electrons and change the atoms to positively charged ions. This principle is utilised in nuclear power stations to generate electricity. However, the resulting highly reactive electrons and ions can disrupt organic compounds in living cells and cause many types of sickness including various cancers. Examples of ionising radiation include X-rays, gamma rays, alpha and beta particles. Non-ionising radiation is the opposite. It is low energy and cannot form ions or knock electrons. Examples of non-ionising radiation include ultra-violet light, visible light and radio waves. Non-ionising radiation may, however, have some effect on people but this is not yet clearly understood.

Key Associated Topics:

Carcinogen; Chernobyl; Eco-catastrophe; Energy; Hazardous Wastes; Nuclear Energy – The Counter-Debate; Plutonium; Precautionary Principle; Radioactive Waste.

Associated Organisations:

Earthlife Africa; Eskom; Koeberg Nuclear Power Station.

RADIOACTIVE WASTE

Radioactive waste covers a spectrum from very low-level waste to very high-level waste and there are different challenges to be faced at the two ends of this spectrum. Low-level waste includes items such as gloves, syringes and cotton wool swabs that have been used in the radiotherapy department of a hospital. By law all such items used in radiotherapy are assumed to be potentially radioactive even if they are not. These items have to be disposed of in a waste disposal process that is strictly controlled by law. Generally there are large quantities of low-level waste produced in a busy radiotherapy department in a hospital, which easily accumulates to a number of drums of waste per day. This whole process presents management and control problems because it is relatively easy for gloves or similar items to be accidentally disposed of outside the official waste process, unless strict control is exercised at all times. A rather different type of challenge faces the issue of very high-level waste. High-level waste is typically spent fuel elements from a nuclear reactor. In contrast to low level waste there are very small quantities of high level waste produced, but the high level waste is very radioactive and if a human being came in direct contact with unprotected high level waste, it would be lethal. High-level waste is also controlled by a disposal process that is strictly controlled by law. A potential problem often cited in relation to high-level waste is its long life. The intensity of a source of radiation is measured by a term called half-life. As radiation intensity dies down it does so rapidly initially and then more slowly as time passes. This is the principle of the Half-Life measurement. A short half-life material (which could have a half-life measured in seconds) could be highly radioactive for a day or two, but can then have died off to a very low intensity after a week. Some radioactive materials have half-lives measured in many years, and they will stay radioactive for thousands of years. As a result of this factor, most countries now have disposal policies that stipulate that any radioactive waste placed in safekeeping in a repository must be able to be picked up and moved to another place in the future, should this become necessary. In years gone by some radioactive waste was sealed into concrete blocks and then disposed of in the sea. This is now considered an irresponsible way of dealing with such waste.

Guest Author: Dr Kelvin Kemm

Key Associated Topics:

Accumulation; Basel Convention; Carcinogen; Chernobyl; Energy; Hazardous Wastes; Life Cycle Assessment; Nuclear Energy – The Counter-Debate; Plutonium; Radiation; Waste Management.

Associated Organisations:

Earthlife Africa; Eskom; Koeberg Nuclear Power Station.

RECYCLED PAPER

Photo: Nampak Recycling

Only about 25% of the world's paper is currently recycled. Experts say that there is no good technical or economic reason why this amount cannot be doubled by the year 2005. (Paper recycling will increase if everyone commits themselves to recycling the paper they use, instead of throwing it away in the rubbish bin.) South Africa recycles approximately 42% of all the paper it consumes. This figure is below that of some European countries but it is believed that it could be increased with further education and heightened awareness. It is estimated that if half of the world's paper consumption were recycled, this would release over 8 million hectares of forest from paper production for other uses. South Africa's paper industry is still desperately short of recycled paper because not enough people are willing to participate in recycling schemes. It is so serious that paper has to be imported from overseas to make up the shortfall.

Key Associated Topics:

Eco-efficiency; Eco-Logical Lifestyle Guide; Recycling; Waste Management.

Associated Organisations:

Nampak Recycling; Sappi; Recycling Forum (The National).

RECYCLING

Recycling is simply the re-use in some form or another of materials that would otherwise be thrown away. Humanity needs to take a lesson from nature where everything is recycled and nothing goes to waste. When plants and animals die, they decompose and the nutrients from the process leach into the soil or water and go to supply food for the next organism. Recycling normally lowers energy consumption, decreases pollution, conserves natural resources and reduces the amount of land needed for landfill sites. Recycling has been shown to be financially successful with Collect-a-Can's top entrepreneur earning almost R15 000 per month, just from collecting tin cans. Other forms of recycling include glass (Glass Recycling Association), plastics (Plastics Federation), and paper (SAPPI).

Key Associated Topics

Agenda 21; Eco-efficiency; Environmental Education; Integrated Pollution and Waste Management; Life Cycle Assessment; Poverty Alleviation; Recycled Paper; Stewardship.

Associated Organisations:

Apple Orchards Conservancy; Bergvliet High School; Biolytics Southern Africa; Canstone; Cart Horse Protection Association; City of Cape Town Solid Waste; Collect-a-Can; Fairest Cape Association; Gauteng Conservancy Association; Hangerman; Hewlett Packard South Africa; Inspire!; Keep Pietermaritzburg Clean; Nampak Recycling; Recycling Forum (National); Responsible Container Management; Thorntree Conservancy; Wales Environmental Partnerships.

REHABILITATION OF LAND

Land rehabilitation has become an important part of the new mining legislation, which stipulates that mines must develop Environmental Management Plans, which must include reference to closure and rehabilitation of the site. Furthermore, the mines are required to put down guarantees to cover the costs of rehabilitation, which cannot be used for anything else. A number of the open cast mining companies have become highly skilful in rehabilitating worked out open cast mines, to such a degree that the previous owner can put his animals straight on the land to start grazing. Amongst one of the important tasks on commencement of mining is the stripping off of the topsoil from the surface of the mine and storing it for later use on the rehabilitated mine. Rehabilitation of land for underground mines is somewhat more difficult but still possible. In recent years, the mining company, ERGO, has managed to completely reclaim old slimes dams and extract additional gold and silver, as well as free the sand for use in building purposes. The land is now ear-marked for office and light factory development.

Key Associated Topics:

Aesthetics; Contaminated Land; Hazardous Wastes; Land Degradation; Landscape Architecture; Opencast Mining; Pioneer Species; Polluter Pays Principle; Radioactive Waste; Soil; Rivers & Wetlands.

Associated Organisations:

African Gabions; Apple Orchards Conservancy; Arcus Gibb; Bapela Cave Klapwijk; Envirocover; Gauteng Conservancy Association; GeoPrecision Services; Groundwater Consulting Services; Kantey & Templer; Morris Environmental & Groundwater Alliances; Richards Bay Minerals; Thorntree Conservancy; Viridus Technologies; Vula Environmental Services; Wales Environmental Partnerships.

Photo: Earthyear Magazine

the view improved
since we picked up.

Collect-a-Can has been driving can recoveries for more than a decade.
Since we started over 500 000 tons of used beverage cans have been
recovered in Southern Africa and recycled, replacing valuable natural resources,
assisting in litter abatement and saving costly landfill sites.

If you can assist, please call us on (+27) (0)11 466 2939.

www.collectacan.co.za

COLLECT-A-CAN

Supported by Iscor and Nampak

RENEWABLE ENERGY AND ENERGY EFFICIENCY

Guest Essay by Glynn Morris ~ AGAMA Energy (Pty) Ltd.

All of us use energy services in our daily lives. Many of these are based on grid electricity, which is essentially very clean and convenient at the point of use in your home or office. However, grid electricity is only one form of energy and it has some significant disadvantages – one of which is that in South Africa it is derived from non-renewable and environmentally problematic sources of energy such as coal. And, as it turns out, our coal has particularly high local and global environmental impacts, being one of the dirtiest coals in the world.

Advantages of conventional grid electricity:

◆ Convenience
◆ Low cost – to the consumer
◆ It is clean at the point of use
◆ Reliability
◆ Wide range of interchangeable appliances available – both new and secondhand
◆ Status and a sense of modernity.

Disadvantages of conventional grid electricity:

◆ High consumption of non-renewable resources such as coal, uranium and gas
◆ High use of water in a water poor country (each unit of electricity generated consumes 1,25 litres of water)
◆ Health and safety issues relating to coal/uranium mining and gas/oil drilling and refining
◆ Land use issues relating to coal/uranium mining
◆ Damaging environmental effects of emissions – greenhouse gases and particulates

Photo: Bruce Sutherland - City of Cape Town

◆ Waste disposal issues relating to ash (from coal) and nuclear waste
◆ Visual impacts of overhead transmission lines across the country and distribution lines in towns
◆ Centralisation of control and the associated dependencies.

What are the alternatives?

Implicit in the question is the awareness that there are different options – other than electricity from Eskom or the local authority – i.e. you actually do have a choice. Secondly, it implies that each option is better or worse than others depending on the criteria – such as cost, convenience, environmental impacts, etc. Modern Renewable Energy Systems and Energy Efficient Options are technically mature, commercially established, readily accessible and they are more or less just as expensive (or cheap) as the conventional option of grid.

Typical renewable energy and energy efficiency options include:

◆ Natural lighting of your building, by allowing sunlight to enter the building structure in a controlled and pleasant manner
◆ Solar heating of your building by orientation and better placement and shading of windows
◆ Solar water heating
◆ Solar and/or wind generated electricity (whether connected to the grid or not).

Some of the benefits include:

◆ Energy savings – you can save energy, and hence natural resources, which benefits everyone.
◆ Financial savings – through using less energy (energy efficiency) and substitution of one form of (non-renewable and expensive) energy to a another (renewable or cheaper) form.
◆ Reduced consumption of our natural resources.
◆ Operational security – renewable energy systems can be designed to provide better reliability than conventional grid systems, such as power for telecommunications, navigational beacons, etc.
◆ Diversity of supply – the diversification of supply means that there is less risk of experiencing a total loss of power, due to the sole source of energy (electricity) going down for some reason.
◆ Reduced or deferred infrastructure costs – including generation, transmission, distribution and maintenance.
◆ Environmental benefits – renewable energy systems do not degrade the environment to the same extent as non-renewable systems.
◆ Social benefits – increased employment opportunities through manufacture, installation and operation of renewable energy systems.

The South African government has developed and published its energy policy in the form of the White Paper on Energy Policy of the Republic of South Africa (December 1998) and a White Paper on Renewable Energy (November 2003). These policy documents officially endorse and recommend the greater use of renewable energy and energy efficiency.

Photo: Department Minerals & Energy

What can you do to begin the process of shifting away from the use of non-renewable energy – Where does one start?

It is always difficult to start on a new path, but one of the benefits of renewable energy and energy efficiency is that you can start small and transform your energy utilization patterns and energy service systems slowly and within the constraints of your budget. The more you implement changes, the more monthly disposable income will become available (from the savings you make) for further investments in your energy transformation.

Introducing energy efficiency measures

Before looking at the ways in which you can transform your use of conventional grid energy, it is important to remember that, in terms of more sustainable energy efficiency practices, there is nothing that can beat the benefits of reducing your need for energy in the first place, through energy conservation and energy efficiency. Any activity which reduces the demand for energy saves our natural resources and also reduces wastes and emissions. Even homes or businesses, which currently use renewable energy, really do benefit from energy conservation and energy efficiency.

◆ Replace incandescent light bulbs with compact fluorescent bulbs in lights which are used more than 3 hours per day.

◆ Install a timer switch and insulation on your geyser or convert your geyser to an instantaneous water heater (with no heat losses from stored water).

◆ Draught-proofing your building to reduce uncontrolled heat losses through air movement.

◆ Insulate your building.

◆ Maintain good fridge habits, including keeping the door seal in good condition; keeping your fridge 75% full all the time, with water bottles, to minimize the amount of warm air which enters the fridge to replace cooled air which "falls" out when you open the door.

Introducing renewable energy systems

The key to the most cost-effective use of renewable energy systems is the matching of the energy services that you require with the capacities and characteristics of available technologies and systems. This is really true for any technology – computers, cars, music systems, etc.

◆ Assess your real energy service needs (after having reduced the consumption of your existing energy supplies to more realistic levels, through energy efficiency measures and increased use of more efficient electrical appliances as above).

◆ Substitute or expand your range of energy supplies to utilize these more effectively, e.g. using solar cooking or bottled gas for cooking rather than electricity.

◆ Start to increase your overall use of renewable energy to reduce your use (and dependence) on non-renewable energy supplies.

◆ Monitor (and then manage) your consumption and the costs of this consumption (including social and environmental costs if you feel up to it) on an ongoing basis.

◆ For access to information and suppliers, the best place to start is by contacting the Sustainable Energy Society of Southern Africa (as listed in The Enviropaedia). The initial questions that you should enquire about include:

  ◆ The initial costs of supply and installation

  ◆ The operating and maintenance requirements (even renewable energy systems need to be maintained)

  ◆ The levels of service which are offered

  ◆ The implications of upgrading (or downgrading) as your needs change

  ◆ The costs of ongoing maintenance

  ◆ The quality assurance for the equipment and for the installation contractor.

We hope you enjoy the experience of taking responsibility for your own energy service needs. Start small and play. The risks are small by comparison with entrusting your (and the Earth's) future to others. Stand up to the fact that one third of all known species on Earth will have been wiped out by 2050, due primarily to global warming as a result of our current practices.

Key Associated Topics:

Agenda 21; Biomass; Cleaner Production; Eco-efficiency; Ecological Footprints; Energy; Fossil Fuels; Geothermal Energy; Nuclear Energy – The Counter Debate; Renewable Resources; Solar Heating; Sustainable Development; Wave Power; Wind Farming; Sustainable Development Section: – Maintaining Profits, or Sustaining People and Planet?

Associated Organisations:

African Wind Energy Association; Agama Energy; Divwatt; Energy Efficient Options; Energy Technoloy Unit – Cape Technikon; ESI Africa Magazine; Eskom; Green Clippings; International Solar Energy Society; Iskhus Power; LC Consulting Services; Oelsner Group; Renewable Energy And Energy Efficiency Partnership; Solardome SA; Solar Engineering Services; SouthSouthNorth; Sustainable Energy Society of SA; Sustainable Energy Africa.

Photo: Bruce Sutherland - City of Cape Town

Photo: Guy Stubbs Photographer

RENEWABLE RESOURCES

These are environmental resources, which are continuously renewing themselves. For example, energy that is harnessed from the sun, wind and waves is renewable. Trees, soil and water are renewable but only in the long term and only if careful plans are laid for sustainable use and re-planting. Soil that is washed away by erosion will take thousands of years to regenerate but if it is conserved, it can be productive for centuries. Therefore, renewable resources need to be conserved and used wisely in order to be truly sustainable. An underlying principle to understanding renewable resources is the recognition that all resources, irrespective of whether they are renewable or non-renewable, must be utilised in a manner that will stretch and extend their usefulness either through conservation or through better and more efficient use. For example, a solvent was used to clean contaminants from machinery. The supplier indicated that once used, the solvent should be disposed of. Some years later, an adviser saw this practice and suggested that the users filter the solvent through a fine gauze filter and try reusing the solvent. The contaminants were filtered out and the solvent was able to be re-used many, many times over.

Key Associated Topics:

Aquaculture; Geothermal Energy; Natural Resources; Organic Farming; Permaculture; Recycling; Renewable Energy; Solar Heating; Wave Power; Wind Farming.

Associated Organisations:

Africabio; Agama Energy; Divwatt; Energy Efficient Options; Energy Technology Unit – Cape Technikon; ESI Africa Magazine; Eskom; Heinrich Boell Foundation; Sustainable Energy Society of SA; Solar Engineering Services; Solardome SA; Wales Environmental Partnerships.

RESPONSIBLE CARE

"Responsible Care" is an international chemical industry programme, operating in 42 countries, which promotes continuous improvement in health, safety and environmental performance. Committed companies sign a formal, public pledge which requires them to: operate in a manner which minimises negative environmental impacts; communicate with neighbours, customers and the public on their activities; promote the principles of Sustainable Development; provide information about their products; and seek to continuously improve their knowledge and performance with regard to health, safety and environment efforts. For an international perspective on Responsible Care, go to: www.cefic.be/activities.

Responsible Care is administered in South Africa by the Chemical and Allied Industries' Association (CAIA).

Key Associated Topics:

Cleaner Production; Eco-efficiency; Ecological Footprint; Industrial Ecology; Interested and Affected Parties; Precautionary Principle; Safety Health & Environment; Stewardship; Sustainable Development; Triple Bottom Line.

Associated Organisations:

Chemical & Allied Industries' Association; Hewlett Packard South Africa; PPC Cement; Richards Bay Minerals.

RIVER CATCHMENT MANAGEMENT

A river catchment or drainage basin is all the land from which rainfall drains into a particular river and its tributaries. Watersheds separate catchments from each other.

Where catchments are covered with natural vegetation, much of the rainfall infiltrated through soils, is stored as groundwater and moves back into the rivers through seepage. Where catchments have become urbanised, the natural vegetation has been removed and replaced by "hard" surfaces such as roofs, roads and parking areas. This "hardening" lessens the amount of infiltration and increases the amount of surface runoff (water that does not infiltrate) discharged to rivers and canals. In the event of a severe rainstorm this increased runoff sometimes causes flooding.

Not only is the quantity of runoff influenced by urbanisation, the quality of the water is also affected. Runoff often contains contaminants such as oils collected from road surfaces, chemicals from dumpsites and litter. Industry often discharges effluent directly into rivers, and some residents use the drainage systems to dispose of waste. In addition, as cities grow, wastewater treatment plants handle increasing volumes of sewage and frequently find it difficult to achieve acceptable standards before discharging treated effluent into rivers.

Catchment management is a philosophy, process and implementation strategy to achieve a balance between utilisation and protection of environmental resources in a particular catchment area. In recent years, catchment management has gained prominence as an approach to managing water resources. In South Africa, The National Water Act (Act 36,1998) is based on this philosophy and proposed the establishment of Catchment Management Agencies to implement the new policies.

The urban drainage catchments in the City of Cape Town are managed according to the catchment management philosophy and deal with the quantity and quality of runoff. This

Photo: Bruce Sutherland - City of Cape Town

involves the integrated management of stormwater reticulation systems, open watercourses, wetlands, groundwater, vleis and estuaries.

Guest Authors: Mark Obree and Randall Adriaans – City of Cape Town

Key Associated Topics:

Alien Invasive Plants; Dissolved Oxygen Content; Eutrophication; Freshwater Ecology; Gabions; Geographical Information Systems; Groundwater; Rivers & Wetlands; Suspended Solids; Urbanisation; Vetiver Grass; Water Quality.

Associated Organisations:

Afrodev Associates; Aqua Catch; Bainbridge Resource Management; Desert Research Foundation of Namibia; Holgate; Hout Bay & Llandudno Heritage Trust; Mondi Wetlands Project; Rand Water; Two Oceans Aquarium; Water Institute of SA; WESSA; Zandvlei Trust.

Photo: African Gabions (Pty) Ltd

RIVERS AND WETLANDS

**Guest Essay by Dr Heather Malan and Dr Jenny Day
~ Freshwater Research Unit, UCT**

Rivers and wetlands provide a range of goods and services for the benefit of people, but need a healthy ecosystem in order to provide these.

People need water in their homes for drinking and washing and for watering their gardens. We also rely on agricultural produce like meat, milk and vegetables, and on industrial products from motorcars to paper and computers. The production of all of these goods requires large amounts of water, which comes from our rivers, and aquifers (underground water sources) and sometimes even from our wetlands. We also use the water that runs in rivers for the generation of electric power, and for diluting and removing much of our waste.

Photo: Copyright South African Tourism

But natural aquatic ecosystems like wetlands and rivers supply us with many more services other than just actual water. They support populations of fish and other resources that we eat; they provide grazing for our cattle and sheep; they provide timber and reeds for building and weaving; in many parts of the world they act as major transport systems. They also cleanse the water that flows in them. Living 'decomposer' organisms – microbes and invertebrates (insects, worms, snails and crustaceans) – break down and thus remove a lot of waste material, so that as water flows down a river it becomes cleansed. Thus water some distance downstream of a source of pollution is a lot cleaner than it was where the pollutant entered it. This is not merely the effect of dilution because rivers that have lost their living organisms because of canalization (see later) or poisoning are unable to clean the water flowing in them.

Another surprising feature of rivers is their importance for the coastal zone in general, as well as for estuaries. The land is much richer in nutrients than the sea is, so the food chain in the shallow coastal zone often depends on nutrients that are brought down rivers, through their estuaries, and out to sea. What is more, the power of the water flow-ing in a river causes erosion of the bed and banks. The eroded particles of soil and sediment are washed downstream; if the flow becomes slower, then those particles are deposited within the bed of the river itself. Some of the material is washed out to sea, however, and contributes to the sand that forms beaches and dunes. If a river stops flowing, it becomes silted up. If a river stops flowing, it also stops providing sand to the beaches near its mouth, although of course currents in the sea continue to erode the beaches. This is why beaches near some rivers are beginning to disappear.

Wetlands, particularly those with dense stands of emergent plants, provide further services for humans. They are effective at controlling floods because the plants slow down the force of floods, forcing the water to spread out and reducing its damaging effects. They also store and slowly release flood waters to river channels, reducing the worst of the flood peak, extending the time taken for the flood waters to drain away, reducing 'flash flooding', and supplying water during the dry season. Because the water is forced to move slowly, it is also likely to drain downwards into local aquifers, thus recharging them. Wetlands are also natural filters, removing sediments, nutrients and bacteria from the water flowing through them. As a result, water emerging from a wetland is often cleaner than the water entering it. The reason is that the plants cause water to flow slowly and thus to drop its silt; at the same time they, and microbes associated with them, extract nutrients for their own growth. Artificial wetlands are sometimes cultivated for the final cleansing of sewage effluent and, in some areas, for flood control. Wetlands also form some of the most productive lands on Earth, providing habitat, food and shelter for an enormous variety of plants and animals, from minute to huge and from aquatic to terrestrial species. The biodiversity of some wetlands approaches that of tropical forests.

Rivers and wetlands can provide us with all of these services only if they are reasonably 'healthy'. So what does an aquatic ecosystem need in order to stay healthy? Firstly, any aquatic ecosystem is affected by its catchment, so a river or wetland is unlikely to be healthy if it receives more than a minimal amount of pollution from its catchment. Then it needs its bed and banks to be left alone (and not be canalized – lined in concrete) so that its water can interconnect with the water upstream and downstream, and below its bed. It is only in this way that the system will be able to acquire (or rid itself of) the sediments and nutrients on which its biota relies.

Despite the many "goods and services" supplied by healthy rivers and wetlands we have abused them in many ways. One of the major threats is by the construction of dams which, in addition to excessive abstraction (by means of pumping), leads to disruption of the natural flow regime and has an adverse effect on the living organisms downstream. For example, the migration and spawning of the large-mouthed yellow fish (Barbus kimberleyensis) in the summer-rainfall region is triggered by high flows in spring. Due to the damming of many rivers the numbers of this fish are dwindling. Reduced or altered flows can have a detrimental effect on aquatic organisms, but can favour others. Flows in the Orange River are extremely regulated – meaning that because of upstream dams and weirs, very high flows and very low flows, which would be the natural regime in the river, no longer occur. These steady flows are considered to be the major cause of black fly (*Simulium* spp.) infestations, which can cause extensive damage to livestock. Dam walls also often pose insurmountable barriers to fish movement – preventing fish from moving upstream to spawn.

Not only do dams control the movement of water, and of animals, but also of sediment. In catchments where there is extensive erosion (often due to over-grazing or inappropriate farming methods), much valuable topsoil ends up in dams, reducing the capacity for storing water. Sediments (and the nutrients bound to them) are often important for downstream floodplains and estuaries. Cessation of the flow of sediments to these systems due to the construction of dams or weirs can lead to catastrophic degradation of these rich ecosystems. A classic example of this is the collapse of the prawn fishery at the mouth of the Zambezi, Mocambique due to construction of the Cahora Bassa dam.

Because rivers are such convenient systems for the disposing of waste, there has been a tendency to pollute them. Sewage (treated and un-treated), industrial waste, salts, nutrients and pesticides have all often ended up in water bodies. This is sometimes due to direct discharge of effluents into rivers and lakes through pipes. However, rainwater flows over the catchment surface soaks into the ground, gradually making its way into groundwater bodies and in so doing, carries pollutants present on the surface or in the soil into the water. Nutrients (phosphorus, nitrogen) and pesticides are often washed from agricultural lands in this manner. Eutrophication is a form of pollution in rivers and lakes, which results from excessive loads of nutrients being washed or dumped into these water bodies (originating largely from fertilizers, sewage and effluent from food-processing plants). Some species of plants thrive in a nutrient enriched environment leading to rapid and excessive growth. The blocking of waterways by reeds and thick smelly scums of phytoplankton on the surface of lakes are just some of the problems caused by eutrophication.

Natural rivers in lowland areas tend gradually to change their channel (banks eroding in some areas and being built up in others). Furthermore, during periods of high flow the flood-plain areas become inundated with water. This is an important ecological process as fish move into the shallow, flood-plain waters to spawn and the protected nutrient-rich backwaters provide an excellent nursery for the young. However in urban situations where land is limited (and developers are greedy) development has been allowed to encroach into the flood-plain areas. Then, because of the risk of flooding or erosion, structural measures need to be put in place to remove flood waters as rapidly as possible as well as to stabilize the banks. Thus many urban rivers are canalised (the bottom and sides encased in concrete). Whilst this may be an efficient engineering solution, the result is not ecologically sensitive since there is no

habitat within the smooth walls for aquatic organisms to live, and few organisms will be found in such systems. More environmentally friendly options are available – for example the use of gabions (metal baskets of stones) that can be arranged to prevent erosion and allow the river to meander). Such options allow plants to grow along the banks and offer some habitat for aquatic organisms (but not a lot of protection from flooding).

Despite the important role of wetlands in cleaning water and regulating water flow in South Africa, it is estimated that at least 50% of our wetlands have already been lost due to draining and filling for agriculture or housing development.

So what can be done about the abuse of our precious rivers and wetlands? South Africa is an arid country in which a large proportion of the population still does not have access to basic amenities such as sanitation and potable water. However, we need to have a balance between protection of aquatic resources on one hand, and development on the other.

Photo: Sappi (*Earthyear Magazine*)

Can I preserve the future?

Picture by Richard du Toit

▶ Canon strives to ensure a future for our natural environment and two ways in which we aim to achieve this is through our association with the World Wildlife Fund (WWF) as a global conservation partner, as well as through the implementation of Canon's Kyosei* Philosophy. We actively promote the conservation of energy and resources in our products and have completed design systems to make our products 100% recyclable. We are also increasing energy efficiency at production sites by 30% to reduce emissions and help prevent global warming.

It is just Canon's way of preserving an environment that you will want to photograph for many years to come.

* Kyosei = Living and working together for the common good - working in harmony between products, customers, partners and in the natural world.

Whatever you can imagine, with Canon you can.

www.canon.co.za

For more information contact Toll Free 0800 004 937

Canon
Conservation Partner

you can
Canon

The New South African Water Act (1998) is remarkable internationally in that it recognizes that the entire aquatic ecosystem needs to be protected because it is the source of all the goods and services provided to people. More than laws will be required however, as there needs to be a greater understanding by all sectors of the community that whilst access to water is a right – that water is precious and our aquatic resources need to be preserved, protected and kept unpolluted. People need to start cutting down on consumption and ensuring that water is not wasted. Some ways in which this can be done include:

◆ Planting indigenous gardens (these use much less water than exotic plants such as kikuyu lawns).

◆ Recycling of "grey water" (e.g. using water from baths, washing machines, etc. for irrigation).

◆ Using dual-flush toilets.

◆ Using porous surfaces for driveways, parking areas, etc. (this allows rain to soak into the ground).

◆ Watering gardens and agricultural fields early evening and mornings (to reduce evaporation).

Pollution can be reduced by using less fertilizer on gardens and disposing of pesticides and oil in a responsible manner (rather than emptying them down the drain). People can campaign against building too close to the flood line of rivers and the filling-in of wetlands.

Although dams have major impacts on rivers they also have an important role in ensuring that people have a reliable water supply. But before building any new dam, there needs to be a very clear understanding of the potential long term costs and benefits, remembering that the impacts may only be felt many miles downstream and many years after completion of the dam wall. Ensuring that we utilize our limited water resources in a manner that will realize the most benefits for all people will be an increasingly difficult challenge in the future. It is however a challenge that needs to be faced if all the people of South Africa are to attain a dignified standard of living.

Key Associated Topics:

Alien Invasive Plants; Biodiversity; Bio-remediation; Dissolved Oxygen Content; Eutrophication; Freshwater Ecology; Groundwater; River Catchment Management; Suspended Solids; Urbanisation; Water Quality; Wetlands.

Associated Organisations:

Afrodev Associates; Aqua Catch; Botanical Society of SA; Bainbridge Resource Management; Desert Research Foundation of Namibia; Holgate; Hout Bay & Llandudno Heritage Trust; Mondi Wetlands Project; Rand Water; Riviera Wetlands And African Bird Sanctuary; Two Oceans Aquarium; Water Institute of SA; WESSA; WWF; Zandvlei Trust.

Photo: Dr J Ledger

SAFETY, HEALTH AND ENVIRONMENT (SHE)

(From S, H and E Management to Integrated SHERQ management systems: latest trends and developments)

Guest Essay by Prof Johan Nel ~ Executive Manager, Centre for Environmental Management (CEM), North West University (Potchefstroom Campus)

Effective and appropriate safety, health and environmental management is a critical component of sustainable business practice. Traditionally fragmented and disjointed safety, health and environmental management (SHE) arrangements are increasingly being integrated into an efficient SHERQms (safety, health, environmental, risk and quality management system). Integrated SHERQ based management systems, or an integrated SHERQms is indeed the strategy of choice of most progressive line managers. Integration however, entails much more than the alignment of traditionally separated SHE programmes.

Business integration

Firstly, it entails integration of traditional SHE programmes that were often perceived as "added on" staff functions, to the actual business or line functions of organisations. Business integration is best achieved by viewing SHE management as supporting business processes that inform key and other supporting business processes.

Risk-based SHE management

Integration of SHE practices with business processes is also enhanced by the adoption of a risk aversive approach to business management, where both business and traditional SHE risks are identified, assessed and managed within an integrated risk-reduction programme. SHE risks are addressed as business risks.

Management system approach

Business integration is also enhanced by the adoption of a management system format to deliver integrated business and SHE programmes. Management systems are designed around the classical Deming Plan-Do-Check-Act (PDCA) management cycle. Line managers readily adopt the PDCA cycle as it forms the basis for almost all line management models or strategies.

Expansion of SHE programmes

Adoption of an integrated business and SHE management approach, within the framework of reducing business risks, often amounts to the extension of SHE programmes to include supporting business processes that traditionally may not have been serviced at all such as: contractor management, purchasing, strategic planning, insurance, design as well as financial management functions and services, etc. The scope of most SHE management programmes is often also extended to include supply chain management as well as product stewardship that may traditionally have received marginal SHE attention. The traditional scope of environmental management is also expanded to include elements of corporate social responsibility.

Quality management

Another addition to the traditional SHE management stable is quality management based on ISO 9001:2000 as SHE management portfolios are also integrated with quality management programmes.

Benefits of integrated SHERQ management systems

Benefits of an integrated and risk-based SHERQms include:

◆ Elimination of wasteful turf wars between line and traditional staff SHE functions
◆ Elimination of costly duplication of fragmented SHE management arrangements
◆ Increased confidence of all stakeholders that all risks are comprehensively identified, assessed and managed
◆ Improved efficiency of all business processes
◆ Enhanced performance tracking and reporting opportunities
◆ Improved SHE performance
◆ Multi-skilling of SHERQ-professionals.

Drivers for the swing towards Integrated SHERQms

◆ Stakeholder demands
◆ Demands for resource efficiency improvement
◆ Corporate governance demands
◆ Increased risk exposures.

Key Associated Topics:

Ecological Footprints; Environmental Auditing; Environmental Management Planning; Environmental Management Systems; Environmental Risk & Liability; ISO 14000 Series; Pollution; Precautionary Principle; Waste Management.

Associated Organisations:

Advantage ACT; Anglo American; BMW; Ceatec; Centre for Environmental Management (CEM) North West University; Chemical & Allied Industries Association; Crystal Clear Consulting; Dekra-its Certification Services; Development Consultants of SA; Dittke Mark; Ilitha Riscom; Imbewu Enviro; Independent Quality Services; Ishecon; Kumba Resources; Metrix Software Solutions; Mining Review Africa; National Air Pollution Services; NOSA; PPC Cement; Price Waterhouse-Cooper; Vula Environmental Services; Wilbrink & Associates.

Photo: **Department Minerals & Energy**

SCOPING

Scoping is a process used particularly when carrying out environmental, or strategic environmental, impact assessments. It is a process, which is intended to narrow the scope of the assessment and ensure that it focuses on the truly important and significant issues or impacts. Scoping does not just narrow down the issues but may also involve adding in issues that were not included before. Additional issues may be raised as a result of consultations and discussions with interested and affected parties. Scoping helps to ensure that all the relevant matters are taken into account and that time and effort is not wasted on matters that are not important or are insignificant or irrelevant. The scoping process is done in consultation with interested and affected parties.

Key Associated Topics:

Environmental Impact Assessment; Environmental Manage-ment Planning; Environmental Management Systems; Geographical Information Systems; Integrated Environmental Management; Land Use Planning; Strategic Environmental Assessment.

Associated Organisations:

Acer; Crowther Campbell; Hilland; Landscape Architects Uys & White; Newtown Landscape Architects-Limpopo; Viridus Technologies; Wilbrink & Associates.

SCRUBBER

A "scrubber" is the common term used to describe a piece of equipment designed to remove or "scrub" toxic gases and other materials from the gases emerging from a chimney. The term is mainly used to describe the use of a liquid into which toxic gases are absorbed or dissolved to prevent them from being dispersed into the air. Sulphur dioxide is often removed from stack emissions using wet scrubbing methods. Dry scrubbing involves the use of materials onto which particles "stick" and are thereby also prevented from going into the atmosphere. Other methods of "scrubbing" these emissions of particles include the use of bag filters (the emissions are passed through filters which trap particles) and electrostatic precipitators. Electrostatic precipitators electrically "charge" the particles so that they are attracted to electrodes and thus prevented from leaving the chimney (or "stack", to use its technical name). Scrubbing is an important method of reducing air pollution from industrial chimneystacks.

Key Associated Topics:

Air Quality; Atmospheric Pollution; Catalytic Converter; Fossil Fuels; Incineration; Integrated Pollution and Waste Management; Particulates; Petrochemicals; Pollution; Sulphur Dioxide; VOCs.

Associated Organisations:

National Association for Clean Air; SI Analytics; Wales Environmental Partnerships.

SICK BUILDING SYNDROME

"Sick building syndrome" describes human illness characteristics that are brought about by air-conditioned buildings that do not introduce sufficient fresh air to the air conditioning systems or filter, or clean the air passing through the air conditioning systems. Symptoms of sick building syndrome include dizziness, headaches, pain over the eyes, nose and throat irritations, flu and cold-like symptoms, and a general feeling of lethargy and heaviness. The shortage of fresh air means that there is a build-up of chemicals and VOCs in the air in the building, which originate from photocopiers and fax machines. There is also a concentration of germs and diseases from the people themselves in the buildings. These germs then breed in the warm, moist conditions in the air conditioning system and are spread throughout the building to all the occupants. Improving designs of air conditioning systems, increasing the quantities of fresh air drawn into the air conditioning system, and disinfecting filtering systems and cooling water tanks can reduce sick building syndrome.

Key Associated Topics:

Air Quality; Environmental Health; Environmental Manage-ment Planning; Green Architecture; Nuisance.

Associated Organisations:

ERA Architects; Greenhaus Architects.

SOCIAL IMPACT ASSESSMENT (SIA)

A social impact assessment is either a stand-alone report or part of an Environmental Impact Assessment. An SIA is about "people impacts" of projects, policies and developments. The SIA tries to analyse how people will be affected socially, whether people and communities will benefit from the actions proposed and, if so, how. The SIA will look at a community and see how strong it is, whether it has good social ties and structures such as clubs, churches, support systems and formal and informal educational structures. From the social study of a community, it tries to come up with suggestions, which may help to integrate the project, policy or action and make it more acceptable or beneficial. It is difficult for an SIA to produce precise results because people are unpredictable and social and community circumstances can change so easily – not just materially but also from the way in which people see and interpret facts. An SIA will try to judge if the decision to implement the project, policy or plan will last the test of time and whether 50 years later, people will say the decision was good or bad. A broad overview guide to Social Impact Assessment can be found at www.iaia.org.

Key Associated Topics:

Communities; Cultural Resources; Environmental Ethics; Environmental Impact Assessment; Environmental Justice; Indigenous Peoples; Interested and Affected Parties; NIMBY; Nuisance; Public Participation; Stakeholders.

Associated Organisations:

Coastal & Environmental Services; Environmental Evaluation Unit – University of Cape Town; Environmental Justice Networking Forum; Groundwork; Hilland.

Photo: Dr J Ledger

SOIL

Soil is formed slowly by the weathering and erosion of rocks over many thousands of years. Almost everything which humans need can be traced back to the soil: food, clothing, medicines, oxygen, etc. As one of the three primary resources of the biosphere, humankind's well-being is inextricably linked to the soil.

South Africa has very limited fertile soil. Less than 7% of South Africa's land area is regarded as high grade, good agricultural production land. It has been said that soil is one of South Africa's greatest exports, as each year over 460 million tonnes of topsoil are stripped from poorly managed land, by rain and wind, thus wasting the limited resource base and reducing the productivity of the land. The loss of protective vegetation through deforestation, over-grazing, ploughing and fire increases the vulnerability of soil to being swept away. The conservation of soil, in particular the precious fertile topsoil, is therefore one of South Africa's most pressing environmental concerns. Soil erosion can be prevented by not exposing bare earth to rain and wind, through the observance of good farming practices (including the use of crop rotations, which include fallow periods under grass or pasture, to allow the soil to recover) and by farming within the carrying capacity of the land. The planting of trees also helps to reduce soil erosion. Conservation of soil requires a commitment from all South Africans – government departments and policy developers, industry, developers, farmers and individual home-owners – who all need to ensure that land-use practices and developments do not further deplete this essential resource base.

Guest Author: Alison Kelly ~ Wildlife and Environment Society of South Africa

Key Associated Topics:

Agriculture; Bio-remediation; Carrying Capacity; Composting; Contaminated Land; Deforestation; Desertification; Fire; Land Degradation; Monoculture; Organic Farming; Permaculture; Pesticide Treadmill; Pesticides; Rehabilitation of Land; Trench Gardening.

Associated Organisations:

Master Farmer Programme; Barlofco; Conservation Management Services; Ecoguard Distributors; Food & Trees for Africa; Food Gardens Foundation; Morris Environmental and Groundwater Alliances; Southern African Wildlife College.

SOLAR HEATING (PRO-ACTIVE)

The sun emits a constant stream of energy towards the Earth. This energy reaches the ground at about one kilowatt per square metre at midday with unobstructed sunlight. The natural sunlight consists of all wavelengths of the visible spectrum plus some ultraviolet (UV) at the short wavelength (blue) end of the spectrum, and some infrared (IR) at the long wavelength (red) end of the spectrum. It is the red end of the spectrum with the infrared that carries heat, while the blue and ultraviolet is the high-energy portion. The blue and ultraviolet is the part of the spectrum that produces electricity from silicon solar cells, but it is the red and infrared part that produces heat. South Africa is particularly fortunate in having generally clear skies and many hours of daylight throughout the year, which makes it an ideal country to exploit solar energy. Solar heating can be used in a number of ways. One is to use solar water heaters placed on the roof of a house. If the house does not have electricity the solar water heater can be directly connected to hot water taps in the house. The temperature of the water will vary depending on the time of the day. At midday the water could be too hot to place one's hands into, and at night it would be cool or cold. One potential addition to the system is to place a storage tank in the roof connected to the solar heater so that hot or warm water can be drawn, even at night. However, if the house has electricity then a solar heater can still be used, but can be connected so that it feeds into the electrical geyser. Again a reservoir will be a welcome addition because it will feed warm water into the geyser, even at night. This saves electricity in electrical heating of the water. Other ways to use solar heating include building houses with windows or roof surfaces facing the sun so that the natural solar heat warms the house directly. Solar heating can also be used for swimming pools.

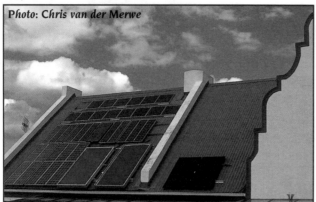

Photo: Chris van der Merwe

A number of developments are under way to utilise solar cookers that can boil water or fry and roast meat. Imagination and ingenuity are still being used to develop other uses of the heat portion of the solar spectrum; it is a resource still waiting for optimum utilisation.

Guest Author: Dr Kelvin Kemm

Key Associated Topics:

Eco-efficiency; Ecological Footprint; Energy; Green Architecture; Renewable Energy; Sustainable Development.

Associated Organisations:

Agama Energy; Energy Technology Unit – Cape Technikon; International Solar Energy Society; Solar Cooker Projekt (Solco); Solar Energy Society of SA; Solar Engineering Services; Solardome SA; Sunstove Organisation; Wales Environmental Partnerships.

STAKEHOLDERS

The term "stakeholder" describes any person or organisation that may affect, or be affected by, a company or industrial process. They are also sometimes called "Interested and Affected parties". Modern companies are trying to identify all the stakeholders affected by their operations so that they can communicate with them and best meet their needs and concerns. Different stakeholders would have different concerns. For example, the main concern of a company shareholder would be to see the company making a profit for him whereas the neighbours to the factory are more concerned about the noise, pollution or aesthetics of the factory. It is important for companies to satisfy their stakeholders' concerns as far as possible and to communicate regularly with them. Current environmental legislation in South Africa requires companies to communicate with stakeholders if they are going to develop or change their activities.

Key Associated Topics:

Activism; Communities; Development of Land; Environmental Impact Assessment; Environmental Justice; Environmental Law; Indigenous Peoples; Interested & Affected Parties; Public Participation; Social Impact Assessment; Women and the Environment.

Associated Organisations:

Environmental Justice Networking Forum; Groundwork; Institute of Directors; National Business Institute; WESSA.

STEWARDSHIP

Whilst a Steward may be defined as "one whose function and responsibility is to manage or serve (on a ship, aeroplane, race-course, estate, etc.)", the concept of stewardship as it relates to the Earth has far wider implications. It certainly means different things to different people as the meaning is also relative to its application. It would motivate a different behaviour if we applied it to the management of the Earth as: a source of food and shelter to survive; a resource to exploit for wealth creation; a biological bank to research and manage for the preservation of its valued components; or as a source of emotional or spiritual upliftment and inspiration. No single definition can therefore adequately express its full significance. However, as a term of increasing importance in the environmental context, it is worthwhile having at least one relative and subjective frame of reference for our human responsibility as Stewards of the Earth. The Vision Statement of The Wilderness Leadership School Trust incorporates several principles that may provide useful points of reference from which to understand our role and responsibilities. These include:

◆ The Universe is a communion of subjects and not a collection of objects.

◆ The Earth is a sacred place and a sacred process and that humans are a part of both this place and process.

◆ The integral Earth community is the primary source of reference and that human well-being is derivative and dependent on the well-being of the whole.

Seen in this context, our function as stewards is not one of purely master or servant, but rather, recognising that we are a co-dependant species within the Earth's greater community of life. Our responsibility therefore is to facilitate a mutually sustaining relationship between the human and non-human Earth communities, in all (physical, emotional and spiritual) aspects of our interaction.

Guest Author: David Parry-Davies ~ Eco-Logic

Key Associated Topics:

African Environmental Tradition; Consumerism; Ecological Intelligence; Eco-Logical Lifestyle Guide; Environmental Ethics; GAIA; Responsible Care; Sustainable Development; Triple Bottom Line.

Associated Organisations:

Conservation International; Conservancy Association; Educo Africa; WESSA – Friends Groups; WESSA; Wilderness Leadership School; WWF.

The Wildlife and Environment Society of South Africa (WESSA) is a leading environmental organization in South Africa. For almost 80 years, the Society has been a motivating force behind many significant environmental actions.

The proclamation of the Kruger National Park, the protection of the St Lucia Wetlands area, the establishment of the Addo Elephant Park, the protection of the Brenton Blue butterfly, pioneering of the MOSS (Metropolitan Open Space System) concept and more recently, lobbying for the protection of the Pondoland Coast, are but a few of the successful conservation initiatives undertaken by the Society.

So, who is WESSA?

WESSA is a society of people who believe in a better quality of life for all. A society of people who speak out, take action, get involved, or simply support the actions of the organisation in whatever way they can.

You too can add your support to the efforts of the society. There are many opportunities for WESSA members and supporters to be actively involved in environmental programmes and projects around the country.

Show you care and join today.

For more information on how you can get involved, visit our website:
www.wildlifesociety.org.za

or contact Membership Services, at (033) 330-3931

People caring for the Earth

WILDLIFE AND ENVIRONMENT SOCIETY OF SA

People caring for the Earth

STRATEGIC ENVIRONMENTAL ASSESSMENT (SEA)

A strategic environmental assessment (SEA) is an environmental assessment that is carried out on one or more strategic actions, policies, plans or programmes. Whereas an EIA studies a physical project, the SEA looks at policies, plans, ideas and programmes, which are more difficult "to touch and feel". The SEA is an important planning tool because it helps planners to understand what will happen to an area if there are different land uses. It will try and provide information and analysis on the consequences of different actions and their environmental impacts in the short, medium and long term. It is a new environmental planning tool in South Africa and has not yet been used very widely. Although not required by law in South Africa, a number of SEAs have been carried out (e.g. for the Cape Town 2004 Olympic Bid, and as a part of the Assessment of Water uses, in particular with respect to forestry). It has already been stated that SEAs will be needed in conjunction with the Land Development Objectives in terms of the Development Facilitation Act, Environmental Implementation Plans in terms of the National Environmental Management Act and Integrated Development Plans in terms of the Local Government Transition Act. A copy of a guide to Strategic Environmental Assessment in South Africa can be found at www.environment.gov.za

Key Associated Topics:

Agenda 21; Ecosphere & Ecosystems; Environmental Management Planning; Interested and Affected Parties; Land Use Planning; Sustainable Development.

Associated Organisations:

Africon; Galago Ventures; Holgate & Associates; Imagis; Morris Environmental & Groundwater Alliances; Settlement Planning Services; Tribal Consultants; Wales Environmental Partnerships.

SULPHUR DIOXIDE

Sulphur dioxide (SO_2) is a colourless, acrid gas formed by the combustion of sulphur. It is an oxidising and reducing agent and is used as a refrigerant, disinfectant, preservative and bleach. It reacts with water to give sulphuric acid, which means that when it is released as a gas through factory chimneys from boilers, it causes serious pollution. "Acid rain" is frequently associated with sulphur dioxide pollution but a significant contributor to acid rain is carbonic acid, which forms when carbon oxides in the atmosphere combine with water. Sulphur dioxide is released from factory chimneystacks but is also released naturally in vast quantities during volcanic eruptions. Sulphur is an important macro-nutrient for plants but as soon as its concentrations reach beyond tolerance levels, it becomes toxic and dangerous to most living things. In very high concentrations, sulphur dioxide is dangerous to health. In low concentrations, it is hazardous to young children, the aged and those with respiratory ailments. Sulphur dioxide mainly aggravates the respiratory tract and can cause secondary health impacts on those suffering from emphysema, asthma, bronchitis and illnesses of the respiratory tract and lungs, and heart disease. Sulphur dioxide is easily detectable and measurable as an air pollutant and is therefore used widely as an indicator for other common industrial air pollutants, including oxides of carbon and nitrogen, and linked photochemical pollutants.

Key Associated Topics:

Air Quality; Atmospheric Pollution; Environmental Health; Environmental Risk and Liability; Pollution; Polluter Pays Principle; Scrubber; Transboundary and Transfrontier Pollution.

Associated Organisations:

Chemical and Allied Industries' Association; South African Institute of Ecologists and Environmental Scientists.

SUSPENDED SOLIDS

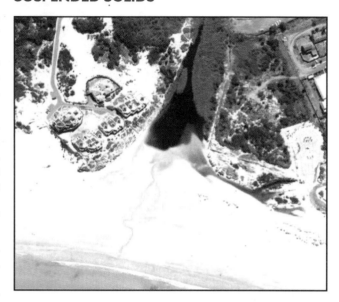

Suspended solids are solids that are found in water and usually originate from soil washed down a river or pollution from industry released into storm water drains. They reflect light and this can result in the water quality deteriorating because organisms cannot photosynthesise effectively. This results in a drop in oxygen and the water begins to stagnate. Fish and other creatures can suffocate through a shortage of oxygen in the water. When fertilisers are washed off the land they can stay in the water as suspended solids. Suspended solid levels can be limited by preventing pollution from entering rivers, by practising good farming methods so that excess fertilisers do not run off the land into the rivers, and also by preventing soil erosion. Suspended solids are a natural occurrence but human activities can contribute substantially to increased levels of suspended solids, which will have an overall negative effect on water quality. It is particularly important to ensure that industrial effluents are properly treated before being discharged into sewers and watercourses, to reduce the levels of suspended solids, thus preventing deterioration of water quality.

Key Associated Topics:

Dissolved Oxygen Content; Eutrophication; Freshwater Ecology; Pollution; River Catchment Management; Rivers and Wetlands; Soil; Water Quality.

Associated Organisations:

Aqua Catch; Groundwater Association of KZN; Rand Water; Water Institute of SA; Umgeni Water; Unilever Centre For Environmental Water Quality.

SUSTAINABLE DEVELOPMENT

The most frequently used definition of Sustainable Development is, "Development that meets the needs of the present without compromising the ability of future generations to meet their own needs." But whilst the commitment to these words may flow simply and sincerely off a chairman's tongue, finding their way into the companies triple bottom line report – the interpretation and practical implementation of these words – is by no means a simple matter.

In 1992 the concept of Sustainable Development emerged, when more than one hundred heads of state met in Rio de Janeiro, Brazil, for the United Nations Conference on Environment and Development (UNCED). The Earth Summit, as UNCED was also known, was convened to address urgent problems of environmental protection and socio-economic development. The assembled leaders signed the Framework Convention on Climate Change and the Convention on Biological Diversity; endorsed the Rio Declaration and the Forest Principles; and adopted Agenda 21, a 300-page plan for achieving sustainable development in the 21st century.

In December 1992 The United Nations Commission on Sustainable Development (CSD) was created to ensure effective follow-up of UNCED, and to monitor and report on implementation of the agreements at local, regional, national and international levels. (It is useful to refer to their website www.un.org

In 1997 the United Nations General Assembly met in special session, for a five-year review of progress since the Earth Summit. This was followed in 2002 by a ten-year review at the World Summit on Sustainable Development (WSSD).

From 28 April-9 May 2003 the 11th Session of the CSD (CSD-11) was held in New York where decisions were made on the Commission's future programme and organisation of work. It was agreed that the CSD's multi-year programme of work beyond 2003 would be organised on the basis of seven two-year cycles, with each cycle focusing on selected thematic clusters of issues. A statement early in 2004 from the new Chairman of the CSD – Minister H.E. Børge Brende of Norway – was bright, positive and optimistic. He said: "The international community has set itself ambitious goals for Sustainable Development. The CSD must help ensure that these commitments are delivered upon. We have no time to lose when it comes to transforming global commitments into action at the local and regional levels."

Whilst The Enviropaedia finds almost universal consensus on the need to implement the principles of Sustainable Development, we have also found such divergence of interpretation and priorities that we have focused this entire edition on the exploration of the term and the issues at stake. We have also added a new section to the publication (The Sustainable Development Section) in which we have invited various business and political leaders to give us their interpretation of Sustainable Development and its implementation within their sector. We hope that this will facilitate constructive debate and build bridges of understanding among those who are empowered to define and determine the implementation path ahead.

Guest Author: David Parry-Davies

Key Associated Topics:

Agenda 21; Cleaner Production; Consumerism; Eco-efficiency; Ecological Footprints; Ecological Intelligence; Economics; Environmental Ethics; Environmental Governance; Environmental Law; Environmental Management Planning; GAIA; Global Compact; Industrial Ecology; Local Agenda 21; Population; Poverty Alleviation; Stewardship; Triple Bottom Line; World Summit for Sustainable Development; as well as All Guest Essays in the Sustainable Development Section.

Associated Organisations:

Absa; Africabio; Allganix Holdings; Beco Institute for Sustainable Development; BMW; Chamber of Mines; Chemical & Allied Industries' Association; De Beers; De Beers Marine; Dept Environmental & Geographical Science – University of Cape Town; Development Bank of SA; Development Consultants of SA; Duard Barnard Associates; Edward Nathan & Friedland; Enact International; Engen; Environmental Ethics Unit – Stellenbosch University; Green Trust; Greenhaus Architects; Hangerman; Hillside Aluminium; Holcim (South Africa); Imbewu Enviro; IUCN SA; Ken Smith Environmentalist; LC Consulting Services; Liz Anderson & Associates; Master Farmer Programme; Metago Environmental Engineers; Metrix Software Solutions; Minerals & Energy Education & Training Institute; Mining Review Africa; Museum Park Enviro Centre; National Cleaner Production Centre; Natural Step The; Oryx Environmental; PPC Cement; Responsible Container Management; Richards Bay Minerals; Solardome SA; SouthSouthNorth; Sustainable Energy Society of SA; Sustainable Energy Africa; Viridus Technologies; Zeri South Africa.

It comes down to respect

The Standard Bank group is a major regional banking force employing more than 37 000 people in its banking and insurance business. Based in South Africa, we have operations in 17 other African countries and niche investment and offshore banking operations in 21 countries outside Africa. Over the past 10 years we have structured ourselves for future growth, with:

- A substantial retail and SME banking presence in sub-Saharan Africa;
- A leading domestic corporate and investment bank, now integrating its activities with our international investment bank focused on other emerging markets; and
- Asset growth from R65 billion in 1992 to R390 billion in 2002.

The international movement towards good corporate citizenship and social responsibility is relatively young. There is still a powerful body of investors demanding companies to stick to the principle of the dominant bottom line. We see it differently. In each society in which we operate as a company and live as individuals, we understand that caring about social and environmental issues is essential. To do this we grasp and embrace the meaning of *ubuntu*– that you can be respected only because of your cordial co-existence with others.

Standard Bank's sustainability report for 2002 is our move to complying with the guidelines released by the Global Reporting Initiative (GRI), a joint initiative of the United States non-governmental organisation, Coalition for Environmentally Responsible Economies and the United Nations Environment Programme. The sustainability report represents a commitment to sustainable development and to comprehensive reporting on it to all stakeholders.

Visit www.standardbank.co.za and click on Investor relations to view our report.

The Standard Bank of South Africa Limited (Registered Bank) Reg. No. 1962/000738/06 SBSA 708529-11/03

SYNTHETIC FUELS

Photo: Sasol (Earthyear Magazine)

South Africa's example of synthetic fuels is the petroleum products that are produced by Sasol at Sasolburg. The process started by converting coal into petrol and diesel using readily available low-grade coal. The technology has been developed further to convert readily available natural gas into liquid fuels using the Sasol Slurry Phase Distillate process.

The synthetic fuels production was the basis and launch pad for a chemical industry that has extended extensively and diversified into a wide range of differing speciality chemicals. When originally developed, the need to inject additional energy levels was not seen as significant because it provided an alternative source of fuel. Furthermore, it made use of low-grade coal, which was available in abundance. Current production has focussed on greater efficiency and cost effectiveness to enable the company to compete aggressively on the global petro-chemical and speciality chemicals markets. The latest advances in the Sasol Slurry Phase Distillates process technology allow for the conversion of natural gas into a high quality diesel fuel with a low sulphur content, low particulate emissions, low aromatics content and high combustion efficiency. These qualities all contribute to improved environmental performance and a lower health impact.

Key Associated Topics:

Energy; Fossil Fuels; Industrial Ecology; Mineral Resources; Non-renewable Resources.

Associated Organisations:

Chemical & Allied Industries' Association; Crowther Campbell; Mintek; Minerals & Energy Education & Training Institute; Oil Industry Environment Committee.

TRANSBOUNDARY OR TRANSFRONTIER POLLUTION

This is pollution that is generated in one country and is carried via air or water to another country where it causes damage, illness or nuisance. The only way to solve this kind of pollution problem is by international agreement and combined environmental education and awareness to reduce pollution levels. It has been demonstrated that much of the pollution that is killing off (and has killed off in the past) many of the forests in Scandinavia, comes from the industrial areas of northern England. The Chernobyl nuclear disaster in 1986 caused clouds of radiation contaminated particles to drift over Scandinavia and has resulted in elevated cancer levels in those countries, as well as the need to slaughter thousands of cattle because of elevated radiation levels in their milk. It has been said that the world has become a global village where communication, trade and access have become close and easy. Transfrontier pollution has always been a problem and requires serious management to prevent pollution generators from moving their problems from one country to the next.

Key Associated Topics:

Air Quality; Atmospheric Pollution; Chernobyl; Eco-catastrophe; Ecosphere & Ecosystems; Global Warming & Climate Change; Groundwater; Kyoto Agreement; Pollution; Waste Management; Water Quality.

Associated Organisations:

National Clean Air Association; Water Institute of SA; Richards Bay Clean Air Association. Unilever Centre for Environmental Water Quality; Aerosol Manufacturers' Association.

TRANSFRONTIER (PEACE) PARKS

The Peace Parks Foundation, a Section 21 company not for gain, was started in February 1997 to facilitate the establishment of transfrontier conservation areas, which are also commonly referred to as peace parks. Transfrontier conservation is a vehicle for sustainable development. This is because the principles underpinning the peace parks initiative are fundamentally correct, as they address poverty, which is caused by massive unemployment and a chronic lack of skills and entrepreneurial training. This often leaves poor people living in or adjacent to conservation areas with very few alternatives but to exhaust the very resource base on which their and our survival depends.

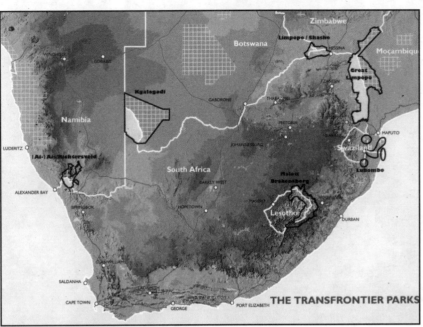

THE TRANSFRONTIER PARKS

The economic potential of peace parks is the mechanism for people to help themselves, by utilising natural resources to their benefit without destroying biodiversity. Only if we ensure sustainable economic growth, based on eco-tourism, which is the fastest growing industry in the world, will the people of Africa and elsewhere have reason to protect their natural assets. A transfrontier conservation area (TFCA), or peace park, usually refers to a cross-border region where the conservation status of the various component areas differs. These areas may include private game reserves, communal natural resource management areas and even hunting concession areas. Fences, major highways, railway lines or other barriers may also separate the various parts. However, they nevertheless border on one another and are managed for long-term sustainable use of natural resources, even though the free movement of animals amongst the various parts may not be possible. The SADC Protocol on Wildlife Conservation and Law Enforcement defines a TFCA as "the area or component of a large ecological region that straddles the boundaries of two or more countries, encompassing one or more protected areas as well as multiple resource use areas."

A transfrontier park is established when the authorities responsible for areas where the primary focus is wildlife conservation and which border on one another across international boundaries, formally agree to manage those areas as one integrated unit according to a joint management plan. These authorities also undertake to remove all human barriers within the Transfrontier Park so that animals can roam freely. The purpose of these parks is to employ conservation as a land-use option to the benefit of local people. At the outset, six TFCAs were identified. Southern Africa's first peace park, the Kgalagadi Transfrontier Park, was formally opened on 12 May 2000 by the Presidents of Botswana and South Africa. On 9 December 2002 the three heads of state of Mozambique, South Africa and Zimbabwe signed an international treaty to establish the Great Limpopo Transfrontier Park. As recent as 1 August 2003, a treaty on the establishment of the |Ai-|Ais/Richtersveld Transfrontier Park was signed in Windhoek by the Presidents of Namibia and South Africa. Agreements to develop further transfrontier conservation areas in southern Africa are under way, with some in the final stages of development.

These successes have resulted in the Ministers for Tourism of the SADC countries, via the Regional Tourism Organisation of Southern Africa (RETOSA), commissioning a detailed feasibility study, funded by the Development Bank of Southern Africa and Peace Parks Foundation, and managed by the latter. The study involves consultations with governments regarding potential TFCAs in their countries, thereby clearly defining possible locations as well as the potential for economic growth and the conservation of biodiversity. It also gives a clear indication of the potential impact future TFCAs will have on the economy and biodiversity of the region as a whole, where they could be established and a cost estimate of doing so. In all, 16 further potential TFCAs have been identified. Should this transpire, it will bring the total area of TFCAs in the SADC to about 120 million hectares, which is roughly the size of France, Germany and Italy combined. They include various major biomes and ecoregions, which means an important contribution to biodiversity. The diversity of wildlife in these areas will attract tourists and form the basis for sustainable economic development. It is estimated that the potential number of tourists that could visit these areas once infrastructure is in place could number 8 million per annum. Compared to South Africa's 1,8 million overseas arrivals last year, the potential benefit to the region becomes obvious.

Guest Author: Prof Willem van Riet ~ Peace Parks Foundation

Key Associated Topics:

Eco-tourism; National Parks; Land Use Planning; Poverty Alleviation; Stakeholders; Wilderness Area; Wildlife; Wildlife Management.

Associated Organisations:

Absa; Conservation International; Imagis; Kumba Resources; Peace Parks Foundation; Wilderness Action Group.

FROM GOATS TO GAME

by Heather Dugmore

There is a highland place where three mountain ranges – the southern Drakensberg, the Witteberge and the Malotis – converge. It is the point where South Africa and Lesotho meet.

The landscape is remote and mystical and rich with San rock art, dinosaur fossils and alpine plant species. Regrettably, much of the land has been overgrazed by cattle, sheep and goats and the more accessible areas have been fenced off as livestock farms. Poachers and hunters have also left their mark and most of the indigenous game hasn't been seen for decades. Cape eland, mountain reedbuck, klipspringer, black wildebeest, Cape mountain zebra, blesbok and oribi … all were once here.

And all will return once the Southern Drakensberg Transfrontier Conservation Area (SDTCA) is established.

This is the dream. This is the dream that is becoming a reality.

The pilot area earmarked for this transfrontier, community-based conservation area is a 100 000-hectare stretch on the border between the Eastern Cape and the southern tip of Lesotho. It incorporates private, communal and state-owned land between South Africa and Lesotho.

Spearheading the initiative is Tenahead Mountain Reserve, a 3 000-hectare private nature reserve on the South African side. Formerly a sheep farm, Tenahead is the first area in the proposed SDTCA to be transformed into a nature reserve, with a small luxury mountain lodge (at 2 700m it's probably the highest in Africa) to accommodate classic safari guests.

"Our goal is to encourage a culture of nature conservation side by side with one of entrepreneurship and ownership of the natural resources for the benefit of all," says Guy Stubbs, one of the directors of Tenahead. So committed is Stubbs to establishing the SDTCA that he sold up his considerable interests in Gauteng last year and moved with his family to Tenahead.

"The conservation area we envisage would see landowners, communities, long-term users and the South African and Lesotho governments joining their various parcels of land together, by contractual arrangement, to create a larger, sustainably managed conservation area," he explains.

"Myself and Tenahead co-director Andrew Weeks, have already had extensive discussions with all the stakeholders – including the community and farmers in the north eastern Cape/Mt Fletcher area (former Transkei), the Lesotho communities and farmers, and the South African and Lesotho governments."

In summary, the goal is to replace the region's small-scale livestock farming economy with a game reserve/eco-tourism economy. For the SDTCA to succeed, it will be imperative that all farmers and communities within the conservation areas (and not simply the chiefs) benefit from it.

Preliminary research suggests that about 5 000 sheep and 300 cattle (between about 35 farmers) are grazed in the upper Sebapale valley on the Lesotho side of the proposed SDTCA pilot area and a further 4 500 sheep and 1 000 cattle (among about 25 farmers) in the Betania/Ulundi grazing areas on the South African side. Assuming an annual net income of R50 per sheep and R100 per head of cattle, rough estimates indicate the livestock represents about R600 000 in annual income for these farmers in the form of meat and wool. This amount would have to be replaced by game-farming and eco-tourism revenue.

Local stakeholders would be assisted in starting small-scale eco-tourism ventures and businesses. For example, the traditional 'xamisi' – stone shepherds' huts – could be transformed into tourist accommodation. The SDTCA management team would assist in upgrading these to an acceptable standard of comfort, adding ablution facilities (including an eco-friendly sewage system) and kitchen facilities.

Additional sites within the reserve would be selected for small, upmarket lodges. Reputable tour and lodge operating companies would be invited to build or manage the lodges that would be owned in partnership with the resident communities. People from the community would be employed in the building and running of each venture.

Another idea, still being developed, is that each participating community would develop a central logistics 'hub' at the entry points to the SDTCA. The hubs would include a restaurant, stables (tourists will be able to explore the reserve on horseback), firewood, first-aid facilities and guest accommodation.

The communities will receive education and training appropriate to eco-tourism and conservation. Activities offered would include hiking, mountaineering, fishing, horse-riding, game-viewing, adventure sports, botanical, palaeontological, geological and rock art tours, cultural experiences, mountain biking and birding.

Of course the creation the SDTCA is far easier said than done as Stubbs explains:

"The local communities and farmers naturally fear the loss of their livestock and grazing lands. This is their livelihood and they don't feel confident about making money from eco-tourism and wildlife conservation because it's foreign to them."

These fears need to be addressed and a significant cultural and educational shift needs to take place, especially in relation to the central role that livestock husbandry plays in Xhosa and Sotho culture. The financial gain and value of nature conservation and wildlife needs to be demonstrated and proved to local farmers and communities. For example, whereas the value of a sheep on the open market is worth about R300, a vaal rhebok (small antelope indigenous to this region) can be sold for R2 600."

Game-farming within a conservation context will provide an important source of income for the SDTCA and all its stakeholders. "Because the project will take a few years to implement, a relief fund will probably be necessary for any-one whose livelihood is negatively affected in the short-term by the creation of the reserve."

In the medium- and long term, with proper management and skills training, the communities and farmers have plenty to gain. The area is currently characterized by extreme poverty and unemployment, illiteracy, stock theft, firearms, drug smuggling, game poaching and severe erosion.

The conservation area will dispense with the serious and persistent problems of stock-theft since no livestock will be allowed in the area and everyone will have a stake in protecting the game. The border fence (which is broken and ineffectual) will be removed and the entire perimeter of the conservation area will be fenced and patrolled by game wardens.

So far, the community response has been extremely enthusiastic and positive and the Mt Fletcher communities have already given their formal consent to the project. "We are now in discussion with the South African and Lesotho governments and the project is looking promising," continues Stubbs.

Subject to the requirements of the South African and Lesotho governments, it is envisaged that a non-profit, section 21 company will be established to manage matters of common concern in the SDTCA.

Ultimately it is hoped the pilot project will be duplicated on the Lesotho side expanding the conservation area from 100 000 hectares to 200 000 hectares – and will fall under the Maloti-Drakensberg Transfrontier Project (incorporating the uKhahlamba Drakensberg Park World Heritage Site).

"This whole region has enormous eco-tourism potential," states Stubbs. "These magnificent mountain ranges should be regarded as a huge asset to South Africa, Lesotho and the communities who live here. The SDTCA's dream is to protect these assets."

The only litter you'll find in the world's greatest animal park.

The harmony between wildlife and man has never looked more promising. Where now, they are coming together as one on this precious land of ours. To achieve this, Absa are proudly working as a conservation partner with the Peace Parks Foundation to break down barriers of the Great Limpopo and Kruger Transfrontier parks. Making such important tracks requires invaluable input from the local communities, creating much-needed jobs and careers along the way. It's a task that involves game ranging, managing the natural beauty of the flora and fauna and above all seeing that our wildlife can live in an environment that isn't a threat to them. Absa will continue to partner the Peace Parks projects. It is after all, in our nature to see that peace and prosperity continues to flourish.

Today Tomorrow Together

TRENCH GARDENING

Trench gardening is a form of subsistence vegetable gardening that has been practised in the townships since the late 1970s. It involves digging a door size trench, about a spade-and-a-half-blade deep. This is then filled with organic domestic waste, plus a few rusty tin cans. The soil is put over the top and is, in turn, covered with mulch (dry organic matter or wood bark). This then creates an ideal and rapid growth media in which the owner can plant vegetables and quickly grow food to eat. This method was developed in response to the township calls that the soil was too weak and poor to support any kind of vegetables. If a group of community members get together and work the land in unison, the quantities of food grown can be increased substantially and everyone can benefit from the gardens.

Key Associated Topics:

Agriculture; Biodegradable; Composting; Organic Farming; Organic Foods; Permaculture; Poverty Alleviation; Recycling; Soil; Urban Agriculture.

Associated Organisations:

Abalimi Bezekhaya; EcoLink; Food and Trees for Africa; Food Gardens Foundation.

Photo: Guy Stubbs Photographer

TRIPLE BOTTOM LINE

The term, "Triple Bottom Line", describes one of the new theories of sustainable development developed by John Elkington and his team at Sustainability Ltd in London. (See www.sustainability.co.uk) This theory suggests that true sustainable development in business must consider not just the financial "bottom line" of prosperity and profit, but also other "bottom lines" such as environmental quality and social equity. Companies, therefore, when they submit their annual reports, should be looking not just at the "financial bottom line" of profit but also at the "Triple Bottom Line" of prosperity, environmental quality and social quality. The triple bottom line represents a level of interdisciplinary thinking that is vital to tackle the complex and varied environmental problems that exist today. The Triple Bottom Line was developed with the aid of industrialists to meet their needs and also to enable the concepts to be easily translated into the financial contexts of business.

Key Associated Topics:

Ecological Footprints; Environmental Auditing; Ecological Intelligence; Economics; Environmental Ethics; Environmental Governance; Global Compact; Poverty Alleviation; Responsible Care; Sustainable Development.

Associated Organisations:

Development Bank of SA; Hillside Aluminium; Holcim (South Africa); HRH The Prince of Wales' Business & the Environment Programme; KPMG; National Business Initiative; Price WaterhouseCooper.

TROPICAL RAIN FOREST

Tropical rain forests appear like a girdle around the equator and occupy approximately 8% of the earth's land surface, yet they comprise about half of all growing wood on the face of the earth and provide habitats for at least two fifths of all the earth's total species. Scientists have still only identified one in six of the estimated two million species that exist in these forests. The rain forests also acts as a "green lungs" for the earth, producing vast quantities of oxygen for us to breathe and survive, and absorbing similar quantities of carbon dioxide as a part of the greater carbon cycle. These forests provide an important balance in managing the air quality of the earth we live on. Temperatures in tropical rain forests usually remain between 20 and 35 degrees Centigrade all year round, and rainfall typically varies from 125 to 510 centimetres per year. The soil in tropical rain forests is normally very thin and low in nutrients so when the canopy cover is removed by, for example, hardwood logging activities, the land becomes seriously degraded very quickly and is lost through being washed away by tropical rainstorms.

Key Associated Topics:

Agenda 21; Air Quality; Biodiversity; Biome; Carbon Dioxide; Deforestation; Ecosphere and Ecosystems; Forestry; GAIA; Land Degradation; Natural Resources; Oxygen.

Associated Organisations:

Botanical Society of South Africa; Conservation International; National Botanical Institute; WWF.

ULTRA-VIOLET LIGHT

Ultra-violet (UV) radiation is between visible light and X rays on the spectrum (4×10^{-7} to 5×10^{-9}). The UV spectrum is further divided by wavelengths into A, B and C bands. Most of UV radiation is absorbed in the ozone layer before it reaches the lower reaches of the atmosphere. In people, it acts on ergo sterol in the skin to produce Vitamin D and is absorbed by melanin in the skin causing it to darken. Within cells, it causes damage through chromosome breakage. Excessive exposure to ultra-violet radiation results in skin cancer. Research studies are beginning to suggest that the weakening of the ozone layer by industrial activities may be having an effect in the southern hemisphere and that the incidence of skin cancer, particularly in South Africa, is rising. Health specialists now agree that it is dangerous to spend time exposed to the sun without some form of protective layers to reduce the effects of ultra-violet radiation. There is no consensus on the value and success of using sun-block lotions to minimise the effects of ultra-violet light on the skin.

Key Associated Topics:

Aerosol; Atmospheric Pollution; Environmental Health; Montreal Protocol; Ozone; Pollution; Radiation.

Associated Organisations:

Environmental & Geographical Science Department – University of Cape Town; South African Institute of Ecologists and Environmental Scientists.

URBAN AGRICULTURE

Cities and towns are not the places where one would normally think that agriculture should be undertaken, but urban agriculture can be a valuable contribution to the development of sustainable cities. Agriculture in urban areas has a number of benefits: 1) Idle land which would otherwise turn into unsightly dump sites can be utilised and beautified through the growing of crops. 2) Using wastewater from drains, storm water drains and other uses (Greywater) means that resources are saved. 3) Household waste can be composted and used to fertilise the crops thus encouraging recycling. 4) Urban communities will benefit through the additional supply of fresh food or added income through the sale of harvested products. Urban agriculture forms one of the parts of a sustainable development philosophy and can be found in most Local Agenda 21 programmes.

Key Associated Topics:

Agriculture; Composting; Greywater; Local Agenda 21; Organic Farming; Organic Foods; Permaculture; Poverty Alleviation; Recycling; Soil; Trench Gardening; Urbanisation; Sustainable Development – In the Built Environment.

Associated Organisations:

Abalimi Bezekhaya; EcoLink; Food and Trees for Africa; Food Gardens Foundation; Greenhouse Project; Permacore; Tribal Consultants.

URBAN GREENING

Urban greening describes the varied initiatives that are being undertaken within the urban areas to "green" and soften the urban landscape and make it more "people friendly". These initiatives include the planting of road verges and islands with indigenous vegetation, the removal of alien species, the upgrading of parks and gardens in inner city areas, the development of urban trails which are planned to traverse through green areas, and the changing of urban landscapes through the use of landscape architects to design or re-design inner city areas to reflect a more natural and green surrounding. A number of South African cities have embarked upon programmes to green specific areas. The Cape Town City Council initiated the very ambitious "Greening the City" programme, which worked up to a point until the pressures of development planning began to cut into the green "islands" and trail areas. Durban revamped its beachfront area to attract more tourists by using natural materials and plants and trees to soften the landscape.

Key Associated Topics:

Aesthetics; Eco-Tourism; Habitat; Landscape Architecture; Local Agenda 21; Permaculture; Rehabilitation of Land; Trench Gardening; Urban Agriculture; Urbanisation; Women and the Environment; Sustainable Development Section: – The Built Environment.

Associated Organisations:

African Gabions; Environmental Education and Resources Unit – University of Western Cape; Green Inc; Newtown Landscape Architects; Newtown Landscape Architects-Limpopo; Rand Water; The Greenhouse Peoples Environmental Centre; Institute of Landscape Architects.

Photo: Bruce Sutherland - City of Cape Town

URBANISATION

Guest Essay by Llewellyn van Wyk
~ Senior Researcher CSIR (Boutek)

The term "Urbanisation" describes the growing trend of populations to gravitate toward living in city or urban environments. The towns and cities of the world are growing at ever increasing rates. Only slightly less than 30 percent of the global population lived in urban areas 50 years ago. Today, three billion people – almost half the global population – live in urban areas. In some parts of the developed world, the percentage is already beyond the 50 percent mark. By 2020, it is estimated that the figure will have reached 57 percent. When the global average breaches the 50 percent mark, a major transformation will have occurred in human society.

Not surprisingly, cities have become the economic powerhouses of national economies, often showing economic growth rates in excess of the national growth rate. Throughout history, cities have been synonymous with economic, social, and cultural growth. Yet, almost one third of today's urban population is desperately poor and live in slums without even the most basic of services.

The burgeoning growth in urban populations has resulted in greater demands for municipal services. In most developing economies it has become a common feature that the demand has substantially exceeded municipalities financial and resource capabilities to deliver such services.

In view of the growing trend toward urbanisation, two questions must be answered when determining the future of the city: Firstly, has the world the capacity to become fully urbanised?

Secondly, what will the effect be on human society?

In response to the first, it would appear that unless urgent and drastic interventions are introduced, it is likely that urbanisation will increase, but with a marked and continuing degradation in the quality of urban life. The second question may however be more difficult to assess, as we must consider whether the anonymity, mobility, impersonality, specialisation, and sophistication that characterises the city dweller, can become the attributes of a stable society, or will the society fall apart?

It seems likely that a completely urbanised world would be significantly different in its social structure from anything we currently know of today. But, for such a technologically advanced and substantially urbanised society to survive and thrive, it will need to adopt very different management strategies from those currently in place. Early trends are already manifesting, indicating that a new urbanised world will be far more integrated, less segregated, more community-orientated, more institutionalised, with ethics, culture and tradition being highly valued.

Key Associated Topics:

Aesthetics; Air Quality; Coastal Zone Management; Consumerism; Environmental Refugees; Green Architecture; Local Agenda 21; Population; Poverty Alleviation; Water; Waste Management; Sustainable Development Section: – The Built Environment.

Associated Organisations:

Built Environment Support Group; SA New Economics (SANE); Department of Geographical & Environmental Science – University of Natal; WESSA.

Photo: Bruce Sutherland - City of Cape Town

USED OIL

In South Africa, there is approximately 90 million litres of recoverable used oil per annum available. Unless recovered, this used oil poses a serious hazard to the environment, and to the country's groundwater resources in particular. The South African oil industry set up a non-profit organisation called the ROSE (**R**ecycled **O**il **S**aves **E**nvironment) Foundation to collect as much used oil for recycling as possible. In the three years that the ROSE Foundation has been operating, it has managed to build up its collection levels to its current rate of 37 million litres per annum. In addition to collecting and disposing of the used oil, the ROSE Foundation also audits and inspects its collection agents to ensure high standards and to prevent leakages at the various sites and depots.

Photo: Department Minerals & Energy

ROSE Foundation is a useful case study on just how successful a used oil collection operation can be. (Visit their website at: (www.rosefoundation.org.za) The case demonstrates the success of good organisation and planning and experts agree that the approach can be applied to any resource recovery initiative.

Key Associated Topics:

Contaminated Land; Hazardous Wastes; Local Agenda 21; Pollution; Polluter Pays Principle; Recycling; Waste Management.

Associated Organisations:

Rose Foundation; Tiger Chemical Treatment; Oil Industry Environment Committee.

VETIVER GRASS

Vetiver grass (*Vetiveria zizanioides*), sometimes called "Mauritius grass", is a coarse grass that can be used as a natural barrier to prevent erosion of soil. The plant is a native of India where it is called "khus" or "khus khus". The plant is very versatile, being able to withstand being submerged in water during floods and also having the ability to withstand the severe pressures of drought. When planted in hedges, it acts as an erosion barrier and can also protect terraces or slopes from damage from heavy rainfall or mud slides. The plant is available in South Africa and has been shown not to be invasive or damaging in any way to the soil. It is known to exist in 24 countries in Africa, 13 countries in Asia and 11 countries in the Americas. The plant is able to function as a natural engineering resource without causing ecological damage or interference with natural systems. For more information on vetiver grass, contact the Southern African Vetiver Grass Network at: www.inr.unp.ac.za/vetiver

Key Associated Topics:

Bio-remediation; Conservation; Land Degradation; Landscape Architecture; Rehabilitation of Land; Soil.

Associated Organisations:

Grassland Society of Southern Africa; Institute of Landscape Architects; WWF.

VOCS (VOLATILE ORGANIC COMPOUNDS)

VOCs are primarily the lighter "fractions" of oil or hydrocarbons, that is, the parts that evaporates easily because they have a low boiling point. Some of these products are used because of their evaporating or "quick drying" characteristics. Most VOCs are carbon-hydrogen compounds (hydrocarbons) but they also include aldehydes and ketones. Emissions come, for example, from households (cleaners, cosmetics, lacquering), metal industry (paints and coatings), printing industry (inks and cleaners), oil refineries (loading, storage and transfer of hydrocarbon products), petrol stations (the pumping of petrol), and the dry cleaning industry (cleaning solvents). Tropospheric ozone forms as a result of photochemical (sunlight) reactions between nitrous oxides and VOCs and this can be hazardous to health in peak concentrations. Various countries (e.g. the European Union, with the Netherlands in the forefront) have begun to control the release of VOCs by setting various reduction targets.

Key Associated Topics:

Air Quality; Atmospheric Pollution; Carcinogen; Environmental Health; Fossil Fuels; Hazardous Wastes; Ozone; Petro-chemicals; Photochemical Smog; Pollution; Polluter Pays Principle; Solvents.

Associated Organisations:

Chemical and Allied Industries' Association; SI Analytics.

WASTE MANAGEMENT IN SOUTH AFRICA

Guest Essay by PF Theron
~ Institute of Waste Management Southern Africa

What is waste?

In The South African White Paper on Integrated Pollution and Waste Management (2000) waste is defined as:

… an undesirable or superfluous by-product, emission, or residue of any process or activity which has been discarded, accumulated or been stored for the purpose of discharging or processing. It may be gaseous, liquid or solid or any combination thereof and may originate from a residential, commercial or industrial area. This definition includes industrial wastewater, sewage, radioactive substances, mining, metal-lurgical and power generation waste.

The exact definition of waste is however the topic of an ongoing debate because of an increasing global trend to reduce, reuse, rework, recycle, recover, so-called "waste" products. One person's waste can now become another person's valuable raw material. Also with changing technologies, availability and cost of original input materials, the demand for, or need to, use recovered "wastes" is also changing. For example, the gold extraction process used a century ago on the Witwatersrand mines was relatively inefficient measured against today's technology. The result is that a great many "waste" sand dumps have now become very valuable assets to remine. The question therefore is – when is "waste" really waste?

A general definition of waste could then be formulated as "something which nobody wants at a particular moment in time and which needs to be disposed of."

Sustainability

Waste generation and disposal is generally viewed as a key indicator of an unsustainable operating society, so waste solutions encompassing a "cradle to cradle" approach (from the source of production beyond the typical "after-life" management – e.g. via disposal – towards a new lease of life) needs to be considered as part of an extended producer liability. Finished products and goods designed in a way that they can be easily de-manufactured and dismantled for material recovery and recycling, results in the reduction of man's reliance on non-renewable resources. This concept is not easy to achieve however, and will require a radical mind shift in our society on the consumer level (demanding waste-wise products) as well as on industrial level (increasing cleaner production technologies). The quest for zero-waste and achieving sustainable waste management can be realised through a process of gradual improvement in production efficiency and consumer waste awareness as outlined in Cape Town's integrated waste management plan. Zero-waste to landfill may not be achieved in the short to medium term, however we must actively pursue this vision as the ultimate goal. The first step is to set achievable targets in reducing the growth of the waste stream and reducing as much waste as possible at source (by preventing it from being generated). We should also be aware and plan for the fact that in aiming for zero-waste, it becomes progressively more difficult and more expensive to achieve the higher percentage targets for reducing waste volumes.

Integrated Waste Management (IWM)

Integrated waste management will become mandatory once the National Integrated Waste Management Bill is promulgated into an Act. The IWM concept recognises the importance of a broad hierarchy of preferred options, which look at the waste stream in a cradle to grave approach namely:

♦ Waste avoidance: The reduction of waste at source. Through a deliberate policy of minimising the creation of waste within an industrial process, many "waste-exchange" opportunities (whereby one company's waste becomes another's raw material) can significantly reduce costs and increase the profitability of companies.

♦ Re-use: The utilisation of a waste product without further transformation (e.g. the use of old newspapers as wrapping material or using old glass jars for storage).

♦ Recycling: The manufacturing of a product that is made from waste materials. This can only be done by a business that is technically equipped to change the properties of a former waste material into a new product (e.g. making plastic pellets out of plastic waste, melting waste glass to make new bottles, melting beverage cans for new steel appliances, etc.). There is a distinction between closed loop and open loop recycling. Closed loop recycling is a process within the same company that generated the waste, whereby the waste materials from one process is "internally recycled" to be used for another process step or to make another product. Open loop recycling means that the waste material leaves the location where it was generated and is sent elsewhere for recycling.

♦ Resource recovery: The recovery or retrieving of recyclable materials out of the waste stream or the collection of recyclable materials before they enter the waste stream – for the purpose of re-use or recycling.

♦ Treatment: The processes of changing the physical and/or chemical properties of a waste product, (e.g. by compaction, incineration, neutralisation of acids and bases and detoxification of poisons).

♦ Disposal: The final and least desirable step in the hierarchy, involving landfilling of wastes in a controlled manner.

Photo: Earthyear Magazine

Domestic Waste Management

Water & Sewage Treatment

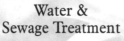

Healthcare Waste Managemnt

Landfill Management

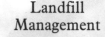

WR
WADE REFUSE
DOMESTIC WASTE MANAGEMENT

SEDIBA
WATER AND SEWAGE

SANUMED
HEALTHCARE WASTE MANAGEMENT

LANDFILL MANAGEMENT
WASTE FACILITY MANAGEMENT

MILLENNIUM
WASTE MANAGEMENT

Millenium Waste Management
Reg. No. 2001/007831/07

Gauteng	: Tel : (011) 456-5597/8	Fax : (011) 453-2241
Western Cape	: Tel : (021) 951-5383/93	Fax : (021) 951-8440
Eastern Cape	: Tel : (041) 466-2741	Fax : (041) 467-1578
KZN	: Tel : (031) 700-3921	Fax : (031) 700-3208
E-mail	: info.gp@millenniumwaste.co.za
Website	: www.millenniumwaste.co.za

PROUDLY SOUTH AFRICAN

Waste Transport

Waste Minimisation

Equipment & Plant

Waste-tech
INDUSTRIAL WASTE MANAGEMENT

PROCESS MANAGEMENT
WASTE REDUCTION, RE-USE AND RECOVERY

CONQUIP HIRE
EQUIPMENT & PLANT HIRE

Waste Technology & Disposal

Hazardous Waste & Emergency Response

Container Reconditioning

Dispose-tech
RESPONSIBLE WASTE DISPOSAL

HAZMAT
SUPPORT SERVICES

ENVIRODRUM
RESPONSIBLE CONTAINER MANAGEMENT

Enviroserv Waste Management
Reg. No. 1990/070417/07

Gauteng	: Tel : (011) 456-5400	Fax : (011) 453-7583
Western Cape	: Tel : (021) 951-8420	Fax : (021) 951-8440
Eastern Cape	: Tel : (041) 466-2741	Fax : (041) 466-2745
KZN	: Tel : (031) 902-1526	Fax : (031) 902-5778
E-mail	: info.ho@enviroserv.co.za
Website	: www.enviroserv.co.za

ENVIROSERV
WASTE MANAGEMENT

PROUDLY SOUTH AFRICAN

Waste generation in South Africa

The estimated total waste generation in the Republic of South Africa, excluding water-borne and atmospheric pollutants, is illustrated in the following table:

Estimated Waste Generation in the RSA in Million Tons/Annum

Source	Qty	% of Total	Hazardous Portion (tons)
Mining	377	80,3	1,05
Industrial	22	4,7	0,81
Power generation	20	4,3	0,01
Agriculture	20	4,3	0,13
Domestic & trade refuse	18	3,8	0,13
Sewage sludge	12	2,6	0,13
Total	469	100	2,26 = +/- 0,5 %

Non-hazardous industrial waste and domestic and trade refuse amount to 38 million tons per year and are classified a "general waste". The per capita generation of domestic waste varies from as low as 0,16 kg per day in informal settlements to over 2 kg per day in the more affluent societies with an average of about 0,9 kg per day, which is roughly equivalent to the world average.

Waste volumes are increasing in South Africa, along with increasing population and increasing living standards. Although the exact increase on an annual basis is not known for South Africa as a whole, over the past three years in Cape Town, it has been estimated to be between 5% and 7%.

Materials recovery and recycling in South Africa

Some 150 000 people are involved either directly or indirectly in the materials recovery process in South Africa. The following table summarises the quantities of material recovered during 2000:

Materials Recovery in South Africa

Type of Material	Total tonnage produced	Tonnage converted into packaging	Tonnage recovered	% of Total
Paper & board	1 990 000	868 000	770 000	38,7
Plastics	923 000	465 000	133 000	14,4
Tinplate	263 000	263 000	121 000	46,0
Glass	521 000	509 000	102 500	19,7
Total	3 697 000	2 105 000	1 126 500	30,5

Waste disposal

The competency conferred on the Department of Environmental Affairs and Tourism by the Environmental Conservation Act (ECA) to issue permits for the landfilling of wastes has been delegated to the Department of Water Affairs and Forestry (DWAF).

DWAF have subsequently developed a trilogy of documents, namely "The Minimum Requirements" to provide standards and give legal effect to the permitting and licensing of waste management facilities in South Africa. The trilogy provides graded standards for:

◆ Document 1: The Handling, Classification and Disposal of Hazardous Waste. This document classifies hazardous wastes into four hazard ratings and prescribes the technical requirements for the receiving landfills.
◆ Document 2: Waste Disposal by Landfill. This document deals with general waste and the requirements for the classes of landfills.
◆ Document 3: Water Monitoring at Waste Management Facilities

The first edition of The Minimum Requirements was published in 1994 and the second edition in 1998. The next in the series (Document 4) is presently being prepared by DWAF, which is due for completion in 2004. Document 4 covers landfill operations (selective aspects) and will set minimum requirements for the upgrading of dumpsites, training in waste management and the auditing of waste facilities.

Some 478 landfills have been authorised and given permits by DWAF at the time of writing, including 36 hazardous waste sites. There are however an estimated 800 landfill sites currently in operation throughout the country.

Legislative, policy and strategy frameworks

Over the past ten years, South Africa has made great strides in addressing key issues, requirements and problems experienced in waste management. Although the Environmental Conservation Act (ECA) addresses issues such as littering, permitting of waste disposal sites and regulatory competency, the Constitution of the Republic of South Africa Act, 1996 (Act 108 of 1996) has, for the first time, guaranteed the right of South Africans to a clean and healthy environment.

This was followed by the Draft White Paper on Integrated Pollution and Waste Management for South Africa, which was published in 1998. The White Paper advocates a shift from the present focus on waste disposal and impact control (i.e. end of pipe) to integrated waste management and prevention as well as minimisation. In terms of legal changes this will entail national government drafting legislation requiring the prevention and minimisation of waste.

The White Paper in turn, gave rise to the formulation of a National Waste Management Strategy and Action Plans, a collaborative effort between DEA&T and DWAF with Danish financial and technical assistance and input by a great many interested and affected parties. At the time of writing, DEA&T was developing a series of user-friendly guidelines called "Working with Waste" to implement needs identified in the strategy. The two volumes completed thus far are Guidelines on Recycling of Solid Waste and Guidelines on Waste Collection in High Density and Classified Areas. Although the strategy is a genuine attempt to develop a holistic and integrated approach to waste management in South Africa, the same cannot, unfortunately, be said of current legislation regarding waste management, which is fragmented, diverse, uncoordinated and administered by a number of different government departments. Mining and radioactive waste are, for example, excluded from the provisions of the ECA. A few of the many separate acts impacting on waste management are:

◆ The National Environmental Management Act, 1998 (Act 107 of 1998)
◆ The National Water Act, 1998 (Act 36 of 1998)
◆ The Hazardous Substances Act, 1973 (Act 15 of 1973)
◆ The Health Act, 1977 (Act 63 of 1977)
◆ The Nuclear Energy Act, 1993 (Act 131 of 1993)

Littering

Many South Africans have been allowed to become litterbugs, which has resulted in a dirty country and the expenditure of huge unnecessary sums of money to clean up the mess (without litterbugs these funds could have been spent on far more productive things such as housing, schooling and help for the poor). While this is a behavioural problem, the causes are diverse and can be traced back to poor education, unemployment and economic problems and even socio-political issues. It will need a concerted effort by all South Africans to change this mindset.

In an effort to curtail littering caused by plastic shopping bags, the Environmental and Tourism Ministry has, with effect from 9 May 2003, banned the use of shopping bags thinner than 30 microns. An agreement reached with a number of retail chain stores has also resulted in these thicker bags having to be purchased by the customer. This has resulted in a reported drop of some 80% in the manufacture of shopping bags and subsequent job losses. Understandably not everybody agrees with these measures but it has, without doubt, severely pruned the "national flower" from the landscape.

Similar efforts to curtail the illegal disposal of tyres and rubber are underway, whilst the glass recycling industry is busy cleaning its own house.

Key Associated Topics:

Agenda 21; Air Quality; Atmospheric Pollution; Cleaner Production; Composting; Consumerism; Dilution; Eco-taxes; Environmental Law; Hazardous Wastes; Incineration; Integrated Pollution and Waste Management; Landfill; Leachate; Methane; NEMA; Pollution; Polluter Pays Principle; Radioactive Waste; Recycling; Scrubber; Water Quality.

Associated Organisations:

Apple Orchards Conservancy; Arcus Gibb; Beco Institute for Sustainable Development; City of Cape Town Solid Waste; Enviro Options; Environmental Risk Management; Enviroleg; Enviroserv Waste Management; Fairest Cape Association; Gauteng Conservancy Association; Greenhaus Architects; Hewlett Packard South Africa; Hillside Aluminium; Icando; Inspire!; Institute of Waste Management of Southern Africa; Jarrod Ball; Keep Pietermaritzburg Beautiful; Millennium Waste Management; NaDeet; Nampak Recycling; National Cleaner Production Centre; Recycling Forum (National); Responsible Container Management; SA Climate Action Network (SACAN); Softchem; Technikon Pretoria; Terratest; Thorntree Conservancy; Umgeni Water Services; Wales Environmental Partnerships.

Photo: Bruce Sutherland - City of Cape Town

WATER QUALITY AND AVAILABILITY

Guest Essay by Dr Eureta Rosenberg

When it comes to water, southern Africa is a region of extremes. Eastern areas have problems associated with too much water, such as flooding, malaria and cholera. Parts of the west are so dry that children grow up without seeing rain and industrial development is difficult. Some South African teachers take water from home to flush the staff toilets; others coach swimming in sparkling blue pools. Some businesses 'sweep' their courtyards with drinkable water, whilst others lack water to wash the fruit they sell by the roadside.

This diversity comes about through rainfall patterns, and through the policies, which determine access to resources. Water availability is also affected by how we use water. One of its more hidden 'uses' is for waste disposal. The resulting pollution lowers water quality which, in turn, affects availability. We start our exploration of water issues here.

WATER QUALITY

Among the many factors affecting water quality is siltation. Deforestation, poor farming and other inappropriate land-use practices contribute to soil erosion. Erosion and construction cause topsoil to wash into rivers and dams. Whilst silt helps to make estuaries fertile, too much of it interferes with water plants and animals, affects fish stocks and shortens the useful life-span of dams, which is a major problem in Africa. Wetlands are invaluable as they remove excess soil (and pollutants) from water. Unfortunately wetlands are now endangered ecosystems, threatened by roads, farming and commercial developments, all of which can reduce water quality. The Okavango Delta, one of the most important wetlands in the region, and a great tourist attraction, is under pressure from mining and agriculture. In South Africa, half our wetlands have already been destroyed.

Another factor lowering water quality is the excessive use of fertilisers to grow crops. Excess fertiliser washes into rivers and lakes causing eutrophication, which suffocates water life. Eutrophication is also caused by human waste, which has become a major cause of water pollution in the region. Not only do thousands of people lack adequate sewage facilities, even expensive resorts and housing estates may allow inadequately treated sewage to flow into the very water that made the place attractive to development in the first place. The sewage treatment plants of many towns and cities are inadequate. Around the world, six million children die of diarrhoea each year, many of these infections being caused by sewage-contaminated water. Major waterborne diseases in the region are spread in this way.

Industrial by-products also affect water quality. Mines, for example, pump vast volumes of polluted water to the surface and, despite efforts to pool and evaporate this effluent, acidic run-off and seepage collect in both surface- and groundwater. Paper mills are amongst the worst polluters in the region. They use chemicals to soften and bleach wood pulp to make paper. This results in organochlorines

Photo: Courtesy of Ithala

like dioxins flowing from the mills into rivers, where they can kill fish or harm their growth and reproduction. Dioxins are amongst the most toxic chemicals known, and are proven carcinogens. Situated below a paper mill, a tributary of Swaziland's Great Usuthu River has been poisoned by these chemicals and clogged with paper fibres. Elsewhere, chemicals in industrial waste dumped in factory yards wash into surface water or seep into groundwater. This is how Maseru's main water supply became contaminated with lead at concentrations six times above World Health Organisation standards, and how the Mukuvisi River, which supplies Harare's drinking water, became highly polluted with phosphorus, nitrogen, sulphate and aluminium. Following recent legal action, a South African steel producer had to buy the land belonging to residents who proved that the producer had been contaminating their groundwater for 30 years. In this area near Vanderbijlpark, as elsewhere in the region, boreholes or wells are the main sources of drinking water. The water has been contaminated by seepage from unlined sludge dams. This has been linked to a range of effects in the surroundings of the steelworks where fourteen families reported cancer, children were diagnosed with stunted growth, fruit trees died and poultry became infertile.

In our mainly dry region, the contamination of groundwater is a serious problem. It is caused by numerous sources of pollution including animal feedlots and kraals, sewage,disinfectants thrown into pit latrines, leaks from bulk storage of chemicals, chemical spills, poorly maintained landfill sites and, as in the case above, mining and industrial effluent.

Printed on Sappi recycled paper 205

WHAT CAN BE DONE ABOUT THE DETERIORATION OF WATER QUALITY?

◆ Industrialists must apply or develop technologies and processes for cleaner production.

◆ If it is impossible to clean up your waste stream, reconsider the nature of your business and explore new opportunities provided by less harmful products.

◆ If the 'polluter pays principle' is used to tax polluting companies, this tax can be used to subsidise the installation of cleaner production technology, especially in small and medium-sized businesses.

◆ Consumers should note the negative impact on our water quality that is created by manufacturers and suppliers of the goods that we so readily purchase.

◆ Local government must take greater responsibility for protecting water quality, improving sewage treatment plants and supporting national authorities in their policing efforts.

As 7,06 million South Africans lack access to safe water and sanitation, inexpensive methods for water testing and purification must be shared widely. We need not wait until government provides everyone with a flush toilet, nor need we resort to the bucket system currently used in some townships. Other systems that are safe, hygienic and easy to build and maintain are available and widely used in, for example, Mexico and Australia. (Contact Share-Net or the Division of Building and Construction Technology at the CSIR for further information.)

WATER AVAILABILITY

A 1987 UNEP report estimated that pollution and population increase could cut the availability of water per person by half in the Third World, where two billion people already suffer chronic water shortages.

Let's now look at the issue of water supply. Supply is determined by the natural limits to southern Africa's fresh water sources; it is also substantially affected by government policy.

The Worldwatch Institute reminds us that water scarcity is linked to food security. They cite the case of Jordan, which grew its own wheat until a drop in groundwater levels forced them to import wheat. This cost the government of Jordan so much that it cancelled its subsidy on bread, resulting in 'bread riots' in 1996. Israel extracts most of the water in the River Jordan. This is a classic example of the political aspects of water distribution.

In South Africa the demand for water exceeds the supply in nine major river catchments, and we are frequently reminded to save water. How often have you heard comments such as "But the water I waste when I'm rinsing cups seems so little compared to what industry uses"? Actually, according to South Africa's *State of the Environment* report (2000), urban and residential consumers use more water than mining and industry together! The biggest and also the least efficient user of water is agriculture, which uses between 70 and 75% of South Africa's water. Only one third of this is used directly to grow crops. Pine, wattle and eucalyptus plantations also consume large volumes of water. Commercial forestry is considered to be the least efficient user in environmental terms. The new Water Act aims to discourage inefficient use, as big water consumers will be charged more per unit than those who use less.

"Does it really matter how much water we use? Surely water makes its way back into the water cycle, via runoff and evaporation, returning to earth as beautifully clean rainwater?" Natural systems do purify and circulate water, but they are increasingly over-stretched. For example, wetland and forest areas that cleanse the water are shrinking, and pollution that makes the water unusable is increasing. Also, rain doesn't always fall where and when you need it and it is expensive, both in economic and ecological terms, to channel clean water to users. Urban areas like Cape Town, Port Elizabeth and Pietersburg are highly water-stressed, and will become even more so as their city populations increase. It is estimated that South Africa's water demand will increase by 52% over the next 30 years!

In the past our answer to the problem of water supply was to build more dams. Now, as the flow in all major rivers has been greatly reduced by dams, which catch more than 60% of the run-off, ecologists warn that we are not leaving enough water in the rivers for these vital ecosystems to survive. We therefore need to find other solutions. Water-saving strategies will have to become a way of life for all of us.

WHAT CAN BE DONE ?

At home: Monitor your household consumption and reduce your usage of water-wasting appliances, such as washing machines and dishwashers. Plant a water-wise indigenous garden, consider buffalo grass if you want a lawn, and install rainwater tanks.

At school: Recycle paper – to save water and wetlands as well as trees! Start a 'water-wise' competition for schools or clubs to develop innovative ways to use water more efficiently and sparingly. In Zimbabwe SCOPE teaches schools to harvest rainwater from their corrugated iron roofs, and grow lush permaculture gardens. Teachers can link this to Technology, Economic and Management Sciences, Natural Sciences and Life Orientation.

In industry and agriculture: Do a water audit to identify where water is used and where it is wasted. Install water-saving devices and processes. Recycle water: this will without doubt reduce operating costs and improve bottom line profits. On the farm, consider a holistic approach to land management through permaculture and other similar strategies.

Local authorities: Consider switching to water-free sanitation. The British flush-toilet system is inappropriate in a dry region; the alternatives can be just as hygienic if well designed and maintained, and will alleviate pressure on sewage treatment plants.

Photo: Guy Stubbs Photographer

Key Associated Topics:

Agriculture; Alien Invasive Plants; Carrying Capacity; Desalination; Desertification; Dilution; Dissolved Oxygen Content; Drip Irrigation; Eco-taxes; El Niño; Energy; Environmental Health; Eutrophication; Forestry; Freshwater Ecology; Greywater; Groundwater; Non-renewable Resources; pH Scale; Pollution; Polluter Pays Principle; Population; Poverty Alleviation; River Catchment Management; Rivers & Wetlands; Suspended Solids; Tropical Rain Forest; Urban Agriculture; Urban Greening; Waste Management; Wetlands.

Associated Organisations:

Aqua Catch; Barlofco; Biolytics Southern Africa; BKS; Digby Wells; Enviro Options; Mondi Wetlands Project; Rand Afrikaans University (RAU); Rand Water; Softchem; Unilever Centre For Environmental Water Quality; Umgeni Water Services; Wales Environmental Partnerships; Water Institute of SA; Water in the Environment Research Group – University of the Witwatersrand.

Photo: Dr J Ledger

Let's all be Water Wise.

Many of us in South Africa today are unaware of just how precious water is. If we did, we'd be much mo[re] vigilant about conserving it, and we'd also be more prepared to pay for it. Let's take a look at the fac[ts] about water and perhaps then we'll change our attitude to water and learn to give it the respect it deserve[s]

Facing the facts about water

The amount of water on earth is constant and cannot be increased or decreased. A country has to make do with the water reserves it has, and learn to manage them. South Africa receives an average annual rainfall of 492mm, almost half the world's average of 985mm. That's why we are classified as a semi-arid country.

There is also an uneven distribution of rainfall across our country. What's more, our hot dry climate causes excessive evaporation. As our population increases, so will the demand for water. All of this puts great pressure on our limited water reserves.

Let's be wise about water in the future

Are you getting a clear picture of just how dire the situation is? What of the future? We could build more dams and water transfer schemes, but we can't afford this type of infrastructure. We could clean up rivers. But these solutions address the symptoms of the problem, when we should be addressing the cause. Only by examining our own attitude towards water, and how we impact on its quality and availability, will we be empowered to make a difference.

To be "Water Wise" means that you will:

- RESPECT water and all life. All life on earth - humans, animals and plants – need water in order to survive.
- Use water carefully and do not WASTE it by using too much or using it inappropriately. Let's start by looking at how we use or abuse water at home or at school.
- Not POLLUTE rivers with liquid and solid waste, making it unclean or dangerous.
- Take ACTION to solve any water problems. Knowing about a problem is one thing – doing something about it is something quite different. We cannot afford to be complacent. If we see an environmental problem we must act.
- CONSERVE water, and thereby conserve the natural environment. By conserving water we are protecting our environment and ensuring the survival of all life on earth.

The quality of water

Water quality is defined as water that is safe, drinkable and appealing to all life on earth. In South Africa, our fresh water – which is already scarce – is decreasing in quality. This is caused by an increase in pollution and the destruction of river catchments, caused by urbanisation, over-population, deforestation, damming of rivers, destruction of wetlands, industry, etc.

Why we must pay for water services

Water comes freely from the sky so we don't need to pay for it, rig[ht] WRONG! Water might be free but gaining access to clean, quali[ty] water is not. We have to pay for the infrastructure and people th[at] bring this water to us, i.e.

- the people who look after the rivers;
- the dams that are built to store the water;
- the purification stations that clean this water;
- the reservoirs that store the water;
- the pumps and pipes that transport the water to the reservoi[rs]
- the pipes that transport the water to our taps;
- the water treatment plant which cleans used, dirty water, and p[uts] clean water back into the river.

Next time you turn on a tap, remember that all life on earth is connecte[d] If humans pollute water then it causes the fish and the insects in t[he] rivers to die. If humans waste water, there won't be enough for t[he] animals and plants that share the earth with us. We all share th[e] planet and its resources, and we must remember that our actions affe[ct] others. Remember water is life and life depends on water. So let's b[e] Water Wise in our daily lives. Let's all act now to respect, conserv[e] and value our precious water!

This information is brought to you by Rand Water WaterWise departme[nt]
Call 0860 10 10 60 for more refreshing informatio[n]

www.randwater.co.za

WAVE ENERGY

The marine environment is the next major frontier for the harnessing of energy, with vast amounts of energy potentially available in the waves (driven by the wind) and tides (driven by the moon) of the seas worldwide. Although records are known dating back to the 12th century of tidal energy being exploited for milling purposes, the development of marine energies has lagged behind other renewables because of the difficulty of installing systems in the sea. However, this is changing as technological developments in marine renewables are picking up pace.

Novel designs range from buoys and other floating devices, from which the energy is captured from their up-down movement in the waves, to seabed-mounted turbines, which rotate due to the action of the current on the blades, but most have yet to proceed beyond the prototype or demonstration stage. The world's first commercial wave power project, the LIMPET (land-installed marine power energy transformer), which captures the energy in shore-breaking waves, is installed on the Scottish island of Islay and began feeding power into the grid in late 2000.

Photo: Dr J Ledger

A study of the wave power off Cape Town was undertaken at the University of Stellenbosch in the early 1980s, finding that the inshore potential along the southwestern Cape coast is as promising as anywhere else in the world. In 2002, as part of Eskom's renewable energy programme, an assessment of the wave power potential of the whole South African coastline was completed, indicating that the potential is greatest off the south-west coast between Cape Town and Mossel Bay. The second phase of the programme, currently underway, is aimed at identifying a suitable technology for use in South Africa. The Oelsner Group has also been investigating wave energy from an independent power producer perspective and is involved in an economic feasibility study of a seabed device that was developed in a University of Stellenbosch study. The Oelsner Group, in conjunction with Interproject Service AB of Sweden, is also involved in developing a project to supply wave power to Robben Island from a moored buoy system just off the Island.

Guest Author: Jonathan Spencer Jones
~ Editor, ESI Africa Magazine

Key Associated Topics:

Eco-efficiency; Energy; Industrial Ecology; Natural Resources; Renewable Energy; Renewable Resources; Sustainable Development.

Associated Organisations:

Oelsner Group.

WETLANDS

Wetlands (vleis, bogs, swamps, sponges or marshes) are vital ecosystems, which have been described as some of the most productive in the world. They help to cleanse water of pollutants, provide habitats for a variety of plant, animal and insect species and regulate floods.

Wetlands act as breeding grounds for important bird and fish species and have an extraordinary diversity of plant species, which contribute to the capacity of wetlands to filter and cleanse water of natural and man-made pollutants.

Wetlands are excellent scavengers of pesticides and toxins, and can remove between 20 % and 100 % of heavy metals. Wetlands also act as vital water regulators, absorbing excess water during rainy periods and releasing water during the dry seasons. However, throughout the world, much effort is being taken to drain these areas and reclaim the land for alternative uses.

There is a great deal of ignorance about the role and importance of wetlands, although in South Africa the new recognition of the importance of water and its costing in legislation is helping to put more focus on the need to protect wetlands both for water supply and purification and ecological and biodiversity purposes.

Until quite recently, South African farmers were paid subsidies to reclaim wetlands and convert them into farmland. In southern Africa, the Okavango Swamp, one of the most important wetlands globally and regionally, is under threat of drainage for mining and agricultural purposes.

Key Associated Topics:

Agenda 21; Alien Invasive Plants; Biodiversity; Bio-remediation; Dissolved Oxygen Content; Eutrophication; Freshwater Ecology; Groundwater; Rivers & Wetlands; River Catchment Management; Suspended Solids; Urbanisation; Water Quality.

Associated Organisations:

Bateleurs Flying for the Enviro; Bergvliet High School; Birdlife of SA; Crowther Campbell; Holgate & Associates; Hout Bay & Llandudno Heritage Trust; Mondi Wetlands Project; National Botanical Institute; WESSA; WWF.

160° 120° 80° 40°

ARCTIC OCEAN

80°

Arctic Circle

ALASKA

Gulf of
Alaska

CANADA

Hudson Bay

40°

UNITED STATES OF AMERICA

NORTH

ATLANTIC

OCEAN

Tropic of Cancer

MEXICO

PANAMA VENEZUELA

0°

P A C I F I C

COLOMBIA

ECUADOR

O C E A N

PERU

BRAZIL

BOLIVIA

S O U T H

Tropic of Capricorn

PARAGUAY

ATLANT

CHILE

OCEA

ARGENTINA

40°

Antarctic Circle

0   2000   4000   6000   8000 Miles (at Equator)

80°

160° 120° 80° 40°

⌂ Holcim

We'd like to thank Holcim for helping us leave our children more of a living planet. If you'd like to help, contact Virginia on 021 888 2861

Global warming means we leave our children less of a living planet.

www.panda.org.za

WILDERNESS AREA

Photo: SA National Parks

Wilderness

Wilderness means different things to many people. The best way to explain what a wilderness area is, is to look at the legal definition of these areas. Under the IUCN's Framework for Protected Areas the highest form of protection an area can achieve is a Wilderness Area. This is described as a large area of unmodified or slightly modified land and/or sea which retains its natural character and influence and which is protected and managed so as to preserve its natural condition.

Why do we need to protect wilderness?

The Earth is a complex relationship between land, air, water and living things. As users of the Earth's resources, humans have a direct impact on these relationships. We need to allow ecological and biological processes to carry on unhindered by man to ensure that there is adequate opportunity for evolutionary adaptation. Wilderness provides protection for many resources which we, as a species, have come to take for granted, such as a supply of fresh, clean water and unpolluted air.

Wilderness offers untold scientific value; knowledge of the Earth gives us a growing understanding of the way global systems work and helps us make the choice about how to use the Earth today and safeguard it for tomorrow.

Wilderness provides exciting recreational activities and a place to experience solitude and freedom.

Guest Author: Andrew Muir ~ Wilderness Foundation

Key Associated Topics:

Antarctica; Biodiversity; Conservation; Ecological Intelligence; Eco-tourism; Environmental Ethics; GAIA; Heritage; Mountains; National Parks; Non-renewable Resources; Stewardship; Transfrontier (Peace) Parks; Wildlife; World Parks Congress.

Associated Organisations:

Bainbridge Resource Management; Bateleurs Flying for the Enviro; Educo Africa; Ezemvelo African Wilderness; Hiking Africa Safaris & Tours; Mountain Club of SA; SA National Parks; White Elephant; Wilderness Action Group; Wilderness Leadership School; WWF.

WILDLIFE

Wildlife forms a key part of the food chain and ecosphere, and provides a series of physiological and logical links with the earth that we depend on. Many argue that animals are of little value and it does not really matter that the Dodo and carrier pigeon are now extinct. The counter argument is that these creatures have an intrinsic economic, scientific, recreational, medical, aesthetic and ecological value, which cannot be ignored. Animals have gone through a unique and structured evolution to reach their present form and the lessons they have learned on the way could well assist us with our problems and concerns. Theologians suggest that humankind has an ethical obligation to protect species from premature extinction because of human activities. They further suggest that humankind has the role of a "steward" to protect and manage each small area. Many different religions and faiths place a high, almost god-like reverence upon animals and wildlife, recognising the importance they have in pointing out food to humans, particularly in hard times.

Key Associated Topics:

Avian Rehabilitation; Birds; Birding; Biodiversity; Biodiversity – The Sixth Extinction; CITES; Conservation; Ecological Intelligence; Ecosphere and Ecosystems; Eco-tourism; Endangered Species; Entomology; Extinction; GAIA; Game Farming; Green/Brown debate; Heritage; Hunting; Stewardship; Transfrontier (Peace) Parks; Wilderness Area; Wildlife Management; World Parks Congress.

Associated Organisations:

Africa Geographic; Africat Foundation (The); Bio Experience; Brousse James & Associates; Cheetah Outreach; Christina Gubic; De Beers; EWT; Ezemvelo KZN Wildlife; Forum For Economics & the Environment; Hiking Africa Safaris & Tours; International Fund for Animal Welfare; Wilderness Leadership School; Elephant Management & Owners' Association (EMOA); Space for Elephants Foundation; White Elephant Lodge; WESSA; WWF.

Photo: Richard du Toit - Londolozi

WILDLIFE MANAGEMENT

Guest Essay by L. D. van Essen
~ Coordinator and Lecturer, Centre for Wildlife Management, University of Pretoria

What is wildlife management?

Wildlife management is what humans do to ensure that wildlife species serve whatever ecological, commercial, recreational, or scientific purposes government or public interest determines. The techniques of wildlife management run the gamut from manipulating wildlife habitat to establishing hunting and fishing seasons and regulations, and from collecting wildlife population data to educating the public about wildlife conservation. Wildlife management is concerned with all wildlife species – both game (those pursued for sport) and non-game. Some wildlife species must be protected because they are rare or endangered. Usually, that's the result of habitat being destroyed or changed by man. At the other extreme, some wildlife species may become so locally abundant that they threaten private property and people's livelihoods. Dealing with those species and conditions can require population or habitat controls.

Ecology forms the primary scientific discipline of wildlife management and therefore embraces the interactions of all organisms with their natural environments. By recognising that humans, as other organisms, have a total dependency upon the environment, it is accepted also that wildlife, in its myriad forms, is basic to the maintenance of a human culture that provides quality living.

What do wildlife managers do?

Wildlife managers are professionals who implement techniques, which help ensure that the objectives of wildlife management are met. While some of what wildlife managers do focuses strictly on wildlife, the challenges are complex and often require that the wildlife manager have expertise in a variety of areas. Often, too, the challenges require specialists in specific areas. Today's wildlife manager is much more than a "game warden". The wildlife manager in the field often counts animals, controls wildlife populations, or recommends seasons, and is a public relations specialist and educator as well. In addition, wildlife managers enforce the laws and regulations, which are designed to maintain optimum wildlife populations.

Wildlife research

Successful wildlife management is based on facts obtained by scientific research. Wildlife biologists perform basic or applied research to obtain facts on such subjects as physiology, genetics, ecology, behaviour, disease, nutrition, population dynamics, land use changes, or pollution.

Wildlife public relations

Nearly all wildlife management work involves some degree of public relations skills. Some wildlife jobs, however, deal almost exclusively with this area.

"My teeth are on edge!"

This expression speaks of the frustrated condition of many wildlife managers who find old ideas presented as discoveries, words misused, and phrases like "stocking rates, edge effect for wildlife, etc." used in nonsensical ways as self-efficacious prophecies or political blocks rather than in conversations that may help solve local, usually unique, problems. The term 'conservation' is a case in point, where it was used in the past decade or two by the general public and scientists alike as a holistic euphemism for "management". In fact both preservation and conservation represents the continuum that is wildlife management. The 'preservationist' approach on the other hand has become the hand puppet of the extreme green movement. Tragically the existing confusion over the two terms the world over is the principal reason why the policies of many wildlife agencies (governmental, multi-national and NGOs) have gone awry.

Wildlife management ignores the emotive rhetoric put out by the protectionist lobby and interprets and acts on the real facts of wildlife/nature issues at hand, such as the improvement in the status of unsafe species and the sustainable utilization of safe wild animal populations to the benefit of mankind.

Wildlife management – as an art and craft, with the solid underpinning of applied science – seeks truly workable solutions to wildlife problems as it recognises the fact that the wildlife resource is a product of the land that can, and should, be used wisely for the benefit of mankind.

Key Associated Topics:

Avian Rehabilitation; Birds; Biodiversity; CITES; Conservation; Ecological Intelligence; Ecosphere and Ecosystems; Eco-tourism; Endangered Species; Entomology; Extinction; GAIA; Game Farming; Green/Brown Debate; Hunting; Stewardship; Transfrontier (Peace) Parks; Wilderness Area; Wildlife; World Parks Congress.

Associated Organisations:

Centre for Wildlife Management – University of Pretoria; Eco Systems; Elephant Management and Owners' Association.

WIND FARMING

Energy from the sun sent to earth every year is equal to more than 15 000 times the commercial energy consumption of the world's population. Almost half of this energy is naturally converted into wind. The theory and practice of using wind energy to do work for mankind is well known and tested. It has been used for centuries, for the purpose of milling, water pumping and many other applications. However, it is only in the last half of the 20th century that people started to explore the use of wind energy to generate electricity. Historically the international interest in electricity generation from wind power was sparked by the world oil crises in 1973 and 1979 when, in addition to the problem of the security of fossil fuel supplies, the focus of developed countries became more environmental. Because of the political focus on environmental issues and, of late, the hot topic of global warming and greenhouse gases, the wind energy market has experienced an enormous growth period over the past 20 years. Commercial wind farms are now being operated on a more competitive basis then ever before. A wind farm typically consists of a group of large wind turbines, which supply the generated electricity directly into the national grid. The movement of air is used to propel three blades mounted on a hub and horizontal axle, transferring the energy to a gearbox and generator where the electricity is generated. The gearbox and generator are housed in the so-called nacelle, which is mounted on top of a tubular steel tower. Wind turbines come in all shapes and sizes and range from 200 Watts to 3.5 Mega Watts. South Africa is blessed with an abundance of wind resources, especially in the coastal regions. There is, however, not yet one large wind turbine installed. The government has declared the 13 MW Darling Wind Farm, north of Cape Town, as a National Demonstration Project which went on line in September 2002. On this wind farm, there will be 10 wind turbines with a generating capacity of 1 300 kW. Each wind turbine has a hub height of 50 m (the same height as a 17 story building). The length of one blade is 32 m (longer than a tennis court). Less than one percent of the land on which the farm is situated is used for the foundations and roads, with farming operations continuing as usual.

The National Demonstration Wind Farm is envisaged to:
◆ Demonstrate to the public and other interested parties not currently informed or educated about wind energy;
◆ Present an opportunity for technology transfer, training and practical experience for the industry and potential wind energy supporting industries and organisations.

During its 25 years of operational life, the anticipated benefits will be:

Savings of national resources
Coal: 450 000 tons • Water: 1 130 000 litres

Avoided release into the atmosphere
Carbon dioxide: 850 000 tons • Sulphur Dioxide: 10 400 tons
Others: 55 500 tons

Guest Author: Dr Hermann Oelsner ~ MD, Oelsner Group

Key Associated Topics:

Eco-efficiency; Energy; Environmental Footprint; Industrial Ecology; Renewable Energy; Renewable Resources; Sustainable Development.

Associated Organisations:

African Wind Energy Association; Divwatt; Solardome SA; Energy Technoloy Unit – Cape Technikon; LC Consulting Services; Oelsner Group; Wales Environmental Partnerships.

WOMEN AND THE ENVIRONMENT

Agenda 21 confirms the importance of women in sustainable development. Women form over half of the world's population yet there are very few women in decision-making positions. Women are also most often the members of the community who are directly involved in the land. Many women are farmers and yet often cannot own the land, which they farm because of patriarchal (male-dominated) systems of land ownership and access to credit. By giving women more rights and decision-making powers, they will be able to have more say in creating more sustainable lifestyles and improving the quality of life for themselves and their families. In South Africa, women and youth are specifically identified as being significant role-players in sound environmental management. Increased environmental awareness, opportunity and environmental education for youth and women are direct means of improving environmental quality in South Africa, and of ensuring greater action to improve the quality of the existing environment.

Key Associated Topics:

Agenda 21; Agriculture; Environmental Education; Population; Poverty Alleviation; Stakeholders; Trench Gardening; Urban Agriculture; Zero Population Growth.

Associated Organisations:

Heinrich Boell Foundation; Art of Living; SA Wildlife College; Southern Health & Ecology Institute; Sustainability United (SUN); Water Institute of SA; Working for Water.

Photo: *Impression by Quinton Lawson - Bernard Oberholzer Architects.*

DEPARTMENT OF MINERALS AND ENERGY

DIVIE - Danida Capacity Building in Energy Efficiency & Renewable Energy

South Africa moves towards sustainable energy and cleaner electricity

ENERGY EFFICIENCY

The Department of Minerals and Energy, South Africa, is proud to announce the drafting of the first 10-year Energy Efficiency strategy. The draft strategy sets a target of a 12% reduction in electricity consumption to be achieved by the year 2014. The draft strategy includes sub-targets for the sub-sectors

- Industry
- Residential sector
- Public and Commercial Buildings
- Sub-targets for the Transport sector and
- Agricultural sector will follow.

To achieve these goal the Department will embark on a range of interventions such as:

- Implementation of technical norms and standards,
- Update building codes,
- Roll-out of an appliance labelling programme,
- Energy management best practice,
- Information campaigns and education, to promote awareness

The strategy is a National Effort to reduce the need to build new power generation capacity and to support South Africa's households and business units to enhance effective utilisation of energy.

Interested partners, investors and Donors are invited to download the draft strategy from our website and support us reaching the goals for a cleaner environment and a better place for all.

RENEWABLE ENERGY

In 2003 the South African Cabinet approved a White Paper on Renewable Energy. The White Paper sets a target of an additional 10,000 GWh of Renewable Energy by the year 2013. 10,000 GWh could be 1200 wind turbines each having an installed capacity of 3 MW.

The Department is formulating a strategy with clear areas of work to be implemented during the next 9 years to achieve this target.

Both the Energy Efficiency strategy and the Renewable Energy Strategy sets forward a number of goals, which all relate to sustainable development.

In the case of Renewable Energy it is envisaged that implementation of the strategy will support

- Social sustainability through improved health, job creation and energisation of rural areas
- Environmental sustainability through reduced GHG emissions and reduced pollution
- Economic sustainability through improved energy security and new capacity.

Implementation of Renewable Energy will primarily happen through grid-connected technologies. Bio-fuels and Solar Thermal will also be in focus through support to R&D.

The market for CO_2 trading is an excellent business opportunities. With a Designated Authority in place to handle CO_2 credits the DME is proud to invite Partners and Investors to join our renewable energy world.

For more information contact: Kevin Nassiep, Chief Director – Department of Minerals and Energy,

Nassiep@mepta.pwv.gov.za or visit our web-site: www.dme.gov.za

WORLD PARKS CONGRESS (DURBAN 2003)

Guest Essay by John Richards
~ presenter SAFM Ecowatch Programme

It will be a long time before we know for certain if the outcomes of the fifth World Parks Congress in Durban in 2003 have born long-lasting, dare it be said, "sustainable" fruit.

That the Congress was huge and impressive was beyond doubt – it represents the high point so far in our species' recognition of the need to conserve and protect. It was the Biggest and Best since World Congresses began. It also consolidated the realisation that we are part of nature, and that people, whether living in areas of great natural beauty or in cities, are interdependent with all living things. Hence the slogan of the Durban Congress, "Benefits Beyond Boundaries".

The Durban Accord was a carefully worded statement that held out encouragement and expressed the urgent need to take conservation seriously. If activists and the faithful were hoping for a revolutionary cry, they were disappointed – inevitably. The IUCN survives because it works in the realms of Realpolitik. But neither was the Accord empty rhetoric, and it's backed by some very concrete outcomes.

These outcomes may be divided into two categories – strong statements without material back-up; and actual recommendations that will have material effects.

Some of the major material announcements include: 3.8-million hectares of the world's most bio-diverse territory proclaimed as protected areas by the Brazil government in the Amazonas, with Conservation International support of at least US$1-million. That's an area of land about the size of Belgium.

The Congress host country South Africa found itself rather embarrassingly below the target of 10% of protected land area. But with measures announced to bring privately-owned conserved land under government protection, and the establishment of five new national parks, the current 6.6% of protected area will reach the magic 10%. These changes are a direct outcome of the Congress, as are the decisions by the Madagascar government to increase the total protected area of that much endangered and diverse set of biomes. The total area will jump from 1.7-million to over 6-million hectares, including marine protected areas and wetlands.

There were also 32 separate recommendation packages resulting from the "streams" and "linkages" of the congress, and here are just a few of the notable ones.

From the "Linkages in the Landscape/Seascape and Governance" Stream comes Recommendation 11 on a Global Network to support the development of trans-boundary conservation – the Peace Parks as they are popularly called. There is no forum, and the situation

should be remedied without delay says the IUCN. There should be a framework and an internationally recognised register of trans-boundary protected areas says the IUCN. Strong recommendations without a material result, but a warning to sit up and take notice.

But you have to be practical – and many parastatal conservation bodies have to justify the public funds and present their true account. For want of a better term it's convenient to speak of 'managing' a wildlife reserve, or conserving biodiversity in a protected area. And here we're doing quite well.

The gross square kilometrage of protected areas as defined by the IUCN has grown exponentially since 1962 and the First World Congress – from 2.4 million km^2 to 4.1km^2 ten years later to double that again in 1982. And in 2004 the tally is 18.8 million km^2. This figure is equivalent to 12.65% of the Earth's land surface, or an area larger than the combined land area of China, South Asia and South East Asia. But the quality and the categories of these areas is of course a different matter.

At last the oceans – and especially the coastal regions – are getting the attention they need. The WSSD emphasised the need to maintain the productivity and biodiversity of important marine and coastal areas, and set target dates of 2012 for the establishment of Marine Protected Areas consistent with international law; 2015 for the restoration of depleted fish stocks; and 2010 for the application of an ecosystem approach to fisheries management. WPC recommendation 22 calls on the international community to increase greatly the marine and coastal area managed in protected areas by 2012, amounting to at least 20% of each habitat. There are further admonitions to governments to set in place monitoring and eco-system approaches, to further strengthen the Ramsar Convention on Wetlands, and limit habitat destruction.

The fifth Congress was a good arena for dialogue on Climate Change. At last scientists laid to rest the critics of the Global Warming model, and consensus was achieved on the fact that Nature is dynamic. Change encompasses many facets, including the growing human population and the drive to development. So in addition to biophysical changes there are socio-economic and political changes, which will have profound implications for protected areas. Climate change and its synergies with the other global changes is a new and unprecedented challenge confronting protected areas.

Many of the impacts on biodiversity will be in tropical regions and developing countries, while the major source of global greenhouse gases are industrialised countries. The equity issues will require new funding mechanisms the IUCN warns. Governments, NGOs and local communities are called upon to identify and designate protected areas that increase representation of species and ecosystems that may be jeopardised by climate change, including (a) all threatened species by 2012; and (b) all species and ecosystems by 2015. There is a call for the IUCN World Commission on Protected Areas to expand partnerships, step up research and monitoring and deepen expertise, anticipate the impacts and adapt management to those changes.

It's always been about money, and if loss of biodiversity is to be significantly slowed by 2010, then Protected Areas (PAs) deserve financial support because of the tremendous benefits they provide – benefits beyond boundaries. But the financial realities belie the good intentions of governments and conservationists to fund PAs. An estimate of the cost of establishing and maintaining these comprehensive PA systems runs at US$20 billion every year; in the mid-1990s PA budgets totalled only about 20 percent of that. Recommendation 7 sends out strong pleas for governments to plug the gap. There's an Initiative on Sustainable Financing which will help administrations address ways to ensure the flow of funds from tourism, trusts and the private sector, and increase grants from the Global Environment Fund (GEF) for developing countries. And of course local and indigenous communities as primary beneficiaries must be granted access to benefits such as tourism revenues, clean air and water and soil conservation.

So, then – we had many nice-to-have sentiments, backed by authoritative voices; many noble ideals and equitable objectives addressed to governments and conservationists (both civil and independent) – backed up by all the science and organisational efficiency of the IUCN and associated NGOs like WWF and the International Professional Rangers' Association.

We can all fervently hope that the policy-makers will be sympathetic and that the financial decision-makers will be creative and generous.

Finally, it needs to be said – and Recommendation 13 puts it well – the cultural and spiritual values of protected areas, are the fundamental sources of meaning in our existence. They serve as fundamental tools for conservation, and are an expression of the highest desires and commitments of humankind for the preservation of life on the planet.

"Many societies, especially indigenous and traditional peoples, recognise sacred places and engage in traditional practices for the protection of nature, ecosystems or species ... they also recognise sacred places as a unique source of knowledge and understanding of their culture, thus providing what could be considered the equivalent of a university." (WPC Recommendation 13.)

Social, scientific, industrial humankind has created and built the cities, transformed the landscape and modified forever the natural world. Now our industrial emissions are contributing even to a changing climate.

The rate of species extinction is way beyond the natural order. The fifth World Parks Congress gave voice to reason and wisdom, and it's up to us to listen and respond.

Key Associated Topics:

Biodiversity; Birds; CITES; Conservation; Ecological Intelligence; Ecosphere and Ecosystems Eco-tourism; Endangered Species; Extinction; GAIA; Green/Brown Debate; Hunting; Stewardship; Stakeholders; Transfrontier Parks; Wilderness Area; Wildlife; Wildlife Management; Sustainable Development.

Associated Organisations:

Conservation International; EWT; SA National Parks; Peace Parks Foundation; Wilderness Leadership School; WESSA; WWF.

Photo: International Institute for Sustainable Development

BMW South Africa
Sustainable Mobility

Sustainability strategy

The BMW Group is committed to the three cornerstones of sustainability: economic success; social responsibility and environmental protection as key to delivering mobility for future generations.

BMW South Africa subscribes fully to this triple bottom line and the need to establish a viable balance between the interests of mankind, nature, technology, business, progress and the right of future generations to an intact environment.

The Group's guiding principles of sustainability were showcased at the BMW Group's EarthLounge on Sandton Square and at an exhibit at BMW SA's Rosslyn Plant, during 2002's World Summit on Sustainable Development. This was not only an opportunity to be directly involved in the global debate, but also to present concepts such as CleanEnergy, which promotes awareness of hydrogen (generated from renewable energy sources) as the fuel of the future.

The Rosslyn Plant has the distinction of being the most environmentally friendly automotive plant in South Africa. Adequate protection of the environment is seen as the prerequisite for a lasting and sustained future, and the company is committed to reducing the impact of its manufacturing processes.

Integrated Environmental Management System

In 1999, BMW SA was the first automotive company in the world to be certified to all three management systems in a form of an Integrated Management System (ISO 9001, ISO 14001 and BS 8800). The system was re-certified in 2002 to ISO 9002, ISO 14001 and OSHAS 18001. The single most important aspect in having a formalised environmental management system is that almost all the environmental improvement projects have improved the efficiency of processes and saved money for the company at the same time.

Over the last five years BMW SA has reduced its water consumption by 50% at the Rosslyn Plant, and water usage is now monitored online. In 2002, a state-of-the-art effluent plant was built as part of a R2 billion investment by BMW Germany. The plant is designed to treat four different waste water streams at source.

More than 80% of solid waste is recycled and the amount of waste going to landfills, where the capacity is becoming increasingly limited, has been reduced. Packaging waste has been reduced by introducing returnable packaging material. Emissions to air have also been reduced due to water-based paint technology which contains 80% less solvents than conventional paint recipes.

Involving suppliers

Suppliers contribute more than 60% of the value of each car to the production line, and supplier non-conformance will therefore impact directly on BMW's production. Because BMW and its suppliers are seen as a single entity in the eyes of the environmental community, it is vital that suppliers are aware of the impact of their processes on the environment. In 2000 BMW SA introduced bi-annual workshops for suppliers, designed to provide them with the support they need to successfully implement environmental management systems which meet the needs of their customers and conform to recognised international standards.

Our people

A content and motivated workforce determines the ultimate success and sustainability of a company. BMW SA recognizes its diverse staff as its most valuable asset.

The company encourages a culture of trust which supports personal responsibility and self-management. This results in greater efficiency, productivity and better business performance – key to the company's success as an organisation competing internationally.

BMW SA's 3000 staff members have access to on-site restaurants and healthcare services, extensive in-house training and subsidised education schemes. BMW South Africa's Early Learning Centres at both Midrand and Rosslyn are available to the children of all BMW employees.

BMW South Africa's commitment to the development and growth of its employees was recognised when it became one of the first eight companies in South Africa, the only motor manufacturer, and the first of the BMW Group to receive accreditation as an Investor in People (IIP) in September 2002.

Employees' health is also a priority, and besides providing a wide range of medical services, the company has developed and implemented an intensive HIV/AIDS workplace programme to create awareness of the pandemic and to establish support structures for its employees. A comprehensive policy was launched in 2001 and BMW continues to counter this threat through ongoing awareness and education campaigns and effective testing and treatment programmes. More than 85% of all employees had already been voluntarily tested for HIV by November 2003.

As part of its strategy for the continuum of care, BMW has expanded its HIV/AIDS programme to its dealers and suppliers

and has established resources in surrounding communities by training local doctors, traditional healers and leaders of religious groups. The company is also providing funding through a public-private partnership with a German agency for a holistic community-based Multipurpose Care Centre, to be built in Soshanguve near Rosslyn.

Social responsibility

BMW South Africa believes in partnerships which result in sustainable community development. The company invests millions of rands every year in empowering surrounding communities, developing sustainable skills and contributing in the long run to a better quality of life.

BMW South Africa invests in the community in six keys areas, namely environmental sustainability, education, science and technology, sport development, the arts and local community development.

One of the company's most successful social investment projects, the SEED (Schools Environmental Education Development) Project, has created awareness around the importance of environmental conservation in over 50 schools and recently also clinics, and has impacted on more than 200 000 people around the country.

WORLD SUMMIT ON SUSTAINABLE DEVELOPMENT (WSSD)

Guest Essay by David Parry-Davies ~ Eco-Logic

The World Summit on Sustainable Development (WSSD) took place in Johannesburg from 26th August to 4th September 2002. This was neither a single nor an isolated event.

The following excerpts 8-9 taken from "The Johannesburg Declaration on Sustainable Development" that was signed at WSSD will put the Johannesburg Summit in context.

"8. Thirty years ago, in Stockholm, we agreed on the urgent need to respond to the problem of environmental deterioration. Ten years ago, at the United Nations Conference on Environment and Development, held in Rio de Janeiro, we agreed that the protection of the environment, and social and economic development are fundamental to sustainable development, based on the Rio Principles. To achieve such development, we adopted the global programme, Agenda 21, and the Rio Declaration, to which we reaffirm our commitment. The Rio Summit was a significant milestone that set a new agenda for sustainable development.

9. Between Rio and Johannesburg the world's nations met in several major conferences under the guidance of the United Nations, including the Monterrey Conference on Finance for Development, as well as the Doha Ministerial Conference. These conferences defined for the world a comprehensive vision for the future of humanity."

In the negotiations preceding WSSD it became clear that even reaching agreement on the "Agenda for discussions" at Johannesburg would in itself be a difficult task due to the divergent political, economic and social interests of the many role-players.

Whilst the environmental community looked forward to a Summit focused on "Saving the Earth" the politicians looked for mechanisms to "Save the People" and business looked for opportunities to help "Develop Prosperity for All".

Even within areas of unanimous common interest, the priorities of developed countries looked far different from those of developing countries.

In the process of the pre-negotiations it was decided to change the name of the forthcoming Johannesburg event from "Rio + 10 Earth Summit" to the "World Summit on Sustainable Development". This signified a clear and distinct shift of focus from that of the previous Rio Earth Summit. Those committed to "people and development" welcomed the change, whilst those focused on protecting "the Earth and our natural resources" lamented, with some declaring the debate lost – before it had even begun.

Some 22 000 statesmen, dignitaries, economists, scientists, activists and representatives of environmental and social interest groups arrived in Johannesburg – all with focused intent and each with a vested interest in the environment and/or access to its resources.

Whilst the central UN sponsored inter-governmental event took place at the Sandton Convention Centre:

◆ A "Global People's Forum" convened, debated and networked at Nasrec;

◆ The IUCN hosted a programme of lectures, panel discussions and book launches;

◆ At St Stithians College discussions and conferences took place in the field of Biodiversity and Bio-piracy;

◆ At the Waterdome there were many panel discussions, debates and presentations, mostly organised by the IUCN and the "water wise" industries;

Photo: Bruce Sutherland - City of Cape Town

◆ The Ubuntu Village showcased green technology, corporate environmental initiatives and Governmental PR stands, in between curio and organic food stands. This was also the venue for initiatives such as the Environmental Education Declaration, which motivated the UN to confirm the proposed decade of "Education for Sustainable Development".

◆ At a former mine site – Shaft 17, the South African Environmental Justice Networking Forum, together with international environmental justice activists, gathered for 5 days of discussions. Issues such as "The capitalist order must be challenged for the sake of environmental justice" and "Sustainable development is used to mean more economic growth" were debated in open forum.

◆ And on the wealthy suburban streets of Sandton demonstrations and parades by social and environmental activists were a common feature. The local residents (like many other urban residents around the world) sat at home overwhelmed by the magnitude of the negotiations, confused by the diverse range of issues, and cynical of any worthwhile results emerging from an event in which political rhetoric and business self-interest appeared to dominate.

However, despite the cynicism, much was achieved and the term "Sustainable Development" emerged as the common and agreed reference point and aspiration for all future economic, environmental, political and social activities. All agreed on this – Sustainable Development is the basis and the glue to bind the divergent interests of People, Planet and Economic Development.

The Johannesburg Declaration on Sustainable Development was signed, with 37 separate points commencing with:

1. We, the representatives of the peoples of the world, assembled at the World Summit on Sustainable Development in Johannesburg, ...reaffirm our commitment to sustainable development.

2. We commit ourselves to build a humane, equitable and caring global society cognisant of the need for human dignity for all.

The Declaration continues with many inspirational and insightful declarations of care and integrity.

Also signed was "The Johannesburg Plan of Implementation" that includes chapters of commitment relating to:

◆ Poverty eradication

◆ Changing unsustainable patterns of consumption and production

◆ Protecting and managing the natural resource base of economic and social development

◆ Sustainable development in a globalising world

◆ Health and sustainable development

◆ Sustainable development of small island developing states

◆ Sustainable development for Africa

◆ Other regional initiatives; sustainable development in Latin America and the Caribbean; Asia and the Pacific; the West Asia region; the Economic Commission for Europe region.

Both The Johannesburg Declaration on Sustainable Development and the Johannesburg Plan of Implementation (JPOI) documents can be found at the United Nations Commission for Sustainable Development: website www.un.org/esa/sustdev/

In response to the JPOI text the Wildlife and Environmental Society of SA (WESSA) commented:

"The official UN Johannesburg Plan of Implementation (JPOI) text will provide the framework for how sustainable development is implemented ... Many of the actions outlined in the JPOI text were in themselves detailed and useful, but unfortunately of a voluntary nature and therefore unlikely to be implemented as effectively as if they had been binding."

The IUCN declared, "Views on the value of the Summit vary widely. To many, WSSD fell short of expectations and thus was a lost opportunity because governments failed to take the sustainable development agenda forward. Others were concerned that the trade liberalisation agenda was being pursued at the expense of sustainable development. In contrast, some argued that the fact that trade was being discussed outside of the confines of the World Trade Organisation (WTO) was a positive sign of the willingness of governments to address trade in terms of sustainable development. While some decried that multilateralism was failing, others felt that the Summit re-affirmed global commitment to sustainable development."

Whilst Richard Douthwaite, an Irish economist said: "Nowhere in the official documentation for the (WSSD) conference is there any recognition that this planet is finite and that there are limits to the scale of human activities which have to be observed."

Other comments from the divergent interest groups included:

"Instead of developing a new momentum at the Summit, civil society has had a very hard job of defending the Summit from a complete take-over by the World Trade Organisation (WTO). The WTO says you have to have economic growth and if it negatively affects society and the environment, that's just tough." "Fighting poverty and saving the environment are in fact the same battle."

Whatever our perception of the WSSD event in terms of its focus, agendas and outcomes, we must return to the observation made at the beginning of this article, that WSSD was part of a process – a current and ongoing process. Nothing is too late to change – everything is at stake, so we conclude with one final quotation from WSSD:

"The real responsibility lives with us – public action and public demand."

Key Associated Topics:

Agenda 21; Cleaner Production; Consumerism; Eco-efficiency; Ecological Footprints; Ecological Intelligence; Economics; Energy; Environmental Ethics; Environmental Governance; Environmental Law; Environmental Management Planning; GAIA; Global Warming; Industrial Ecology; Local Agenda 21; Non-renewable Resources; Pollution; Population; Poverty Alleviation; Stewardship; Triple Bottom Line; Sustainable Development; + See All Guest Essays in the Sustainable Development Section.

Associated Organisations:

Art of Living; Department of Environmental Affairs & Tourism; Department of Water Affairs and Forestry; Department of Trade and Industry; Department of Minerals and Energy; Earthlife Africa; Environmental Monitoring Group; EWT; Gondwana Alive; Heinrich Boell Foundation; Iskhus Power; National Botanical Institute; National Business Initiative; Oryx Environmental; Wales Environmental Partnerships; WESSA; WWF.

Photo: Bruce Sutherland - City of Cape Town

Bringing envi

management

Telkom's stand against safety, health and environmental risks garners world-class certification

Already known for its ground-breaking programmes to ensure sound practices in the environment in which it operates, Telkom has turned its vision inwards to ensure a safe and healthy working environment for its many employees. Achieving ISO 14001 and OHSAS 18001 certification status this year has vindicated the information and telecommunications giant's home-brewed strategy in this regard, and the company is once again blazing a trial along which other South African corporates are bound to follow.

Due to the nature of their activities, members of Telkom's technical workforce face a wide array of safety, health and environmental risks. Hazards regularly encountered include noxious and flammable gases present in manholes, unsafe power reticulation encountered in many informal communities and, of course, the many risks that come with being suspended from masts and towers of various descriptions.

Instead of accepting these risks as an unavoidable feature of the arena in which it does business, Telkom decided in 1999 to pull out all the stops to reduce incident frequencies and the overall number of incidents.

Reducing risks, saving lives

This decision made sense on two very important levels. Firstly, Telkom's stated value system includes acting as a responsible corporate citizen, and this extends not only to its external environment, but also to the human capital it so highly values. Secondly, reducing the cost involved in days lost due to injury and other incidents would obviously hold financial benefits for the company and its employees.

Faced with what it regarded as an unacceptably high incident rate and frequency, as well as days lost due to incidents, the company investigated the best the world had to offer at that time in terms of systems to manage its occupational environment and formulated its own Safety, Health and Environmental (SHE) management system.

ronmental
home

Getting down to basics

The central engine powering Telkom's SHE efforts throughout the organisation is its Incident Prevention Plan (IPP), an integrated management system for managing safety, health and environmental impacts and risks. This system was developed by Telkom itself in accordance with world-class standards, and focuses primarily on reducing incidents and costs incurred due to non-compliance and unsafe working practices.

Clearly, uniform application of the IPP would not be possible in a company such as Telkom with its wide diversity of functions, all of which represent different levels of risk potential. A multifaceted approach was called for, and to that end sections were grouped together in three categories: high-risk service organisations, medium-risk service organisations and moderate-risk service organisations.

The rationale behind this division was that it would allow SHE specialists to ensure compliance with essential requirements (such as ensuring statutory compliance) in all three levels, and then to focus on the higher risk areas where the real need lay. High-risk sections such as Access Networks Operations (ANO), Core Network Operations (CNO), Data and Special Services (DSS) and Technology and Network Services (TNS) were targeted for immediate and sustained safety performance improvements. Pockets of excellence established in the process would serve to show what could be achieved and the lessons learnt in these areas would be applied elsewhere in the organisation.

Knowledge is power

Putting new safety, health and environmental measures into place is one thing; getting the message through to the people who matter – the workers – quite another. To ensure information saturation to all levels and areas within the organisation a number of communicative measures were employed. The most important means of communication is probably via the Internet – IPP is web-based, and the relevant information can be accessed by all employees. Other successful measures include workplace posters conveying safety messages, newsletters, onsite presentations and, most important, employee training in sound safety, health and environmental principles.

An enviable report card

Improvement in all targeted areas has been immediate and dramatic, as is borne out by the following trends identified by Telkom's SHE division in a recent study:

- The number of incidents reported per month has dropped by more than 60% in the period 2000 to 2003.
- Incident frequency rates have shown an equally impressive decrease over the same period.
- The number of days lost as a result of incidents has shown a steady decrease.
- The comparative estimated cost to the company regarding days lost as a result of incidents has reduced progressively. This cost is at present the lowest recorded to date, as is the estimated cost per employee.

Independent accolade

Impressive though these figures are, Telkom felt that independent scrutiny of its SHE management system would serve as a further assurance that world-class quality was indeed evident throughout the organisation.

Dekra ITS, an international leader in the field of certifying safety, health and environmental management systems, was appointed to conduct an independent review of Telkom's integrated SHE systems. Dekra ITS's audit focused specifically on measuring whether the company met the exacting standards contained in ISO 14001 and OHSAS 18001.

ISO 14001 is the cornerstone of the ISO 14000 series of standards. It provides the specification of an environmental management system as well as requirements that an organisation must meet for certification purposes. Its overall aim is to support environmental protection and prevention of pollution in balance with socio-economic needs.

In the same way, Occupational Health and Safety Standard (OHSAS) 18001 evaluates the relevant organisational systems against international standards for quality and environmental management systems respectively. This certification further signifies that Telkom is able to control occupational health and safety risks across a wide spectrum.

Towards zero tolerance

System success and certification notwithstanding, Telkom's efforts to highlight areas of systems improvement are ongoing. The lessons learnt in the various pockets of excellence are applied throughout the company to ensure improvement in all risk areas, and Telkom's SHE specialists continue to measure progress against targets in the company's various divisions.

In the end, it's all about people – the people who work for a company, the people in the communities served by the company and, of course, the environment shared by all. In raising the bar concerning a safer working environment Telkom hopes that it is setting the tone for good corporate citizenship, and that major South African companies will follow suit to ensure a better working and living environment for all our people. ■

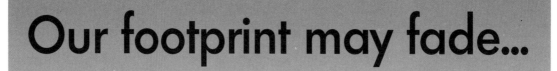

Our footprint may fade...

...but our legacy remains

At Richards Bay Minerals (RBM) we believe in building assets that will remain for future generations. In addition to the infrastructure created by our operation, such as roads, electricity and telephones, we have enabled new assets to be created from the minerals we mine.

As a world leader in the supply of the heavy minerals ilmenite, rutile and zircon, we have a significant impact on the economy, providing more than R5-billion annually to the South African gross domestic product (GDP). In addition, by providing about a quarter of the world's demand for these products, we generate billions of rands in taxes and foreign exchange.

Our internationally acclaimed social investment programme encompasses education, health, community safety, economic empowerment and rural development. These initiatives continue to improve the lives of thousands of people living in the surrounding area and provides a sustainable solution to the empowerment of communities.

Another essential component in our operation is its impact on the environment. To this end we spend over R10-million a year on our rehabilitation programme which aims to return the dunes to a similar state prior to mining. These rehabilitating forests are also a valuable outdoor laboratory for scientists and some 35 independent studies have been undertaken and results are published in local and international journals.

RBM
Creating a sustainable future for all

For more information about our operations
visit our website at www.rbm.co.za
Tel: +27 (035) 901-3441   Fax: +27 (035) 901-3442

SUSTAINABLE DEVELOPMENT SECTION

The concept of sustainable development is built on three pillars: economic growth, ecological balance and social responsibility.

Whilst sustainable development has become widely accepted as a critical basis for decision-making, the manner in which it is understood and interpreted varies quite significantly according to the needs and priorities of the different sectors (pillars).

It is therefore almost inevitable that situations will arise in which individuals, from different sectors, may take an entirely different course of action that may be in direct and challenging opposition to the other – whilst each considers their own actions to be fully justified "in the context of sustainable development."

In the interests of building bridges of understanding, *The Enviropaedia* has therefore invited high profile, key decision-makers to present their own interpretation of what sustainable development means and how they believe that it should be practically applied within their own area of activity.

Authors were invited to address:

◆ Strategies – how to implement the principles of sustainable development.

◆ Examples of "best practice".

◆ Benchmarks for measuring the success of policies and strategies.

◆ Current and future trends.

◆ Advantages of and obstacles to implementing sustainable practices.

In coming to understand the motivation behind contrasting viewpoints, role-players may thus be assisted in finding common ground and mutually acceptable solutions.

SUSTAINABLE DEVELOPMENT:

SUSTAINABLE DEVELOPMENT ~ MAINTAINING PROFITS, OR SUSTAINING PEOPLE AND PLANET?

Guest Essay by Dr Eureta Rosenberg

A tricky concept

Sustainable development is widely advocated as the way to deal with the issues which fill the pages of this publication. The concept is however hard to apply in practice, and easy to manipulate. These days many promote 'sustainability', including some who are only interested in sustaining that which benefits them directly. But to make the world a better place for all, current practices must change. The term 'sustainable development' implies that some forms of development cannot be sustained (continued indefinitely).

Here we look at why we need a different kind of development, what sustainable development is about, and how to apply the concept in practice.

The greatest taboo

There is a powerful taboo against questioning development. You just don't dare to. If you do, you may be labelled a rich greenie or 'to the far left' with 'pie in the sky' views, or some other derogatory term suggesting that you are either foolish or selfish with no concern for the poor.

Why this taboo? Our world is in the grip of a powerful development myth. In its simplest form, this is a belief that simply increasing economic growth will create more jobs and wealth for more people, and everyone and the planet will be better off.

The dominant model of development is usually associated with these assumptions:

◆ Poverty is best addressed by growing economies through accumulating wealth in the form of profits, which will eventually 'trickle down' to benefit the poor.

◆ Growing economies through competitive capitalism is the only way to develop, as proven by the failure of socialism.

◆ Wealth is best accumulated through industries and mechanisation; so industrialised countries are called 'developed' and those which are not (yet) industrialised and energy-intensive, are called 'developing' ('Third' World).

◆ Economic growth should be measured in terms of how much we produce and consume, and what we destroy in the process need not be included in the calculations. (See ECONOMICS).

Should we hold on to these beliefs? A quick look at the REALITY CHECK Box suggests not.

REALITY CHECK – The Track Record of Development As Measured by Economic Growth

Globally, there has been great economic growth. According to the UNDP consumption of goods and services in 1997 was twice that in 1975, and six times more than in 1950. Economic activity is now at an estimated $30 trillion annually. To what extent has this reduced poverty, inequality, crime or unemployment, and improved quality of life?

Poverty: An estimated 1 billion people still do not have the means to meet their basic needs. This includes 12 million children living in poverty in the United States of America, and 47 million in OECD countries. In India grain production increased by 149 million tons from 1951-1999, but in 1995, while granaries overflowed, 350 million people could not afford the available food. In 75 of 100 'developing' countries, the average person's income today is lower than it was 10 or even 30 years ago. In SA, 33% of households are now poverty-stricken, compared to 28% in 1995.

Equality: Worldwide, the income gap between rich and poor is increasing. The assets of the world's three richest men are greater than the combined national product of the 48 poorest countries. In Nigeria 80% of revenues generated from oil stay in the country, but only a small elite benefits from them. In SA the richest 10% of households still account for almost half of total consumer spending; the poorest 10% of households account for just over 1%.

Crime: Higher crime rates, including violent crimes, are associated with wider income gaps, in the West and elsewhere. America's prison population has increased 800% over the past 30 years. All South Africans' quality of life is affected by the effects of rampant crime.

Jobs: Economic growth is not solving unemployment; the global trend is towards 'jobless growth'. Market economies do not have a natural tendency to create jobs, and with government control of economic affairs waning, and technological innovations favouring smaller numbers of skilled workers, job creation suffers. Trans-national companies control over 33% of the world's productive assets, but tend to focus on capital-intensive industries, so they employ only about 5% of the global workforce. In SA the number of unemployed rose with 2,3 million to 4,2 million between 1995-2002, a period characterised – with the exception of 1995 – with better than expected growth.

Health: Despite medical advances we cannot stop the spread of HIV/AIDS, tuberculosis or malaria. There is a rise in diseases – ranging from allergies and diabetes to cancer – related to pollution, 'rich' diets and increasingly chemical environments. At the same time, more than half of all deaths among children can be attributed directly to under-nutrition.

Quality of life: Billions of people have no shelter, no adequate sanitation, no safe water supply. The lack of basic, relatively inexpensive facilities is still widespread in South Africa, despite it being one of the most developed countries on the continent. Some countries dump food in the sea to keep markets stable, while elsewhere in the world children die of hunger. And those with adequate resources do not necessarily enjoy a good quality of life. Spending on tranquillisers in the North matches total published expenditure on health in the 67 poorest countries.

Questioning development

Despite significant economic growth around the world, poverty, disease, unemployment and crime persist. This is reported in many national and international forums. Those who are already well-off do gain from capitalist economic growth, but very little 'trickles down' to benefit people in need.

At the same time, many of those activities which we call 'development' gobble up land, water and the diversity of life, and spew out pollution, sick people and degraded places. These unintended outcomes worsen poverty and create new health risks. Even those who benefit financially from economic growth find their quality of life spoiled by its 'shadow side' – crime, pollution and stress.

Development-as-growth has been sustained for as long as it has, by excluding the majority of the world population from its benefits, even as it used these (first slaves, then 'cheap labour' and ready markets, as well as the natural resources of the Third World) to create wealth.

Our current model for development is clearly not delivering the goods. And yet we cling to it … probably because it does (in some ways) benefit the 'consumer classes' in both the 'developed' and 'developing' parts of the world. There has also not yet been a historically proven, effective alternative. So the old "economic growth" myth has remained as the favoured model for everyone to follow.

One of the most dangerous aspects of our current model is that it is so resource-intensive that it cannot be multiplied on a big enough scale to benefit all people. We are already consuming more resources than the Earth can renew, and are actually overshooting its capacity by an estimated 30%. The bulk of this over-consumption is enjoyed by only 20% of the world's population, who consume up to 80% of its resources. Scientists say that if we were to extend this consumption and the pollution that goes with it to the rest of the world (to those billions who are currently excluded from development benefits) we would need to have five Earths to provide all the resources and absorb all the pollution.

We have only one Earth!

Having only one Earth does not mean that inequality and poverty should persist. Clearly, we need new ways to develop, that would stay within the limits of our planet's capacity, at the same time bringing benefits in a more fair and equitable manner to all.

Among those who dared to question development-as-growth were environmentalists who called for it to slow down and scale down, within the limits of the Earth's capacity to renew the resources that sustain both life and development. (An example is the 1972 Club of Rome report, *Limits to Growth*.) Others said that if we are to protect and improve life on Earth, we actually need MORE development.

Enter Sustainable Development

So the idea of 'sustainable development' was introduced, first in the World Conservation Strategy of 1980, and then in the Brundtland Report, produced by the World Commission on Environment and Development in 1987.

The authors of these documents were among those who called for MORE development, to pay for environmental protection and reduce poverty. And here a great muddle started,

because linking the two concepts – sustainability and development – opened the door for some to interpret the call for action as a call for more of THE SAME development. They argued that where there is poverty and suffering, there is simply not enough economic growth. The track record of development-as-growth was ignored – along with its unsustainable environmental costs. So the focus of the idea of sustainability shifted – from sustaining living resources, to sustaining DEVELOPMENT.

There are other ways of thinking about development – as a kind of development that would sustain (nourish) people, including the poor, and at the same time not overshooting the Earth's capacity to renew ecological resources. In *Caring for the Earth, A Guide to Sustainable Living* a coalition of environmental agencies described sustainable development as "improving the quality of human life while living within the carrying capacity of supporting ecosystems." The Brundtland Report talked about "development that meets the needs of the present without compromising the ability of future generations to meet their own needs." These ideas really require us to think in a different way – more long-term – about the needs of the present, especially the plight of the most needy, but also beyond them, about the future. In this view, what we need to focus on is how to sustain people and the planet – not following the old misleading path of "economic growth" as we know it.

Having gained popularity at the 1992 Rio Earth Summit (the United Nations Conference on Environment and Development) the term sustainable development is now widely used – and misused.

Applying the concept

South Africa's Department of Environmental Affairs and Tourism (DEA&T) built the idea of sustainable development into the National Environmental Management Act. DEA&T identified three components to sustainable development: environmental, social and economic sustainability. This means a form of development that sustains the natural environment; looks after people; and ensures that economic welfare can be maintained. It would be to the benefit of People, Planet and Prosperity – which was our slogan for the World Summit on Sustainable Development, held in Johannesburg in 2002.

All three of these 'principles' must be part of every development decision. The politician who agrees to turn a nature reserve into a golf course because of the economic benefits, is ignoring the ecological sustainability of his decision. If privatisation of water resources helps us use the natural resource more sustainably, but users cannot afford to pay, it is not socially sustainable. An industry which hides its health impacts on workers, even as it educates children about nature, is simply trying to sustain itself and 'business as usual'.

So how do we move towards a kind of development that will sustain people and planet and help all of us to prosper? There are no easy answers, but there are several worthwhile directions to consider.

ECOLOGICAL SUSTAINABILITY

A first principle for sustainable development is that any development activity should help to sustain (and not harm) our natural resources. Both scientists and ordinary people have noticed that the Earth's life-support systems are taking strain – once fertile areas can no longer sustain people; global

Sustainable Development
is about protecting our resources for future generations

GREEN Clippings
a weekly digest of environmental & conservation news

Keeping up to date on environmental issues is vital for sustainable development.

As more organisations begin to consider the environmental impact of their operations, staff need to be more informed about the complex issues that arise. It also becomes vital to be aware of the public image that is being projected in the media.

Green Clippings will keep you and your organisation informed via our weekly e-mail news service, containing brief reports of the most critical environmental news stories affecting Southern Africa.

Green Clippings weekly news service is followed closely by government, business, educators, religious groups, environmental organisations and conservation groups, lawyers, students, journalists, health practitioners and the general public.

Visit our website today, or contact us to subscribe to Green Clippings weekly update of environmental & conservation news

www.greenclippings.co.za
gc@greenclippings.co.za
Tel 021 8833366

fishing stocks have been depleted; the atmosphere, water courses and our food supplies have been polluted. It is difficult to determine how much of this kind of development the Earth can take. It is also hard, for each activity, to predict its impact on ecosystems. For this reason, we need to apply the "precautionary principle" which states: Don't proceed with a development until we are reasonably sure that it won't have negative impacts; if we cannot be sure, then don't proceed! There is also much that we can do to restore our environments, which in turn opens up many new development opportunities. Examples include organic farming, de-contaminating the soil and re-growing forests. On a macro-scale, if we are to sustain the planet, we must reduce population growth and change our patterns of consumption (see Topics POPULATION and CONSUMPTION).

SOCIAL SUSTAINABILITY

This second principle implies that a majority of people must benefit from development – not just a lucky few. It calls for fairness in the access to and benefits from the Earth's resources.

As we saw in the REALITY CHECK Box, our world is far from fair. The majority of the global population does not have access to resources; these same people suffer disproportionately from pollution, resource depletion and land degradation. How do we work towards greater equity, environmental justice and social sustainability?

South Africa has taken a huge step forward by abolishing unfair apartheid laws and instating a democracy which recognises equal rights However, inequalities remain in the way people participate in the economy and benefit from it. On a global scale there are calls for fairness in international trade regulations, which currently benefit the wealthiest nations and discriminate against less powerful ones.

But there are also economic injustices within each country. We therefore need to review our economic policies and practices in terms of their 'people impact'. To what extent do they reduce the appalling income gap in South Africa? What best supports the livelihoods of the majority? A few capital-intensive projects or many smaller, job-intensive ones? A weaker rand or a stronger rand?

Of course, people's well-being is not only dependent on jobs or income – social sustainability also involves education, health and a healthy environment, security, opportunities for relaxation and spiritual renewal, and people's right to participate in decisions which affect them. This includes the right to information about the environmental health impacts of development activities, and the right to legal action if such activities prove to be harmful.

ECONOMIC SUSTAINABILITY

This principle should be about economic activities that sustain people and planet – NOT about maintaining an economy, especially one based on development-as-growth which, as noted, harms the planet and fails to benefit the majority of South Africans.

'Sustainable Development' calls above all for reforms in the manner that we conduct our economic activities. Numerous measures have been proposed including:

- Removing unfair trade barriers – Economists estimate that this would allow poorer countries to generate a total income three times the sum of official development aid;

- Removing government subsidies which harm the environment and the poor;

- Upholding the polluter pays principle – those who do harm, must pay for redress;

- Instituting clear paths of responsibility and liability – for example, a CEO should be liable for the accuracy of a company's environmental reporting;

- Shifting the tax base from labour to resource use – in other words, rather than taxing us on what we earn through the work we do, tax us on our impacts and what we consume;

- Price products not only on what value has been added to them, but also in terms of what value they have deducted from the common natural resource base;

- Increase resource productivity – create wealth with ever fewer resources.

On a micro-scale, the question is not what constitutes a 'sustainable business' – how to sustain 'business as usual' is taught in any business school and does not require a new concept. The question is – what kind of business will sustain people and planet? Hard, truthful answers will require many companies to change either the nature of their business, or the way in which they go about it. Examples include more responsible waste management, cleaner production, recycling and energy efficiency. (See CONSUMERISM.)

The Jo'burg Summit (WSSD) and beyond

In their policies, decision making and on platforms such as WSSD it appears that the South African Government's interpretation of sustainable development focuses on generating greater economic growth. In this they are supported by business and industry. "Poverty alleviation" is frequently put forward as the basis for decisions, ranging from how we manage protected areas to what we do about waste.

It is certainly critical to recognise the scale of human hardship in our country, and finding sustainable ways to help the needy to build a better life, should indeed be a priority. But such efforts may be futile, if government and other powerful players:

- Choose to ignore the need to sustain our ecological resources, as if this were separate from and secondary to the process of sustainable development.

- Fail to consider the disappointing track record of the development-as-growth model, from the point of view of the poor.

We noted at the start that sustainable development is a tricky concept, hard to define and easy to manipulate by vested interests. If we do try to pin it down, we find it very challenging to put into practice, for it involves a new ethic, radical changes and long-term thinking. Most of us, including our politicians and partners in business and industry, focus primarily on short-term interests. It is hard to think of future generations if their interests conflict with one's current power base and ecosystems don't vote. Sustainable development will also require political sustainability – more sustaining forms of public and corporate governance that will prioritise the welfare of the least powerful amongst us, and of the only planet we can offer our children.

DEPARTMENT OF ENVIRONMENTAL AFFAIRS AND TOURISM

OUR VISION IS THE CREATION OF A PROSPEROUS AND EQUITABLE SOCIETY LIVING IN HARMONY WITH OUR NATURAL RESOURCES

MISSION

To lead sustainable development of our environment and tourism for a better life for all through:

- Creating conditions for sustainable tourism growth and development
- Promoting the conservation and sustainable development of our natural resources
- Protecting and improving the quality and safety of the environment
- Promoting a global sustainable development agenda
- Transformation

VALUES

- Sustainability: *efficient use of resources, walking the talk*
- Performance: *quality of products, impact of our work in society, energy driven*
- Professionalism: *accuracy, punctuality, knowledgeable, corporate governance*
- People: *team building, Batho Pele, capacity building*
- Integrity: *ethics, honesty, non-corruptive*
- Innovation: *face challenges and offer new solutions, pushing frontiers*
- Diversity: *richness of culture, ideas, a fabric of our organisation and nation.*

We are continuing to transform our society and create a better life for all. In doing so, we remain guided by the principle of sustainable development and the sustainable use of our natural and cultural resources. The environment is not only our source of life, but the protection and efficient management of the environment, is core to our tourism growth and development strategy.

The responsibility to grow tourism through sound environmental management, extends beyond that of Environmental Affairs and Tourism, and therefore integrated delivery by government as a whole is essential if we want to achieve our objectives. The implementation of the Tourism Growth Strategy will be a major focus over the next few years. The unprecedented growth in tourism in South Africa over the last year alone, is an indication of the potential in this sector to contribute to job creation and economic growth.

Sustainable development has become a thread that runs through all our work. South Africa's successful hosting of the World Summit on Sustainable Development in September 2002 in Johannesburg, clearly demonstrated that the programmes of our Government have a lot to contribute to the global dialogue on sustainable development. South Africa has successfully chaired the eleventh session of the Commission on Sustainable Development (CSD) in 2003, which set out the CSD programme of work for the next 10 years, following the adoption of the Johannesburg Plan of Implementation.

South Africa hosted the World Park Congress in September 2003. This was the single most important event of its kind in the world dealing with issues of conservation and protection of the environment. South Africa showcased our best practice in conservation management, the sustainable utilisation of our natural and cultural resources, as well as our successes in community-based natural resource management.

Addressing the pollution and waste issues of our country will be another major focus for the coming year. Issues such as air quality management will be addressed and polluting companies will be taken to task should they not comply with approved permit regulations.

All of our efforts are to make sure that the economic value of both tourism and environment are utilised in a sustainable manner to ensure that there is firstly a better life for all in South Africa, but also a better life for all in the world.

DEPARTMENT OF ENVIRONMENTAL AFFAIRS AND TOURISM

Private Bag X447 Pretoria 0001

Tel +27-12-310 3911 • Fax +27-12-310 3457 • www.deat.gov.za

Welcome

RECOUNTING JOHANNESBURG: THE WAY FORWARD FOR SUSTAINABLE DEVELOPMENT

Guest Essay by Mohamed Valli Moosa

(Minster of Environmental Affairs and Tourism at Johannesburg WSSD and Chairman of the 11th Session of the United Nations Commission on Sustainable Development)

Introduction

The Johannesburg World Summit was the biggest United Nations conference ever held. It was attended by 17 000 delegates, including 105 heads of state and government. Altogether 180 countries were officially represented at the Summit. In addition, 500 parallel events took place in Johannesburg and elsewhere in the country. It is estimated that the total number of international delegates attending the Summit and its parallel events was 37 000.

The World Summit on Sustainable Development has opened the way for the world to take new strides in the foremost challenge of our time – the eradication of poverty and closing the gap between rich and poor, combined with protection of the environment. It constituted a huge victory for human development and for the environment and it fulfilled a number of key objectives:

◆ The Summit created the correct balance of the three pillars of sustainable development, which are, social development, economic growth and the protection of the environment. This is a decisive shift from the predominantly wrong perspective over the past decade that sustainable development equals the protection of the environment.

◆ The Summit emphatically pronounced that sustainable development cannot be achieved separately from the quest to eradicate poverty, and that the growing gap between rich and poor is one of the biggest threats to sustainable development. Among the decisions in this regard is the decision to establish a world poverty fund.

◆ The Summit introduced a major shift from the donor-recipient paradigm to one that focuses on the obstacle to economic growth in poor countries posed by the unfair global economic system. While there is agreement to increasing aid from rich to poor countries, there is, more importantly, an acknowledgement that by far the biggest obstacle to poverty eradication is lack of market access and the anti-poor trade system.

◆ The cause of the African continent was greatly advanced with the practical focus on the New Partnership for Africa's Development.

◆ The Summit served to advance the cause of multilateralism during this troubled time in the world. It asserted the centrality of the United Nations and called for democratic global governance.

◆ The Summit brought a global focus on the state of the environment, and renewed high-level commitment to environmental protection.

The decisions we took

The Johannesburg Plan of Implementation (JPOI), which was adopted by consensus, includes programmes to deliver water, energy, health care, agricultural development and a better environment for the world's poor.

New targets will have enormous impact on the global agenda:

◆ In addition to the already agreed target of halving the number of people unable to access safe drinking water by 2015, it was agreed also to halve the number of people without basic sanitation by 2015.

◆ Countries agreed to reverse the trend in biodiversity loss by 2010 and to restore collapsed fish stocks by 2015.

◆ Chemicals with a detrimental health impact will be phased out by 2020.

◆ Energy services will be extended to 35% of African households over the next 10 years.

We believe that the Johannesburg Summit shifted the focus of world leaders from policy debates to the real task of "making it happen" and achieving high-level commitments by heads of state and leaders from business and civil society to meet the goals set. As testimony to this, many concrete actions, partnerships and funding targets were announced by countries and stakeholders.

Who could have thought that as skepticism hovered above our quest to make the world a better place, a commando of development cadres would emerge from the United Nations with a plan to ensure that the outcomes of the World Summit on Sustainable Development are set for global action and national implementation?

Organisation of work for the future

The new organisation of work of the UN Commission on Sustainable Development (CSD) is practical, and aimed at enhancing implementation. We have found a way to give practical effect to some of the innovations of Johannesburg: the 2-year implementation cycle, and the role of regions in implementation. The first year of the cycle will be the Review Year. This session will evaluate progress in implementation, identify constraints and share lessons learned and best practice. A key feature of this year will be the regional and sub-regional inputs in addition to contributions from major groups, UN agencies and the traditional core inputs in the form of the Secretary-General's State of Implementation Report and national reports.

The second year, the Policy Year, will take decisions on practical measures to expedite implementation.

Each CSD cycle will undertake an evaluation of progress in implementing Agenda 21, the programme for the Further Implementation of Agenda 21 and the Johannesburg Plan of Implementation, while focusing on identifying constraints and obstacles in the process of implementation with regard to the selected thematic cluster of issues for the cycle.

Each cycle will also address poverty eradication, changing unsustainable patterns of consumption and production, protecting and managing the natural resource base of economic and social development, sustainable development in a globalizing world, health and sustainable development, sustainable development of SIDS, sustainable development for Africa, other regional initiatives, means of implementation, institutional framework for sustainable development, gender equality, and education as cross cutting issues.

The programme of work agreed to has the following thematic clusters:

- Cycle 1 (2004/5): Water and Sanitation, and Human settlements
- Cycle 2 (2006/7): Energy for sustainable development, Air pollution/Atmosphere, Climate Change, and Industrial Development
- Cycle 3 (2008/9): Agriculture, Rural Development, Land, Drought, Desertification and Africa
- Cycle 4 (2010/11): Transport, Chemicals, Waste Management, Mining, 10 year Framework of Programmes on Sustainable Consumption and Production Patterns
- Cycle 5(2012/13): Forests, Biodiversity, Biotechnology, Tourism, and Mountains
- Cycle 6 (2014/15): Oceans and Seas, Marine Resources, SIDS, Disaster Management and Vulnerability

In 2016/17, there will be an overall appraisal of Agenda 21, the programme for the Further Implementation of Agenda 21 and the JPOI.

Reporting and UN system co-ordination

We need an effective system of reporting, to enable the review, evaluation and monitoring of progress in the implementation of Agenda 21, the programme for the Further Implementation of Agenda 21 and the JPOI. These reporting systems must be streamlined and easy to use.

The United Nations agencies and programmes, the Global Environment Facility and the relevant financial and trade institutions have been invited to participate actively in the work of the CSD. The Secretary General will report on activities to promote inter-agency co-operation and co-ordination on implementing the JPOI and other major UN conferences.

Major groups and partnerships

The involvement of major groups in the implementation of the decisions we took in Johannesburg must be strengthened, and we must ensure that major groups actively participate in interactive dialogues and we should collectively strive for better balance and representation of major groups in the implementation of these global commitments.

The important role of partnerships in implementation has been acknowledged. This is complementary to, and not a substitute for, the intergovernmental commitments. A number of useful criteria have been agreed on, to strengthen the link between partnerships and the targets we set in Johannesburg. Reporting on the progress of these partnerships should enable us to share the lessons learned and the best practice developed in partnerships.

Follow-up in South Africa

Our government has put together the building blocks necessary for the successful national and global follow-up of the Johannesburg Summit decisions.

In ensuring the appropriate follow-up of the decisions taken at the Johannesburg World Summit, and building on our successful chairing of the 11th Session on the UN Commission on Sustainable Development, which adopted a programme of work for the implementation review of the Johannesburg Plan of implementation, a coordinating task team led by the Department of Environmental Affairs and Tourism and the Department of Foreign Affairs has finalised our country's response strategy for meeting our country's commitments to the Johannesburg decisions.

The response strategy commits in detail all government departments to integrate the implementation priorities of the Johannesburg targets and the Millennium Development Goals in government's programme for the next ten years.

It further outlines the need to focus on the following as essential for the successful implementation of a global sustainable development agenda:

1. Integrated follow-up and planning in the implementation of outcomes of major conferences and summits in the economic and social fields.

2. Building of strong partnerships with civil society and our development agencies for the implementation, monitoring and evaluation of progress.

3. Raising awareness, capacity building and mobilising new and additional resources for implementation.

It is government's view that sustainable development should not be regarded as an add-on, but should be an organic part of what we do. We must avoid the danger of making sustainable development the responsibility of just one organ of state. Sustainable development is very much about integrated government.

In Johannesburg, as we did in Rio, we entered into a solemn pact with future unborn generations not to destroy beloved planet Earth. We also entered into a deal with the poor and hungry and with oppressed women to ensure social and economic development.

The poor watch and wait to see whether hunger, disease and global warming will be tackled with the same vigour displayed by some on the military front. The poor watch and wait to see whether the rich will say that they have as much resources and determination to prevent the early deaths among the children of the poor, from water-borne diseases.

> *"Fighting poverty and saving the environment are in fact the same battle."*
>
> ~ World Summit on Sustainable Development 2002

SUSTAINABLE DEVELOPMENT: THE UN VISION

Guest Essay by Minister Børge Brende ~ Chairman of the UN Commission on Sustainable Development and Norwegian Minister for Environment

At the World Summit in Johannesburg in 2002 and at the Millennium Summit in 2000, the international community set itself ambitious goals for sustainable development – the Johannesburg Targets and the Millennium Development Goals (MDGs). The UN Commission on Sustainable Development (CSD) has been challenged to oversee the implementation of the ambitious goals into viable action for our common future – regionally, nationally and locally. We have no time to lose.

As chair of the upcoming session of the CSD – CSD12 – I will make it a priority to uphold the political momentum from Johannesburg, with particular emphasis on water, sanitation and human settlements. These are the topics to which the Commission has decided to devote most of its attention over the next two years. Progress in these areas will help reach goals in other important areas such as health, education, gender equality and biodiversity – as well as poverty eradication.

Great expectations rest on the shoulders of world leaders and the international community.

We have promised to:

◆ Reduce by half the proportion of people without sustainable access to safe drinking water by 2015.

◆ Achieve significant improvements in the lives of at least 100 million slum dwellers by 2020.

◆ Halve the number of people without access to improved sanitation by 2015.

It is time to stop being defensive. What is needed is political courage.

◆ More than 1.2 billion people around the world lack basic water supplies.

◆ More than 2.6 billion people do not have access to adequate sanitation.

◆ The global goals on water require that safe drinking water be delivered to another 270 000 people every day for the next 12 years. The sanitation goal means that basic sanitation must be made available to another 370 000 people every day for the next 12 years.

Although a daunting task, I believe it can be done. During the Water Decade in the 80s, we managed to reach such figures. We can do it again. But we have to learn from the mistakes and the setbacks that we experienced. This time, we have to do it in a sustainable manner. Our ambition is to enhance the CSD's role in monitoring actual implementation of the global targets. Focus must be on identifying obstacles and providing clear and concise recommendations for further action. We must ask ourselves what works and what doesn't, and learn from best practices. And we must bear in mind that this is a global agenda, and the specific needs and challenges of each

region must be duly reflected in the way that we seek adequate solutions.

To succeed we should draw on the strengths of the CSD, addressing the crosscutting dimensions of sustainability and mobilising active participation of all relevant stakeholders. Even though the targets are global in character, they must be implemented where people live and shelter and services are required – at village, community and city levels.

I agree that the numbers themselves are overwhelming. But more important – what do they mean in practical terms?

◆ They mean that at any given moment, almost half the population of the developing world are sick from unsafe water and sanitation.

◆ They mean that a child is killed every eight seconds from water-borne diseases.

◆ They mean that every year, around four million people – corresponding to the total population of my home country Norway – die from water-related diseases.

◆ They mean that it is crucial to improve the situation of the urban slum dwellers around the world.

◆ They mean that health is the most important reason for investing in water, sanitation and hygiene.

This is the reality that we constantly need to keep in mind.

Water and sanitation

Focusing on water and sanitation is very timely as it serves as a prerequisite for reaching other MDGs. There is also a strong linkage between the state of the environment, of freshwater resources in a country, and its capacity for development and poverty eradication.

Water is the key factor in economic growth, contributing to improved living conditions through employment and increase in gross national product. When focusing on water and sanitation, we need to address two important issues – governance and financing.

To bring about change – mindset also has to be changed. We need to empower the poorest and listen to their concerns and needs. We should introduce clear mechanisms for accountability between the poorest and their governments. We need to include water and sanitation in poverty reduction strategy papers, and we need to integrate poverty into water resources management and efficiency plans at national level. According to the Johannesburg Targets, all countries are expected to develop such plans by 2005. By reaching this target we can effectively set the tone for our future work.

Today, many governments spend a lot of money on water, but in an uncoordinated manner, and without sufficient attention to basic services. In many cases, governments therefore need to spend money more efficiently, and with greater focus on results. At the same time, there is an imminent need for additional resources. We can say neither "only more ODA" nor "only governance reforms". We need to strike a rational balance, avoid dogmatism, and develop a sensible framework for evaluating which countries need which approaches. However, the largest share of investment will need to be met from public and private resources in each country. Developing countries need investments, not just in basic infrastructure and production capacities, but also in goods and services.

Human settlements

Safe and better housing conditions will provide people with dignity and security, reduce severe health risks, often linked to bad environmental conditions, and improve the future prospects of the young.

This can only be achieved if we address three core challenges:

◆ We must focus on good governance and political leadership. We can achieve improved quality of life for all citizens by developing sound strategies and focusing on good governance at the national and local levels – and by giving political priority to the upgrading of existing slums.

◆ We must address the lack of access to land and to secure tenure, which today is a major obstacle to development. In South Africa, the right to housing is included in the Constitution, and since 1994, more than 1.5 million houses have been constructed, providing close to 6 million people with a home. This is a remarkable achievement in a short time, in a country with 40 million people.

◆ And we must mobilise financial resources – at domestic and international level. A key problem for the poor is to get access to financial resources to improve their living conditions. In order to provide such access, micro financing has been developed as an instrument. One example is the Self-Employed Women's Association in India. It started in 1974 on the savings of 4 000 women, all very poor members of a labour union. By 1995 it had given 22 500 women the opportunity to own tools and other means of production, to earn a higher income and thereby be integrated in the economy.

It's happening. It's do-able.

A revitalized UN Commission on Sustainable Development

With only three main topics on its 2004-2005 agenda, the CSD can become more focused and solution-oriented in its approach. The 2004 Session will act as a review session, undertaking an evaluation of progress in implementing Agenda 21, the Programme for the Further Implementation of Agenda 21 and the Johannesburg Plan of Implementation. The job is to identify constraints, obstacles, successes and lessons learned in the process of implementation. A good review is necessary to secure a common platform for the subsequent policy session in 2005, mobilising the necessary political and financial support to ensure continued efforts at global, regional and local level to meet the specific targets, including the overall goal of poverty eradication.

The Earth Summit in Rio in 1992 was the culmination of a political momentum that had been building up over several years. In retrospect, it is easy to see that too little emphasis was put on an effective follow-up. As chair of the CSD, I will do my best to ensure that last year's World Summit in Johannesburg will come to be seen as the beginning of a process, rather than the end of one – a process that will be marked by one word: implementation.

In our every deliberation, we must consider the impact of our decisions on the next seven generations.

~ Iroquois Native Americans.

Everyone has the right :

a) to an environment that is not harmful to their health or well being; and

b) to have the environment protected for the benefit of present and future generations through reasonable legislative and other measures that –

i) prevent pollution and ecological degradation;

ii) promote conservation; and secure ecologically sustainable development and the use of natural resources while promoting justifiable economic and social development.

~ The South African Bill of Rights

SUSTAINABLE DEVELOPMENT
A CORPORATE RESPONSIBILITY

Guest Essay by Tony Dixon
~ Executive Director, Institute of Directors South Africa

Spotlight on capitalism

A number of spectacular corporate failures both here and abroad have put the spotlight on capitalism and the free enterprise system, with many commentators questioning whether or not the system might have lost its way.

An increased focus on the principles of sound corporate governance was a natural reaction and the Institute of Directors (IoD) in southern Africa responded by initiating the King Committee on Corporate Governance.

For South Africa, which has just come through the most remarkable period of transformation, the challenge is even greater.

Credibility of free enterprise system at stake

The vast majority of people in this country probably have a pretty jaundiced view of business and at stake is the credibility of the free enterprise system. Commerce and industry will ultimately be judged not only on their ability to create much-needed jobs and wealth but also on their contribution and relevance to society. In short, the ability of South African business to generate sustainable growth and long-term relevance is the only real hope this country has for the long-term prosperity of its citizens.

Its ability to deliver meaningful value into the future is further compounded by the huge challenge of AIDS and the whole question of corporate transformation which, of necessity, demands more focus here than elsewhere.

Ability of directors to meet enormous challenges ahead

The ability of directors to deliver on the enormous challenges ahead probably needs a change in emphasis and a move away from the traditional leadership style which, sadly, has too often been driven by short-term goals, greed and self interest.

A more encompassing leadership style that focuses more on the common good and long-term sustainability is what current wisdom dictates.

The move towards sustainability reporting is therefore welcome in that it encourages a more positive response to some of these vitally pressing issues.

Triple bottom line reporting

The King Committee's recommendations of a move towards triple bottom line reporting were acclaimed worldwide as a precedent in addressing the need for more inclusive management and accountability. That it has since been adopted in many other codes of governance around the world is a tribute to its authors.

The importance of triple bottom line accounting is highlighted in King II but one of the real benefits is that it encourages a far more integrated approach to managing an enterprise and requires a more positive demonstration of performance.

Align interests of individuals, corporations and society

Sir Adrian Cadbury summed it up rather aptly in a Corporate Governance Overview for the World Bank where he stated: "Corporate Governance is concerned with holding the balance between economic and social goals and between individual and communal goals ... the aim is to align as nearly as possible the interests of individuals, corporations and society."

There is an increasing belief that commitment to sustainable development enhances an organisation's reputation and reduces risk. That good governance makes good business sense is now more apparent than ever.

Correlation between good governance and long-term profitability

Research from both McKinsey and the Deutsche Bank provides conclusive evidence that there is a strong, positive correlation between good corporate governance, shareholder value and long-term profitability. The premium in share price is substantial and a demonstrable level of consistently higher earnings makes a compelling argument for the adoption of sound principles. There is little doubt that over time additional linkages between shareholder value and sustainable performance will be forged.

Those who ignore the bugle call for sound corporate governance and the need for sustainable development do so at their peril.

Increased shareholder activism is a symptom of the malady

More often than not an unhealthy preoccupation with short-term profit leads to value destruction rather than value creation. Shareowners are already demanding greater accountability from Directors and this trend is likely to get worse before it gets better. Increased shareowner activism is a symptom of the malady and provides tangible evidence of the dissatisfaction with the status quo.

The past lack of transparency and the tendency to pay lip service to the demands of sustainability reporting have

Think of what sustainability will do for Africa...

Our commitment to sustainable development is unwavering.

The DBSA cares for the planet through funding programmes that are environmentally sound and sustainable.

DBSA
Development Bank
of Southern Africa

Development Bank
of Southern Africa

1258 Lever Road
Headway Hill, Midrand

PO Box 1234
Halfway House
Midrand, 1685

Telephone:
+27 (0)11 313 3911
Facsimile:
+27 (0)11 313 3086

www.dbsa.org

been disappointing, illustrating that Corporate South Africa has been slow to buy into the benefits of a more integrated and holistic approach to management.

A KPMG annual survey, conducted in 2002, on sustainability reporting of the top 100 companies, revealed that only one produced a sustainability report, only four produced social reports while 12 produced reports on one or more aspects of Safety, Health or Environmental issues. Hardly inspiring!

Demands for social reform higher than anywhere else

Also disappointing was that in an international survey conducted in the same year, when it came to corporate accountability for sustainable development, South Africa was ranked last out of 19 countries. In a country where the demands for social reform are perhaps higher than anywhere else this was unacceptable.

Happily, the response to this criticism has been positive with a noticeable improvement in adherence to the principles recommended in King II. KPMG's 2003 survey shows that 85% or more of the top 100 companies met the baseline recommendation of King II in that they reported in some form or another on the nature and extent of their sustainability policies and practices.

Unfortunately many of these reports appeared superficial in nature with the vast majority providing only partial disclosure. Very few provided the full disclosure recommended in King II in respect of social, transformation, ethical, safety, health, and environmental management practices. The trend however is a positive one.

While the war is not yet won, this move towards fuller and more responsible disclosure is encouraging. The Johannesburg Stock Exchange (JSE) has entered the fray with the introduction of the FTSE/JSE SRI Index, which it is hoped will encourage even further participation amongst the top 160 companies on the bourse.

The initiatives of the IoD through King II, the JSE through the SRI Index and its listing regulations together with ongoing pressure from concerned stakeholders should contribute to even further compliance in the future.

Move away from tick box compliance mentality

What is needed in the interim however is a commitment by management to move away from a tick box compliance mentality to a more comprehensive form of reporting ,which sees substance over form as its hallmark.

If business fails to respond to the challenge, it could find the demands of the World Summit on Sustainable Development (WSSD), which called for enforced corporate accountability, becoming a reality. This would be unfortunate.

*The world we have created today
has problems which cannot be solved
by thinking the way we thought
when we created them.*

~ Albert Einstein

*Never believe that a few caring people
can't change the world.
For, indeed,
that's all who ever have.*

~ Margaret Mead

SUSTAINABLE DEVELOPMENT IN MANUFACTURING

Guest Essay by Tony Phillips, CEO ~ Barloworld

The populist image of the corporate leader is a caricature human being whose sole focus is maximising their personal wealth with little or no care for the social or environmental consequences of their actions. As so often in life, the facts get in the way of a good story. On closer examination corporate leaders are, in my experience, as much a cross section of society as any other arbitrary slice of the employed population. What is really insightful is to observe the way in which, through their actions, an ever increasing number are moving their organisations at an ever increasing pace, in the direction of improved social and environmental performance.

14 years ago at the Rio Earth Summit, business was invisible when NGOs and Governments argued and debated in search for the solutions to the world's problems. A decade later at the Johannesburg Summit over 700 leaders of the world's largest organisations regarded the issues as sufficiently important and relevant to their businesses that they made the trek to a small country at the bottom of the planet to participate in the heart of the sustainability debate.

Cynics will argue that their motivation was primarily to be seen on the side of the angels – a publicity stunt that had nothing to do with what their organisations were actually doing in their everyday operations.

The extraordinary variety of sustainability initiatives on display in Johannesburg suggest a much more profound transformation. These initiatives ranged from high-level corporate commitment to generic principles such as the Global Compact through to an explosion of small businesses founded on innovative use of re-cycled materials and every imaginable approach in between. Few attending Rio in 1992 would have imagined that not one, but two auto manufacturers would present hydrogen powered cars to the world as commercially viable products in 2002.

I believe the driving force behind this transformation of sustainable development from the avant-garde to the corporate mainstream is the increasingly compelling evidence that it makes sound commercial sense. This is vitally important since a modern chief executive, no matter how passionately he or she feels the personal wish to lead a business into more socially and environmentally sustainable patterns of behaviour, is bound by a primary need to make the enterprise economically sustainable (profitable).

This is not some corporate excuse, it is the reality that failure to achieve a foundation of profitability means, for most leaders, the end of their leadership role and the end of any possibility that they might play a role in leading their organisation to higher levels of sustainability.

Sustainable development bears many similarities to risk management in the different ways that it is seen by companies within the manufacturing sector. In the early stages of adoption, it is often seen as imposition of another layer of cost on the business. Typical reactions amongst managers revolve around "another series of things we have to do that take our eye off the ball and our focus away from getting the real job done." However as the systems for measurement of social and environmental impact are established invariably they identify opportunities for financially beneficial improvements in the business. In the environmental arena manufacturers that measure and target lower levels of electricity and water consumption inevitably lower their cost base. On the social front companies that become involved in comprehensive HIV/AIDS programmes find they can reduce the incidence within their employee base and increase the effectiveness of treatment and hence the period in which sufferers are able to remain effective members of the workforce.

A key element in sustainable development is the engagement of, and reporting to stakeholders. By actively, openly and transparently reporting on an organisation's performance, pressure is created for continuous improvement. By their very nature companies are competitive. This competition manifests of course in comparison to others but also in comparison to itself in prior periods. Thus companies which annually present a public assessment of their progress in the sustainability arena typically seek to be both better than their industry peers, and to better their own performance last year.

The alacrity with which manufacturing companies are adopting the Global Reporting Initiative (GRI) framework for triple bottom line reporting is a remarkable feature of the first four years of the twenty first century. The power of the GRI lies in both its comprehensive approach and the flexibility that companies can evolve the complexity of their reporting at a pace that suits them and makes sense for their business.

In the environmental arena the implementation of sustainable development strategies is encouraging companies to use FEWER resources and to seek to avoid the use of harmful products. The virtual eradication of the use of CFCs (key contributors to the hole in the ozone layer) has been

achieved largely due to the willingness of companies to adopt alternative propellant gases in their aerosol products, in fridges and cooling systems and in packaging. In the paint industry in a very few years the industry has switched from predominantly solvent-based to water-based technologies.

In the social context, the manufacturing industry has made significant strides in the reduction of child labour and improving the quality of life of factory workers in developing nations. This has occurred because companies in sectors such as shoe and garment manufacture, which have outsourced their manufacturing to suppliers in those lower cost developing nations, increasingly make the same demands on their suppliers as they would impose (or have imposed) on themselves if they were manufacturing in their developed world home bases. They not only make the demands, but back up their words with action and resources, employing staff to not only monitor supplier conformance but also to provide expertise to assist suppliers to improve performance.

All this activity is most powerful when propelled by enlightened self-interest and the naturally competitive instincts of successful global corporations responding to an idea that is already understood as the only way to successfully do business in the twenty first century.

Sustainable development must become an inherent part of the culture of any company if it is to thrive over the long term. In our organisation we have our own language to describe it – it is called Value Based Management and within it, we set our goals to create value simultaneously for all our stakeholders: our shareholders, employees, customers and the social and environmental "communities" in which we operate. The key word is simultaneously and not, by implication creating value for one stakeholder at the expense of another. It is the kind of philosophy that has stood our organisation in good stead for 102 years and is the foundation on which we are building our future.

For the next century, the challenge is to implement substantial increases in natural resource productivity, to become effective and systematic in doing more with less.

~ John Bruton, Prime Minister of Ireland

I recognize the right and duty of this generation to develop and use our natural resources, but I do not recognize the right to waste them or to rob by wasteful use, the generations that come after us.

~ Theodore Roosevelt

" ...bad environmental practice produces costs, whatever the stage of development. Countries like China and Brazil are increasingly showing that conservation and development go hand in hand."

~ World Summit on Sustainable Development 2002

SUSTAINABLE DEVELOPMENT IN THE RETAILING INDUSTRY

Guest Essay by Tessa Chamberlain
~ General Manager, Corporate Marketing, Pick 'n Pay

Just as the World Summit for Sustainable Development in 2002 identified three interdependent elements of "sustainable development" (people, planet and prosperity), companies within the retail sector need to recognise the interdependence of their own businesses, the communities they operate within as well as their environment at large.

Supermarkets and other retail outlets play an increasingly important role in our lives and our interactions. They, and the ever increasing number of malls in which they belong, change the function and appearance of cities, provide a natural focus for the community, and replace the town square as a forum for entertainment and human interaction. If supermarkets comprehensively and objectively review their influence on people's behaviour they will quickly come to the conclusion that they have an important role to play, both in providing their customers with the best service they can offer, but also as role models for improving social and environmental conduct.

The bottom line principle of sustainable development within the retail sector is accountability. Whether in a retail environment or otherwise, accountability not only forces retailers to allocate resources to environmental projects, but also forces them to take responsibility to make sure that these projects actually work, and have a positive, ongoing impact. The concept of triple bottom line reporting plays a key role in sustainable development.

In addition to accountability, the following points demonstrate what can and should be done by retailers to minimize their environmental impact.

◆ **Primary packaging of goods.** Consumerism places a huge strain on environmental resources; firstly on our raw materials and energy resources, and secondly by pushing up the amount of waste that has to be dealt with. Obviously it would be detrimental for retailers to discourage consumerism, but they should:

a) use their influence to force manufacturers into the use of more environmentally-friendly packaging, and

b) educate consumers to take an active role in environmental protection by seeking out products that have less environmental impact, and by being responsible with their waste.

Recycling is one of the base principles of sustainability. Most Pick 'n Pay Hypermarkets have created Recycling Centres for a variety of materials. Customers are encouraged to deposit used plastic, glass bottles, plastic carriers, cardboard, paper and cans at the recycling centres. The Centres are supported by local communities, and also by the recycling companies responsible for removing the items collected. In the 1990s Pick 'n Pay introduced a range of "green" products ranging from paper to cleaning products. Together with the manufacturers, these ranges were developed as an environmentally friendly range focussing not only on packaging but the ingredients used as well. We continue to offer our customers these products as an alternative choice.

◆ **Carrier bags.** Plastic bags have become a symbol of waste and garbage in South Africa. This litter is seen wherever you go – next to our highways, our beaches, in our harbours and river mouths. And sadly their damage goes way beyond just the fact that this litter looks ugly – the damage to flora and fauna is well documented. Whilst the Government-led initiative is a huge step forward towards better management of the impact of plastic on the environment, there are still many issues that need to be addressed to truly make progress in this area. The crux of the legislation currently in place relates to the thickness of carrier bags (a minimum of 30 microns). The aim of this is to enable easier recycling, and greater opportunity for re-use. The retail sector has also looked for alternatives. After extensive research on consumer trends, environmentally friendly alternatives and international successes, Pick 'n Pay has introduced the 'Green Bag' as an alternative to the government-regulated bags. The 'Green Bag' is ergonomically designed and is made of environmentally friendly, non-woven polypropylene fabric, making it fully washable and long lasting. The Bag is sold at cost and 20% of the price is set aside for environmental projects. In addition, the durable "Bag for Life" remains on sale in most stores. At the end of the day though, the real solution to the plastic bag problem lies in consumer education and in changing the consumers' habits regarding re-use of bags.

◆ **Refrigeration.** The use of CFCs has been a long-standing environmental thorn. At one stage they were widely used in the coolants used in refrigerators and freezers that are so important to the delivery of high standards of freshness by food retailers, and in the air-conditioners that are so widely used by all retailers. The Montreal Protocol of some years back stipulated a deadline for a complete phase-out of coolants containing CFCs, and retailers need to ensure that they are in compliance with this. This has been a long and expensive process for Pick 'n Pay. In some instances this has entailed converting equipment and in others it has meant the purchase of completely new equipment.

- **Usage of electricity.** Bright lights and refrigeration are two essentials of the retail trade, especially in the food sector, leading to high levels of energy consumption by stores. Higher energy consumption means greater use of raw materials to create that energy. Pick 'n Pay has taken steps to reduce this consumption, by introducing automatic power saving switches in every store, and in many stores the hot water is warmed by energy-saving systems to minimize the use of electricity.

- **Impact on environment of new stores.** Because consumerism is a core driver of environmental decay, the building of new retail stores and malls brings additional risk to an area. Waste levels increase, energy requirements increase, raw materials are consumed, water is used and abused, flora is destroyed, traffic and pollution increases, and so the list goes on. Retailers need to be responsible about the development of new stores by incorporating environmental impact studies into their planning process.

- **Distribution.** As with refrigeration and electricity, distribution of product is a primary process within the retail sector – more so within the food sector because of the frequency with which deliveries need to take place. The environmental impact of this process can be reduced by introducing low noise and low pollution vehicles.

- **Waste.** Owing to the volume of foodstuffs in food stores, and the amount of secondary packaging that is removed prior to products being packed on shelves, retailers generate a considerable amount of waste. Hence waste management becomes an important area where retailers can and must be active in environmental protection. In Pick 'n Pay's case, the aim is twofold: firstly to minimise the amount of waste, and secondly to work with contracted recycling agents to ensure that nothing is "dumped". For example, agreements are in place with the cardboard packaging companies to collect and recycle all boxes, which are compacted at store level. Similar agreements are in place for the removal of food waste. It makes sense for retailers to work with partners – organisations that focus specifically on the environment and have the expertise at hand and to tap into their knowledge, rather than trying to reinvent the wheel.

- **Poor consumer awareness.** Bad habits and practices, usually through lack of appropriate education, is perhaps one of the most substantial problems. Looking beyond the specifics of where retail practices place our environment at risk, perhaps one of the biggest contributions that a retailer should make is communication. Because shops provide a hub for our modern lives, they also provide a platform for communication. Along with the media, retailers and retail communication affords a huge, and respected, forum for educating consumers and demonstrating good environmental practice that brings effect to sustainability. Through effective communication and education, Pick 'n Pay seeks to create an awareness of environmental issues amongst its employees, customers and local communities with its Enviro Fact Sheets and campaigns. All employees are exposed to the company's philosophy with regard to the environment to encourage their buy-in. This helps ensure that environmental responsibility is seen as part of each employee's job, and not an additional function. Most stores have staff environmental groups, who work together to set goals for the improvement of their work and home environments, with an emphasis on creating a positive impact on communities in need of upliftment.

Returning to the explosion of consumerism, to a very large extent it is the retail sector that is driving that phenomenon, and retailers must take responsibility for their actions. Subscribing to this viewpoint simultaneously implies a realisation that ensuring a more efficient use of natural resources is in the long-term interest of the business. Sustainable development calls for economic and social development that meets the needs of the present generation without compromising the ability of future generations to meet their own needs. And retailers are very dependent on future generations!

At the end of the day, if one considers that accountability is such a fundamental principle of sustainable development, companies need to be absolutely committed to doing what's right for their own industry, whether the cost entails direct investment, such as was the case with refrigeration in the food retail sector or the contribution of funds and resources to specific projects.

This commitment must be led from above, and be entrenched in the company's vision and mission. This is the only way it can have real impact. The responsibility Pick 'n Pay has assumed in positioning itself as an environmentally aware company is laid out in the form of a policy based on principles held by the International Chamber of Commerce and the Business Charter for Sustainable Development, and has its roots in the building blocks of the retailer that were laid out more than 36 years ago. Considering the areas of risk that retailers expose to the environment, they hold a great responsibility to nurture it, but this must be entrenched in the philosophy of the business, just as Pick 'n Pay has done with its own Environmental Code of Practice. A clearly defined code of practice, coupled with a formal reporting procedure, can only lead to better and more sustainable environmental activity.

We do not inherit the earth from our forefathers, we borrow it from our children.

~ His Majesty King Sobhuza II

SUSTAINABLE DEVELOPMENT IN THE MINING INDUSTRY

Guest Essay by Karin Ireton
~ Group Manager, Sustainable Development, Anglo American plc

It is not surprising that engineers, accountants, marketing people and artisans don't have a blueprint of the principles of sustainable development at the forefront of their minds.

The term sustainable development has become shorthand for a debate that has raged for over 20 years through a variety of core disciplines including environmental conservation, social development and economics, collecting bits along the way from the human rights agenda, governance and democracy debates. It encompasses everything from politics and philosophy to the most exact sciences.

The nearest we can get to an international consensus is contained in voluminous UN policy documents. And the truth is, there isn't a "one-size fits all" response to the issue of sustainable development. There are, however, practical ways of addressing the challenges of sustainable development such as those being taken in the mining industry.

The Mining Industry's response

Through the International Council on Mining and Metals leading companies have created a framework of principles around which they will concentrate their efforts on sustainable development. The ICMM members include a number of companies operating in South Africa – Anglo American plc, Anglo Gold, BHP Billiton, Placer Dome and Rio Tinto. The framework includes principles clustered around a number of core issues and includes: corporate governance and ethical business practice; the incorporation of sustainable development principles into business decision-making; human rights and respect for the cultures, customs and values of employees and others; risk management and continual improvement in health, safety and environmental performance.

Other commitments are to contribute to the social, economic and institutional development of the communities in which we operate; contributions to integrated land use planning and biodiversity conservation; and effective, transparent communication. A sectoral guideline for independently verified reporting is also being developed with the Global Reporting Initiative.

A further framework for Sustainable Mining is provided by the Mining Charter in South Africa. This addresses human resources, development issues, employment equity, migrant labour, community and rural development, living conditions, procurement, ownership, beneficiation and reporting.

But how?

In the past, individual professionals operating often quite separately addressed operational level challenges such as those of worker health and safety, the biophysical environment, the economic viability of the mine. Their main focus was on short-term deliverables against performance criteria and budgets. Sustainable development now adds new challenges and requires that multi-disciplinary approaches are developed. The time frames are also far longer. For example, a sustainable development challenge for mine planners is to conceptualise and plan for post-mining scenarios in which communities can create sustainable livelihoods for themselves. These future scenarios may be many years away and are therefore very difficult to plan and anticipate This highlights the need to work in partnership with government and society, as the mining industry alone cannot be expected to achieve all the desired goals. However, real efforts are being made by the mining industry to create the tools to assist operational staff to give daily effect to these commitments.

One of these is a joint project between the ICMM members and the World Bank to create a set of tools, which will assist in community engagement and designing public participation programmes, assessing community needs for development, and identifying regional government approaches and methodologies for conflict resolution. The final project will be published in the second quarter of 2004 and has drawn on southern African regional expertise and experience.

Of course, for some the debate is not how we should mine, but whether we should. Many of the remaining resources are found in less developed parts of the world, and those countries have a sovereign right, acknowledged in the Rio Declaration, to utilise their resources. The WSSD reinforced the Rio principles and also explicitly recognised that mining played an important catalytic role in the development of many economies.

However, putting the benefits that accrue to the State from those mining activities to work in creating sustainable futures is essential. This will require transparency and accountability of all parties. The Extractive Industries Transparency Initiative is being negotiated between key governments and leading industry players that will set out a framework for disclosing payments to governments for the extraction of natural resources.

Many of the metals and minerals that are mined are finite but not rare, and a significant proportion is almost infinitely recyclable. The end products of mining are in household, industrial and infrastructural use around the globe and

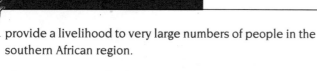

provide a livelihood to very large numbers of people in the southern African region.

There are man-made alternatives to most mining products but they too have environmental, economic and social impacts and consequences.

Addressing the problem

The challenge, I believe, is to remain in ongoing debate with society and in partnership, to address a number of key issues, including:

◆ Continuously refining mining practices and improving economically viable recycling and reuse patterns, thereby ensuring multiple lives for products from mining.

◆ Ensuring that the revenue that a country or region earns from mining is used for building human and social capacity and developing infrastructure.

◆ Ensuring that mining contributes to the social, human, manufactured and financial capital in countries where it operates, including fighting poverty through employment and other means.

◆ Planning for the post-mining phase at specific sites and at a strategic level in areas where resources are becoming less economically viable.

◆ Finding the best mix between job retention for low-level workers and technology improvement, together with the migration of jobs up the skills ladder, which will make the mining industry a safer and healthier working environment.

One of the sustainable development challenges within southern Africa is the issue of migrant labour. As in most things in this debate, a simplistic response will just move the problem along the line, not resolve it.

Current repatriation of money to neighbouring countries by miners working in South Africa by far exceeds the aid monies received by those countries, according to the South African Chamber of Mines economists. As we are regularly reminded by the Zimbabwe debate, having stability in neighbouring countries is critical to our own well-being.

While many proponents of sustainable development focus strongly on the need to hire local labour from neighbouring communities, the mining scorecard requires that companies do not discriminate against foreign migrant labour.

The major mining companies in this country have a proud record of community development projects and philanthropic work and the Anglo American Chairman's Fund has for a number of years been voted by development NGOs as the corporate fund most attuned to the needs of sustainable development. Key focus areas are education and youth, health (including HIV/AIDS), and community

development as well as a number of core biodiversity and cultural history related projects.

A new and important depth was added to our initiative to fight HIV/AIDS through a ground breaking public-private partnership between Anglo American, LoveLife, the Department of Health, the Nelson Mandela Foundation, the Henry Kaiser Family Foundation and the Global Fund to fight AIDS, TB and Malaria. Anglo American will provide R30 million over a period of 3 years for the development of youth-friendly community clinics.

This is in addition to voluntary counseling and testing, awareness raising efforts, wellness programmes and appropriate treatment for employees. About 1 000 employees are on anti-retroviral treatment at present. More than 90% of them are well enough to return to normal working duties and the compliance with the treatment regimes has been impressive.

Efficiency in the use of resources such as energy and water and programmes to prevent pollution are also part of everyday mining practice. Management of biodiversity on our properties is a key issue being addressed through biodiversity strategies and action plans.

A great deal of work is being undertaken in the major companies at an operational level on rehabilitation and restoration work and also in seeking creative land use alternatives at the post mining stage, at industry level and through international programmes such as the sponsorship of the Millennium Seedbank and the World Conservation Monitoring Centre.

Conclusion

Sustainable development is a shared responsibility. Society, industry and government must all contribute and work together to achieve meaningful results. This is a long and winding road – a journey, not a destination – but leading mining companies have taken the initiative, are making progress and are committed to being part of a sustainable future.

"Sustainable Development is a security imperative. Poverty, environmental degradation and despair are destroyers of people, of societies, of nations."

~ World Summit on Sustainable Development 2002

DEPARTMENT OF MINERALS AND ENERGY
REPUBLIC OF SOUTH AFRICA

Mining and the Environment

National Phepafatso Strategy

A holistic integrated and co-operative approach for mine environmental management and pollution control In South Africa

Energy and the Environment

My fire works best, in 5 easy steps!

1 2 3 4 5

● gives less smoke ● burns longer
● Heats quicker ● is much safer

Basa njengo Magogo

Spare the environment. Spare your health.

SUSTAINABLE DEVELOPMENT IN SOUTH AFRICAN TOURISM

Guest Essay by Cheryl Carolus ~ CEO, South African Tourism

Sustainable development of the tourism industry lies at the very heart of the challenges facing South African Tourism in marketing and positioning South Africa as a world-class, value for money preferred tourist destination and in delivering on its mandate to increase tourist volume, spend and length of stay, improving geographic spread and seasonality and encouraging industry transformation.

The vision embodies a vitally balanced compromise between the conservation of natural and cultural resources, tourism and the tangible involvement of, and economic benefit to, local communities.

This vision is the very essence of the Government's White Paper on Development and Promotion of Tourism in South Africa adopted in 1996. Inspired by the very real contribution that tourism can make towards poverty alleviation, the paper stated as its vision '… to develop the tourism sector as a national priority in a sustainable and acceptable manner, so that it will contribute significantly to the improvement of the quality of life of every South African. As a lead sector within the national economic strategy, a globally competitive tourism sector will be a major force in the reconstruction and development efforts of the government.'

Cognisant of the harmful impact uncontrolled tourism growth can have on societies and the environment (as demonstrated by the increased pressure on the natural, cultural and socio-economic environments of many international tourism destinations), South African Tourism has fully supported the Department of Environmental Affairs and Tourism in remaining committed to emphasising the importance of responsible tourism development in South Africa.

After extensive work, the Department released provisional National Responsible Tourism Development Guidelines in March 2002. The guidelines define the responsible tourism development concept as 'enabling local communities to enjoy a better quality of life, through increased socio-economic benefits and an improvement in environment. It is also about providing better holiday experiences for guests and good business opportunities for tourism enterprises.'

Importantly, the guidelines identify three core areas of responsible tourism development behaviour; namely economic, social and environmental responsibility. The essence of these focus areas is as follows:

Economic responsibilities

Economically-speaking, the South African tourism industry has grown significantly towards playing a much larger role in the South African economy.

Having realised that the formal tourism sector can assist with creating opportunities for the informal sector, South African Tourism has begun to work on maximising the local economic benefits which tourism can bring to an area. Initiatives include creating a more diversified tourism product and marketing a wider range of experiences, activities and services to tourists in line with its new customer-focused approach. The emphasis is placed on historically disadvantaged groups operating small guesthouses, shebeens and restaurants with local cuisine; offering community tours, or traditional music, arts and crafts; and support services such as transportation, laundry and gardening.

Clearly, this strategy does more than weave the African cultural tourism experience into the fabric of mainstream South African tourism. It also provides opportunities for emerging entrepreneurs to contribute increasingly to the growth of the tourism industry, and to further encourage the incidence of domestic tourism that currently comprises 67% of all travel within South Africa.

Indeed, South African Tourism has begun to pay formal tribute to these newcomers through its Emerging Tourism Entrepreneur of the Year Award (ETEYA), launched in 2001. Designed primarily to encourage and sustain the development of SMMEs in the tourism and hospitality sectors, the award also contributes to economic empowerment and transformation through job creation and has the potential to help realise South African Tourism's industry growth objectives.

Social responsibilities

Aiming to ensure that all South Africans have equal access to tourism services, both as consumers and providers, the responsible tourism development guidelines detail the importance of encouraging communities in tourism planning, decision-making and development processes, helping to overturn the former myopic focus of the industry.

By definition, this means increasing the involvement of historically disadvantaged communities in the tourism industry, whether as tourists, employees or entrepreneurs. Naturally, members of these communities must first be made aware of the opportunities in tourism. Only then can they be trained and educated to maximise these opportunities within a realistic commercial framework.

A key challenge in this regard is a general lack of appreciation for the inherent value in local culture as a source of diverse tourism products. Yet cultural offerings such as the township tours have been phenomenally successful. Many similar opportunities exist whereby communities can be socially (and economically) empowered to market their cultural traditions and products.

The key to developing these opportunities lies in responsible tourism practices. For instance, potential negative social impacts of any tourism venture (whether mainstream or ethnic in nature) must be identified and minimised where necessary.

Environmental responsibilities

Responsible and sustainable tourism implies a proactive approach by the tourism sector to the environment. This is particularly important where the focus of tourist activity is the natural environment, as is the case with wildlife viewing, hunting and marine-based tourism.

Indeed the guidelines maintain that the environmental responsibilities of any tourism enterprise cannot be divorced from the process of managing the business; and that environmental impacts must be taken into consideration during the design, planning, construction, operation and decommissioning of any venture.

Looking at the bigger picture, the guidelines recommend that all tourism enterprises contribute to environmental sustainability by exercising care in purchasing decisions; indeed, responsible producers and suppliers should be sought out and supported. Likewise, conscious effort must be made to use local resources in a sustainable manner, for example, avoiding waste and over-consumption, and minimising pollution.

World Summit parallels

Interestingly, the three core areas identified in the National Responsible Tourism Development Guidelines For South Africa corresponded exactly with the three pillars of the 2002 UN World Summit on Sustainable Development (WSSD), held in Johannesburg in August 2002 – namely 'People, Planet, Prosperity'.

"The gap between the poor and the rich is widening by the day," said South Africa's Minister for Environmental Affairs and Tourism, Mohammed Valli Moosa. "This poses a great threat to all nations as the instability, conflict, disease and environmental degradation associated with poverty threaten the overall socio-economic fibre of society. The interaction that exists between the economy, social structures and the environment cannot be denied."

In other words, we need to play smarter all round if we are to achieve sustainable growth in tourism and its contribution to the GDP. We've adopted the principles of responsible tourism as route markers in the transformation of the South African tourism industry, particularly through the development of skills and products. By creating sustainable employment in the tourism sector, we will be able to redistribute tourism spend and the accruing benefits throughout South African society, effectively contributing to meeting the challenge of alleviating poverty in our country.

> *To cherish what remains of the Earth and to foster its renewal is our only legitimate hope of survival.*
> ~ Wendell Berry

> *It is not for him to pride himself who loveth his country, but rather for him who loveth the whole world. The earth is but one country and mankind its citizens.*
> ~ Baha'u'llah

> *"...Today more than ever before life must be characterised by a sense of Universal responsibility, not only nation to nation and human to human, but also human to other forms of life..."*
> ~ Dali Lama

The **dti** recognises the environmental industry's contribution to competitiveness through energy and resource efficiency and industry symbiosis

POLLUTION MONITORING, ANALYSIS, ASSESSMENT AND MANAGEMENT
- ❑ **Air pollution abatement**
- ❑ **Contaminated soil and water remediation**
- ❑ **Marine pollution abatement**
- ❑ **Noise and vibration abatement**
- ❑ **Waste minimisation, trading, recycling, re-use and management**

CLEANER TECHNOLOGIES AND PRODUCTS
- ❑ **Instrumentation and technology for monitoring, analysis and assessment**

RESOURCE MANAGEMENT
- ❑ **Energy management including energy efficiency and promotion of renewable sources of energy**
- ❑ **Resource accounting**
- ❑ **Water management**

There are an estimated 1500 environmental enterprises in SA, providing about 500 000 employment opportunities

the dti

Environment is a new economic area that poses many business opportunities.

NATIONAL CLEANER PRODUCTION CENTRE

SOUTH AFRICA

Responsible Care

FEE
FORUM FOR
ECONOMICS
AND
ENVIRONMENT

U K U S A
WASTE MANAGEMENT

- On-site management
- Collections
- Recycling
- Refuse removal
- Other Waste Services

South African Waste Management Employers Association

SECTOR PARTNERSHIP FUND

Projects Africa
www.ProjectsAfrica.com

NEDBANK

GREEN AFFINITY

www.nedbank.co.za

MOLEPI
RECYCLING cc.
Reg. No. CK 2000/024790/23

PROVIDING TRANSPORTATION & COLLECTION SOLUTION FOR YOUR RECYCLABLES

RCMASA

BECO
Institute for Sustainable Business

Business Council for
Sustainable Developmen
South Africa

The Waste Group
Managing Waste For A Friendlier Environment
Reg. No.: 96/15131/07 The Waste Group (Pty) Ltd.

ldo Leopold Institute
Environmental Management Training

SUSTAINABLE DEVELOPMENT
IN TRANSPORT

Guest Essay by Prof Gordon Pirie
~ Dept Geography and Environmental Studies, UWC

Around the world transport has been allowed to develop without regard to its effect on the environment. Transport users have wanted convenient and fast transport, and infrastructure and vehicles have been constructed to meet that demand. Only in the last three decades has the real cost of unprecedented mobility become apparent. Motor vehicles of all kinds produce toxic CO_2 emissions. Wrecked and abandoned vehicles have to be disposed of somehow and somewhere. Roads and motorways are built on agricultural land or where habitats and ecosystems are affected adversely by changes to vegetation and drainage.

Airport noise may have abated by the use of new generation aircraft engines, but engine emissions at high altitude still deplete ozone causing long-term climate change. Inspection of the sea-worthiness of ocean-going tankers might have reduced the risk of accidents and oil and chemical spills round the South African coast, but errors and oversights occur, and irresponsible operators can still wash out ships holds far out at sea undetected. Technically speaking, transport that does little damage to the environment is within our grasp. It is quite possible now to design and build vehicles that are propelled by an energy source that makes no unsustainable demands on fossil fuels that will be difficult to replenish.

Non-polluting engines can be designed. Legislators could easily enact measures that place a moratorium on more motorway construction and that confine heavy freight to rail transport that is slower, less convenient, but environmentally less harmful. It would be possible to legislate a surcharge on every passenger air ticket sold so as to discourage people from flying when they could use other modes of transport. The leaders of South Africa's biggest cities could follow London and Singapore and charge vehicles for entering the central city. But what is technically possible is not always politically feasible, economically sensible or socially fair.

In South Africa, as elsewhere, some compromise has to be made between transport that is environmentally perfect and transport that is practical and affordable. Zero vehicles, zero infrastructure and zero mobility might be best for the environment, but human life requires transport for commerce, trade, sociability and recreation. We need forms of transport that will not prevent those activities in the short term and that will not destroy our planet in the longer term. Transport is needed to give people access to

work and basic services like education and healthcare. The transport sector also provides many jobs. There are various ways of defining sustainable transport. To varying degrees they entail promoting clean transport that does not harm human health, energy-efficient transport that does not unduly deplete natural resources, and environmentally sensitive transport that does not damage ecosystems. Various strategies may be adopted to reach these ideals. On the one hand, providers of transport vehicles can be required to design, construct and operate environmentally appropriate vehicles and infrastructure. Manufacturers of energy sources, lubricants, car bodies, batteries, glass, engine blocks and tyres can be required to improve the environmental compatibility of their products and the product chains that involve onward selling of used items. On the other hand, transport users should be made aware of the environmental consequences of their travel behaviour, and persuaded to change it as necessary.

A mix of self-interest, consumer pressure, international activism, and national legislation has obliged multinational motor vehicle manufacturers and petroleum companies to attend to the environmental effect of their products. Cleaner fuels and cleaner burning fuel combustion systems that produce less lead and sulphur emissions are used in new vehicles. But aging motor fleets still contribute to pollution and inefficient combustion, especially in poor districts and communities where most public transport vehicles are second-hand. In those settings it would be hard to deny improved mobility to people who previously did not have access to motor transport. In addition to requiring use of cleaner road vehicles, the disposal of vehicle carcasses and parts that have reached the end of their life remains a major problem, and manufacturers should be encouraged to use materials that can be recycled easily or destroyed without using considerable energy. Waste is a significant by-product of transport.

Supplying transport that is environmentally acceptable is difficult, but it is perhaps easier than modifying the use of transport by millions of motor vehicle and aircraft users. Individual awareness of how unrestricted mobility can damage the environment does not necessarily change transport behaviour. A combination of prohibition, penalties and persuasion needs to be used, for example, to stop people using private motor cars unnecessarily. But application of the "polluter pays principle" lacks purpose if there are no alternatives to the transport that people and companies currently use. People can be urged to walk whenever possible, and to use public transport for daily work trips and school trips which are predictable and

PRETORIA

JOHANNESBURG

KIMBERLEY

BLOEMFONTEIN

RICHARDS BAY

DURBAN

EAST LONDON

SALDANHA

PORT ELIZABETH

CAPE TOWN

Our system also carries minerals, fuel and foodstuffs essential for your being.

There is another system that's as critical to life as that complex network of blood vessels that nourishes your body. And just like it, our system is also pumping every second of the day, every day of the year.

As one of the largest freight railway companies in the world, Spoornet carries the foodstuffs that end up on our tables; the minerals, chemicals and raw materials that feed our industries; as well as the fuel that fires our power stations.

And through our world-renowned heavy haul operations, we also earn billions in foreign exchange.

The bottom line? When it comes to the economy, job creation and growth – issues that affect every single South African – Spoornet's at the heart of it all.

SPOORNET
At the heart of it all.

can be served by shared vehicle. But even local shopping trips are not possible on foot if neighbourhood shops close because of retail competition at out-of-town shopping centres that offer capacious free car parks. And work trips on public transport are impossible if the service is erratic, expensive and unpleasant. The projected Pretoria-Johannesburg "Gautrain" should relieve commuter congestion on the intercity motorway by providing an effective alternative, but other traffic may grow to fill the vacant road space.

Fuel taxes, car licensing charges and road tolls that make car drivers think twice about using cars privately can help curb unnecessary car use. Road toll profits used to finance additional road provision simply create more space for transport to expand. Creative ways need to be sought to subsidise new public transport provision out of state revenue sources like fuel taxes, traffic fines, company car tax, and tax on privately provided car parking spaces. Measures need to be introduced not just to limit car use, but to cut back on the need to use cars. Land use planning that minimises the demand for travel, or that provides safe walking and cycle options (and secure cycle storage facilities), is essential. Residential "densification" is environmentally preferable to low-density suburbanisation that obliges household units to own one or two cars. "Telecommuting" that enables employees to work from home for at least part of the working week is environmentally positive. So too is "teleconferencing" that reduces corporate travel. "Flexitime" work arrangements that spread peak hour traffic, and ease congestion and pollution, make less radical changes to standard office routine.

In 1998 the transport sector in South Africa consumed 17.5% of total primary energy, and accounted for 17.5% of total carbon emissions. If road vehicle use had peaked in South Africa, anxiety about the environmental impact of transport would be serious enough. Unfortunately, road vehicle use is predicted to increase in a growing population where car ownership is still below saturation point and is still a status symbol, and where alternative public transport is poorly developed and associated with inferior service. Some relief from environmental stress in the transport sector might occur as a result of the taxi recapitalisation project that will put larger and more modern taxis on the road. Similarly, railway reinvestment might be expected to make better use of a significant but deteriorating national asset that has the potential to divert passenger traffic (especially leisure travellers) and freight traffic (especially dangerous cargo) away from roads and airways. The so-called "legacy transport" of apartheid which maximised the distance between the workplaces and homes of millions of Black people is slowly being overturned. Welcome as it is for the economy generally, and for airlines and for car-hire firms particularly, booming tourism contributes to increased transport use.

In poor communities the alternatives to car use involve extensive use of non-motorised transport that is affordable and energy efficient. Cycle rickshaws and animal-drawn carts, for example, compromise on speed, but are manoeuvreable and can operate on mixed quality surfaces. Non-motorised vehicles can be built from recycled motorcar parts, produce little noise and pollution, and do not consume non-renewable energy. Small repair industries provide unskilled work opportunities and are a nursery for business talent. Such informal transport also tends to be socially inclusive. In these ways, environmentally sustainable transport is part of the jigsaw of sustainable development generally.

If transport in South Africa is to become environmentally sustainable, regulatory and institutional reform is required. There is a need for stricter vehicle licensing and inspecting, and for more effective enforcement of minimum operating standards. Transport subsidies that conceal environmental costs need reconsideration. There is also a need to build institutional capacity to specify, monitor and regulate environmental codes in transport. Information needs to be collected on a regular basis so as to enable diagnosis of environmental conditions due to transport, and quick assessment of the financial and other cost of putting different policies in place.

"...think globally, but act locally..."

~ René Dubos

Most of the important things in the world have been accomplished by people who have kept on trying when there seemed to be no hope at all.

~ Dale Carnegie

Your real **opportunity**
to **participate** in
Progress

Reaching
corporates,
Government, **NGOs**
and **civil society**
via **targeted**
distribution
channels and select
bookstore
availability.

You can't afford
to be left behind.

SUSTAINABLE DEVELOPMENT
LAND REFORM

(An example — The Franschhoek Empowerment and Development Initiative)

Guest Essay by Willem Steenkamp ~ Community leader, former ambassador and presently CEO of FRANDEVCO (Pty) Ltd.

From model Apartheid village to a model for sustainable development – that is the story of Franschhoek, the historic village nestled in the mountain valley where the French Huguenots were settled in the 17th century.

Despite its scenic beauty, its rich cultural heritage and its renowned wine estates and restaurants, Franschhoek under Apartheid was a town in conflict. The community was split into separate "white" and "coloured" municipalities, held apart by a buffer zone. No black Africans were allowed to live here. A substantial squatter problem arose, with a serious backlog in low and medium cost housing. The Group Areas Act land claims awaited resolution, and a pressing need emerged to afford the disadvantaged access to the mainstream of the economy, consisting of agriculture and tourism.

In line with the unfortunate reality encountered in many parts of Africa and the Third World, these socio-economic and political issues impacted negatively on the environment. Squatting, in particular, was at the root of contaminating water sources through inadequate waste disposal facilities, leading as well to the destruction of trees through unavailability of electricity for cooking and heating purposes. Disease was rife, and crime as well as anti-social behaviour occurred all too regularly. Regular shack fires posed a serious threat to life and limb, and caused numerous veld fires. In the spheres of heritage conservation and aesthetics, the informal settlement along the main road right at the entrance to the village, certainly did not promote the image it wished to portray to tourists.

In terms of the "triple bottom line" underpinning sustainable development, the lack of attention to human well-being caused not only hardship and resentment, but negatively impacted on environmental integrity and economic efficiency. In the same way, land reform and poverty alleviation must be seen as being among southern Africa's top environmental issues, because without attention to human well-being, desperate humans will impact on their environment simply in order to survive.

In 1998 the Franschhoek community chose co-operation over confrontation and signed a ground-breaking Social Accord. At the heart of this initiative was the understanding that a private/public/community partnership had to be established, and that the communally-owned land of the village had to be optimised and utilised in a manner that would allow cross-subsidisation of the upliftment and conservation actions required, but that would also look beyond immediate capital needs to the continued sustainability of the initiative.

The most valuable such asset was the municipal commonage land of some 102 hectares, which had previously been misused and abused as sewerage farm, rubbish dump, sand mining area and rifle range, and which had become overgrown with invasive alien vegetation. The choice for private sector development partner fell on a grouping of local and international investors who had adopted the "triple bottom line" approach to sustainable development, who committed themselves to ISO14001 environmental management standards, and who proposed a holistic solution to Franschhoek's challenges, synergising with community and conservation bodies and with UNESCO's MaB programme.

A development plan was drawn up and workshopped with the community, who gave it unanimous support. The project consists of three major components:

◆ A wine and olive estate of some 80 hectares, with 19 residential werfs of one acre (4000 sq m) each, enjoying the Cape Winelands farm ambience.

◆ A culture and eco-tourism commercial node being established on some 15 hectares.

◆ An urban residential development of 14 plots sites on the banks of the La Cotte stream.

The developers took transfer of the commonage land in December 2001. Since then, the land has been cleared of alien vegetation in co-operation with the "Work for Water" programme of the Department of Water Affairs and Forestry, and was then systematically rehabilitated. With all planning authorisations in place, the site was handed over to the contractor in mid-September 2003 for the installation of civil infrastructure such as sewage, roads, electricity and storm water. It is noteworthy that, throughout the entire development authorisation process, not a single objection or appeal was lodged by the public. On the contrary, a large number of expressions of support were received. With grant funding from the Development Bank of Southern Africa an exhaustive set of research studies was commissioned on all environmental and societal concerns, leading to the project being described by the Western Cape Directorate: Environmental Affairs as a model for such ventures.

The transfer of the erstwhile commonage permitted the implementation of the Social Accord through generating capital for cross-subsidisation. This included the allocation of 2 hectares of prime development land to the local Group Areas Act land claimants, plus R1,2M for its development, which is a local contribution in addition to what they received from the State in December 2002. Most importantly, the first phase of a low-cost housing project of one thousand units could be completed, in terms of which the beneficiaries received 771 homes worth at least R38K each, entirely free (as opposed to the usual R18K for the typical RDP house).

A striking feature of the commonage development is the cross-subsidisation being achieved, through wealth creation that funds upliftment and conservation. The wine estate being developed on it will be home to a select group of very wealthy investors, who seek the ambience of a wine farm without having to farm it themselves. The farming venture as such, will be co-owned by the historically disadvantaged as a business enterprise, giving them meaningful access to the agricultural sector. Similarly, a "culture and eco-tourism incubator node" is being established on a section of land bordering the main road, scenically situated below the Franschhoek pass, between the historic village and the Mont Rochelle nature reserve. The infrastructure of this node, such as a luxury boutique hotel, vinotherapy spa, conference/exhibition centre, plastic surgery clinic, art gallery, artists & crafters village, together with a Man and Biosphere information centre, will also be co-owned by the historically disadvantaged. Most encouraging has been the support of the business sector, as evidenced by the active participation of the likes of the V&A Waterfront Company, and the funding by Absa Bank of related community and environmental projects.

In conservation terms the project is gathering increasingly widespread recognition. Intended as the "entrance portal" to the proposed Boland Biosphere Reserve, a project delegation headed by Dr. Franklin Sonn was received in September by UNESCO, in Paris. The project also financially benefits the Franschhoek Empowerment and Conservation Trust (FREMCO), headed by John Samuel, a Franschhoek resident and MD of the Nelson Mandela Foundation. The Mont Rochelle Nature Reserve Board appoints two trustees to the FREMCO board. A joint management agreement between the Fransche Hoek Estate Development and the Reserve will ensure that they seamlessly integrate. The project also ties conservation to tourism, with a breeding initiative for near-extinct local fish species such as Witvis, to be re-introduced as sport fish for fly-anglers. A successful fynbos propagation nursery has been set up, and the historic "Cats se Pad" hiking trail has been rehabilitated and re-opened to the public.

As a result of the Franschhoek experience a number of "keys to success" have been identified, which need to be employed if land reform initiatives are to prove sustainable. These are:

◆ First and foremost, craft a community consensus regarding the basic vision, goals and methodology, then document this consensus and have it signed by all stakeholder groups;

◆ Once such a social accord is signed, don't as a consequence of having it in place for one moment stop consulting with the community, who need to be constantly kept informed and on board;

◆ Government, at whatever level, can only create enabling environments, not initiate or execute such projects for and on behalf of communities, who have to engage with their own business sector as experienced wealth creators to serve as locomotive;

◆ Although empowerment and conservation are essentially moral causes, they cannot be properly served without adequate funds. It is therefore imperative that business must be incentivised to serve these very essential causes by linking them to high-value business opportunities, thereby creating the necessary opportunity/obligation for cross-subsidisation;

◆ By bringing in responsible business that actively subscribes to the "triple bottom line", huge capacity, capital, experience, expertise and energy can be tapped into, whilst also gaining another very important benefit, which is to have projects properly packaged in "bankable" documents. Most project proposals in South Africa seem to fail not because of lack of potential funding, but because of an inability to unlock these funds through lack of demonstrable "bankability" i.r.o. the project's presentation.

In preparing such an initiative, every minute spent at the outset on careful consultation, research and conceptualisation saves hours in the later stages when authorisations and development rights have to be negotiated. Go as slow as you must in the beginning, in order to be able to go as fast as you can at the end.

Each project will be unique, its content determined by local circumstance. It must essentially be owned and driven by locals, with a motivated "champion", not by people who can be perceived as outsiders.

Unless poverty can be alleviated and land reform be creatively implemented to create wealth capable of cross-subsidising essential upliftment initiatives, the natural environment will be perceived by desperate humans as being the only resource available to them with which to sustain their meagre existence. Franschhoek's experience has shown that even a very small village can muster the means with which to achieve significant results, through actively engaging with the private sector, adopting a businesslike, wealth creating approach, fostering consensus and actively searching for win-win solutions. Thus the private sector has learnt that conservation and empowerment are good for business, just as the disadvantaged community has learnt that business is good for conservation and empowerment, and environmentalists have learnt that business and empowerment are good for conservation.

More information on this model initiative is available on its web-site: www.fedi.co.za

A *sustainable society is one that shapes its economic and social systems so that natural resources and life support systems are maintained.*

~ Lester Brown

SUSTAINABILE DEVELOPMENT IN THE BUILT ENVIRONMENT

Guest Essay by Etienne Bruwer – Greenhaus Architects

A new ethos of people making the manmade world

In talking about southern African literature, esteemed professor Eskia Mphahlele once observed that we live in a culture obsessed with place. Our indigenous literature reveals a yearning for 'place of being', the rare ingredient of 'at-homeness' in our experience. With an obsession, we hang on to 'there whence we came', our mythical places of wholeness and belonging, places that remember us, and honour our humanity in its fullness. In our stories, we 'read' one another by association with these remembered places of origin, allocations of place serve as metaphors of identity. Prerequisite to 'at-homeness' is a combination of life-enhancing social, economic and environmental factors that we have come to call 'sustainability'.

Over the last seventy years – the period commonly known as the Modernist era – much of architecture has become 'containers-for-profit-behind-billboards' – anonymous and bland, 'buildings' that are computer-generated and ex-catalog, cloned and repeated ad infinitum. So too, much of that which previously became the public realm has either been privatized/annexed for profit or otherwise disowned, becoming 'non-places', characterised by environmental apathy, social estrangement and extreme economic disparity. Areas zoned for profit exclude participation by most, and the leftovers – the no-man's lands twixt counting house and securi-zoned consumer-compound, contain little or nothing of us except our taxes in tar. As urbanity gives way to mall-and-motorway 'culture', memory of the liveable city has faded; so too has the practice of civility that our cities once afforded us. Add to that the combined effects of virtual reality (the placelessness of the p c) and loss of propinquity (place-boundness), and the odds stacked against civic life and environment alike become evident. Our obsessional longing for 'at-homeness', for a homecoming and healing of our godforsaken state, is a direct consequence of our attitude to shaping the environment.

We participate in the manmade environment in two ways; we engage with it through our experience of built architecture, and more profoundly, we 'own' it by participating in its making and transformation. Both are prerequisite to achieving sustainability. We 'belong' by finding something of ourselves in the world and by 'making our mark', by transforming environment. More plainly put, we tend to maintain and add value to environments that we find worthy reflection of our humanity, and we tend to disown, neglect and vandalise that which we cannot recognise, respect or 'own'. At best, we delight and relate to the qualities of human scale, proportion, texture, etc. as found in the organic morphology (correct membering of parts to whole) of living, life-giving architecture. At worst, when the fundamental human need to relate and participate in the making and transformation of the manmade environment is no longer perceived (and hence neither educated for nor

invested in) we will have arrived in the industrial wilderness. On our current trajectory, we are literally, 'running out of space'.

Although environmental aspects (correct physical processes/sources/materials, etc.) and economic aspects (embodied/used energy, life cycle planning, demonetarised energy exchange systems, voluntary simplicity, etc.) of sustainability in the built environment are more or less known, the social aspects are not. In the first decade of transformation in South Africa, even as the world computerised and globalised, the right relationship between people, work and the making of the built environment was sought for by many agencies, but largely still awaits discovery. We see this problem across the board – from managing nature reserves, to building inner cities, to achieving equitable land reform, but most acutely in the procurement of housing, where perhaps apartheid housing was only different from post-apartheid housing in perhaps having delivered more durable buildings.

We are still 'counting the units' rather than chalking up sustainable nodes, clusters and neighbourhoods. Given our history and the absolute primacy of social development and creating job opportunities on our current agenda, this is a huge missed opportunity. Half the world's energy is used by the building industry, yet this sector has remained remote from people, and as such has not utilized its potential and capacity as a social transformation tool, a mass employment source and broad wealth creator. The building industry is still perceived as 'industrial' and hence remains a de-humanised business, in which 'environment-making' is measured as 'labour cost'.

It would seem that we have yet to come to the fundamental realisation that development comes 'from within', that human dignity is 'self-made' and can only develop through participation and co-ownership of the process of making environment.

Given the state of the environment (the globe), the contraction of wealth and economic disparity (between the hemispheres) the social polarities and divides (at a local/regional level) achieving sustainability is the most important challenge of this new century. People are also now realizing that in order to achieve sustainability, in addition to institutional change, action at an individual level will be required – the kind that in turn will require personal change (mainly, re-directing material aspirations – such as altering lifestyle and relinquishing conformism). With so little tangible evidence of success and reward, the paradigm shift to sustainability will be neither easy nor comfortable. It also cannot be bought for all the money in the world, as the neo-liberalists would have us believe.

We call the connective tissue between identifiable/isolatable processes, which are working in series and sequence 'SUSTAINABILITY'.

Consisting as it does in the quality of the links between the parts of wholes, defining the sustainability of 'things' in a process as complex and open-ended as the built environment is not really appropriate or useful. At best, one could list key indicators of a sustainable process:

1. Sustainable processes always have an aesthetic dimension; along with Commodity and Firmity, the presence of Beauty and Variety is always essential. This is prerequisite to communication across our many divides, because words tend to divide, whereas shared cultural goods unite.

2. Proportional to sustainability will be the extent of human involvement. Human presence is maintained throughout planning, designing, negotiating, building, finishing, fitting, and maintaining sustainable built environments. We have been de-humanising the manmade world since the advent of the industrial revolution, developing machine-made 'products', which in turn have produced ever more costly, unhealthy and soulless environments void of life. Designed in the image of the machine as they are, indus-trial-aesthetic spaces cannot achieve sustainability, because the energy embodied in these intentionally limit human-involvement in favour of factory-made goods.

3. Integrated process is characteristic of sustainable practice. All parts belong, forming part of a greater whole. The goodness and the number of the links between the parts determine stability. Recursive and integrative practices broaden interdependency, spread risk, multiply fall-back positions, use energy better, lessen dependency on money. It is good for business but also not bad for anything else.

4. Within a sustainable manmade environment, nothing is designed or built that could harm the environment or that uses energy squanderously. Nothing is made harder than needed for its purpose. Sustainability in the making of the manmade environment must incorporate values for: Social Capital (by rating levels of participation); Economic Capital (by rating the amounts of embodied energy); and for Environmental Capital (by chalking up the scope of the life-enhancing practices and elements used such as renewable resources, toxicity, recycling, source purity, remanufacture, transport economy, etc.)

5. Adopting a "source economy" (using what you have to hand) is prerequisite to sustainability. In nature, sources are abundant, in the city, the waste stream and alien vegetation are endless. The closer the proximity of sources to 'products', the greater the economy and most probably the purer the source. In the Cape region for example, abundant clay and sand underfoot and 'free' alien vegetation cannot be beaten as 'triple bottom line' sources – in terms of transport, embodied energy, and positive environmental impacts (alien clearing, water consumption, etc.). Also, when buying, sustainability dictates that the sources themselves must be sustainable.

To conclude, a few examples of influential initiatives in the sphere of sustainability. In the United States, land of plenty, (also junk and debt) the re-manufacture of building materials and establishment of stockyards for distributing re-useable building materials is a growth industry. Cradle-to-cradle planning, in which building components are made and fitted to facilitate easy alteration and re-use at the inception/design stage (Prof Charles Kibert, University of Florida). Buildings 'begin' before they get built and 'go on' long after they are decommissioned.

Often working in concert, the global Permaculture movement (Bill Mollison et al), the Urban greening and food producing movement (Abalimi Bezekhaya, Trees for Africa, Soil for Life, etc.), the Ecovillage and New settlement movement (Kuthumba/Plettenberg Bay, Crystal Waters/Australia, the Camphill Movement, Bethel/Lesotho, Hotshoj/Aarhus, Denmark, etc), the inner city "green housing cooperative movement" (Flintenbreite/Hamburg, etc.), the new energy development and awareness centres (Greenhouse People's Project in Joubert Park, Gauteng, Agama-EDG green office centre in Westlake, CapeTown); green building materials (Real Goods Centre, Ukiah, California) are the thinktanks and laboratories providing showcases for visionary ideas and better practices in re-creating our manmade environment.

The remarkable work of Christopher Alexander in Central and South America (Favela/Bairro Housing Movement) and Hassan Fathy in Egypt (natural building), and Victor Lerner (mayor of Curitiba/Brazil – integrative processes) form part of the spectrum of a global 'organic' architecture and civic processes with community as its core. Younger talented 'junk architects' like Jan Korbes in Den Haag (works with Romany Communities in Rome) are emerging and being recognised as highly creative waste-stream/remanufacturing experts.

Global networks, conferences and exhibitions such as the World Organic Architecture Exhibition (Berlage Bourse, Amsterdam, 2003) and South Africa's own bi-annual Strategies for Sustainability in the Built Environment Conference Series jointly convened by leading local institutions and professionals (UCT Engeo, CSIR Boutek, etc.) are the prime venues for exchange.

Organisations like Craterre in Grenoble and the van Leer Foundation working out of Holland are making a measurable difference. Their work evidences understanding of the links between unemployment, social unrest and housing and the many difficulties of foreign interests and aid and governments operating in a global blue-suit economy.

Active involvement of communities in managing their environment must be the order of the day. Equality, access, accountability, transparency and sustainable living must be our watchwords.

~ Nelson Mandela

Environmental Partnership Opportunities

Sustainable development technology is a priority growth area for Wales and is high on the Welsh Assembly Government's agenda. Wales has over 1000 companies with a wealth of experience and cutting-edge technological know-how in the environmental sector.

Effective partnerships between Wales and South Africa are already providing:

- *Growth and development opportunities for environmental goods and services companies in both countries*
- *Transfer of technology, helping to build capacity and competitive edge*
- *Improved skills in environmental protection and resource management*

To find out what partnership opportunities you could benefit from, contact us today:

Michael Sudarkasa, Africa Business Direct, PO Box 413586, Craighall 2024

Tel: 011 445 9460 • Fax: 011 445 9459 • Email: msudarkasa@africabusinessdirect.com

Sustainable development technology is supported in Wales by:

www.walestrade.com • www.wda.co.uk

For further details we are listed under "Wales Environmental Partnerships" in the Enviropaedia Directory Section.

SUSTAINABLE DEVELOPMENT
~ IS IT ACHIEVABLE
UNDER OUR CURRENT LEGAL FRAMEWORK?

**Guest Essay by Cormac Cullinan ~ CEO of EnAct International
and a director of the specialist environmental law firm,
Winstanley Smith and Cullinan Inc.**

South African society and economy, like the global economy, is unsustainable in its present form. This is recognised in several White Papers and Acts, which refer to the need to achieve "sustainable development" or "ecologically sustainable development." The Constitution even gives everyone the right "to have the environment protected … through reasonable legislative and other measures that secure ecologically sustainable development and use of natural resources while promoting justifiable economic and social development."

Despite these provisions, in many cases we are not only failing to achieve the goal of ecological sustainability, but accelerating away from it. This is commonly attributed to a failure to implement and enforce environmental laws properly. What we overlook is that the overall effect of our current legal system is to legitimise and encourage unsustainable human practices, rather than to prevent them.

The magnitude of the "environmental crises" that faces us at the present time is difficult to overstate. Human pressures on Earth's life-supporting systems are causing them to deteriorate rapidly and possibly irreversibly. Global warming and climate change is proceeding rapidly and the number of disasters such as droughts, windstorms and floods have been increasing significantly for years now. Scientists predict that relatively small changes in average temperatures are likely to have a devastating impact on South African agriculture.

A heavy social price is also being paid for the "progress" and "development" achieved by industrialised societies. The gap between rich and poor continues to widen. Furthermore, even dramatic economic achievements such as the seven-fold expansion of the world economy between 1950 and 2000 and the explosion in world trade, have been achieved by squandering the Earth's "natural capital."

A crisis of governance

The fact that our species is consuming its habitat can only mean that our systems for governing ourselves are seriously dysfunctional. What we call the "environmental crisis" is really a governance crisis. The crucial task is to regulate human behaviour so that it contributes to, rather than undermines, the Earth's systems. This means not merely adjusting our legislation to restrict the most environmentally-harmful activities, but completely rethinking our legal and political systems.

The roots of the problem

Our governance systems are defective because they are based on a false understanding of how the Universe functions, and of our role within it. The core falsehood is that we humans are separate from our environment and that we can flourish even as the health of Earth deteriorates. In fact, we have convinced ourselves that our health and well-being depend on exploiting the Earth as fast as possible.

The exact opposite is true: we have evolved within, and remain an inextricable part of, the community of life on Earth. Desolation, dysfunction and disease are the consequences of believing that human fulfilment is attainable outside of this "Earth Community" or that it can be achieved at the expense of the health of the Community as a whole.

This illusion of independence is exacerbated by the myth that we are the "master species" whose destiny it is to run this planet for our own benefit. The dominant cultures in our world are as convinced of the superiority of our species over others and of our right to rule and exploit the planet as most white South Africans once were about their right to oppress other South Africans.

Laws have been used to "hard-wire" these mistaken beliefs into the structure of our societies. For example, the law has reduced all other aspects of the Earth and all other creatures to the status of objects for the use of humans. Other living creatures have no rights, and for as long as they are legally defined as "things" (as slaves once were), the law will regard them as incapable of having rights. Even the National Environmental Management Act 107 of 1998 ("NEMA") and the draft Biodiversity Bill currently before Parliament, do not affirm the fundamental principle that all forms of life have intrinsic value, regardless of their usefulness to human beings, despite the fact that this has been recognised by many international instruments such as the 1992 Convention on Biological Diversity.

It is important to appreciate that the defects in our govern-ance system cannot be solved merely by legislative reform. The problem is not simply that our laws must be refined to make them more effective. In fact our laws, by and large, do accurately express the defective worldview on which they are based. Our governance systems, including our legal and political establishments, perpetuate, protect and legitimise the continued degradation of Earth by design, not by accident. We must now seek to ensure that new laws (including the new Bills on Biodiversity, Protected Areas, Air Quality, Waste Management, and the Coastal Zone that are due to be considered by Parliament during 2004) reflect a new underlying philosophy.

Earth Governance

If the societies that dominate the world are ever to turn away from the destructive and potentially fatal direction in which they are headed, we must change not only our understanding of the role of humans within the Earth Community, but also our governance systems to ensure that humans act in accordance with this new worldview.

We humans are an integral and inseparable part of the Earth Community. In the same way that a liver must regulate itself so that it not only functions efficiently internally, but also contributes to the health of the whole body, so we humans must govern ourselves in a manner that ensures that the pursuit of human well-being does not undermine the integrity of the Earth, from which our well-being is derived.

In developing this new "Earth Governance" approach, we must recognise that we are but one part of the larger Earth Community and consequently our governance systems must be designed to be consistent with the larger order of the Universe, and of Earth in particular. In order to do this we must develop laws, policies, institutions and political structures that:

(a) Place a greater emphasis on our responsibilities towards the rest of the Earth Community than on our rights to exploit it.

(b) Focus on strengthening the relationships, which create the Community and on emphasising balance, reciprocity and restorative justice in relationships both within human communities, and between humans, other species and the ecosystems within which we exist.

(c) Recognise the important role of non-human members of the Earth Community and restrain humans from unjustifiably preventing them from fulfilling those roles.

(d) Require us to expand our understanding of governance and democracy to embrace the whole Earth Community and not just humans.

Conclusions

In order to achieve ecological sustainability, we need to make a dramatic philosophical shift from a worldview, which places human beings at the centre of the Universe, to a view that sees the maintenance of the integrity of the whole Earth system as the overriding concern. This will also require us to expand our understanding of community to include non-humans, and to base our decision-making on what is good for the Earth Community as a whole. This may sound fanciful, but we should remember that this more comprehensive understanding of the nature of community has prevailed for most of human history, and is supported by the great wisdom traditions of many cultures. It is also an understanding that is inherent in African cosmologies and customary law.

The initial changes in our governance systems are unlikely to be so dramatic. However, as more and more people accept that in the long-term human well-being can only be achieved within the context of a healthy Earth Community, they will begin to re-orientate their lives, business and eventually, our societies, toward achieving this purpose. The challenge for us, and the next few generations, is to find novel and practical ways of drawing inspiration from the rich diversity of human experience as well as modern scientific insights in order to establish effective means of governing human behaviour to ensure that we contribute to the flourishing of the whole Earth Community instead of destroying it. Government of the people by the people is no longer sufficient. We need governance of the people by the people for Earth.

The views in this article are expressed more fully in his book *Wild Law* (*Siberink*, 2002), which lays the foundations for a new ecocentric approach to law and governance.

If we go on the way we have, the fault is our greed and if we are not willing to change, we will disappear from the face of the globe, to be replaced by the insect.

~ Jacques Cousteau

"The real responsibility lives with us - public action and public demand."

~ World Summit on Sustainable Development 2002

SUSTAINABLE DEVELOPMENT IN THE BANKING AND FINANCE SECTOR

Guest Essay by Ruan Kruger

~ Environmental Analyst: Development Bank of Southern Africa

Background

According to the International Finance Corporation (IFC 2003), developed countries launched the modern environmental movement during the 1960s. This was followed by the emergence of sustainable development in the 1980s, a movement that forced a country to examine its pursuit of economic growth while addressing environmental protection and social development. The 1990s heralded the concept of globalisation and there was a growing awareness that companies had a role to play in sustainable development and that their activities had global ramifications.

By the new millennium, most major companies responded to the call for social responsibility by implementing internal environmental management systems, reporting on their environmental activities, adopting voluntary environmental and social codes/guidelines, and /or joining public-private partnerships. Financial institutions also undertook these efforts, but, as relatively "clean" industries, their internal environmental and social goals were fairly straightforward. All of these companies have one thing in common. They realised that they had to make a choice – be proactive and undertake these voluntary efforts or remain unprepared for the requirements of globalisation and/or future legislation.

While the financial sector has been dealing with the reality of environmental risk directly affecting its core business practices, it also has had to adapt to the treatment of environmental and social issues in a changing world. According to the International Finance Corporation (2003) the 1980 United States Superfund Act forced US banks quickly to become aware of the serious risk it posed to their bottom line – it meant that contaminated land might be worth a fraction of its former value or, worse, might represent a major liability. Banks responded by incorporating formal environmental risk management procedures into their lending policies to reduce their lender liability and default rate. Although their potential liability was eventually reduced, banks built on their environmental experience by reducing their internal ecological footprint, undertaking environmental audits and introducing environmental conditions into loan agreements. In turn, this drove the insurance industry to exclude certain types of environmental risk from their standard policies and to introduce new environmental impairment liability insurance products.

Potential benefits of applying sustainability principles

The IFC (2003) further states that financial institutions have been pursuing efforts that not only reduce environmental risk and improve their ecological footprint, but also add value via new products/services. Their experience demonstrates that adhering to sustainable development allows them to: uncover latent, potentially harmful downstream risks; save money;

respond to the increasing call for socially responsible behaviour; upgrade their reputations; strategically position them as market leaders; access new markets; generate revenue; and perhaps most importantly, it has become standard practice for involvement in the international financial community.

Within this community, multilateral/bilateral institutions were the first to include environmental and social requirements as part of the financing terms. The World Bank Group (including IFC), the European Bank for Reconstruction and Development, the Asian Development Bank, the Inter-American Development Bank and the Development Bank of Southern Africa all have such policies/procedures in place. As the largest financiers in emerging markets, the inclusion of environmental and social loan conditions can significantly influence financial institutions' contribution to sustainable development.

Whatever the initial motive of financial institutions to incorporate social and environmental aspects into their business operations, numerous benefits typically emerge. Those frequently cited include enhanced management, governance, communications, and stakeholder relations.

Application of sustainable development principles

Private sector banks took the initiative in 2003 and drafted an agreement called the Equator Principles. Under this agreement, the banks agreed to adopt the IFC's social and environmental rules for sustainable development. According to the Equator Principles, the banks will agree "not (to) provide financing to projects where the borrower will not or is unable to comply with our environmental and social policies and processes." A recent press report mentioned that 15 banks have already agreed to implement the Equator Principles.

In addition to the above-mentioned ground-breaking initiative in the banking and finance sector, the Global Reporting Initiative (GRI) was started approximately three years ago to encourage all business organisations to report voluntarily on their individual success in implementing steps to become sustainable. The GRI (2002) states that:

… accountability, governance, and sustainability are three powerful ideas that are playing a pivotal role in shaping how business and other organisations operate in the 21st century. Together, they reflect the emergence of a new level of societal expectations that view business as a prime mover in determining economic, environmental and social well-being. These three ideas also point to the reality that business responsibility extends well beyond the shareholders to people and places both near and distant from a company's physical facilities. Defining, measuring, and rigorously reporting on these economic, environmental, and social issues lie at the core of the mission of the Global Reporting Initiative (GRI).

The guidelines complement and strengthen traditional financial reporting by providing critical, non-financial information that helps users assess the current and future performance of the reporting organisation. Whereas financial reporting primarily targets one key stakeholder – the shareholder – sustainability reports have a wide audience, reflecting the diverse groups and individuals with a stake in high quality information. Financial analysts, employees, customers, advocacy groups, trade unions, communities and others are all part of GRI's audience.

Performance indicators, both qualitative and quantitative, are the core of a sustainability report. The performance indicators are grouped under three sections covering the economic, environmental, and social dimensions of sustainability. In each area, GRI identifies core indicators (required for reporting in accordance with the guidelines) and additional indicators (used at the discretion of the reporter to enrich a report).

GRI in collaboration with the United Nations launched the United Nations Environment Program – Financial Institutions (UNEP FI) initiative two years ago, and followed up the work with a finance sector working group. This group was tasked to develop indicators that will be specifically applicable to the finance sector. The work started in September 2003 and the first drafts will be available towards the end of 2004.

The potential difficulties in implementing sustainable development principles

The main difficulty that remains is that the implementation of sustainable development principles is still a voluntary option to all businesses. But the positive angle to this issue is that big business is increasingly forcing the implementation of this approach through not doing business with organisations that do not subscribe to the same philosophy. This power that business can use in their trading and financing operations is an immense force that can change the future of sustainable development dramatically.

A problem that is being experienced in the developing countries/emerging markets is that these governments are often so eager to attract foreign direct investment that environmental and social legal requirements are not made strict enough. This situation is referred to in the IFC document, and emphasises the important role that international financial institutions should play in the emerging markets.

Emerging markets are often faced with such pressure from their electorate to provide basic infrastructure like housing, water and sanitation, as well as the creation of job opportunities, that the required attention is not given to long-term strategies like implementing sustainable development principles. These situations can only be solved by consistent awareness raising campaigns through global summits and active involvement by developed world financial institutions in the developing economies.

Additional opportunities are being created for international financial institutions to play a positive role in the emerging markets by financing and participating in initiatives like the Prototype Carbon Fund, launched recently in South Africa through a partnership between the World Bank and the Development Bank of Southern Africa. This will enable cleaner technologies to be transferred from the developed world to the developing countries, and will include a financial benefit for the developing countries for using the cleaner technology. Difficulties experienced are the cost associated with the preparation of baseline information and the monitoring of environmental information to claim the required credits from the Fund.

Anticipated future trends that will affect the application of sustainable development principles

Financial institutions have until recently largely escaped the wrath of environmental groups, which have tended to shower their criticism on international institutions such as the International Monetary Fund and World Bank, or manufacturers accused of operating sweat shops in developing countries (Financial Times, April 7 2003). But at this year's World Economic Forum in Davos, over 100 advocacy groups signed the so-called Colleveccio Declaration, which called on financial institutions to implement more socially and environmentally responsible lending policies.

According to the IFC (2003), developed country financial institutions have undertaken most of the sustainable development efforts in emerging markets, while the emerging market financial institutions are lagging behind. There are several exceptions, but embracing even basic environmental and social efforts has yet to reach a critical mass. Most emerging market financial institutions still do not believe that these issues are their problem to solve, but perversely, it is in their marketplaces that environmental and social challenges for business and society as a whole are most profound. Another problem is a dearth of in-country capacity. In recent years, development agencies such as IFC have begun to bridge this knowledge gap by training financial institutions on environmental and social issues.

Trends that are starting to influence financial institutions – and will increasingly do so in the future – are:

◆ "No place to hide" – NGOs will continue to expose financial institutions' activities.

◆ Investors increasingly will scrutinise companies for their transparency, governance, and environmental and social efforts.

◆ International financial institutions will continue to include environmental and social conditionality for emerging market projects.

◆ Weak emerging market regulations/enforcement will spawn more informal environmental and social regulators.

◆ Best practices will become more documented.

◆ Financial institutions will be pressured to apply environmental and social standards to their entire operation.

> *There are no passengers on spaceship earth. We are all crew.*
> ~ Marshall McLuhan

SUSTAINABLE DEVELOPMENT INVESTMENTS

(The JSE Socially Responsible Investment Index)

**Guest Essay by Nicky Newton-King ~ Deputy CEO, JSE Securities Exchange
South Africa Johannesburg Stock Exchange (JSE)**

Over the last few years there has been an increasing awareness of and need to measure corporate social responsibility. As an emerging market, South Africa's companies are doing much already and deserve to be recognised for their corporate social responsibility ("CSR") efforts. The global nature of capital markets is forcing South Africa to fight for every investment cent.

The JSE's proposed socially responsible investment index ("SRI Index") has been a work-in-progress for some time and is expected to be launched during the first quarter of 2004. The Index will be a market driven mechanism that will aim to focus the South African financial sector in identifying and addressing the global socially responsible investment ("SRI") imperatives taking the spotlight. In South Africa in particular, where there is a recognised premium for good corporate governance, companies have had to address labour, affirmative action and health-related issues with more focus than internationally and hence the demands of broader SRI should not be unfamiliar to our companies.

Companies will need to demonstrate that they meet the Index Criteria, which will reflect widely accepted sustainability and socially responsible practices and are designed to measure the integration of triple bottom line principles relating to the economy, environment and society into companies' business activities. Companies that obtain the requisite score will be included in the Index.

The principle of environmental sustainability recognises that Africa has a rich and critical resource base, which must be wisely used if it is to provide any sustainable support to the development of South Africa and its people. This need has been recognised by the South African government by its commitment to many international conventions that relate to sustainability and also its pledge along with other African leaders to the imperatives set by the New Partnership for Africa's Development ("NEPAD").

Evaluation of environmental sustainability will take cognisance of, amongst other things, environmental risk identification and management, protection and rehabilitation of the environment, reporting and auditing of performance, and the production and supply chains.

Social sustainability is premised on the fact that, as recognised in the King Code on Corporate Governance, a company is a key component of modern society, often forming a more immediate presence to citizens than government or civil society. Companies need to engage wider than only with their shareholders, developing positive relationships with all stakeholders, including staff and the community. Balancing

performance and compliance while taking account of the expectations of all stakeholders is the key challenge facing companies in this respect today.

Economic sustainability is often regarded as a synonym for financial performance. While this is certainly an aspect of economic sustainability, it does not necessarily reflect long-term growth. Economic sustainability will be evaluated with reference to, amongst other things, corporate governance principles, identification and management of economic risk, the economic impact of HIV/AIDS will have on business activities, accounting and employee satisfaction.

International practice often sees funds in the SRI area exclude companies who operate in areas of particular controversy, as they assume that commonly excluded industries are by the very nature of their business embarking on activities which are not socially responsible. The inevitable consequence of exclusion is that those excluded companies are then free from scrutiny and remain unhindered in their activities without regard for the three strong pillars of sustainability principles. In the development of the JSE SRI Index the exchange believes it important to include as many industries as possible to ensure economy-wide embracing of principles.

The Index has invited voluntary participation from the 160 companies that form part of the FTSE/JSE All Share Index. The JSE expects the introduction of the Index to create an aspirational benchmark for both listed and unlisted companies, which in itself will encourage participation in the screening process and the Index. Over time, it is also expected that the generally accepted practice in relation to good corporate citizenship and hence the Criteria, will become more demanding and reward those companies in the Index with greater investor confidence and a stronger profile.

If the Index Criteria does become widely accepted it would greatly facilitate the development of capacity and encourage the awareness of sustainability matters amongst stakeholders of companies. It is anticipated that the Index will catalyse the debate on what constitutes sustainability and social responsibility in the South African context and how this can be leveraged to influence behaviour. If its measure of good (and bad) company behaviour is an accurate reflection of South African realities, then the Index's contribution to aligning company behaviour with socio-economic and environmental realities could be significant.

The local SRI Index will be a tool not only to crystallize what good SRI and CSR practices are, but also to identify companies with good SRI and CSR practices. It will enable investors to utilise their capital to reward companies that demonstrate excellent practices. Ultimately, investors will be empowered to influence company behaviour in a way that makes it more responsive to South Africa's development needs.

The good, the bad and the ugly

What's happened to the planet you call home this year? If you answer "I don't know", maybe it's time you picked up a copy of *Earthyear*. Because if you have read this far, it probably means that you are concerned about the welfare of your planet — just like most people you'll find on the pages of this award-winning magazine.

What have you done for the Earth this year?

Earth YEAR
JOURNAL FOR SUSTAINABLE DEVELOPMENT

Editor: Fiona Macleod • **Advertising:** Dean Wright • **Subscriptions:** Amanda Chetty
Published by **M&G Media, Media Mill, 7 Quince Road, Milpark 2092** • **PO Box 91667, Auckland Park 2006**
Phone: **(011) 727-7000** • Fax: **(011) 727-7111** • E-mail: **earthyear@mg.co.za** • Website: **http://www.earthyear.co.za**

SUSTAINABLE DEVELOPMENT
ACCOUNTING FOR NATURE'S SERVICES

Guest Essay by Prof. Maartin de Wit ~ Department of Geological Sciences, UCT

In recent years, ecologists and economists alike have provided empirical documentation of the direct dynamics between the rural poor and the natural environment. A deep understanding of this relationship between poverty and Nature begins with an appreciation of the basic environmental functions on which human life depends. Sustainable development thus demands that we address Nature's future needs also; and Nature's hidden expenses are considerable, yet seemingly difficult to quantify. How, for example, does one put a price on a future need of a natural resource that is currently consumed free of charge? How will a commodity like clean air be costed when its consumption is restricted by pollution? Will it go the way clean drinking water did towards the end of the 20th Century and become part of a competitive global bottling industry? "Oxygen bars" already thrive in some Asian cities.

Costs of these and other common natural resources, like ecosystems, are not addressed seriously at present because they have no equivalent monetary value and therefore remain outside the day-to-day economic market structures. Government and business can easily hide these costs because their accountants are not instructed to incorporate any figures for Nature's services, despite the fact that globally, Nature continuously submits invoices for damage control, and warns of looming environmental crisis and extinctions. This is a contemporary accounting scandal of unprecedented scale. Enron and other large corporations have been found out to be doing a scandalous job of corporate accounting, yet we are all party to far more corrupt accounting on a global scale. This global fraud will continue to undermine any attempts at sustainable development until Nature's costs are recognised and paid for by those who use and abuse them. But how? Presently, a new breed of ecological economist is trying to explore a new road towards sustainable development, by providing a practical way to acquire a better idea of just how valuable Nature is, and how we can learn to incorporate its benefits into more robust economic activity and analyses.

Economic wisdom of contemporary western society teaches that Nature provides no returns to the investor until it is harvested. If money in banks grows faster than trees in forests, the owners should chop them down and invest their profits elsewhere. This advice was given to many young budding economists and entrepreneurs at the burgeoning new schools of business administration and of resource economics throughout the developed world over the last half of the 20th Century. Assuming that their resources would last forever, the captains of economic growth and consumerism put their newly learned skills of discounted cash flows into practice with vigor and confidence. As fishermen, foresters, miners and energy extractors, they harvested everywhere with a lust spurred on by optimisation theory and magic formulae of maximum sustainable yield.

This rate of harvesting would ensure their profitable harvests forever. Mathematically it made good economic sense (and excellent profits), for example, to slaughter all the existing whales in the southern oceans, and to continue relentless fishing of cod, anchovies and halibut on the world's rich fishing grounds, to cut more forests, and to indiscriminately dump mine waste in lakes and rivers in exchange for more copper and gold. Then, early in the 1970s, the Canadian mathematical bioeconomist Colin Clark and Nobel economist Robert Solov revisited and highlighted what economist Harold Hotteling had calculated nearly 50 years earlier. Clark, Solov, and soon others, warned of the dangers of this approach to maximise monetary profits without considering its incremental side-effects. Yet few listened with the attention that their analyses deserved, and preventive action was delayed until cumulative consumer damages of global dimensions were brought sharply into focus during the last decades of the 20th Century. Declining fish stocks closed the Newfoundland cod-fishing industry off the Grand Banks (and is now threatening to do likewise in European waters); accelerated deforestation clashed with rubber-tappers and indigenous ways of life, reduced biodiversity and increased tropical diseases in South and Central America, throughout southeast Asia and central Africa; coal and oil burning fueled acid rain in Europe and North America; dumping of industrial waste reached far into Africa; soils were swept to sea from Madagascar to China; water and air became toxic; the ozone layer thinned; climates changed.

How had society been so hoodwinked into squandering gifts of Nature?

Why had commerce students for so long been taught "the price of everything but the value of nothing?"

And who will pay for the collateral damage to Nature and to people?

Towards the start of the new millennium people all over the world woke up to realise that economists and industrialists had led us astray by failing to admit into their natural resource equations many important social and environmental costs (externality costs) that the profit-makers were passing on to them. By choosing to ignore these "invisible costs", the practices of natural resource accountancy by industry were being mismanaged deliberately.

The mining industry, for instance, had made no attempt to estimate the long-term health and pollution costs associated with extracting and concentrating lead, uranium, or asbestos; timber and fishing industries had not tried to quantify the damage they cause to ecosystems and the continuous extermination of species. The costs of cleaning up the atmosphere or hydrosphere were not discounted from the enormous profits of the oil, coal or gas industries. Yet today, these so-called externalities, that include the uncalculated side-effects of their relentless pursuit of economic growth and globalisation, are still largely ignored: the current economic growth process continues to place the burdens of social and environmental costs on the weakest, and on future generations. This means that we continue to support a corrupt natural resource accountancy that verges on a crime against humanity.

Unless we insist on immediate action to stop and change this behaviour, we will, through our apathy, allow the profit makers to ravage our planet until there is nothing left to plunder.

Resources economists are now attempting to address externality costing in two different ways. On the one hand, traditional resources accounting has continued to concentrate on incrementally increasing the complexity of economic equations. By introducing the effects, for example, of improved exploitation technology; resource substitution; depletion and intergenerational equity; environmental abatement and more recently anthropogenic-induced global warming and ozone depletion, they hope to avert confrontations with existing capitalist market structures.

This approach of internalising many factors has clarified much about natural resource rents and their solutions guide natural resources accounting and legislation of many developed nations, as well as developing nations under the watchful eyes of organisations like the World Bank. But exploration of this economic avenue has achieved very little in protecting the delicate structures of Nature and ecosystems, nor seriously unravelled the link between rural poverty and degradation of the natural environment.

More recently, on the other hand, ecological economics has taken a different, almost inverted approach to evaluate natural resources capital. Partha Dasgupta, Robert Costanza, Gretchin Daily, Paul Ehrlich, Stuart Pimm, Karl-Goran Maler, and many others that understand the importance of a holistic approach, have started to address the costing issues of Nature in a more innovative, indeed revolutionary way. In doing so, they have created an exciting new field of study which views ecosystems as capital assets and Earth's geological processes as a global service industry that provides us with fresh water, clean air and fertile soils to grow our crops.

The key then is to figure out the costs of this natural service industry. What is the price tag of ecosystems; how much should we pay for pollinating services of bees and other insects; what should scavengers be charging for mopping-up; and how much does it cost Nature to produce our fresh water and clean air? Most of Earth's services are hard to quantify. Nevertheless this new avant-garde of eco-economists has attempted to calculate this for Earth's ecosystems.

US$33 trillion/year is a minimum estimate of the annual value of global ecosystems, nearly twice the Gross World Product (GWP).

The annual cost of globally consumed fresh water provided by Nature's hydrological cycle alone is estimated to be at least US$3000 billion (or about 12% of the GWP) and soil formation a whopping US$17 trillion. Although these numbers are widely questioned, and the methodology criticised, by assigning Nature's worth in dollars and cents, ecological economics has motivated people to start paying attention where there was none before. Clearly vast sums of money are invested by Nature to keep our backyard clean and to provide us with fresh water, mineral deposits and other basic life-support services and enjoyment of Nature.

We must learn now to pay our dues, retrospectively, and balance our industrial expense account. But at least three quarters of the world population cannot afford Earth's services, because they are too far in debt, and without poverty relief global pollution will increase unabated. Continuous ecosystem destruction is almost inevitable, and environmental health care remains in danger of collapse unless we take alternative actions now.

Rachel Carson's *Silent Spring* (1962) and the initial Club of Rome reports (1972) are milestones that initiated and influenced international debate about alternative futures.

By the mid-80s these debates took on growing momentum, during which scientists started, collectively, to influence political discourse in earnest: first through general environmental impact statements; and then more specifically through global climate-change forecasts.

The 1997 study of Robert Constanza and other ecological economists has provided another quantum leap towards the possibility of achieving some form of sustainable development. They show clearly that to protect our natural resources and our life support systems, we must now implement a new accounting system that incorporates natural capital and rural environmental assets.

> "...Earth provides enough to satisfy every man's need, but not every man's greed..."
>
> ~ Mohandas K Gandhi

Environmental Management in the City of Cape Town

VISION

The vision of Environmental Management in the City of Cape Town is to ensure that sustainable and equitable development is combined with sound environmental practice for a healthy environment, which sustains people and nature, provides protection for our unique resources and results in an enhanced quality of life for all.

The environment touches all aspects of our lives:

 The Natural Environment

 The Built Environment

 The Socio-Economic Environment

In order to guide our various projects and programmes we have developed an Integrated Metropolitan Environmental Policy, or IMEP.

IMEP
Integrated Metropolitan Environmental Policy
City of Cape Town

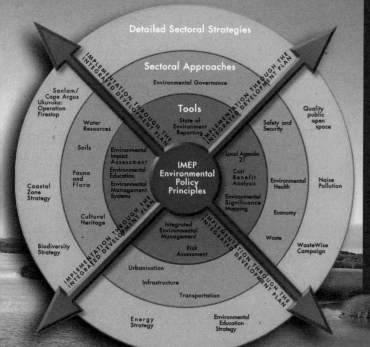

Detailed Sectoral Strategies

Sectoral Approaches

Environmental Governance

Tools

State of Environment Reporting

IMEP Environmental Policy Principles

Sanlam/ Cape Argus Ukuvuka: Operation Firestop

Water Resources

Soils

Environmental Impact Assessment
Environmental Education
Environmental Management Systems

Fauna and Flora

Coastal Zone Strategy

Cultural Heritage

Biodiversity Strategy

Integrated Environmental Management

Risk Assessment

Urbanisation

Infrastructure

Transportation

Energy Strategy

Environmental Education Strategy

Safety and Security

Quality public open space

Local Agenda 21

Cost Benefit Analysis

Environmental Significance Mapping

Environmental Health

Noise Pollution

Economy

Waste

WasteWise Campaign

IMPLEMENTATION THROUGH THE INTEGRATED DEVELOPMENT PLAN

Guidelines for Outdoor Advertising

Natural Heritage Management and Monit...

World War II Structures on Blouberg

Environmental Management Systems

PHOTO : JOHAN VAN PAPENDORP

Management of our Cultural Heritage

Local Agenda 21

Environmental Impact Assessments

Environmental Education (EE)

Environmental Management Programmes

Moddergat Rehabilitation

Smitswinkel Bay - False Bay Coast

Metropolitan Open Space System

Coastal Zone Strategy

If you would like to find out more about the City's environmental programmes, please contact:

Blaauwberg Region
Tel: (021) 550 1096
Cape Town Region
Tel: (021) 400 3912
CMC Region
Tel: (021) 487 2284
Helderberg Region
Tel: (021) 851 6982

Oostenberg Region
Tel: (021) 980 6250
South Peninsula Region
Tel: (021) 710 8134
Tygerberg Region
Tel: (021) 918 2007

CITY OF CAPE TOWN
ISIXEKO SASEKAPA
STAD KAAPSTAD

www.capetown.gov.za/emd
enviro@capetown.gov.za

SUSTAINABLE DEVELOPMENT IN LOCAL GOVERNMENT

**Guest Essay by Nomaindia Mfeketo: Executive Mayor
~ City of Cape Town**

"TOGETHER WE CAN MOVE MOUNTAINS"

Table Mountain looms large and majestic in the minds of every Cape Town resident and visitor. It is an awe-inspiring symbol of our human frailty compared with the greatness of nature. It is also a constant reminder of our enormous responsibility as guardians of this exceptionally beautiful corner of planet Earth.

South African poet and visionary, Credo Mutwa, tells of a popular folklore around this world-famous icon that daily defines the lives of the millions of people living around it.

"Tixo, the god of the Sun, and Djobela, the Earth goddess, made love and Qamata was born. When Qamata created the world, he was attacked by the Great Dragon of the Sea, who was jealous and wanted to stop the work of creation. In the battle with the Great Dragon, Qamata was crippled. In order to help Qamata, Djobela created a number of mighty giants to guard the world. She positioned one at each of the four corners of the earth – in the East, the West, the North and – the biggest giant – in the South."

"After many angry battles with the Great Dragon of the Sea, the giants were killed, one by one, but they asked Djobela, the Great Earth Mother, to turn them into mountains so that even in death they could guard the world. And so, the greatest giant of all, Umlindiwelingizinu, became Table Mountain, the Watcher of the South".

This tale reminds us of our power to influence our natural environment. We can unwittingly destroy it through short-sighted decisions or we can conserve it for future generations by means of visionary strategies.

For many months, our environmental management managers have sat with the very many people who are passionate about our environment and deliberated the best options for a sustainable management plan based on IMEP – our very own integrated environmental policy.

In the spirit of the World Summit for Sustainable Development theme of 'from Agenda to Action', the City is ready to put words into action with the launch of our first implementation strategies. These are actions, which will make a tangible difference to the lives of the people of Cape Town.

This is a major milestone for environmental management in the country and sets the ball rolling for local government to become one of the main protagonists of sound conservation practice.

Recognising that Cape Town's environment is an asset with vast economic potential, the City of Cape Town (CCT) has adopted a three–tiered approach to sustainable development based on Local Agenda 21:

◆ A strategic vision for the City and environmental policy framework

◆ Implementation partnerships, tools and priority programmes for service delivery

◆ Information systems to streamline administration and empower residents and decision-makers by providing feedback on progress.

The City's IMEP is an overall strategic policy, which includes a 2020 vision for the environment of Cape Town and general policy principles for sustainable development, implementation tools and detailed approaches to a range of 14 environmental, social and economic themes.

A leadership pledge to IMEP has been publicly signed by myself and the City's 200 Councillors, committing Cape Town to sustainable development.

Six priority environmental issues were identified from residents' input:

◆ Litter and illegal dumping

◆ Air pollution

◆ Biodiversity

◆ the Coastal Zone

◆ Noise pollution, and

◆ Quality open spaces for disadvantaged communities.

Strategies and targets for these are being prepared and three strategies (Biodiversity, Coastal Zone Management and Environmental Education) have been approved by Council.

Biodiversity Strategy

CCT proposes the creation of a Biodiversity Network, which will consist of a set of ecologically representative conservation areas connected through a network of corridors and nodes.

Core biodiversity nodes have been identified and a Biodiversity Network has been mapped to include all 37 Cape Flats core flora conservation sites identified by the Botanical Society of SA.

High priority sites will be managed primarily for biodiversity whilst other mixed-use areas will form part of the network system of corridors. The prioritisation of network areas will inform future processes of land acquisition, open space management and negotiation with landowners.

The strategy also proposes a plan for the False Bay Coast, which identifies social and economic opportunities for local communities, and aims to connect the False Bay Coast with the Cape Flats via key areas such as the Edith Stephens Wetland Park and Driftsands.

Coastal Zone Strategy

Cape Town is recognised as one of the world's top destinations for kite surfing, surf skiing, kayaking, windsurfing and sailing. However vandalism, crime and safety concerns are posing major threats to the quality of our coastline.

A coastal recreation plan is being developed to address these problems as well as the poor standard of facilities and services such as ablutions and lifesaving.

CCT has identified 15 beaches for potential Blue Flag status. This is an international 'eco label' awarded to beaches where environmental protection is a high priority.

Three sustainable coastal management plans are being developed for Muizenberg, Hout Bay and Sea Point, which will define a new holistic approach to integrate natural, economic, social and administrative aspects.

An Informal Beach Trading Policy is in place, setting clear guidelines for the licensing of vendors. A policy is also being developed to ensure that the City's resorts are well run, financially viable, safe and accessible for all.

Environmental Education and Training Strategy

The CCT's Youth Environmental School (YES) programme represents one of its largest projects, involving 20 000 learners at 470 primary schools.

This is supplemented by the Schools Environmental Policy (SEP) Project, which assists schools to form action-oriented environmental committees. The City also publishes a "State of the Environment (SoE) Report and workbook to help primary school teachers with the formulation of environmental lessons in Curriculum 2005.

Other educational initiatives include the Themba the Edutrain, which provides 12 000 less-privileged learners with a unique classroom-on-wheels experience; the annual Cheetah's Challenge to raise awareness for endangered species; and the publication of an annual Environmental Resource Directory.

Support programmes for CCT employees include an analysis of their environmental training needs; the preparation of a database of projects and resource materials; and the development of an evaluation tool for environmental education and training projects.

Cape Town hosted the international Cities Energy Conference in November 2003 and showcased its own Energy Strategy – the first for an African city.

Annual State of Environment (SoE) reporting allows both officials and communities to monitor progress in key areas. Our SoEs are even being integrated into the Western Cape schools science curriculum.

Progress since the adoption of IMEP has been encouraging and a number of programmes and projects have been completed. Whilst there are some success stories, our City will not shy away from highlighting the enormous challenges that remain and the need to build further partnerships for sustainable and sustained progress.

The way forward, which ensures that a responsible approach to our environmental assets is combined with our commitment to alleviate poverty, will not be easy, but we shall succeed, for "Together we can move mountains."

Sustainable development is not something governments or international bodies do to people. It is something people do for themselves, and for their children.

~ Cielito F Habito

Secretary of Socio-Economic Planning, the Philippines

The Department of Water Affairs and Forestry

OUR VISION

We have a vision of a democratic, people-centred nation working towards human rights, social justice, equity and prosperity for all.

We have a vision of a society in which all our people enjoy the benefits of clean water and hygienic sanitation services.

We have a vision of water used carefully and productively for economic activities, which promote the growth, development and prosperity of the nation.

We have a vision of a land in which our natural forests and plantations are managed in the best interests of all.

We have a vision of a people who understand and protect our natural resources to make them ecologically stable and safeguard them for current and future generations.

We have a vision of a Department that serves the public loyally, meets its responsibilities with energy and compassion and acts as a link in the chain of integrated and environmentally sustainable development.

We have a vision of development and co-operation throughout our region; of playing our part in the African Renaissance.

OUR MISSION

The mission of the Department of Water Affairs and Forestry is to serve the people of South Africa by:

Conserving, managing and developing our water resources and forests in a scientific and environmentally, sustainable manner in order to meet the social and economic needs of South Africa, both now and in the future;

Ensuring that water services are provided to all South Africans in an efficient, cost-effective and sustainable way;

Managing and sustaining our forests, using the best scientific practice in a participatory and sustainable manner;

Educating the people of South Africa on ways to manage, conserve and sustain our water and forest resources;

Co-operating with all spheres of Government, in order to achieve the best and most integrated development in our country and region;

Creating the best possible opportunities for employment, the eradication of poverty and the promotion of equity, social development and democratic governance.

OUR VALUES

The Department of Water Affairs and Forestry is a loyal servant of the Government and the people of South Africa.

As public servants, our skills will, at all times, be used for the benefit of the people and for the reconstruction and development of our country in the spirit of Batho Pele (People First).

As management, our responsibility is to provide high quality transformational leadership and a disciplined work ethic and to promote a working culture for motivated, accountable and committed teamwork.

As citizens of the African continent, we are dedicated to long-term integrated regional security and co-operation, and to the spirit of the African Renaissance.

Our working environment is governed by the principles of representivity, equality, mutual respect and human development.

CORE VALUES FOR TRANSFORMATION

We recognise that people are the cornerstone of the Department's success. Diversity is valued as a source of strength. We strive for a Department that fosters personal and professional growth and achievement.

We have the courage to change.

DEPARTMENT: WATER AFFAIRS
AND FORESTRY

CONTACT DETAILS
Chief Directorate Communication Services
185 Schoeman Street / Private Bag X313 • **PRETORIA** • 0001 • Tel: (012) 336 7500

DIRECTORY ACTIVITY ICONS

Agriculture

Animal

Avian

Botanical

Certification

Commerce Trade and Industry

Conservation

Consultancy Services

Eco-Tourism

Environmental Education and Training

Environmental Management & Planning

Environmental Media

Environmental Networking

Environmental Research & Development

Environmentally-Friendly Products

Environmental Ethics

Government Bodies

Healthcare

Land Management & Planning

Marine

Mining

NGO, CBO or Non-Profit Company

Professional Associations & Institutions

Environmental Risk or Liability Management

Rural Management & Development

Sustainable or Renewable Resources

Waste Management & Recycling

Water

Wildlife

Specialist Services

NETWORKING

DIRECTORY

SECTION

50/50 – SABC2

SABC Artillery Road AUCKLAND PARK Gauteng
Tel: 011 714 6621 • Fax: 011 714 6844
Email: 5050@sabc.co.za
Website: www.5050sabc.co.za
Contact Name: Danie van der Walt – Executive Producer

SABC2's top-class Environmental Television programme which is broadcast from 17:30 to 18:30 on Sundays.

ACTIVITIES:

ABALIMI BEZEKHAYA

PO Box 44
OBSERVATORY 7935 W Cape
Tel: 021 447 1256 • Fax: 021 447 1256
Email: abalimi@iafrica.com
Website: www.abalimi.org.za
Contact Name: Rob Small – Director

ABALIMI ("the people who plant") is an independent non-denominational affiliate of Catholic Welfare and Development (CWD). ABALIMI incorporates Abalimi Bezekhaya ("planters of the home") and The Cape Flats Tree Project. ABALIMI is a community-based environmental development NGO based at two People's Garden and Environmental Centres in Khayelitsha and Nyanga. ABALIMI resources, capacitates and partners individual survival gardeners, community market garden groups and community greening initiatives. Abalimi supports the development of leading best-practice community-owned Model Projects, which build democracy, renew and conserve the indigenous urban-environment and generate permanent informal jobs. The Model Projects influence broader policy formulation in open space development and catalyse the emerging environmental and ecological urban agriculture movement in the grassroots communities of the Cape Flats.

ACTIVITIES:

ABSA

17th Floor Absa Towers
160 Main Street GAUTENG
Tel: 011 350 4182 • Fax: 011 350 5800
Email: hammondm@absa.co.za • Website: www.absa.co.za
Contact Name: Hammond Mtembo – Brand Consultant

The Absa Group is one of South Africa's largest financial services organisations serving both individual and business clients in South Africa, the United Kingdom, Germany and the United States, China, Singapore, Hong Kong and Africa. Absa employs a customer-centric business model with targeted business units serving specific market segments. The Group offers from the basic products and services for the low-income personal market to customised solutions for the corporate market. Absa is a leading player in the home loan, installment finance as well as the debit and credit card markets. With a comprehensive e-business strategy, the Group continues to launch innovative services.

The Group has an extensive social investment programme focusing primarily on job creation, education and the battle against HIV/Aids. Absa has more than 32 500 staff, assets of R269 billion, 611 full and subsidiary outlets, more than 3 311 self-service outlets and 400 000 internet banking customers – the largest in SA.

ACTIVITIES:

ACER (AFRICA) ENVIRONMENTAL

Management Consultants
PO Box 503 MTUNZINI 3867
KwaZulu-Natal
Tel: 035 340 2715 • Fax: 035 340 2232
Email: info@acerafrica.co.za
Website: www.acerafrica.co.za
Contact Name: Dr R D Heinsohn – Managing Director

ACER (Africa) is an environmental consultancy firm, which adopts an holistic, multi-disciplinary and integrated approach to human endeavour in order to contribute to responsible decision-making and sustainable growth and development. ACER is proud of its reputation for excellence and client service. ACER provides customised Integrated Environmental Management consultancy services in an African context. In all its efforts, ACER actively supports, complies with, and contributes to international principles, approaches and standards. ACER is able to meet its private and public sector clients' needs in a customised manner in two key areas of operation:

1. Environmental Management
   - ◆ Strategic Environmental Assessments
   - ◆ Social/Biophysical Impact Assessments
   - ◆ Environmental Management Systems
   - ◆ Resettlement Planning and Implementation
   - ◆ Monitoring and Evaluation
   - ◆ Project Management

2. Communications Management
   - ◆ Public Participation Programmes

The ACER team comprises a range of specialists who address fully the requirements and needs of clients across a wide sectoral range. ACER places primary importance and full value on honesty and integrity in all its activities and dealings.

ACTIVITIES:

ADVANTAGE A.C.T.

Advantage House 38
Gordon Verster Road
The Willows PRETORIA 0041
Tel: 012 807-3503 • Fax: 012 807 1539
Email: ben@advantageact.co.za
Website: www.advantageact.co.za
Contact Name: Ben Fouché – Director

Constantly rising international Occupational Health, Safety and Environmental (SHE) standards and the challenges of competing in a global economy have forced companies to re-assess and align their SHE programs and systems in order to stay competitive.

At Advantage ACT we believe that an organisation's Environmental Management System should add value to the business and contribute to the ability to compete.

In line with this philosophy, Advantage ACT provides our clients with following world-class SHE training, consulting, auditing and personnel placement services:

- Advantage ACT is a SETA 23 and City & Guilds accredited training institution, providing the complete range of outcomes-based and customised SHE courses and seminars that fit our clients' specific circumstances.
- We consult on, and assist with the implementation of ISO/OHSAS-based management systems and provide knowledgeable and well-qualified expertise at board-level SHE committees (King II Report).
- The available auditing expertise spans the whole range of existing management systems from ISO to NOSA.
- Our unique resources and industry knowledge assist us when headhunting suitable candidates for your organisation's SHEQ-related positions.

"Experience the Difference"

ACTIVITIES:

AEROSOL MANUFACTURERS' ASSOCIATION

PO Box 9630
CENTURION 0046 Pretoria
Tel: 012 663 2999 • Fax: 012 663 2998
Email: ama@uskonet.com
Contact Name: Mr Mike Naude – Executive Director

The Aerosol Manufacturers' Association and the aerosol industry – a partnership of over 40 years – dedicated to a better life and a better product for producers and consumers alike.

ACTIVITIES:

AFRICABIO

15 Stopford Road Irene CENTURION 0062
Tel: 012 667 2689 • Fax: 012 667 1920
Email: africabio@mweb.co.za • Website: www.africabio.com
Contact Name: Prof. Jocelyn Webster – Executive Director

ACTIVITIES:

AFRICA BIRDS AND BIRDING

P O Box 44223, CLAREMONT 7735, W Cape
Tel: 021 762 2180 • Fax: 021 762 2246
Email: wildmags@blackeaglemedia.co.za
Website: www.africa-geographic.com
Contact Name: Peter Borchert – Publisher/Editor

As an independent magazine reporting on and about the continent, Africa Birds & Birding strives to offer balanced and comprehensive coverage of the natural history, identification, distribution and conservation status of the continent's abundant avifauna. The standing objective of Africa Birds & Birding is to entertain, inform and generally to foster awareness of birds as important indicators of general environmental health throughout the cities, towns, villages, farms and wilderness areas of Africa. In so doing Africa Birds & Birding will encourage the active involvement of readers in broader environmental and conservation challenges of our time, consistently advocating the need for the widest use of natural resources in a manner that involves and is of real benefit to the people of Africa.

ACTIVITIES:

AFRICA CO-OPERATIVE ACTION TRUST (ACAT)

PO Box 747 UMTATA 5100 E Cape
Tel: 047 536 0147/9 • Fax: 047 536 0148
Email: acatec@pixie.co.za
Contact Name: Mr Boyd Cekeshe – Director

ACTIVITIES:

AFRICA GEOGRAPHIC

P O Box 44223, CLAREMONT 7735, W Cape
Tel: 021 762 2180 Fax : 021 762 2246
Email: wildmags@blackeaglemedia.co.za
Website: www.africa-geographic.com
Contact Name: Peter Borchert – Publisher/Editor

As an independent magazine reporting on and about the continent, Africa Geographic strives to offer balanced and comprehensive coverage of the environmental and conservation challenges facing Africa so that our readers may thoughtfully and actively respond. The standing objective of Africa Geographic is to entertain, inform and generally to foster an awareness of important issues, consistently advocating the need for the wisest use of natural resources in a manner that involves and is of real benefit to the people of Africa. Africa Geographic is well respected by non-government organisations involved in conservation and the environment. In particular it enjoys the support and endorsement of WWF-SA and the Peace Parks Foundation.

ACTIVITIES:

Achieving sustainable economic growth will require the remodelling of agriculture, energy use, and industrial production after nature's example.

~ Jessica Tuchman Matthews

AFRICAN GABIONS (PTY) LTD/ MACCAFERRI

PO BOX 133 KYA
SANDS 2163 Gauteng
Tel: 011 708 1102
Fax: 011 708 3230
Email: enviro@africangabions.co.za
Website: www.africangabions.co.za
Contact Name: Ronel Suthers – Environmental Manager

Aim: To provide market leadership in environmentally engineered solutions,through professional service, technical expertise and quality products.

Specialising in the manufacture and supply of environmentally friendly products used by the Civil Engineering and Construction Industries.

Environmental solutions for:
- Biotechnical Engineering
- Soil Erosion Control
- Reinforced Soil Structures
- Embankment Stabilisation
- Earth Retaining Works
- Revetment/Channels
- Bridge Abutment/Pier Protection
- Culvert Inlet/Outlet Protection
- Low-level River Crossings
- Wetland and River Rehabilitation
- Weir/Drop Structures
- Coastal Protection

Services:
- Free Design Service
- On-Site Installation Assistance
- Training on Technical Software
- Product Training
- Quality Assurance System
- Design and Build packages
- Design and Supply

Resources:
- Qualified Civil Engineers
- Design Office with Professional Civil Engineers
- Professional Indemnity Insurance
- SABS & BBA Certification on Products
- BVQI Accredited - ISO 9001:2000
- Soil Erosion Control Specialists (120 Years)
- Biotechnical Engineering Department
- Ongoing Research and Development

African Gabions has offices in Durban, Johannesburg, Cape Town, Madagascar and Malawi. Worldwide we are known as Maccaferri and have over a century of experience in providing Environmental Solutions.

Please contact us if you would like copies of our literature or software.

ACTIVITIES:

AFRICAN GAMEBIRD RESEARCH, EDUCATION AND DEVELOPMENT TRUST – (AGRED)

PO Box 1251
FOURWAYS 2055 Gauteng
Tel: 011 465 4391 • Fax: 011 467 2118
Email: agred@netdial.co.za
Contact Name: Mr Rob Baillie – Director of Marketing, Education and Development

AGRED undertakes research, education and development of gamebirds in SA.

ACTIVITIES:

AFRICAN HEALTH & DEVELOPMENT ORGANISATION

PO Box 680
STELLENBOSCH 7599 W Cape
Tel: 021 883 3366
Email: ahdo@ahdo.org.za • Website: www.ahdo.org.za
Contact Name: Richard Weeden

African Health & Development Organisation is an NGO dedicated to informing and educating Africans on issues that affect our long-term sustainability. We focus on health & environmental issues, and how they affect the everyday lives of ordinary South Africans.

ACTIVITIES:

AFRICAN INSTITUTE OF CORPORATE CITIZENSHIP

PO Box 37357
BIRNHAM PARK 2015 Gauteng
Tel: 011 643 6604 • Fax: 011 642 6918
Email: info@aiccafrica.org • Website: www.aiccafrica.org
Contact Name: Paul Kapelus – Director

African NGO undertaking research, advocacy and network development in the field of banking and investing, corporate governance, accountability and reporting and responsible competitiveness.

ACTIVITIES:

AFRICAN NODE OF THE MOUNTAIN FORUM (ANMF)

C/o ICRAF
PO Box 30677
NAIROBI Kenya
E-mail: k.atta-krah@cgiar.org
Contact Name: Dr K. Atta-Krah

A pan-African network of mountain communities, individuals and institutions living, working or deriving their livelihoods in the mountain areas of Africa.

ACTIVITIES:

AFRICAN WIND ENERGY ASSOCIATION (AfriWEA)

PO Box 313 DARLING 7345 W Cape
Tel: 022 492 309 • Fax: 022 492 3095
Email: afriwindea@wcaccess.co.za
Contact Name: Dr Herman Oelsner – Chairperson

AfriWEA will promote and support wind energy development on the African continent by facilitating the exchange of political and scientific information, expertise and experience in the wind energy sector. It aims to further the wind energy interests of Africa in particular and developing countries in general.

ACTIVITIES:

AFRICAT FOUNDATION (THE)

PO Box 1889 OTJIWARONGO Namibia
Tel: 09264 67 304566
Fax: 09264 67 304565
Email: africat@natron.net
Website: www.africat.org
Contact Name: Carla Conradie

The AfriCat Foundation is a Namibian non-profit organisation committed to the long-term conservation of large carnivores in Namibia. This is achieved through creating awareness, preserving habitat, research, education and animal welfare.

ACTIVITIES:

AFRICON

PO Box 905 PRETORIA 0001
Tel: 012 427 2476 • Fax: 012 427 2390
e-mail: thomasv@africon.co.za
Website: www.africon.co.za
Contact Name: Mr T.R. van Viegen – Director

Africon Environmental Services is an environmental business consultancy practice with professional staff comprising a range of expertise in the fields of integrated environmental management and planning ensuring compliance with environmental legislation. Africon specialises in providing services and environmental solutions in the African context in the following environmental fields:

- Spatial Environmental Resource Management Systems (ERMS) designed to assist local government and large-scale property developers in meeting their environmental obligations by integrating environments into all planning processes
- Environmental Impact Assessments (EIA) for specific development applications
- Environmental Management Programme Reports (EMPR) for the mining and classified activities, including borrow pits
- Environmental Management Plans (EMP)
- Environmental Management Frameworks
- Environmental and ecological planning
- Environmental auditing and monitoring

- Local authority and environmental policy development
- Landscape Design and planning
- Environmental management systems
- Environmental risk assessments

ACTIVITIES:

AFRIDEV ASSOCIATES (PTY) LTD

14 Klipkers
PROTEA VALLEY 7530
Tel: 021 913-1785
Email: afridev@direcway.com
Contact Name: Cheryl Wither or Jack Hennessy (Director)

AfriDev Associates specializes in the management of major projects in the environment, rural development and agriculture sectors, as well as in HIV/AIDS impacts, particularly:

- All sectors of biophysical environmental assessment
- Ecological (instream) river flow requirements
- Socio-economic assessment
- Community participation and facilitation
- HIV/AIDS impact analysis and intervention
- Rural infrastructure
- Water resources management

AfriDev seeks to provide the highest quality expertise to address issues of the environment both natural and social; and to enhance and maintain sustainable solutions to the parallel demands of social and economic development as well as biophysical conservation. The management of AfriDev Associates has worked in a variety of countries throughout Africa, the Caribbean, Central and North America, and with many donor agencies.

ACTIVITIES:

The international community ought to give special attention to promoting the transfer of environmental technologies, a pivotal task in environmental cooperation.

~ Kim Young Sam,

President of the Republic of Korea

AGAMA ENERGY (PTY) LTD

PO Box 606
CONSTANTIA 7848 W Cape
Tel: 021 701 7052
Fax: 021 701 7056
Email: mark@agama.co.za • Website: www.agama.co.za
Contact Name: Mark Harris – Business Manager

AGAMA Energy provides green energy solutions. Our clients and customers benefit by being able to deliver high-quality products and services using cleaner, more sustainable energy inputs thereby demonstrating their commitment to better public accountability and corporate governance.

Other services include:

- ◆ Energy research and consultancy
- ◆ Engineering design and supervision of renewable energy systems
- ◆ Environmentally conscious design of buildings and development projects
- ◆ Development and implementation of energy management plans

The technical team comprises Glynn Morris, (MSc Eng - Energy Studies) - founder and Managing Director, who brings over seventeen years of experience in the field; Mark Harris (BA LLB CFP) - Business Manager; Greg Austin (MSc Eng) – Projects Engineer, Rendani Kharivhe (BSc Eng) – Field Engineer.

Former and current projects include:

- ◆ Implementation of a baseline CDM project for the City of Cape Town for energy interventions in low-income houses in Kuyasa, Khayelitsha
- ◆ Regional co-ordination for the Renewable Energy and Energy Efficiency Partnership
- ◆ The supply of 845MWh of Green Power for the WSSD

ACTIVITIES:

AGENCY FOR CULTURAL RESOURCE MANAGEMENT (ACRM)

PO Box 159
RIEBEEK WEST 7306 W Cape
Tel/ Fax: 022 461 2755
Email: acrm@wcaccess.co.za
Contact Name: Jonathan Kaplan – Director

Established in 1989, ACRM specialises in:

- ◆ Archaeological impact assessments and heritage resource management
- ◆ Development, implementation and monitoring of heritage site management plans
- ◆ Development of community-based heritage tourism products.

ACTIVITIES:

AGRICULTURAL RESEARCH COUNCIL

PO Box 8783
PRETORIA 0001 Gauteng
Tel: 012 427 9700 • Fax: 012 430 5814
Email:mpurnell@arc.agric.za • Website: www.arc.agric.za
Contact Name: Mariana Purnell

ACTIVITIES:

AGRICULTURAL RESOURCE CONSULTANTS

PO Box 3474 PARKLANDS 2121 Gauteng
Tel: 011 486 2254
Contact Name: Jim Findlay – Principal

Pesticide Development Specialists.

ACTIVITIES:

ALBANY MUSEUM

Somerset Street GRAHAMSTOWN 6139 E Cape
Tel: 046 622 2312 • Fax: (046) 622 2398
Email: l.webley@ru.ac.za
Website: www.ru.ac.za/albany-museum/
Contact Name: Dr Lita Webley – Acting Director

Curate, conserve and research collections including entomology, herbarium, palaeontology, archaeology, freshwater invertebrates, freshwater ichthyology and history. Provides an educational service to township schools and mobile museum services to farm schools.

ACTIVITIES:

ALDO LEOPOLD INSTITUTE

PO Box 35453
MENLO PARK 0102 Pretoria
Tel: 012 346 5446 • Fax: 012 460 2989
Email: aldo.Leopold@envirolaw.co.za
Website: www.envirolaw.co.za
Contact Name: Christa Barnard – Manager

The Aldo Leopold Institute provides advanced and practical environmental management training for environmental practitioners and the planning professions who require mid-career reskilling or training on an advanced or specialised level. Two types of courses are offered to cater for different training needs: customised corporate programmes and standard courses, which are presented upon demand. The Institute's short Learning Programmes are accredited by the Project Management Chamber of the Services SETA. They also meet international training requirements as set by ISO 14 001 and ISO 14012.

ACTIVITIES:

ALLGANIX HOLDINGS (PTY) LTD

Unit I The Vineyards Building Adam Tas Road
STELLENBOSCH 7600 W Cape
Tel: 021 886 7566 • Fax: 021 886 7564
Email: leonard@allganix.com
Contact Name: Leonard Mead – CEO

Allganix manufactures and trades high-quality organic and natural foods. We're also active in fresh produce, from box schemes to exports, as well as fresh and long-life beverages.

ACTIVITIES:

AMPHIBION TECHNOLOGIES

30 Constantiaberg Business Park DIEP RIVER W Cape
Email: rudy@wassertec.co.za • Website: www.amphibion.co.za
Cape Town: Contact: Rudy Regenaas
Tel: 021 705 9090 • Fax: 021 705 9092
Johannesburg: Contact: Ian Wright
Tel: 011 476 4862 • Fax: 011 476 5414

Manufacture, sales and service of high-quality ozone generators for use in industrial air and water disinfection and sterilisation, also for potable and effluent water treatment.

ACTIVITIES:

ANGLO AMERICAN

Anglo Operations Limited
PO Box 61587
MARSHALLTOWN 2107 Gauteng
Tel: 011 638 4261 • Fax: 011 638 8521
Email: jstacey@angloamerican.co.za
Contact Name: Julie Stacey
– Manager Safety, Health and Environment Policy Unit

Anglo American plc, with its subsidiaries, joint ventures and associates is a global leader in the mining and natural resource sectors. It has significant and focused interests in gold, platinum, diamonds, coal, base and ferrous metals (producing copper, nickel, zinc), industrial minerals (producing zircon and rutile (titanium dioxide) and forest products, as well as having financial and technical strength. The group is geographically diverse, with operations and developments in Africa, Europe, South and North America and Australia. Anglo American represents a powerful world of resources.

ACTIVITIES:

ANIMAL ANTI-CRUELTY LEAGUE

PO Box 7
ROSETTENVILLE 2130 Gauteng
Tel: 011 435 0672
Email: aacljhb@iafrica.com • Website: www.satis.co.za/aacl
Contact Name: Heather Cowie – Public Relations Officer

Established 1956, the Animal Anti Cruelty League is SA's second largest independent animal welfare organisation. **Branches:** Johannesburg, PE, Bredasdorp, Durban, Pietermaritzburg and Ladysmith. **Mission:** Care and protection of all animals.

ACTIVITIES:

ANTARCTIC AND SOUTHERN OCEAN COALITION (ASOC)

PO Box 13371 MOWBRAY 7705 W Cape
Tel: 021 422 5594 • Fax: 021 422 5594
Email: asoc-safrica@mweb.co.za • Website: www.asoc.org
Contact Name: Anton Boonzaier – Southern African campaigner

ASOC is an international coalition of numerous environmental NGOs from all over the world. Its various activities are focussed on the protection and conservation of the environment of Antarctica and the Southern Ocean.

ACTIVITIES:

APPLE ORCHARDS CONSERVANCY

PO Box 409 DE DEUR 1884 Gauteng
Tel: 011 949 1517 • Email: roy@thorntree.co.za
Contact Name: Roy Venter – Chairman
 ◆ Create a clean and healthy environment
 ◆ Conserve and increase indigenous plants
 ◆ Conserve birds, animals, insects and reptiles
 ◆ Create a sense of community

ACTIVITIES:

AQUA CATCH CC

PO Box 6675 KRAAIFONTEIN NORTH 7572 W Cape
Tel: 021 987 2566 • Fax: 021 987 2566
Email: gale@aquacatch.co.za • Website: www.aquacatch.co.za
Contact Name: Dr Barbara Gale

Aqua Catch provides consultancy services related to aquatic ecology, aquatic environments, aquatic conservation, integrated catchment management and water demand management. Aqua Catch is qualified to undertake general environmental impact assessments (usually with some element of aquatic impact) as well as specialist aquatic studies to determine current ecological state and make preliminary management recommendations related to potential impacts on current state. Aqua Catch can also assist with the development of catchment management plans, and construction and operational management plans for developments impacting on aquatic environments. Aqua Catch has access to all equipment necessary for on-site assessments and can outsource for detailed chemical and microbiological analyses and specialist hydrological assessments. Aqua Catch is firmly committed to the wise utilisation of South Africa's scarce aquatic resources.

ACTIVITIES:

ARCUS GIBB (PTY) LTD

PO Box 2965
CAPE TOWN 8000 W Cape
Tel: 021 469 9100
Email: jmcstay@gibb.co.za • Website: www.arcusgibb.co.za
Contact Name: Jon McStay – Director: Environmental and
Remediation Services

ARCUS GIBB is a multi-disciplinary consulting company that
offers Integrated Environmental Management Services;
including:

- ◆ Environmental Impact Assessment
- ◆ Environmental Management Plans
- ◆ Environmental Management Programmes
- ◆ Environmental Management Systems
- ◆ Environmental Auditing (including Due Diligence
  and Compliance Auditing)

In addition, we offer Environmental Remediation and Waste
Management Services, including:

- ◆ Remediation of contaminated industrial sites
- ◆ Pollution prevention
- ◆ Environmental Monitoring Protocols
- ◆ Asbestos Surveys
- ◆ Risk assessment and groundwater assessment
- ◆ Integrated Waste Management Plans
- ◆ Permitting of landfill sites (incl. site selection,
  design, EIA, permit application)
- ◆ Closure of landfills
- ◆ Hazardous waste management
- ◆ Leachate treatment and landfill gas monitoring

ACTIVITIES:

ART OF LIVING FOUNDATION (THE)

PO Box 1714
GLENVISTA 2058 Gauteng
Tel: 012 374 4949
Fax: 012 374 4949 • Cell: 082 322 0117
Email: kirtidab@telkomsa.net • Website: www. artofliving.org
Contact Name: Kirtida Bhana – National Co-ordinator

The Art of Living Foundation, founded in 1982 by Sri Sri
Ravishankar, is an international Non-Governmental
Organisation (NGO) in special consultative status with the
Economic and Social Council (ECOSOC) of the United
Nations. The Foundation has accredited representatives at
the United Nations in New York, Geneva and Vienna.

"This current crisis in our global environment is only a
projection of our limited minds, clouded in the stress and
strain of separation from one another and our common source.
We begin to solve the problem by returning to this common
source and the values which surround it." From a statement
by the Art of Living Foundation given to the U N World
Summit on Sustainable Development, Johannesburg, 2002.

The 5H Programme – Health, Homes, Hygiene, Human
Values and Harmony in Diversity (www.5H.org) is a joint effort
of the Art of Living Foundation in India and International
Association for Human Values (IAHV).

It is a massive social and empowerment programme with the
aim of uplifting individuals and communities. Projects are
currently being implemented in South Africa, Poland, Kosova,
USA, Latin America, Indonesia and India.

The Breath Water Sound Workshop is an integral part of the
5H Programme. This programme is offered to economically
and socially challenged sections of society and is aimed at
developing a self-reliant community.

The Art Excel Course (All Round Training in Excellence) for
children and teens challenges the learners to go beyond their
limited perspectives to consider the world at large with all its
diversity. The true measure of success is a happy, healthy,
well-adjusted child who is able to deal with life's challenges
(www.artexcel.info).

ACTIVITIES:

ASSOCIATION FOR WATER & RURAL DEVELOPMENT (AWARD)

Private Bag X483 ACORNHOEK 1360 Mpumalanga
Tel: 015 793 3991 • Fax: 015 793 3991
Email: tessa@award.org.za • Website: www.award.org.za
Contact Name: Tessa Cousins – Director

Award's vision is that the Sabi Sand catchment will stand as
a model of integrated development and sustainability
(social, environmental and economic). To this end, we have
adopted an approach that integrates projects that focus on
the sustainability of the water resource base (the Save the
Sand Project) and water-based livelihoods (community water
supply).

ACTIVITIES:

@ LAND LANDSCAPE ARCHITECTS & ECOLOGICAL PLANNERS

@ L A N D
LANDSCAPE ARCHITECTS
& ECOLOGICAL PLANNERS

PO Box 206
MENLYN 0063 Gauteng
Tel: 012 366 0124 • Fax: 012 366 0111
Email: wvrla@mweb.co.za • Website: www.atlant.co.za
Contact Names: Johan Goosen / Gwen Breedlove – Directors

Landscape and Urban Design: Design and management of
infrastructure design, landscape architectural design, graphic
& brochure design, construction documentation and
construction management.

Strategic Planning and Feasibility Analysis: Prepare
evaluation of the critical factors affecting development,
including environmental analysis, physical site analysis, real
estate market overview, community socio-economic analysis,
and public policy analysis.

Environmental Planning and Assessment: Impact analysis
including gathering and coordination of all environmental
consultants' information, permit requirements, and
preparation of applications, compilations and criteria for
review of information, and impact statements for both natural
and built environments.

Tourism planning: Large scale game park analysis and planning to include environmental evaluation, capacity studies, lodge and infrastructure locations and optimisation of aesthetic characteristics. Lodge placement to include site layout, amenities design, facilities design and cost of construction determinate. Planning and design of adventure and leisure facilities to include construction package and development cost.

Geographic Information Systems: Informed large-scale planning tool for decision making from various national and provincial databases as well as interpretation/queries/analysis of data /building of custom-made databases and map/image creation.

ACTIVITIES:

AUDI OF SOUTH AFRICA

P O Box 80 UITENHAGE 6230
Port Elizabeth
Tel: 041 994 5528
Email: brand@vwsa.co.za • Website: audi.co.za
Contact Name: Chayne Brand – Brand Manager

Environmental protection and preservation is an important aspect of Audi's global philosophy. The Audi Terra Nova Awards were developed in this spirit and focus on known and unsung heroes who have been tirelessly working in the environmental sphere without due recognition. The primary objectives of the Audi Terra Nova Awards include the following: firstly, the recognition and awarding of individuals and organisations for their dedication and work done to protect and conserve our environment; secondly, to motivate people to become more environmentally concerned and involved; and thirdly, to initiate public support for the work done by nominees. The Audi Terra Nova Awards cover the full environmental spectrum – Air, Land, Water and all Living Creatures, which reflects the four rings of Audi.

ACTIVITIES:

BABOON MATTERS

3 Balmoral Road PLUMSTEAD 7860 W Cape
Tel: 021 797 2751 • Fax: 021 797 2751
Email: nprm@netactive.co.za
Contact Name: Leigh Sax or Jenni Trethowan – Directors

Sustainable management of baboons on the Cape Peninsula and Baboon Tours.

ACTIVITIES:

BAINBRIDGE RESOURCE MANAGEMENT

314 Alexandra Road PIETERMARITZBURG 3201 KwaZulu-Natal
Tel: 033 386 9133 • Fax: 033 386 9133
Email: wrbainbr@iafrica.com
Contact Name: W.R. Bainbridge – Chief Consultant

Environmental consultant, specialising in management and planning of protected area and natural areas associated with commercial aforestation, environmental impact assessments and aforestation applications.

ACTIVITIES:

BAPELA CAVE KLAPWIJK CC

PO Box 11651 HATFIELD 0028 Gauteng
Tel: 012 362 4684 • Fax: 012 362 0394
Email: cka@mweb.co.za
Website: www.cka.co.za
Contact Name: Alan Cave – Director

Proficient in the application of geographic information systems to inform planning and design and visual impact assessment studies. Experienced in the environmental impact and mitigation of linear developments, such as roads and pipelines. Opportunities and influences beyond the project are always considered.

ACTIVITIES:

BARLOFCO CC

10 Queens View Place
UMGENI HEIGHTS 4051 Kwazulu-Natal
Tel: 031 564 0848 • Fax: 031 564 0848
Email: enviro@barlofco.co.za • Website: www.barlofco.co.za
Contact Name: Ian Jackman – Marketing Manager

Agricultural & environmental research equipment, training & professional advice – for sampling testing & monitoring of:

- ◆ Soils & sediments
- ◆ Surface and groundwater – including Diver® water quality & quantity monitoring instruments
- ◆ Crops & plants
- ◆ Agro climate & meteorology
- ◆ Physical survey & aerial map interpretation
- ◆ Waste sludge & slurry

Unique selection of affordable solutions – supported by certified trained personnel – and widest range of durable, reliable, quality instruments for environmental research. Trusted worldwide because you get results! Specialist training will reveal the secrets of better sampling & monitoring. FREE detailed user instructions supplied with every set.

Users include soil scientists, agricultural & environmental specialists, mining & industrial environmental officers, hydrology & geohydrologist professionals, meteorologists, engineering technicians & consultants, university & technikon departments, agricultural colleges, government departments.

Barlofco is the Certified Distributor for Eijkelkamp Agrisearch Equipment in Southern Africa – ALL IT TAKES FOR ENVIRONMENTAL RESEARCH.

ACTIVITIES:

BATELEUR RAPTOR REHABILITATION

PO Box 46147
ORANGE GROVE 2119 Gauteng
Tel: 011 640 1577 • Fax: 011 640 1057
Email: eaglelady@wol.co.za
Contact Name: Lorna Stanton

ACTIVITIES:

BATELEURS (THE) – FLYING FOR THE ENVIRONMENT IN AFRICA

9 Woolston Road
WESTCLIFF 2193 Gauteng
Tel: 011 646 0175 • Fax: 011 486 2238
Email: info@bateleurs.org • Website: www.bateleurs.org
Contact Name: Nora Kreher – Chair

We are a non-profit, non-governmental, Section 21 organisation, privileged to have dedicated volunteer pilots who want to help the environment and conservation by using their time, flying skills and aircraft, free of charge, to fly stakeholders over areas under threat in order to give them an aerial perspective on the problem or situation they are assessing.

Airborne pilots, like eagles, are all-seeing and have the opportunity to report anything amiss within their ranges. Bateleur pilots offer their services to help halt malpractices by reporting them, showing them to the right people who can do something about them. We also fly missions for other NGOs, media, government departments and individuals deeply concerned with the environment, wildlife, conservation and wilderness.

ACTIVITIES:

BEAUTY WITHOUT CRUELTY

PO Box 2332
CLAREMONT 7735 W Cape
Tel: 021 671 4583 • Fax: 021 671 4583
Email: beautywc@netactive.co.za
Contact Name: Ms Beryl Scott – Managing Trustee

Beauty Without Cruelty is an educational organisation concerned with cruelty to animals in vivisection, the fur and ivory trade and factory farming.

ACTIVITIES: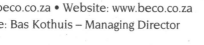

BECO INSTITUTE FOR SUSTAINABLE BUSINESS

BECO Institute for Sustainable Business

PO Box 12485 MILL STREET 8010
W Cape
Tel: 021 689 7117 • Fax: 021 689 7116
Email: info@beco.co.za • Website: www.beco.co.za
Contact Name: Bas Kothuis – Managing Director

We consult to both the private and public sectors taking a systems view of sustainability. BECO has extensive expertise and proven methodologies in Pollution Prevention and Waste Minimisation in both industrial and commercial enterprises, as well as government strategies and policies.

BECO Institute for Sustainable Business is a research and consultancy institute specialised in strategic environmental management issues. Our philosophy is based on the principles of "prevention is better than cure", and our experiences are based on taking a systems view of sustainability.

BECO has extensive experience in cleaner production and products:

◆ Prevention of wastes & emissions (Cleaner Production, Waste Minimisation, Pollution Prevention)
◆ Integrated Waste Management
◆ Benchmarking and Eco-efficiency indicators
◆ Sustainability scans
◆ Improvement of energy, raw material and water efficiency
◆ Environmental Management Systems (environmental action plans, ISO 14000 & EMAS)
◆ Monitoring environmental performance
◆ Annual corporate environmental reports
◆ Management of, and assistance for Waste Minimisation Clubs
◆ Pollution prevention through environmental permitting
◆ Preventive environmental chain management
◆ Environmentally-sound product development (eco-design) and product-oriented environmental management, including Life Cycle Assessments (LCA)
◆ Training and workshops for (local) authorities and companies (SAQA accredited)

BECO has offices in Johannesburg and Cape Town and employs 7 consultants, all with an academic degree. Besides its wide international network, BECO-ISB has an associated partner in The Netherlands and Belgium: BECO Group BV, with more than 40 consultants.

A broad consultancy capacity is ensured by the diverse backgrounds of the consultants: business administration, management of technology, product development, chemistry, chemical technology, biology, process technology, agricultural science, etc.

ACTIVITIES:

BERGVLIET HIGH SCHOOL

Firgrove Way BERGVLIET 7945 W Cape
Tel: 021 712 0284 • Fax: 021 715 0631
Email: tim@bhs.wcape.school.za
Website: www.wcape.school/bhs
Contact Name: Tim le Mesurier – HOD Environmental Education

Bergvliet High School has been a leader in environmental education for more than 15 years, with its annual Grade 10 Conservation Camp firmly established as part of its curriculum. Approximately 250 students and teachers spend three days of hands-on environmental discovery in the Betty's Bay area on the southern Cape coast.

Bergvliet High has also been a leader in recycling of paper, bottles, tins and plastic since 1987. The recycling depot, operated by a volunteer team of parents and supporters of the school, earns the school approximately R60 000 each year.

Bergvliet High is active in campaigning against the destruction of local environmental assets. Hacks and other forms of alien vegetation control are undertaken regularly in the Silvermine Nature Reserve and the local wetland. School projects are aimed at local environmental issues.

Bergvliet High would welcome being able to share experiences with other groups or bodies with similar interests.

ACTIVITIES:

BIO-EXPERIENCE

31 Stilt Avenue TABLE VIEW 7441 W Cape
Tel: 021 557 4942 • Fax: 021 556 7976
Email: bioexperience@absamail.co.za
Website: www.bioexperience.org
Contact Name: Marlise Pentz – Office Manager

Bio-Experience acts as a travel agency, but with a difference. Clients volunteer at various wildlife and conservation placements around South Africa as an alternative to conventional travel. Because we charge for placements we are able to donate much needed monies to the various venues to which we send volunteers. In addition, our venues get the hands to assist them with their daily tasks as they are often understaffed. Volunteers may choose between various venues including working with lions, vervet monkeys, sharks, penguins, buffaloes and baboons. Some of our non-wildlife related placements include conservancies, private nature reserves and game lodge management.

We even offer various community development projects including working with HIV orphans and street children. Our clients benefit by getting work experience, giving something back to the environment, enjoying South Africa from a very different perspective and have the opportunity to work up close with our wildlife. It's a win-win situation for volunteer and venue alike.

ACTIVITIES:

BIOLYTIX SOUTHERN AFRICA

PO Box 129
LYNEDOCH 7603 W Cape
Tel: 021 880 2340 • Fax: 021 880 2341
Email: info@biolytix.co.za • Website: www.biolytix.com
Contact Name: Theunis Duminy – CEO

Biolytix supplies and installs environmentally-sound sewage treatment and re-use systems. Suitable for developments or the individual house, these systems and environmentally sensitive approach to resources, transform your waste into a homegrown water and nutrient resource. By re-using 'waste' water generated in the house for irrigation purposes in the garden, the technology allows for more sustainable gardens in our arid country. This common sense application of technology helps to reduce pressure on our water resources, increase the lifespan of existing water and sewage infrastructure AND saves both money and water.

ACTIVITIES:

BIOWATCH SOUTH AFRICA

PO Box 13477 MOWBRAY 7705 W Cape
Tel: 021 447 5939 • Fax: 021 447 5974
Email: biowatch@mweb.co.za • Website: www.biowatch.org.za
Contact Name: Nicci van Noordwyk – Office Manager

Biowatch is committed to facilitating South Africa's adherence to international and national commitments on biodiversity through research, monitoring, advocacy and informing people about their rights.

We specifically concentrate on genetic engineering in agriculture and the impact on the above.

ACTIVITIES:

BIRDLIFE SOUTH AFRICA

PO Box 515
RANDBURG 2125 Gauteng
Tel: 011 789 1122 • Fax: 011 789 5188
Email: info@birdlife.org.za • Website: www.birdlife.org.za
Contact Name: Dr Aldo Berruti – Director

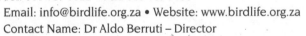

BirdLife South Africa promotes the conservation, enjoyment and understanding of birds and habitats through a network of 48 branches and affiliates, and is growing. BirdLife manages 13 national conservation and education programmes:

- ◆ Primary School Education for Sustainable Living
- ◆ The Global Seabird Programme, which aims to reduce seabird mortality, largely caused through adverse longline fishing techniques
- ◆ Species and sites conservation (part of a global programme) including Blue Swallow, Cape Parrot, Rudd's Lark, White-winged Flufftail, Spotted Thrush, Lappet-faced Vulture and Bald Ibis Working Groups
- ◆ Sappi-WWF-SA Wakkerstroom Training & Environmental Centre
- ◆ The Sasol Guide-training Programme
- ◆ The Oppenheimer Trust/de Beers Programme
- ◆ Richards Bay Rio Tinto Avitourism Development Programme
- ◆ BirdLife Travel and Birder Friendly Establishments
- ◆ The Bedford Wetland Park

- ◆ The Development of Local Guides in South Africa – Job Creation and Capacity Building for Ecotourism Development and Biodiversity Conservation
- ◆ Birding routes and avitourism development programme
- ◆ Building on Experience NGO and CBO Management and Skills Development Programme.

BirdLife South Africa is the local partner organisation of BirdLife International.

This coalition of more than 100 nationally-based organisations, with more than 2 500 000 members worldwide, strives to conserve birds, their habitats and global biodiversity, working with people towards the sustainability of natural resources. BirdLife South Africa publishes a National Newsletter, a section in African Birds and Birding, Ostrich (a scientific journal) and other publications including the Eskom Red Data Book for Birds of South Africa, Lesotho and Swaziland; Atlas of Birds of Southern Africa and the Important Bird Areas of Southern Africa. Branches have exciting and varied programmes of indoor and outdoor activities for their members – come and join the growing number of people who care for our environment and derive great social and recreational pleasure from the rich diversity of birdlife in South Africa.

ACTIVITIES:

BKS (PTY) LTD

PO Box 3173 PRETORIA 0001
Tel: 012 421 3854 • Fax: 012 421 3501
Email: martinv@bks.co.za • Website: www.bks.co.za
Contact Name: Martin van Veelen – Director: Environmental Management

Founded in 1965, BKS is a dynamic, multi-disciplinary organisation with specialist expertise in engineering and management, including managing the legally required environmental processes for successful project implementation.

ACTIVITIES:

BMW SOUTH AFRICA (PTY) LTD

PO Box 2955 PRETORIA 0001
Tel: 012 522 2755 • Fax: 011 805 2775
Email: odette.green-thompson@bmw.co.za
Website: www.bmw.co.za
Contact Name: Odette Green-Thompson – Corporate Communications Manager

BMW South Africa's operations centre around its head office in Midrand and a manufacturing plant and vehicle distribution centre in Rosslyn near Pretoria.

Plant Rosslyn accommodates the company's main production functions such as Logistics, the Body Shop, the Paint Shop and the Assembly Line. The plant, which employs some 2 500 employees in a two-shift operation, is currently geared to manufacture over 55 000 3 Series (4-door, right and left-hand drive) vehicles per year of which 80% are exported.

Substantial investments by BMW Germany since the mid 90s have resulted in Rosslyn becoming one of the most modern automotive manufacturing plants in the world.

This investment in the latest technologies, along with proven world class production processes, has secured Plant Rosslyn's position in the global BMW manufacturing network and the opportunity to produce the successor of the current 3 Series.

Plant Rosslyn received the highly prestigious J.D. Power and Associates' European Gold Plant Quality Award in 2002.

ACTIVITIES:

BOHLWEKI ENVIRONMENTAL (PTY) LTD

PO Box 11784 VORNA VALLEY 1686 Gauteng
Tel: 011 466 3841 • Fax: 011 466 3849
Email: bohlweki@pixie.co.za • Website: www.bohlweki.co.za
Contact Name: Carol Volbrecht – PA to MD

As a specialist environmental consulting company, Bohlweki Environmental provides expert services to both public- and private-sector organisations. As a multi-skilled consultancy, we offer solutions to a wide range of environmental management needs, and provide a comprehensive integrated environmental management service.

Our understanding of the need to integrate development and the needs of the environment ensures that environmental management solutions offered by Bohlweki Environmental has identified appropriate approaches to projects, offering minimised environmental impact, and practical and achievable mitigation measures. Bohlweki Environmental have extensive experience in undertaking environmental impact assessments. As part of this experience, Bohlweki Environmental offers expertise in related environmental management requirements, including feasibility studies, environmental planning, risk assessment, the development and implementation of environmental management plans and management systems, auditing, training and public involvement (participation) processes.

ACTIVITIES:

BOOK LEAGUE

PO Box 52 DARLING 7345 W Cape
Tel : 022 492 3153 Fax : 022 492 3153
Email : bookleague@telkomsa.net
Contact Name : Wendy MacIlrae – Owner

Offering a wide selection of mail-order environmental books.

ACTIVITIES:

BOTANICAL SOCIETY OF SOUTH AFRICA

Private Bag X10 CLAREMONT 7735 W Cape
Tel: 021 797 2090 • Fax: 021 797 2376
Email: info@botanicalsociety.org.za
Website: www.botanicalsociety.org.za
Contact Name: Dr Dave McDonald – Deputy Director

The Botanical Society of South Africa has close to 100 years of commitment to the development of the National Botanical Gardens, the conservation of the indigenous flora of the country and the promotion of indigenous gardening.

The society has over 15 000 members worldwide with thirteen branches in South Africa, seven of which are directly associated with a national garden. Member volunteers assist in the gardens in activities such as tour guides, education programmes and horticulture activities. In addition, they run events to raise funds for the gardens such as special plant sales, concerts, and craft markets. All branches run excursions and outings to interesting wildflower places and many promote flora conservation in their area. The Society has produced ten regional field guides suitable for use by the layperson and has a very successful annual environmental education project that includes a poster depicting a particular biome each year. The poster and associated workbook is made available to thousands of school children annually. The conservation arm of the Society runs projects, which address conservation priorities outside protected areas especially in the world-renowned Cape Floral Kingdom.

ACTIVITIES:

BROUSSE–JAMES & ASSOCIATES

PO Box 13885 CASCADES 3202 KwaZulu-Natal
Tel: 033 347 0322 • Fax: 033 347 0323
Email: brousse@sai.co.za
Contact Name: Barry James – Partner

Game Ranch Planning and Management; Botanical/Ecological Surveys and Research; Environmental Impact Assessments; Customised Biological Computer Programming and Database Management; GIS; Writing/ Editing of Geographic Information Systems; Editing and Environmental/ Adventure Journalism.

ACTIVITIES:

BUILT ENVIRONMENT SUPPORT GROUP (BESG)

University of Natal DURBAN 4041 KwaZulu-Natal
Tel: 031 260 2267 • Fax: 031 260 1236
Email: besg@wn.apc.org
Website: www.usn.org.za/member/besg.html
Contact Name: Mrs Lizell Bouwer

The Built Environment Support Group (BESG) works with people in urban areas who are disadvantaged in their access to resources, to demonstrate, implement and promote participatory, innovative and sustainable ways of improving living environments. The diverse skill base of its 34-member staff component strengthens BESG. This includes 21 professionals with experience and qualification in: town and regional planning, architecture and urban design, project management, engineering, construction management, social, economic and environmental sciences, organisation building and development, financial and general business management, adult education, policy analysis and research.

ACTIVITIES:

CAMERON CROSS INCORPORATED

PO Box 8867 CENTURION 0046 Gauteng
Tel: 012 671 3720 • Fax: 012 671 3560
Email: mail@cameroncross.co.za
Contact Name: James Cross – Director

Cameron Cross Inc. specialises in Environmental Law. Our vision is to consistently provide environmental legal advice of the highest standard, to add value by ensuring innovative legal solutions in relation to environmental issues, and to contribute to a sustainable future through the protection of the environment. Cameron Cross Inc. assists corporates in the mining and heavy industrial sectors with identification and management of environmental legal obligations and associated risks. The firm has unique experience in rendering environmental legal advice within the context of multi-disciplinary teams requiring the application of environmental principles to technical and scientific scenarios. We furthermore incorporate Environmental Law into various other legal disciplines such as planning and corporate law. Services include:

◆ Environmental Legal Risk Assessment
◆ Due Diligence Investigations
◆ Compilation of Legal-Environmental Registers, Guides and Software
◆ Legal Opinions
◆ Prospecting and Mining Law
◆ Litigation

ACTIVITIES:

CANON

Canon

PO Box 1782
HALFWAY HOUSE 1685 Gauteng
Tel: 011 265 4900 • Fax: 011 265 4954
Email: k.whittall@canon-sa.com • Website: www.canon.co.za
Contact Name: Kathryn Whittall – Senior Marketing Assistant

Canon South Africa (Pty) Ltd is a wholly-owned subsidiary of Canon Europa in Netherlands and is responsible for the marketing, sales and distribution of Canon's full range of Consumer Imaging and Business Solutions products in South Africa, as well as a number of sub-saharan African countries. Canon SA's consumer imaging line-up includes: Bubble Jet printers, scanners, faxes, personal copiers, multifunctional devices, calculators, printer and fax consumables, multi media projectors, as well as a full range of photo and video products.

ACTIVITIES:

*I believe in God,
only I spell it Nature.*
~ Frank Lloyd Wright

CANSTONE

PO Box 707 WELLINGTON 7654 W Cape
Tel: 021 864 1212 • Fax: 021 864 1557
Contact Name: Alex Eichmann – Owner

Alex Eichmann, a Swiss engineer and inventor of the Canstone, wanted to produce low-cost bricks which would help thousands of homeless South Africans to build their own houses.

Canstone is a sturdy, lightweight cement brick which incorporates two empty cans. This innovative idea reduces the amount of cement required for a standard brick and utilises empty cool drink or beer cans to strengthen the brick and recycle unsightly litter. The result is a cheap, lightweight cement brick that is extremely easy to work with.

Two workers can easily produce 200 Canstones per day. Because of their light weight and size the Canstone brick is easy to handle, accelerates the building process considerably, and reduces costs. Each Canstone brick will save more than 50 cents of materials and does not require electricity to produce. In tests carried out by The SA Bureau of Standards it was found that a Canstone can withstand pressures of up to 8 000 kg.

On his Wellington Farm, Eichmann has used Canstones to build all his outbuildings and pump houses. Farmers around Wellington have already ordered moulds to start making bricks on their own farms for labourer houses, stables and other buildings. The cans in the stone make an opening approximately 60 mm in diameter for the installation of electrical pipes or water pipes, etc.

ACTIVITIES:

CAPCORV RECYCLING

37 Keswick Road GERMISTON 1400 Gauteng
Tel: 011 873 3447 • Email: thammo@netactive.co.za
Contact Name: Thammo Schultz

Principle Products Recycled: PP and PVC. Principle Products Produced: PP Granuals for the flower pot market. PVC cable is compounded into shoe soles.

ACTIVITIES:

CAPE ACTION FOR PEOPLE AND THE ENVIRONMENT

A partnership for sustaining life in the fynbos and adjacent shores.

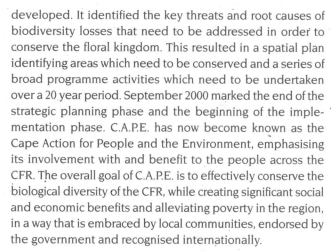

Private Bag X7 , CLAREMONT 7735 W Cape
Tel: 021 799 8790 • Fax: 021 797 3475
Email: info@capeaction.org.za • Website: www.capeaction.org.za
Contact Name: Trevor Sandwith – Co-ordinator

C.A.P.E. is an innovative programme of the South African Government developed to protect the rich biological heritage of the Cape Floristic Region (CFR). Its development was made possible with an initial grant from the Global Environment Facility (GEF) in 1998. In this period, the Cape Action Plan for the Environment, referred to as the C.A.P.E. Strategy, was developed. It identified the key threats and root causes of biodiversity losses that need to be addressed in order to conserve the floral kingdom. This resulted in a spatial plan identifying areas which need to be conserved and a series of broad programme activities which need to be undertaken over a 20 year period. September 2000 marked the end of the strategic planning phase and the beginning of the implementation phase. C.A.P.E. has now become known as the Cape Action for People and the Environment, emphasising its involvement with and benefit to the people across the CFR. The overall goal of C.A.P.E. is to effectively conserve the biological diversity of the CFR, while creating significant social and economic benefits and alleviating poverty in the region, in a way that is embraced by local communities, endorsed by the government and recognised internationally.

C.A.P.E. has three broad areas of focus:

◆ Conserving biodiversity in priority areas: establishing an effective reserve network, enhancing off-reserve conservation, and supporting bioregional planning;

◆ Using resources sustainably: developing methods to ensure sustainable yields, promoting compliance with laws, integrating biodiversity concerns into catchment management, and promoting sustainable ecotourism; and

◆ Strengthening institutions and governance: strengthening and enhancing institutions, policies, laws, cooperative governance, and community participation.

Please contact us for further information and to register as a C.A.P.E. partner.

ACTIVITIES:

CAPE BAT ACTION TEAM

c/o Department of Zoology
University of Cape Town RONDEBOSCH 7701
Tel: 021 650 4011 • Fax: 021 650 3301
Email: djacobs@botzoo.uct.ac.za • Website: www.capebat.co.za
Contact Name: David Jacobs

Cape Bat was founded to assist in conserving our local bats, many species of which are on the brink of extinction and to provide education in the environmental importance and economic value of bats.

ACTIVITIES:

CAPE ENVIRONMENTAL TRUST (CAPTRUST)

PO Box 231 BERGVLIET 7864 W Cape
Tel: 083 338 9319 • Contact Name: Frank Wygold

CAPTRUST was formed to act as a forum for the exchange of information and experience in matters pertaining to the sustainable use and conservation of the natural and man-made environment and, where appropriate, for common action in dealing with environmental issues.

ACTIVITIES:

CAPE TOWN (CITY OF) – ENVIRONMENTAL MANAGEMENT

P O Box 16548
VLAEBERG 8018 W Cape
Tel: 021 487 2283 • Fax 021 487 2255
Email: keith.wiseman@capetown.gov.za
Website: www.capetown.gov.za/emd
Contact Name: Keith Wiseman – Manager Policy Research and Review

The Environmental Management Department assists with planning sustainable development in the City of Cape Town so that it is healthy and suitable for the human, natural and built needs of the area. We aim to ensure that our children and ourselves will live in a safe and healthy environment that is able to support a better life for all.

We do this by:

◆ implementing sound planning and development of policy and guidelines;
◆ building awareness and understanding through special educational and communication projects, and undertaking monitoring and assessment of development projects.

Our Vision: "To ensure that sustainable and equitable development is combined with sound environmental practice for a healthy environment, which sustains people and nature, provides protection for our unique resources and results in an enhanced quality of life for all."

Our Websites: Environmental Management Department – www.capetown.gov.za/enviro/emd/

Integrated Metropolitan Environmental Policy (IMEP) – www.capetown.gov.za/imep
State of Environment Report (SoE) – www.capetown.gov.za/soe

Our Needs: Looking for partnerships (technological, financial and personal) for projects involving environmental policy, strategy, research, communication and implementation in the fields of air quality, biodiversity, climate change, coastal zone, education, energy, public transport and waste.

ACTIVITIES:

CAPE TOWN (CITY OF) – SOLID WASTE: DISPOSAL

9th Floor, 38 Wale Street
CAPE TOWN 8000 W Cape
Tel: 021 487 2477 • Fax: 021 487 2476
Email: saliem.haider@cmc.gov.za
Website: www.capetown.gov.za
Contact Name: Saliem Haider – Area Manager: South

Waste Wise is an initiative of the City of Cape Town's Solid Waste Management Department designed to encourage all communities to take ownership and responsibility for their environment by raising awareness on waste issues through outcomes-based education.

ACTIVITIES:

CAPE TOWN HERITAGE TRUST

103 Hout Street HERITAGE SQUARE 8001 W Cape
Tel: 021 424 9591 • Fax: 021 424 3159
Email: ctht@heritage.org.za • Website: www.heritage.org
Contact Name: Laura Robinson – Director

A leading non-governmental organisation tasked with the conservation and promotion of the built environment, cultural and natural landscape of metropolitan Cape Town.

ACTIVITIES:

CART HORSE PROTECTION ASSOCIATION (CHPA)

PO Box 846 EPPINDUST 7475 W Cape
Tel: 021 535 3435 • Fax: 021 535 3434
Email: chpa@mweb.co.za
Contact Name: Megan White – Fundraiser

Over the past seven years, CHPA has established regular clinics, introduced the sale of subsidised feed, a farrier service and veterinary care. With 70% of income going directly to benefit the horses, the enforcement of relevant provisions of the Animal Protection Act and education of carthorse owners, CHPA has successfully improved the condition and standard of many carthorses on our roads today. We have grown from providing services at two morning static clinics, to nine clinics held at sites adjacent to scrap metal yards where horses congregate to offload their burdens.

ACTIVITIES:

CEATEC

CEATEC

PO Box 211 UMZUMBE 4225
KwaZulu-Natal
Tel: 039 699 2293 • Fax: 039 699 3129
Email: chris@ceatec.co.za • Website: www.ceatec.co.za
Contact Name: Chris Early – Owner

Established 14 years ago as suppliers of environmental test instruments, CEATEC have expanded their range to include sound level meters, octave analysers, noise dosimeters, lux meters, heat stress monitors, dust measurement equipment, both optical and gravimetric, breath alcohol testers and weather station equipment. As South African agents for Casella, CEL, UK, CEATEC are able to import and offer state-of-the-art instruments at competitive prices to meet the growing demands of Environmentalists, Meteorologists and the Occupational Health Industry. In recent years, the firm's activities in the control of alcohol abuse in the workplace have grown with the importation of hand-held and wall-mounted breath alcohol testers from AME (Pty) Ltd, Australia. These instruments are being successfully operated by South African coal mines where alcohol consumption at work can present serious safety risks.

CEATEC are able to offer cost-effective solutions to most environmental test problems and the firm prides itself on a history of excellent customer care.

ACTIVITIES:

CEBO ENVIRONMENTAL CONSULTANTS CC

PO Box 11945 UNIVERSITAS
Bloemfontein 9321 Free State
Tel: 051 430 2001 • Fax: 051 430 2011
Email: cebo1@icon.co.za • Website: www.cebocyber.co.za
Contact Name: Neil Devenish – Director

CEBO Environmental Consultants CC is a professional consultancy based in Bloemfontein.

Due to the central location of the firm, a large number of projects have been conducted throughout the country. The firm provides a variety of services in the field of Environmental Impact Assessments (EIAs), Environmental Management Plans (EMPs), Land Use Planning, Game Farm Planning, Environmental Education and Vegetation Studies. A wide client base consists of the telecommunication industry, engineering industry, transport industry, private developers, authorities and educational/academic institutions. Specialists in the firm address fully the diverse requirements and needs of clients across a wide sectoral range. The staff places primary importance and full value on honesty and integrity in all their activities and dealings.

Working for an environmental sustainable South Africa based on progress, biodiversity, equality and liability.

ACTIVITIES:

CEDARA COLLEGE OF AGRICULTURE

Private Bag X6008 HILTON 3245 KwaZulu-Natal
Tel: 033 355 9309 • Fax: 033 355 9303
Email: lutgeb@dae.kzntl.gov.za
Contact Names: Bernd Lutge – Vice Principal

Cedara College offers a two-year Higher Certificate and a three-year Diploma in Agriculture specialising in Animal and Crop Production.

ACTIVITIES:

CEN INTEGRATED ENVIRONMENTAL MANAGEMENT UNIT

36 River Road WALMER 6070 E Cape
Tel: 041 581 2983 • Fax: 041 581 2983
Email: steenbok@aerosat.co.za.
Contact Name: Dr Mike Cohen

ACTIVITIES:

CENTRE FOR CONSERVATION EDUCATION

9 Aliwal Rd WYNBERG 7800 W Cape
Tel: 021 762 1622 • Fax: 021 762 8690
Email: postmaster@cce.wcape.school.za
Contact Name: Ms Sigi Howes

The Centre for Conservation Education is a unique environmental educational institution under the auspices of the Western Cape Education Department.

Its aim is to generate positive action through creating environmental awareness and by influencing attitudes among learners and teachers. The teaching programme is outcomes-based; cuts across all learning areas; and supports the National Environmental Education Programme (NEEP). 'Environment' is interpreted in its widest sense and covers natural, man-made and social issues. National and world environmental days are incorporated into the programme annually.

A wide variety of related services includes:
- ◆ Quarterly newsletter
- ◆ Annual exhibition for International Museum Day
- ◆ Schools' anniversary exhibitions
- ◆ Education Museum on the history of education
- ◆ Reference library with resource material on environmental matters
- ◆ Research material on the history of education
- ◆ Illustrated lectures to adult interest groups
- ◆ Networking with other state and provincial departments, NGOs, museums and environmental organisations

ACTIVITIES:

CENTRE FOR ENVIRONMENTAL MANAGEMENT(CEM) NORTH WEST UNIVERSITY

Private Bag X6001 Internal Box 231
POTCHEFSTROOM CAMPUS 2520 North West
Tel: 018 299 2725 • Fax: 018 299 2726
Email: aokml@puknet.puk.ac.za
Website: www.cem.puk.ac.za
Contact Name: Madel Lottering – Senior Administrator

The Centre for Environmental Management (CEM) is positioned to deliver high-quality training programmes and finding innovative solutions to environmental challenges that facilitates change in your organisation on virtually any topic related to the environment and environmental management. An extensive network of supporting specialists, state of the art facilities and equipment as well as dedicated, professional staff, ensure high-quality service delivery.

The CEM is respected locally, as well as internationally, as a Centre of Excellence for its leadership role in finding innovative, environmental management solutions and to facilitate change towards a more sustainable future.

Our mission is to support, research, analyse and build capacities on a transdisciplinary, value creating and enabling basis with respect to:
- ◆ Greening development
- ◆ Greening business in all sectors
- ◆ Greening governance at all levels

These are done within a developing country context, taking cognisance of international developments, regional trends and local requirements.

The core activities are to build capacity and facilitate change by:

- ◆ Conducting applied research in environmental management, sustainable development and related issues, and finding innovative solutions to environmental and sustainability challenges
- ◆ Developing and conducting training programmes in environmental management and related fields
- ◆ Rendering advisory services in environmental management and related fields

In addition to the above-mentioned, CEM is serious about rendering community service, and is involved in many non-profit initiatives aimed at professional development of previously-disadvantaged individuals into skilled environmental professionals, provision of assistance to local government towards greener governance, and much more.

Enquiries:

Mrs Dydré Greef   (018) 299 2714
                      aokdg@puknet.puk.ac.za

Mrs Madel Lottering   (018) 299 2725
                      aokml@puknet.puk.ac.za

ACTIVITIES:

CENTRE FOR ENVIRONMENTAL STUDIES, UNIVERSITY OF PRETORIA

University of Pretoria 0002 PRETORIA
Tel: 012 420 4048 • Fax: 012 420 3210
Email: mdobson@zoology.up.ac.za
Website: www.up.ac.za/academic/centre-environmental-studies/
Contact Names: Prof Willem Ferguson; Ms Marinda Dobson – Director / Assistant

Trans-disciplinary Masters and Doctoral degree options in Environmental Studies are offered in Environment & Society, Environmental Ecology, Environmental Economics, Environmental Management, Environmental Education, Water Resource Management, and Environmental Engineering. These are one year full-time and two years part-time.

ACTIVITIES:

CENTRE FOR RURAL LEGAL STUDIES

PO Box 1169 STELLENBOSCH 7599 W Cape
Tel: 021 883 8032 • Fax: 021 886 5076
Email: postmaster@crls.org.za
Contact Name: Sharron Marco-Thyse – Director

ACTIVITIES:

CENTRE FOR THE REHABILITATION OF WILDLIFE (CROW)

PO Box 53007
YELLOWWOOD PARK 4011 KwaZulu-Natal
Tel: 031 462 1127 • Fax: 031 462 9700
Email: info@crowkzn.co.za • Website: www.crowkzn.co.za

Contact Name: Joanne Hardy – Public Relations Officer

CROW is a non-profit organisation that prides itself in taking care of South Africa's indigenous wildlife, both injured and orphaned. CROW is one of KZN's main rehabilitation centres.

We never refuse assistance to any indigenous creature within, and sometimes outside KZN. We are open for admissions 24 hours a day, 365 days a year. CROW's policy is not to charge for services rendered by the Centre. We derive our income from donations, memberships, public relations functions, bequests and income from investments.

CROW is only open to the public on the last Sunday of every month. From Monday to Friday we conduct educational tours. A rehabilitation centre has a unique opportunity to educate children through visual contact and tactile experiences.

After rehabilitation, an animal is released back into the wild. CROW has a release committee, which monitors all releases. Released animals are monitored for a period of up to 6 months.

ACTIVITIES:

CENTRE FOR WILDLIFE MANAGEMENT UNIVERSITY OF PRETORIA

Hatfield Experimental Farm, South Street
HATFIELD 0002 Pretoria
Tel: 012 420 2627 • Fax: 012 420 6096
Email: lvessen@wildlife.up.ac.za
Website: www.wildlife.up.ac.za/centre
Contact Name: Mr L D van Essen – Lecturer

After 38 years of qualifying leaders in the field of conservation and wildlife management the Centre for Wildlife Management, University of Pretoria is a globally respected provider of exceptional quality wildlife management training, relevant and applicable consultancy, as well as research services in wildlife management.

Wildlife management and conservation experience have been developed in the private, as well as the public sector at international, national, regional and local levels.

The Centre aspires in developing appropriate and cost-effective wildlife management solutions of the highest international standard, within the African and developing country context.

The Centre for Wildlife Management continuously strives to:

- ◆ be recognised, locally and internationally for teaching and research on wildlife management and biodiversity conservation
- ◆ maintain a focus relevant to Africa and southern Africa in particular, with regard to biodiversity conservation in the context of sustainable human development.

ACTIVITIES:

CERTIFICATION BOARD FOR ENVIRONMENTAL ASSESSMENT PRACTITIONERS (THE INTERIM)

PO Box 1749 NOORDHOEK 7979 W Cape
Tel: 021 789 1385 • Fax: 021 789 1385
Email: eacertify@intekom.co.za • Website: www.eapsa.co.za
Contact Name: Jeanette Walker – EAP Certification Co-ordinator

The Interim Certification Board for Environmental Assessment Practitioners of South Africa aims to advance the practice of, and promote quality in, environmental assessment through a process of (presently) voluntary certification. Registered professionals would have pledged to uphold a defined code of conduct, and to act in the best interest of the environment and sustainable development.

ACTIVITIES:

CHAMBER OF MINES OF SOUTH AFRICA

PO Box 61809 MARSHALLTOWN 2107 Gauteng
Tel: 011 498 7661 • Fax: 011 498 7429
Email: nlesufi@bullion.org.za • Website: www.bullion.org.za
Contact Name: Mr Nikisi Lesufi – Environmental Adviser

The Chamber of Mines of South Africa provides strategic support and advisory input to its members. It facilitates interaction among mine employers to examine policy issues and other matters of mutual concern in order to define desirable stances and joint initiatives. It does not promote the mining industry at all costs, but works for the good of South Africa generally. A key activity is the Chamber's representation of the formalised policy position of its membership to South Africa's national and provincial governments, and to other relevant policy-making and opinion-forming entities inside the country, and internationally. The Chamber is active on all major environmental policy issues affecting the South African mining industry, including national and provincial environmental management, waste, water, climate change and trade issues. It recently published guidelines for cyanide management and for public participation, both of which were developed by stakeholder task teams.

ACTIVITIES:

CHEETAH CONSERVATION FUND (CCF)

PO Box 1755 OTJIWARONGO Namibia
Tel: 09264 67 306225 • Fax: 09264 67 306247
Email: cheeta@iafrica.com.na
Website: www.cheetah.org
Contact Name: Dr Laurie Marker – Executive Director

Nestled amidst the panoramic Waterberg Plateau, CCF's Research and Education Centre is dedicated to achieving best practice in the conservation and management of the wild cheetah and their ecosystems.

ACTIVITIES:

CHEETAH OUTREACH

PO Box 116 LYNEDOCH 7603 W Cape
Tel: 021 881 3352 • Fax: 021 881 3352
Email: cheetah@intekom.co.za
Website: www.cheetah.co.za
Contact Name: Dawn Glover – Education Officer

Cheetah Outreach is a non-profit organisation, established in 1997 on land donated by Spier Wine Estate, Stellenbosch, to create awareness of the plight of the free-ranging cheetah through eco-tourism and education; and support in-situ conservation efforts in Namibia (Cheetah Conservation Fund) and South Africa (National Cheetah Management Programme). The Global Cheetah Forum, for which Cheetah Outreach currently acts as the Education Co-ordinator, recognises education as one of the five key issues that will ensure the cheetah a place in the future. Cheetah Outreach holds 80 class presentations (±14 000 learners) a year, develops and distributes resources using the cheetah as a learning tool, holds annual teacher-training workshops to introduce resources and train teachers on the incorporation of the environment into the Revised National Curriculum Statement and offers annual two-month Environment Education Teacher Fellowships in the United States. Cheetah Outreach not only trains education ambassadors for the National Cheetah Management Programme, but in the future will also take up the position of Education Co-ordinator nationally for this initiative. Captive-born, handraised ambassador cheetahs greet visitors at the Estate as well as travel to local schools, hotels, community groups and public events. Since our inception our ambassadors have taken the plight of Africa's most endangered Big Cat to more than a million people as well as reaching wider national and international audiences through TV documentaries and other media coverage. Merlin, the project's Turkish Anatolian Shepherd, helps promote the use of Livestock Guarding Dogs as a successful method of Non-Lethal Predator Control on Namibian farmlands. Cheetah Outreach invites members of the public to have a personal encounter with a cheetah, 7 days a week, 365 days a year. Teachers are welcome to contact us for more information regarding workshops and school visits.

ACTIVITIES:

CHEMICAL AND ALLIED INDUSTRIES' ASSOCIATION (CAIA)

PO Box 91415,
AUCKLAND PARK 2006 Gauteng
Tel: 011 482 1671 • Fax: 011 726 8310
Email: caiainfo@iafrica.com Website: www.caia.co.za
Contact Name: Dr MD Booth – Director Information Resources

The Chemical and Allied Industries' Association (CAIA) was established in 1994 to promote a wide range of interests pertaining to the chemical industry. These include promoting the industry's commitment to a high standard of health, safety and environmental performance.

CAIA is the South African custodian of the international Responsible Care initiative, which has been adopted by 47 countries worldwide. Through this initiative, companies make a formal public commitment to steadily improving performance in health, safety and environmental management. Another key component of this initiative is the promotion of transparency when dealing with the public. In South Africa, those companies, which are signatories to the initiative, account for over 90% of chemical production.

Each year CAIA produces a report on the industry's performance in implementing Responsible Care and sets targets for future developments. Performance data are submitted every two years to the International Council of Chemical Associations, which produces a status report for the United Nations.

ACTIVITIES:

CHEMICAL EMERGENCY CO-ORDINATING SERVICES CC

PO Box 119, WESTVILLE 3630 KwaZulu-Natal
Tel: 031 266 9035 • Fax: 031 266 5613
Email: wilbrink@icon.co.za
Contact Name: Frans Wilbrink – Owner

Chemical Emergency Co-Ordinating Services CC was registered during 1998 at the request of some clients who required such service due to the introduction of the Major Hazard Installations Regulations of the Occupational Health & Safety Act. During the last couple of years the company has experienced a gradual increase in the number of clients who have retained its service for such emergencies.

By joining this service, you will ensure that you and your clients have 24 hours a day access to a well-equipped emergency co-ordinating service. It will free you and your staff from having to attend to these emergencies.

We guarantee immediate response, eliminating the delays that are normally associated with an emergency, thus reducing the negative impact it may have on the environment and the public.

ACTIVITIES:

CHENNELLS ALBERTYN

PO Box 58 OBSERVATORY 7935 W Cape
Tel: 021 448 2333 • Fax: 021 448 0209
Email: chenstel@iafrica.com
Contact Name: Johan van der Merwe – Partner

A firm of attorneys specialising in human rights law, which developed from administrative law actions against the State and the Police, to Labour Law, and now into Environmental Law as a means to ensure the protection of human rights and the right to an environment that is sustainable and healthy as now provided by the Constitution.

ACTIVITIES:

CHITTENDEN NICKS DE VILLIERS

PO Box 10211 CALEDON SQUARE 7905 W Cape
Tel: 021 461 6302 • Fax: 021 461 6466
Email: planning@cndv.co.za /
landscape@cndv.co.za
Website: www.enviro-plan.com
Contact Names: Carmen Stout-Smith

Founded in 1987, CNdV has developed into one of South Africa's leading integrated environmental planning and design practices. This award winning consultancy is committed to excellence in the fields of urban and environmental planning, landscape architecture and urban design.

The practice operates across the full planning and design spectrum including strategic environmental assessments, regional, city and local area planning, urban and environmental design, tourism and resort planning, visual impact assessment and project facilitation.

CNdV's integrated approach is particularly well suited to projects where a strong overlap of skills is demanded. We have established a solid reputation for being at the cutting edge of our rapidly changing urban and natural environment, with its increased emphasis on sustainability, accessibility and urban quality.

ACTIVITIES:

CLOTHING & TEXTILE ENVIRONMENTAL LINKAGE CENTRE

15 Lower Hope Street ROSEBANK 7700 W Cape
Tel: 021 685 6126 • Fax: 021 689 1726
Email: pfoure@mweb.co.za
Website: www.ctelc.co.za
Contact Names: Pat Foure – Director

The Centre is funded by Danida and DTI and supports cleaner products in the clothing and textile pipeline. It provides information and raises environmental awareness in clothing and textile production.

ACTIVITIES:

COASTEC COASTAL AND ENVIRONMENTAL CONSULTANTS

PO Box 370 RONDEBOSCH 7701 W Cape
Tel: 021 685 5445 • Fax: 021 685 5445
Email: coastec@mweb.co.za
Contact Names: Barrie Low / Uschi Pond – Owner / Associate

COASTEC specialises in botanical analysis of aquatic and terrestrial ecosystems, environmental impact assessments and management plans. Activities are supported by an extensive indigenous plant database and 25 years' field experience.

ACTIVITIES:

COASTAL AND ENVIRONMENTAL SERVICES (CES)

PO Box 934 GRAHAMSTOWN 6140 E Cape
Tel: 046 622 2364/7 • Fax: 046 622 6564
Email: lisl@cesnet.co.za • Website: www.cesnet.co.za
Contact Name: Lisl Griffioen – Marketing Manager

Coastal & Environmental Services (CES) specialises in Environmental Impact Assessment and Management. Established in 1990, CES is one of the largest specialist environmental consulting companies in South Africa, and has extensive experience in the management of all aspects of the Environmental Impact Assessment (EIA) process.

CES has a highly qualified staff complement, most of whom hold PhD or Masters degrees, with specific expertise in terrestrial, marine and estuarine ecology, coastal zone management, ecological baseline assessments, social impact assessments and public involvement, heavy mineral mining EIAs, ecological water requirements, rehabilitation, strategic environmental assessments (SEA) and environmental management plans (EMP).

CES has successfully completed a range of environmental studies in South Africa, Mozambique, Madagascar, Kenya, Swaziland, Lesotho, Malawi, Namibia, Angola and Zimbabwe.

ACTIVITIES:

COASTCARE

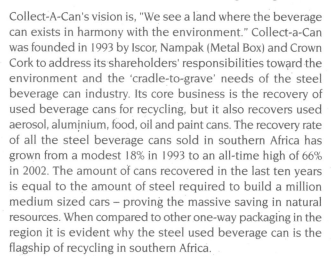

Private Bag X2 ROGGEBAAI 8012 W Cape
Tel: 021 402 3911 • Fax: 021 402 3009
Email: czm@deat.gov.za
Websites: www.deat.gov.za www.sacoast.uwc.ac.za
Contact Name: Dr D E Malan

The Department of Environmental Affairs and Tourism provides national leadership for promoting sustainable coastal development in South Africa. This is primarily achieved through CoastCare, a partnership programme involving the private and public sectors. The policy guiding this joint initiative is presented in the White Paper for Sustainable Coastal Development in South Africa.

CoastCare aims for:
- ◆ Coastal economic development that makes the best use of available resources
- ◆ Coastal development that promotes social equity through improved livelihoods of poor coastal communities
- ◆ A healthy coastal environment for the benefit of current and future generations

CoastCare provides financial and technical assistance for:
- ◆ Coastal development projects
- ◆ Institutional capacity building of coastal management organisations
- ◆ Legal development to support policy
- ◆ Awareness education and training initiatives
- ◆ Coastal resource planning

- ◆ Applied research
- ◆ Coastal information management projects

ACTIVITIES:

COLLECT-A-CAN (PTY) LTD

PO Box 30500 KYALAMI 1684 Gauteng
Tel: 011 466 2939 • Fax: 011 466 2927
Email: shabeer@collectacan.co.za
Website: www.collectacan.co.za
Contact Name: Shabeer Jhetam – Marketing Manager

Collect-A-Can's vision is, "We see a land where the beverage can exists in harmony with the environment." Collect-a-Can was founded in 1993 by Iscor, Nampak (Metal Box) and Crown Cork to address its shareholders' responsibilities toward the environment and the 'cradle-to-grave' needs of the steel beverage can industry. Its core business is the recovery of used beverage cans for recycling, but it also recovers used aerosol, aluminium, food, oil and paint cans. The recovery rate of all the steel beverage cans sold in southern Africa has grown from a modest 18% in 1993 to an all-time high of 66% in 2002. The amount of cans recovered in the last ten years is equal to the amount of steel required to build a million medium sized cars – proving the massive saving in natural resources. When compared to other one-way packaging in the region it is evident why the steel used beverage can is the flagship of recycling in southern Africa.

ACTIVITIES:

CONSERVANCY ASSOCIATION

PO Box 1552
WALKERVILLE 1876 Gauteng
Cell: 084 590-2312 • Fax: 084 505-4922
Email: conservancy@conservancies.org
Website: www.conservancies.org
Contact Name: Dr A Kruger – Chairman

A conservancy is: 'The voluntary co-operative nature and environmental management of an area by its community and its users and in respect of which registration has been granted by the relevant Provincial Authority.' There are four types of conservancies: Rural conservancies, Urban conservancies, Urban conservancies in townships, Industrial conservancies.

What Are Conservancies About ?
Conservancies originated in KwaZulu-Natal in 1975 in the Balgowan area where poaching was a major problem. Nick Steele, a nature conservator, developed the idea of co-operative management of the game in Balgowan and the Parks Board agreed to give training to the game guards of this area. Today there are more than 200 conservancies in Kwa-Zulu-Natal, covering 1 200 000ha. The Free State followed, establishing the first conservancy in February 1985 and soon there were 18 in the Eastern Free State. Since January 1993 their conservancies have increased to 106, covering about 500 000ha. The Free State was also the first to promote the conservancy concept in townships and at this stage these

conservancies have "urban rangers" who are trained in basic environmental management. There are now 600 registered conservancies in South Africa, covering more than double the areas managed by SANParks and official conservation bodies.

Gauteng's 33 registered conservancies took the initiative in forming the Gauteng Conservancy Association (GCA) in February 2003. In August 2003 the GCA hosted a National Conservancy Conference, which was attended by 170 representatives from all nine provinces. At this conference a National Conservancy Association (NCA) was established and a steering committee put in place. On this steering committee there are representatives from all nine provinces. KZN and Gauteng are the only provinces that have Conservancy Associations and it was stressed at the conference that such associations should be promoted in other provinces.

"Our bio-diversity is legendary but becoming increasingly threatened," says Mark Botha of the Botanical Society. "The network of formal reserves is not enough to protect the species and habitats under pressure. With 80% of South Africa's land in private hands, conservation strategies must involve private land-owners."

If you would like to know more about the Conservancies in you area, the contact persons and representatives on the National steering committee are:

Chairman: Dr A Kruger – Email: at.kruger@conservancies.org
1. Eastern Cape: Tim de Jongh • tbone@eetmind.ecape.gov.za
2. Free State: Duart Hugo • Email: duart@global.co.za
3. Gauteng Conservancy Association: Ivan Parkes ivan.parkes@conservancies.org
4. KwaZulu-Natal Conservancy Association: Jean Lindsay lindsayjd@mweb.co.za
5. Limpopo: Annemie de Klerk deklerka@finptb.norprov.gov.za
6. Mpumalanga: Rob Vollet • lamaisonvol@mweb.co.za
7. Northern Cape: Dr Vicky Ahlmann • ahlmann@yebo.co.za
8. North West: Pierre Weinberg • Tel 053 927 0431 or Cell 083 298 1527
9. Western Cape: Chris Martens • chriscip@maxitec.co.za

ACTIVITIES:

CONSERVATION INTERNATIONAL

Private Bag X7 CLAREMONT 7735 W Cape
Tel: 021 799 8655 • Fax: 021 762 6838
Email: t.mildenhall@conservation.org
Website: www.conservation.org
Contact Name: Tessa Mildenhall – Manager: Communications & Development

CONSERVATION INTERNATIONAL

Conservation International's SA Hotspots Programme is committed to facilitating conservation activities in the Succulent Karoo, Cape Floral Kingdom and Maputo-Pondoland, while developing existing and new local expertise and institutions.

ACTIVITIES:

CONSERVATION MANAGEMENT SERVICES

4 Chestnut Street Heather Park GEORGE 6529 W Cape
Tel: 044 870 8472 • Fax: 044 870 8472
Email: consken@cybertrade.co.za
Contact Name: Ken Coetzee – Owner

Providing innovative planning, design, advice for wildlife/ habitat management; natural resource inventories; habitat rehabilitation, resource utilisation and sensitive eco-tourism development to landowners, authorities and the eco-tourism industry.

ACTIVITIES:

CONSERVATION PARTNERSHIPS

Botanical Society Private Bag X10
CLAREMONT 7735 W Cape
Tel: 021 797 2284 • Fax: 021 761 5983
Email: paisley@nbict.nbi.ac.za
Website: www. Botanicalsociety.org.za/ccu
Contact Name: Wendy Paisley – Administrator

The Botanical Society's Conservation Unit focuses on catalytic, collaborative projects in four interlinked areas: conservation planning, conservation stewardship, biodiversity best practice in business, and law and policy.

ACTIVITIES:

CONSERVATION SUPPORT SERVICES (CSS)

PO Box 504 GRAHAMSTOWN 6140 Port Elizabeth
Tel: 046 622 4526 • Fax: 046 622 4526
Email: mpowell@cssgis.co.za • Website: www.cssgis.co.za
Contact Name: Mike Powell – Member

CSS is a specialist natural environment GIS company. CSS is actively involved in precision DGPS mapping, 3-D and spatial analysis and remote sensing.

ACTIVITIES:

" … The ecological crisis has assumed such proportions as to be everyone's responsibility…"

~ Pope John Paul II

CONTACT TRUST

10th Floor Dumbarton House No.1
Church Street CAPE TOWN 8001
Tel: 021 426 1413 • Fax: 021 426 1446
Email: alison@contacttrust.org.za
Website: www.contacttrust.org.za
Contact Name: Alison Bullen – Information Services Manager

Established in 1997, the Contact Trust focuses on working towards enhancing the capacity and opportunities for business and state partnerships, dialogue and co-operation in the development, monitoring and implementation of national policies and legislation initiatives relating to sustainable development. This project understands that while there is a need for enhanced dialogue and partnerships, the capacity within both sectors is often inadequate.

Contact Trust:
- tracks key policy and legislation developments across a variety of sectors, and emails the latest developments to subscribers
- assists organisations in developing lobbying strategies around particular issues or policy and legislation
- monitors various parliamentary committees, producing reports which are sent out to organisations across South Africa

Contact Trust also provides an Online library of current policy and legislation documents and an Online archive of parliamentary committee minutes. Contact's key services provide you with the information and knowledge you need to keep ahead of developments. Our services are structured around immediate, customised, information distribution.

ACTIVITIES:

COUNCIL FOR GEOSCIENCE

Private Bag X112 PRETORIA 001 Gauteng
Tel: 012 841 1911 • Fax: 012 841 1076
Email: pzawada@geoscience.org.za
Website: www.geoscience.org.za
Contact Name: Dr P K Zawada

The Council for Geoscience provides a variety of earth science related services and expertise, which can greatly benefit the EIA process and land-use planning issues.

ACTIVITIES:

C.R.H. CLANAHAN & ASSOCIATES

PO Box 4271 WHITE RIVER 1240
Mpumalanga
Tel: 013 750 2697/8 Fax: 013 750 2699
Email: canavit@mweb.co.za
Contact Name: Robin Clanahan

Specializes in the management of water resources development and dam safety, with associated expertise in fields of environment, agricultural and rural development, ecological (instream) river flow requirements, socio-economic assessment, community participation and facilitation, rural infrastructure, water resources management, river hydraulics, flood hydrology, and associated disaster management.

ACTIVITIES:

CROWTHER CAMPBELL & ASSOCIATES CC

Unit 35 Roeland Square
30 Drury Lane CAPE TOWN 8001
Tel: 021 461 1118 • Fax: 021 461 1120
Email: crowcamp@iafrica.com
Website: www.crowthercampbell.co.za
Contact Name: Jonathan Crowther

Crowther Campbell and Associates (CCA) is a professional environmental consultancy that was established in 1993. Based in Cape Town, the firm has been involved in a large number of projects in the Western Cape region and throughout South Africa. A wide client base ranges from manufacturing and petroleum exploration industries through to private development proponents and authorities. A variety of environmental services are offered in the fields of infrastructure, rivers and wetlands, waste management, oil and gas industry, environmental education, tourism and resort development. In all project involvement the firm aims to achieve a desirable balance between conservation and development. Studies are guided by the principles of Integrated Environmental Management and, where applicable, are conducted in accordance with the relevant environmental legislation. An equal opportunity employment policy has ensured the employment of a highly professional team.

ACTIVITIES:

CRYSTAL CLEAR CONSULTING & MERCHANTS (PTY) LTD

PO Box 1982 BRAMLEY 2018 Gauteng
Tel: 011 882 3368 • Fax: 011 882 3395
Email: info@crystalclear.co.za • Website: www.crystalclear.co.za
Contact: Milton Buchalter – Director

Crystal Clear has been providing Environmental Training and ISO 14001 Software to the South African market for the past 9 years.

Environmental Training
- Executive Briefing
- International IEMA Approved Environmental Auditor Courses
- Introduction to ISO 14001 for Supervisors
- Internal ISO 14001 Auditor Course
- ISO 14001 Software

The 'isotop' Integrator Software, is a complete ISO 14001 documentation package.

This package ensures efficient implementation of ISO 14001, it also provides easy integration of your company's Environmental, Quality, Health & Safety documentation into one software programme. For more information see www.isotop.net

The ISO 14001 Awareness Training Software is an interactive training CD with sound. This easy-to-use Training and Induction Programme, explains what the environment is, how pollution can occur and your staff's role in ensuring pollution prevention and legal compliance.

ACTIVITIES:

CSIR WATER, ENVIRONMENT & FORESTRY TECHNOLOGY (ENVIRONMENTEK)

PO Box 395 PRETORIA 0001
Tel: 012 841 3680 • Fax: 012 841 2597
Email: pmanders@csir.co.za
Website: www.csir.co.za/environmentek
Contact Name: Dr Pat Manders – Business Development

CSIR Environmentek provides technologies for environmental assessment and management, terrestrial resources, forestry and forest products, water resource management and coastal development.

ACTIVITIES:

DATA DYNAMICS

P O Box 940 WESTVILLE 3630 KwaZulu-Natal
Tel: 031 262 8240 • Fax: 031 262 8158
Tel: 011 880 5860 • Fax: 011 8806496
Email: ddyn@ddyn.com Website: www.ddyn.com
Contact Name: Auryn Phoenix – Account Facilitator

Data Dynamics Electronic Publishers of S.A. Legislation has started our new venture into producing Smart Books. We have decided that we will produce products that will make a difference. Our first offering is GEM – a Guide to Environmental Management systems.

GEM is an electronic route planner to understanding, developing and implementing an Environmental Management System. It is a map to the internationally accepted environmental management system standard, ISO 14001.

Whether you seek ISO 14001 certification or just want to understand and manage your environmental risks, GEM will lead the way through key features and benefits that include:

◆ The knowledge and tools to ensure complete environmental management system implementation
◆ An electronic version of the ISO 14001 Standard.
◆ An environmental legal reference library for a trial period
◆ An aspect and impact register developer
◆ Tools for risk assessment
◆ An ISO 14001 self-assessment score-card
◆ A document manager to facilitate ongoing management of system documentation

◆ An EMS manual developer tool
◆ Template system procedures for customisation to the needs of your business

Data Dynamics Electronic Publishers and Metago Environmental Engineers have collaborated to produce a CD that contains the information and tools that will ensure that your journey into a complex subject is an easy one. With many years of experience in their respective industries, this team approach offers intelligent support for the product. The up-market product priced specifically for SME's and BBE's.

To find out more about GEM, simply visit the GEM web site at www.ddyn.com/GEM or call one of our sales teams to begin your GEM journey.

ACTIVITIES:

DAVIES LYNN & PARTNERS (PTY) LTD

PO Box 586 KLOOF 3640
KwaZulu-Natal
Tel: 031 764 7335 • Fax: 031 764 7385
Email: dlpdbn@dlp.co.za
Contact Name: Chris Ross – Director

DLP offer consultancy services in Groundwater, Water and Soil Engineering, GIS, and specialist EIA studies, and are committed to implementing appropriate, effective and sustainable projects.

ACTIVITIES:

DE BEERS

DE BEERS
A DIAMOND IS FOREVER

PO Box 350 CAPE TOWN 8000 W Cape
Tel: 021 409 7222 • Fax: 021 419 0022
Email: patti.wickens@debeersgroup.com
Website: www.debeersgroup.com
Contact: Patti Wickens – Environmental Manager

For more than a century, the name "De Beers" has been synonymous with diamonds. The company leads the world in diamond exploration, mining, recovery, sorting, valuation and marketing. De Beers produces about 45% by value of the total annual global diamond production from its mines in South Africa, and through its 50:50 partnerships with governments of Botswana and Namibia. De Beers' gem mining operations span every category of diamond mining – open pit, underground, alluvial, coastal and under sea – while its exploration programme extends across six continents. De Beers also manages a number of wildlife conservation areas, many of which are associated with the various operations. Through its selling arm, the Diamond Trading Company based in London, De Beers markets some two thirds of global supply, and has conducted a renowned diamond advertising and promotion campaign for over half a century under the banner of "A Diamond is Forever."

ACTIVITIES:

DE BEERS MARINE

PO Box 87 CAPE TOWN 8000 W Cape
Tel: 021 410 4259 • Fax: 021 410 4257
Email: lesley.roos@debeersgroup.com
Contact Name: Lesley Roos – Environmental Management
Co-ordinator

De Beers Marine (Pty) Ltd is a contracting company providing global marine mineral resource solutions through the provision of innovative products and services. We offer best practice capabilities in high-resolution geophysical surveys and interpretation, seabed sampling and resource estimation, as well as micropalaeontological research.

De Beers Marine embraces the concept of sustainable development and aims to strike a balance between its economic, social and environmental responsibilities. Commitment to the highest standards of environmental management extends to all aspects of the company's operations, which are managed through an ISO 14001 certified Environmental Management System.

ACTIVITIES:

DE TWEEDESPRUIT CONSERVANCY

PO Box 117
CULLINAN 1000 Mpumalanga
Tel: 012 732 0093 • Fax: 012 732 0093
Email: akruger@csir.co.za
Contact Name: Dr A Kruger – Chairman

A rural area (11 057 ha) in north-east Gauteng where the landowners voluntarily established a conservancy with the aim of jointly looking after the environment within and beyond their own farm boundaries.

ACTIVITIES:

DE WILDT CHEETAH AND WILDLIFE TRUST

PO Box 1756
HARTBEESPOORT 0216 Mpumalanga
Tel: 012 504 1921 • Fax: 012 504 1556
Email: cheetah@dewildt.org.za
Website: www.dewildt.org.za
Contact Name: Vanessa Bouwer – Assistant Director

The Centre's mission is to ensure the long-term survival of predators, specifically the cheetah and wild dog in their natural environment. This is achieved through breeding, release, education, research into wildlife disease and a conservation programme for free roaming cheetah and wild dog.

An extensive educational programme has been developed, which includes an informative walking trail for children and an accessible sensory trail for disabled visitors.

The Centre now boasts a luxury lodge where visitors and donors can enjoy unique cheetah experiences.

The De Wildt Cheetah and Wildlife Trust has spearheaded an initiative to protect free roaming cheetah in South Africa. Field officers and researchers work in the areas that cheetah occur to lessen conflict between landowners and the cheetah.

ACTIVITIES:

DEKRA–ITS CERTIFICATION SERVICE (PTY) LTD

PO Box 523 HALFWAY HOUSE 1685
Gauteng
Tel: 011 315 1607 • Fax: 011 315 7649
Email: dekraits@metroweb.co.za
Website: www.dekra-its.com
Contact Name: Tamsin Jolly – Deputy Managing Director

DEKRA-ITS Certification Services is a division of the International DEKRA Group of companies. Through joint agreements with Intertek Testing Services (USA) and Semko AB (Sweden), certification and product approvals are offered worldwide.

Established in 1994 the local subsidiary has firmly entrenched itself in southern Africa and the Indian Ocean Islands, providing the highest quality services to both industry and commerce. Our customers range from small private enterprises to government bodies and large national and international corporations.

We offer:

◆ Government accredited and internationally recognised certification of QM systems to ISO 9001:2000, QS 9000, EN 46000, EM systems to ISO 14001, Occupational Health and Safety Management Systems to OHSAS 18001 and HACCP.
◆ Training courses
◆ Product approval

In the belief that "our customer's success is our success and their failure our failure," we offer a personalised service which provides the highest value for money.

ACTIVITIES:

DELTA ENVIRONMENTAL CENTRE

Private Bag X6
PARKVIEW 2122 Gauteng
Tel: 011 888 4831 • Fax: 011 888 4106
Email: delta.enviro@pixie.co.za
Website: www.deltaenviro.org.za
Contact Name: Mrs D.V. Beeton – Executive Officer

Delta Environmental Centre is an environmental education and training organisation that also offers an environmental consultancy service. The main focus is educators, learners and community-based structures.

ACTIVITIES:

DENDROLOGICAL SOCIETY AND FOUNDATION

PO Box 15277 SINOVILLE 0129 Gauteng
Tel: 012 567 4009 • Fax: 012 567 0008
Email: dendron@iafrica.com
Contact Name: Jutta von Breitenbach

The Dendrological Foundation publishes the *National List of Indigenous Trees* and the *Tree Atlas*. It provides tree labels. It offers a Correspondence Tree Knowledge Course.

ACTIVITIES:

DENEYS REITZ INCORPORATED

DENEYS | REITZ
ATTORNEYS

Deneys Reitz Inc.
1984/003385/21

P O Box 784903
SANDTON 2146 Gauteng
Tel: (011) 685-8500 / (031) 367-8800 / (021) 418-6900
Fax: (011) 883-4000 / (031) 301-3346 / (021) 418-6900
Website: www.deneysreitz.co.za
E-mail: Tim de Wet (Jhb) - tjdw@deneysreitz.co.za
Penny Lee (Jhb) - pl@deneysreitz.co.za
Tina Costas (Durban) - tc@deneysreitz.co.za
Contact Name: Penny Lee – Professional Assistant

As practising attorneys, the members of the Deneys Reitz environmental law department will give you constructive advice and assistance on environmental protection issues, including the law and legislation, due diligence investigations, drafting and reviewing contracts, environmental and regulatory requirements, policies, liabilities, certification, conservation, eco-tourism, environmental management and planning, liaison with government bodies and marine, water and waste management issues. We also have an expert mining law department dealing with mining-related environmental issues.

ACTIVITIES:

DEPARTMENT OF ENVIRONMENTAL & GEOGRAPHICAL SCIENCE – UNIVERSITY OF CAPE TOWN

Private Bag RONDEBOSCH 7701 W Cape
Tel: 021 650 2873 • Fax: 021 650 3791
Email: sadams@enviro.uct.ac.za
Website: www.egs.uct.ac.za
Contact Name: Mrs S Adams – Administrative Assistant

Environmental and Geographical Science at the University of Cape Town is characterised by an integrated approach to the study of human-environment relations. Academic and associated research staff have an international reputation for teaching and research in the fields of Environmental Management, Physical and Human Geography, Atmospheric Science, Climate Science and Disaster Risk Science. The department offers courses that contribute to a range of undergraduate programmes spanning several Faculties. We also offer a range of postgraduate studies leading to Honours, Masters and Doctoral degrees in Environmental and Geographical Science and the Mphil programme in Environmental Management. The Environmental Evaluation Unit and the Disaster Mitigation for Sustainable Livelihoods

Project (DIMP) are integral to departmental activities. Staff of these units consult on a wide range of practical environmental issues and problems, and also contribute to the teaching and research activities of the academic department.

ACTIVITIES:

DEPARTMENT OF NATURE CONSERVATION POLYTECHNIC OF NAMIBIA

Private Bag 13388, WINDHOEK Namibia
Tel: +064 61 207 2141 • Fax: +064 61 207 2143
Email: mdeklerk@polytechnic.edu.na
Website: www.polytechnic.edu.na/natural
Contact Name: Mrs M de Klerk – Head of Department

The Department offers certificate, diploma and BTech degree programmes in Nature Conservation. Graduate students are employed at various Government Ministries, NGOs, game ranches, tour guide and safari organisations throughout Namibia.

ACTIVITIES:

DESERT RESEARCH FOUNDATION OF NAMIBIA (DRFN)

DRFN

PO Box 20232 WINDHOEK Namibia
Tel: 09264 61 229855 • Fax: 09264 61 230172
Email: johnp@drfn.org.na / andreb@drfn.org.na
Website: www.drfn.org.na
Contact Names:
John Pallett – Scientific writer, Communications Unit /
Andre Botes – Manager, Support Services

"The DRFN is dedicated to furthering understanding and competence to appropriately manage arid environments for sustainable development." The DRFN is involved with numerous short- and medium-term projects that contribute to this mission. All projects have elements of training, research and awareness creation, and contribute to sharpening our focus on application of environ-mental understanding to development in Namibia. Although we are mainly active in Namibia, our focus has broadened to encompass the SADC region.

The DRFN offers:

◆ Facilitation services for community-based environmental projects
◆ Training programmes towards sustainable development
◆ Environmental writing, publication design and editing service
◆ Environmental assessment consulting service
◆ Expertise in environmental policy analysis
◆ Opportunities for research
◆ Opportunities to work in partnership with other groups aiming towards sustainable development

ACTIVITIES:

DEVELOPMENT ACTION GROUP (DAG)

101 Lower Main Road OBSERVATORY 7925 W Cape
Tel: 021 448 7886 • Fax: 021 447 1987
Email: dag@dag.org.za
Contact Name: Anthea Houston – Director

DAG is an NGO in the Western Cape that works in partnership with communities to plan and implement housing and other development projects. DAG promotes energy efficiency in house design and household consumption.

ACTIVITIES:

DEVELOPMENT BANK OF SOUTHERN AFRICA (DBSA)

PO Box 1234 HALFWAY HOUSE 1685 Gauteng
Tel: 011 313 3911
Email: letsholon@dbsa.org
Website: www.dbsa.org
Contact Name: Nonnie Letsholo – Manager:
Corporate Communications and Marketing

The DBSA strives to be a leading change agent for accelerated and equitable socio-economic development in southern Africa.

It is committed to:
 ◆ Sensitivity to the need of the poor and responsiveness to the demand of clients
 ◆ Alignment of operations to national policies, regional programmes and local priorities
 ◆ A network of collaboration and partnership with the public and private sectors institutions
 ◆ Implementation of a best practice policy in all activities, including social and environmental practices
 ◆ Adherence to principles and practices of good government

The DBSA cares for people and the environment and takes seriously issues that relate to environmental, social and economic justice and human rights. The DBSA is concerned with the sustainability performance of policies, programmes, projects and organisations. It is also deeply committed to building a just and healthy society – one that works in harmony with nature and upholds the integrity of ecosystems. The DBSA provides services for sustainable development activities as a lender, development partner and as an advisor, with a focus on the provision of infrastructure.

ACTIVITIES:

DEVELOPMENT CONSULTANTS OF SA (PTY) LTD

PO Box 782278 SANDTON 2146 Gauteng
Tel: 011 807 4229 • Fax: 011 807 4230
Email: info@dcsa.co.za • Website: www.dcsa.co.za
Contact Name: Kosta Babich – Managing Director

Development Consultants of SA (Pty) Ltd – promoting development and investment in industry, mining, tourism and agriculture. Facilitates company finance, take-overs, rationalisation as well as environmental issues such as ISO 9000, ISO 14000 and Environmental Impact Assessment. The oldest established consultancy specialising in Government Incentive schemes. A proud reputation with foundations built on strong corporate governance.

ACTIVITIES:

DIAL ENVIRONMENTAL SERVICES CC

PO Box 6377 WESTGATE 1734 Gauteng
Tel: 011 958 0136 • Fax: 011 958 0136
Email: dionmesh@iafrica.com
Contact Name: Dion Marais

DiAL Environmental Services is an Approved Inspection Authority providing specialist consultancy in:
 ◆ Design of customised occupational hygiene systems and procedures
 ◆ Legal compliance audits
 ◆ Occupational exposure hazard risk assessments
 ◆ Ventilation, noise, illumination, airborne pollutant sampling, lean and asbestos exposure assessments (accredited asbestos monitoring facility), vibration, ergonomic and water quality surveys and isokinetic stack sampling
 ◆ Environmental impact assessments for noise, dusts, fibres, gases and associated contaminants
 ◆ Rehabilitation of asbestos contaminated sites
 ◆ Design, installation and commissioning of industrial and mine ventilation and air filtration systems

Training in occupational hygiene (both for practitioners and for members of health and safety committees) and asbestos monitoring.

ACTIVITIES:

DIETER HOLM

PO Box 58 HARTBEESPOORT 0216 Pretoria
Tel: 012 371 9584 • Fax: 012 371 9584
Email: dieterholm@worldonline.co.za
Contact Name: Prof. Dieter Holm – Director

Sustainable Development in the Built Environment: Design and assessment of Energy Efficiency and Renewable Energy in Buildings, mainly in subtropical areas. Local Integrated Resource Planning. Strategic Planning.

ACTIVITIES:

DIGBY WELLS & ASSOCIATES

Private Bag X10046
RANDBURG 2125 Gauteng
Tel: 011 789 9495 • Fax: 011 789 9498
Email: info@digbywells.co.za
Contact Name: Mr Ken van Rooyen

Digby Wells & Associates was founded in 1995 from the old Rand Mines Environmental Department as a result of a perceived gap in the marketplace relating to the provision of environmental services to industry. An increased awareness of the need to protect our environment by all stakeholders in a global village has further highlighted the need to provide these specialist services. The company aims to provide practical and cutting edge environmental solutions, while shaping a desired future. Accordingly, we implement an integrated approach toward Environmental Management, which includes the compilation of Environmental Impact Assessments, Environmental Management Programme Reports, Environmental Audits, Feasibility Studies, Liability Assessments and Environmental Reviews.

The company prides itself on core values, which comprise a sense of purpose beyond just making money. These relate to integrity, openness, friendliness, and having fun at being the best. Every day our team seeks to harness a fluid exchange of knowledge, ideas, resources and technologies, aimed at improving the quality of solutions, reputation and culture of the company. Most importantly, we strive to be an integral part of the client's team and adding value to their core ideology and daily operations. Although we are all passionately committed to fulfilling our work responsibilities, time is always invested in those that are around us and who form a critical part of the milieu of Digby Wells.

Our clients in return provide value to our everyday tasks, yet provide warmth, and the formation of close friendships often occur. We firmly believe that this approach enables us to contribute to the advance of economic growth and creation of social responsibility in addition to improving our environment, all of which leads to the desired goal of sustainable development.

ACTIVITIES:

DITTKE (MARK) ATTORNEY AT LAW

61 Peninsula Road
ZEEKOEVLEI 7945 W Cape
Tel: 021 705 2943 • Fax: 021 706 6622
Email: mdittke@xsinet.co.za
Website: www.attorneys.co.za/dittke
Contact Name: Mark Dittke

powered by nature

Our firm practices almost exclusively in the field of Environmental as well as Health and Safety legislation. We regularly advise both large and medium sized companies and corporations on these two issues. In addition, we were involved in several by-law drafting and review projects for local authorities. We have wide experience in conducting Legal Audits, Legal Compliance Assessments, Due Diligence matters and Legal Reviews. We also compile and maintain Electronic Legal Registers for clients, especially for ISO 14001 purposes. Moreover, our firm offers an Electronic Legal Updating Service, the EnviroLaw Updater, covering Safety, Health and Environmental legislation,

coupled to a newsletter informing subscribers of both actual and pending changes to the above fields. We also have a very good working relationship with consulting firms, local government departments, and certification authorities. Since we are fluent in German we also assist German companies.

ACTIVITIES:

DIVWATT (PTY) LTD

313 Boundary Road Northriding
RANDBURG 2190 Gauteng
Tel: 011 794 3825 • Fax: 011 794 1337
Email: jpdev@divwatt.com • Website: www.divwatt.com
Contact Name: Jean-Paul de Villiers – General Manager

DIVWATT

Divwatt (Pty) Ltd was established in February 1993 with the express purpose of developing and manufacturing high quality solar specific equipment.

Our first undertaking was the SOLASTAR project, a research and development programme into a high efficiency submersible pumping system. After two short years of research and development the fruition of the SOLASTAR project was realised, and in 1996 the SOLASTAR solar powered submersible pump received the "Product of the Year" award from the Institute of Engineers. Many of our other innovative designs and products have also received various accolades; the WOZAMANZI hand pump received the same "Product of the Year" award from the Institute of Engineers a year later. Other completed research and development programmes include the TERRA and POWERSTAR, both products are set to revolutionise the inverter and peak power tracking solar charge controller markets respectively.

We've had numerous of our products tested and approved by independent authorative organisations in various countries. This includes the Solar Research Division of the University of Reading, U.K., the University of New Mexico, U.S.A., and the University of New South Wales, Australia. We are currently working with the University of Shanghai, China, and the University of Port Elizabeth, South Africa.

Our products have also been accepted as the international standard. Our digital pump controller the SOLAR POWER OPTIMISER (SPO) has been, and is currently exported to numerous countries around the world, including the U.S.A., India, China and Australia.

Due to the fact that our expertise lies in the areas of research, development, and manufacturing, we seek to join forces with reputable companies in various parts of the world, capable of handling the installation and distribution of our technically superior products. Through our partners we are able make our products and expertise available for the benefit of all concerned.

ACTIVITIES:

DOLPHIN ACTION PROTECTION GROUP (THE)

PO Box 22227
FISH HOEK 7975 W Cape
Tel: 021 782 5845
Fax: C/O Zip Print 021 782 6223
Email: mwdapg@mweb.co.za
Contact Name: Nan Rice – Secretary

Originally founded in 1977 with the express aim of campaigning for the protection and conservation of Dolphins and Whales (Cetaceans), DAPG has since then broadened its role and activities and has run many successful national educational and fundraising campaigns including: SAVE THE WHALES; SAVE ANTARTICA; DOLPHINS SHOULD BE FREE and SAVE OUR SEA LIFE – PREVENT PLASTIC POLLUTION. DAPG has worked effectively and has been a catalyst in introducing protective legislation in South Africa for Cetaceans and the Great White Shark. Education is a priority under the DAPG Constitution and thousands of educational pamphlets are distributed each year through schools and libraries and also to fishermen and merchant vessels to prevent dumping of plastics at sea. DAPG has also become involved in the issue of high seas pelagic drift-netting and played an important role in phasing out this destructive and rapacious fishing method in the southern Indian and Atlantic oceans.

The success and effectiveness of DAPG is significantly due to the tireless work of a small number of dedicated people and the wise and efficient use of limited resources, which come mainly from donations and bequests. DAPG has trained a team of volunteers who can help at mass strandings when they occur.

ACTIVITIES:

DOUG JEFFERY ENVIRONMENTAL CONSULTANTS (PTY) LTD

PO Box 44 KLAPMUTS 7625 W Cape
Tel: 021 875 5272 • Fax: 021 875 5272
Email: dougjeff@iafrica.com
Contact Name: Doug Jeffery – Director

Doug Jeffery Environmental Consultants (DJEC) is a professional, committed, independent environmental consultancy firm, specialising in Environmental Impact Assessments (EIA), Environmental Management Programmes (EMP) and Baseline Studies. Operating since 1987, DJEC have gained extensive understanding as EIA practitioners on a variety of projects, such as housing developments (low and high density), resort developments, wine cellars, pipeline and road alignments, service stations and river/wetland rehabilitation programmes. We consult with a variety of role players ranging from private individuals, NGOs and governmental bodies. We believe in a fully inclusive, hands-on approach to all our work.

ACTIVITIES:

DRIFTERS ADVENTOURS CC

PO Box 48434
ROOSEVELT PARK 2129 Gauteng
Tel: 011 888 1160 • Fax: 011 888 1020
Email: drifters@drifters.co.za • Website: www.drifters.co.za
Contact Name: Andy Dott – Managing Director

DRIFTERS – The leading Safari eco-tour operator in southern Africa with guaranteed tours of 5 to 30 days throughout South Africa, Botswana, Zimbabwe, Namibia, Mozambique, Zambia, Malawi and Tanzania.

ACTIVITIES:

DUARD BARNARD & ASSOCIATES

PO Box 35453
MENLO PARK 0102 Gauteng
Tel: 012 460 8982 • Fax: 012 460 2989
Email: duard@envirolaw.co.za
Website: www.envirolaw.co.za

Contact Name: Duard Barnard

- ◆ Project management of the environmental component in the planning, commissioning and operation stages of projects
- ◆ Acquiring environmental authorisations and ensuring environmental legal compliance
- ◆ Environmental quality control in the planning, commissioning and operation stages of projects
- ◆ Integration of environmental considerations into business management structures including strategic environmental management plans
- ◆ Analyses of environmental impacts and the establishment of detailed management structures that accommodate the impacts within the prescribed legal parameters
- ◆ Environmental law consultancy services

ACTIVITIES:

DURBAN PARKS DEPARTMENT (ETHEKWINI MUNICIPALITY)

P O Box 3740 DURBAN 4000
KwaZulu-Natal
Tel: 031 201 1303 • Fax: 031 201 7382
Email: sandig@prcsu.durban.gov.za
Contact Name: Garth Kloppenborg

The Department's objectives include the administration, development and maintenance of:

- ◆ All parks and open spaces covering over 4 500 hectares, over 500 children's playgrounds and 180 recreation playing fields
- ◆ Over 60 prime parks of horticultural and floral attraction
- ◆ Freeway and main arterial road landscapes and tree plantings

- Six major natural system parks and several smaller nature reserves, also education resource and visitors centres and an indigenous and medicinal plant nursery
- The Durban Metropolitan Open Space System (D'MOSS) which is a planned integrated system to ecologically link and incorporate over 1 416 hectares of openspace from nine parks utilising existing parkland, open space river valleys and coastal land
- The 14,5 hectare Durban Botanic Gardens including guided tours and horticultural education programmes
- The production of over 5 million trees, shrubs and groundcovers annually
- The cutting of overgrowth of over 4 million square metres of verge in the City

ACTIVITIES:

EAGLE BULLETIN

Private Bag X1017
HILLCREST 3650 KwaZulu-Natal
Tel: 031 767 0244 • Fax: 031 767 0295
Email: arend@eagleenv.co.za • Website: www.eagleenv.co.za
Contact Name: Arend Hoogervorst – Editor and Publisher

Eagle Bulletin provides concise focused environmental information for the busy professional. Arend Hoogervorst has edited and published the authoritative Eagle Bulletin since inception in 1990 and is also the Founding Author of *The Enviropaedia*.

ACTIVITIES:

EAGLE ENCOUNTERS

Spier Estate Lynedoch Road
Stellenbosch
PO Box 152
LYNEDOCH 7603 W Cape
Tel/Fax: 021 842 3684
Email: eagles@telkomsa.net
Contact Names: Hank and Tracy Chalmers

Eagle Encounters is dedicated to the conservation and preservation of birds of prey. It has been designed as a multi-faceted organisation, equipped to deal with a range of activities including: the rehabilitation of young and injured raptors; conservation projects; breeding projects; research; educational outreach programmes to schools and farming communities; on-site interactive educational talks and interactive falconry demonstrations.

ACTIVITIES:

EAGLE ENVIRONMENTAL

Private Bag X1017 HILLCREST 3650
KwaZulu-Natal
Tel: 031 767 0244 • Fax: 031 767 0295
Email: arend@eagleenv.co.za
Website: www.eagleenv.co.za
Contact Name: Arend Hoogervorst – Director

Eagle Environmental is an environmental consultancy, which provides strategic environmental support, advice and information to commerce, industry, the academic sector and consultants. This is done in the form of advice and assistance with Environmental Impact Assessments and environmental audits, a wide range of formal and informal customised environmental training, and documentation, manuals, books, journals and national and international environmental support material accumulated from over 20 years of environmental practice and field experience. Eagle Environmental's Director, Arend Hoogervorst, has extensive experience in environmental matters from working in central and local government, commerce and industry and in private practice. He has assisted many companies and organisations in conceptualising their environmental issues and concerns and then helped them, practically, to deal with the issues or plan and develop strategies and action plans to ensure that environmental management becomes a financial saving – not a financial loss.

ACTIVITIES:

EARTHLIFE AFRICA

PO Box 11383 JOHANNESBURG Gauteng 2000
Website: www.earthlife-ct.org.za

Earthlife Africa is an association of voluntary activists working for environmental justice, founded in 1988. Some key current campaigns are: The halting of proposed additional nuclear reactors in South Africa, with the associated fuel plants and transport of radioactive material; the halting of a proposed radioactive waste smelter at Pelindaba near the World Heritage Site; genetic engineering; anti-incineration campaign; promoting sustainable energy and response to climate change; toxic waste; the promotion of Zero Waste; water and dams issues; promotion of food and energy security for all.

For information about environmentally-friendly products, visit the Green House Website and look in the Glad Files.

BRANCHES:

Johannesburg:	Patrick Johnson	
	084 424 1234	patrick@earthlife.org.za
Durban:	Bryan Ashe	
	031 201 1119	bryan@earthlife.org.za
Namibia:	Bertchen Kohrs	
	+264 61 227 913	earthl@iway.na
Cape Town:	Richard Weeden	
	021 883 3366	green@streetkids.org.za

ACTIVITIES:

EARTHYEAR

PO Box 91667
AUCKLAND PARK 2006 Gauteng
Tel: 011 727 7000 • Fax: 011 727 7111
Email: earthyear@mg.co.za • Website: www.earthyear.co.za
Contact Name: Fiona Macleod – Editor

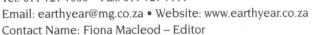

Earthyear is a glossy magazine that endeavours to ensure that sustainable development and enlightened environmentalism are essential components of everyone's lives. It presents problems and constructive solutions in an accessible, entertaining and credible format so that everyone, from school pupils in informal settlements to government ministers and corporate magnates, can understand the issues and be inspired to bring about a shift in their thoughts and actions. It provides inspiration and practical guidelines in a format that is attractive without being decorative, that is up-market but affordable. Launched as a one-off publication in 1990, it has grown over the years and in 2004 is being published as a bi-monthly. Earthyear is distributed to subscribers and is available in CNAs, Exclusive Books and selected bookstores.

ACTIVITIES:

ECO–ACCESS

PO Box 1377
ROOSEVELT PARK 2129 Gauteng
Tel: 011 477 3676 • Fax: 011 477 3675
Email: info@eco-access.org • Website: www.eco-access.org
Contact Names: Rob Filmer – Director
Julie Filmer – Deputy Director

Eco-Access celebrates diversity in humanity and nature through twinning disabled and non-disabled children at Diversity Challenges, Camps, Outings and Eco-Schools. Participants learn respect for themselves, others and the environment through twinned nature-based activities.

ACTIVITIES:

ECOAFRICA TRAVEL (PTY) LTD

PO Box 6281 UNIEDAL 7612 W Cape
Tel: 021 809 2180 • Fax: 021 809 2187
Email: katharinavg@ecoafrica.com
Website: www.ecoafrica.com
Contact Name: Katharina von Gerhardt – Content Manager

EcoAfrica Travel (Pty) Ltd is the inbound tour operator behind www.ecoafrica.com, which pioneered web promotion of ecotourism to Africa back in 1995. Since then, we have arranged over 9 000 tours to the African bush for over 20 000 delighted travellers. We focus on true ecotourism, defined by the International Ecotourism Society as: responsible travel to natural areas that conserves the environment and sustains the well-being of local people.

In this light, we promote destinations and tours that offer the traveller an authentic experience of Africa's wilderness, augment resources for environmental conservation, contribute to sustainable community development and host no more than 30 people at a time.

Conservation and development projects are monitored by environmental management consultants ERM, whose Managing Director is also EcoAfrica Travel's Director for Environmental Ethics. EcoAfrica Travel does not promote the luring, taming or killing of wildlife for entertainment.

We have recently launched www.ecoBotswana.com and www.ecoCape.com

ACTIVITIES:

ECOBOUND ENVIRONMENTAL & TOURISM AGENCY

PO Box 10274 GEORGE 6530 W Cape
Tel: 044 871 4455 • Fax: 044 871 2274
Email: ecobound@pixie.co.za
Website: www.george.co.za/ecobound
Contact Name: Wikus van der Walt – Director

EcoBound Environmental and Tourism Agency was established in 1996 and is a leading environmental consultancy in the Southern-Cape Garden Route area. We specialise in the fields of eco-tourism, environmental impact assessments (EIA), integrated environmental management (IEM), environmental management plans, and associated services.

ACTIVITIES:

ECO-CARE TRUST

PO Box 54131 NINAPARK 0156 Gauteng
Tel: 012 546 1423 • Fax: 012 546 1392
Email: chamara@netdial.co.za
Contact Name: Chamara Pansegrouw – Marketing Manager

ACTIVITIES:

ECOCITY TRUST

Postnet Suite 117 Private Bag X65 HALFWAY HOUSE 1685
Tel: 011 407 6726 • Fax: 011 403 7904
Email: annie@ecocity.org.za • Website: www.ecocity.org.za
Contact Name: Annie Sugrue – Managing Director

The Trust has been created to alleviate poverty, create environmental awareness, sustain the utilisation of natural resources in an equitable manner and secure the ecology for future generations.

ACTIVITIES: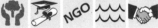

ECO-FONE

PO Box 30440 KYALAMI 1684 Gauteng
Tel: 011 691 6500 • Fax: 011 466 3574
Email: shelley@gg-assist.com
Contact Name: Shelley Whitfield – Executive Assistant

Specialists in the refurbishment and correct disposal of cellular phone accessories and related components – maintaining the balance between mobile telecommunications and the environment.

ACTIVITIES:

ECOLINK

PO Box 727 WHITE RIVER 1240 Mpumalanga
Tel: 013 751 2120/ 750 1067 • Fax: 013 751 3287
Email: eco.link@mweb.co.za
Website: www.lowveld.com\ecolink
Contact Name: Liezl Geldenhuys – Administrative Assistant

Ecolink is a training and development NGO, founded in 1985 by Dr Sue Hart, Executive Director. Ecolink's mission is to serve the disadvantaged people of South Africa by empowering individuals and community groups to achieve development and self-sustainability by providing appropriate education and training.

ACTIVITIES:

ECOREX

PO Box 1848 WHITE RIVER 1240 Mpumalanga
Tel: 013 7513491 • Fax: 013 7513491
Email: ecorex@iafrica.com
Contact Name: Mr Graham Deall

ECOREX provides an environmental consulting and advisory service to forestry, mining, agriculture and industry. Our major focus is on natural resource surveys and environmental impact assessments (EIAs).

ACTIVITIES:

ECOSENSE

PO Box 12697 DIE BOORD 7613 W Cape
Tel: 021 886 4056 • Fax: 021 887 2654
Email: info@ecosense.co.za
Website: www.econsense.co.za
Contact Name: Mark Sasman – Director

Professional services in Environmental Impact Assessment, Scoping Applications, Construction and Operational Environmental Management Plans, Conservation Planning, Environmental Site Management, Audit and Audit checklist by a very experienced team.

ACTIVITIES:

ECOSYSTEM PROJECTS T/A EKO WILD

PO Box 235 THABAZIMBI 0380
Mpumalanga
Tel: 014 777 1814 • Fax: 014 772 3601
Email: ekowild@global.co.za • Website: www.ekowild.co.za
Contact Name: Dr Wilhelm Schack – Director

The company EKO WILD was established in 1991 under the name Ecosystem Projects cc.

Our core activities revolve around the following three categories:

1. Wildlife Veterinary Services

Game capture and translocation (esp. rare and difficult to handle species like rhino, buffalo, sable, roan, giraffe).
Also: Rare species breeding projects, genetics, nutrition, disease control, risk analysis.

2. Game Reserve Management Services

Ecological analysis: vegetation, soil, water, carrying capacity, species composition, stocking rates, harvesting quotas and methods.
Management planning, infrastructure, roads, fire control on game farms.

3. Wildlife Marketing

Sourcing of healthy and superior breeding stock.
Risk analysis and health inspections before purchases and translocation.

Our offices and quarantine station are located in Thabazimbi in the Limpopo Province of South Africa. The game capture unit operates country wide under direct control of a wildlife veterinarian.

ACTIVITIES:

ECOTRAINING

PO Box 19113
NELSPRUIT 1200 Mpumalanga
Tel: 013 744 9639 • Fax: 013 744 0953
Email: ecotrain@mweb.co.za
Website: www.ecotraining.co.za
Contact Name: Jenny Avery

ECOTRAINING is a company that specialises in the training of young people who wish to follow a career in eco-tourism with a particular emphasis on guiding.

ACTIVITIES:

EDUCO AFRICA

7 Dalegarth Road
PLUMSTEAD 7800 W Cape
Tel: 021 761 8939 • Fax: 021 797 5292
Email: contact@educo.org.za • Website: www.educo.org.za
Contact Name: Ms M Goodman – Executive Director

Educo Africa develops leadership and personal mastery through outdoor- and wilderness-based experiential education. Programmes work towards social change, valuing of diversity, and respect for the natural environment.

Specialised focus on youth at risk or in need.

ACTIVITIES:

EDWARD NATHAN & FRIEDLAND (PTY) LTD

EDWARD NATHAN & FRIEDLAND
CORPORATE LAW ADVISERS & CONSULTANTS

PO Box 783347
SANDTON 2146 Gauteng
Tel: 011 269 7600 • Fax: 011 269 7899
Email: gfj@enf.co.za / ros@enf.co.za • Website: www.enf.co.za
Contact Names: Francois Joubert / Robyn Stein – Directors

Environmental Law and Biotechnology are growth areas in the legal arena. Edward Nathan & Friedland has seen the potential and has established a dynamic and innovative Environmental Law and Biotechnology Department (ENF Envirolaw).

The department has experience in a wide range of matters including environmental, water, mining and biotechnology regulatory issues, public-private partnerships, sustainable development issues, due diligences, reviewing contracts, drafting policy and legislation and corporate and social responsibility reporting.

ACTIVITIES:

EKOGAIA FOUNDATION

PO Box 222 NOORDHOEK 7979 W Cape
Tel: 021 789 1751 • Email: ekogaia@iafrica.com
Contact Name: Glenn Ashton

- Investigation of environmental management and holistic systems planning for development of environmentally sustainable projects of any size
- Research, networking and compilation of reports related to environmental planning
- Investigative research and writing of articles and reports

Specialising in wetlands, fisheries, agriculture, land use planning, genetically modified organisms (GMO), Gaian systems theory and practice and many other issues; please enquire for further details in southern Africa.

ACTIVITIES:

ELEPHANT MANAGEMENT AND OWNERS' ASSOCIATION (EMOA)

PO Box 98 VAALWATER 0530 Mpumalanga
Tel: 014 755 4455 • Fax: 014 755 4455
Email: mgarai@pop.co.za • Website: www.emoa.org.za
Contact Name: Dr Marion Garai – Chairperson

EMOA incorporates elephant owners and managers, wildlife biologists and interested persons into an organisation that promotes, monitors and advises on all aspects of elephant management, conservation and welfare.

EMOA
- advises and assists elephant owners on the appropriate management of their elephants
- advises and assists prospective owners of the feasibility of acquiring elephants
- provides information to suppliers of elephants on the feasibility of a proposed translocation

- initiates and promotes research and data management on elephants
- acts as link between private owners, national and international conservation agencies, Universities and other NGOs involved in elephant management
- promotes and facilitates the management of small populations as part of a meta population
- advises and promotes welfare issues of elephants
- advises on the sustainable utilisation of elephants according to specific guidelines

Members play an active role and have input into policies and strategies and participate at regular workshops.

ACTIVITIES: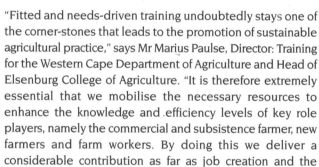

ELSENBURG COLLEGE OF AGRICULTURE

PO Box 54 ELSENBURG 7607 W Cape
Tel: 021 808 5450 • Fax: 021 884 4319
Email: mariusp@elsenburg.com
Website: www.elsenburg.com
Contact Name: Mr Marius Paulse – Director Training

"Fitted and needs-driven training undoubtedly stays one of the corner-stones that leads to the promotion of sustainable agricultural practice," says Mr Marius Paulse, Director: Training for the Western Cape Department of Agriculture and Head of Elsenburg College of Agriculture. "It is therefore extremely essential that we mobilise the necessary resources to enhance the knowledge and efficiency levels of key role players, namely the commercial and subsistence farmer, new farmers and farm workers. By doing this we deliver a considerable contribution as far as job creation and the struggle against poverty in rural communities are concerned."

Elsenburg brings training to the people
Decentralised training strives to bring training closer to especially resource-poor clients. For this purpose training centres are planned at George, Oudtshoorn and in the West Coastal region. Where training was previously offered only in Afrikaans, short courses such as Agricultural Management, Marketing and Financial Management are now, for the first time, being offered in English and Xhosa.

Elsenburg keeps tread with requirements
To fit the arsenal of courses offered at Elsenburg in line with the needs of the clients, new study fields are constantly explored. Courses in subjects such as life skills, communication, food security, mechanisation planning, organic- and greenhouse cultivation of crops, currently enjoy attention.

New "Degree" in prospect
An agreement which made extensive co-operation between Elsenburg and the University of Stellenbosch possible, was recently concluded and it is envisaged to be implemented in 2004. The advantages for both organisations, as well as prospective students are legion. An extension of the degree- and diploma training programmes increase the options for students considerably.

ACTIVITIES:

EMPOWERMENT FOR AFRICAN SUSTAINABLE DEVELOPMENT

PO Box 165 GREEN POINT 8051 W Cape
Tel: 021 434 6012 • Fax: 021 434 6134
Email: dmacdev@icon.co.za / info@easd.org.za
Website: www.easd.org.za
Contact Name: Mr David MacDevette – Director

EASD is a private non-profit corporation working to promote environmentally sustainable development in Africa and globally. EASD provides information products, training and facilitation services and conducts projects in support of sustainable development in Africa and globally. Our focus areas are: (1) state of the environment reporting, (2) regional and national environmental strategies and programmes, (3) the development of environmental information systems, development of knowledge management systems, technology for sustainable development, project and programme evaluation and organisational development (design of environmental organisations).

ACTIVITIES:

EMPR SERVICES (PTY) LTD

Private Bag X12 ROSETTENVILLE 2130 Gauteng
Tel: 011 435 0372/0 • Fax: 011 436 2689
Email: brian.dawson@iafrica.com
Contact Names: Brian Dawson, Doug Edgar – Directors

EMPR Services (previously Chamber of Mines Vegetation Unit) are rehabilitation contractors, specialising in mine dump, open-cast mine, roadside, quarry and degraded land rehabilitation, and land management.

ACTIVITIES:

ENACT INTERNATIONAL

76 Strand Street, CAPE TOWN 8001W Cape
Tel: 021 425 7068 • Fax 021 425 7065
Email : enact@law.co.za wscinc@law.co.za
Contact Name : Cormac Cullinan – Chief Executive Officer

EnAct was started in 1997 in Cape Town as a sister organisation to EnAct International of London, founded in 1994. EnAct specialises in the legal, policy and institutional aspects of environmental and natural resources management. The firm has worked with governments, international organisations, non-governmental organisations and companies in more than 20 countries world-wide.

EnAct's primary expertise is in the reviewing and drafting of environmental policy and legislation at an international, regional, national, provincial or local level. It also provides advice on the designing and strengthening of institutional structures and the training of government officials in order to enhance the implementation of environmental laws.

EnAct employs three senior and 4 junior consultants on a full-time basis. Each of these lawyers has, or is in the process of obtaining, a Masters degree in environmental law. In addition,

Enact has associates in various countries, including Canada, Chile, Portugal and the United Kingdom. The organisation routinely works in multi-disciplinary teams with other environmental specialists.

ACTIVITIES:

ENCHANTRIX

D9 Prime Park Mocke Road
DIEP RIVER 7900 W. Cape
Tel: 021 706 9847 • Fax: 021 706 7642
Email: info@enchantrix.co.za • Website: www.enchantrix.co.za
Contact Name: Anthea Torr – Member

There is no getting away from it, that an acute state of emergency exists on Earth. There are over 13 millions tons of toxic chemicals released into the environment every day, causing havoc with the air we breath, the environment and Earth's inhabitants. The petro-chemical companies sell their toxic waste for use in every day cleaning and personal care products and the majority of these chemicals have not been tested in the cocktail in which they are used. 60% of what you put on skin gets absorbed. Cancers, heart disease, allergies, skin disorders to name a few are epidemic, and the children are suffering the most as their organs are not developed enough to excrete the toxins effectively. There are between 5 – 10 millions cases of poisoning reported every year in America – most of them are fatal, most of them children and most of them originate from the cleaning/chemical products kept at home for daily use! ENCHANTRIX is offering a workable solution. You can look after your body, your clothes, and your home without any chemicals, it is a much gentler and kinder solution for Earth and those living on her. Living consciously and sustainably is not going to be an option – it is going to be essential if we are to continue living on Earth.

ACTIVITIES:

> *Biodiversity is the tool with which you play the game of promoting global stability. But it also consists of the organisms that give wonder and beauty and joy to the world, and that provides the context in which we evolved.*
>
> ~ Peter H Raven

ENDANGERED WILDLIFE TRUST (THE)

Private Bag X11 PARKVIEW 2122 Gauteng
Tel: 011 486 1102 • Fax: 011 486 1506
Email: ewt@ewt.org.za
Website: www.ewt.org.za
Contact Name: Dr. Nick King – Director
Yolan Friedmann – Conservation Manager

The Endangered Wildlife Trust is a non-government, non-profit, conservation organisation, founded in 1973. The EWT conserves threatened species and ecosystems in southern Africa by:
- Initiating research and conservation action programmes
- Implementing projects which mitigate threats facing species diversity
- Supporting sustainable natural resource management

The EWT communicates the principles of sustainable living through awareness programmes to the broadest possible constituency for the benefit of the people of the region. Projects are undertaken by EWT Working Groups and Strategic Partners (listed separately) which harness the energy and enthusiasm of numerous participants. Head-quartered in Johannesburg, South Africa, the EWT is membership based and has a regional office in Mozambique (*a registered national NGO, the Forum para a Natureza em Perigo*).

ACTIVITIES:

EWT – AIRPORTS COMPANY OF SOUTH AFRICA

Africa Strategic Partnership
Private Bag X11
PARKVIEW 2122 Gauteng
Tel: 011 486 1102 • Fax: 011 486 1506
Email: acsabirds@ewt.org.za • Website: www.ewt.org.za
Contact Name: Albert Froneman

The Airports Company of South Africa Strategic Partnership bird strike avoidance project of the Endangered Wildlife Trust is dedicated to minimising bird strikes and other interactions between wildlife and airport facilities at South African airports by applying environmentally sensitive management techniques.

EWT – AFRICAN WATTLED CRANE PROGRAMME

Private Bag X11 PARKVIEW 2122 Gauteng
Tel: + 27 11 706 3823 • Fax: + 27 11 463 4094
Cell: + 27 (0) 82 493 1991 • Email: lindy.r@global.co.za
Contact Name: Lindy Rodwell

The African Wattled Crane Programme aims to conserve Wattled Cranes and their habitats by promoting cooperation in and among African nations in partnership with people who depend on these same habitats. This is to be achieved through conservation programmes that include:
- Research

- Management
- Capacity building
- Education and awareness

EWT – BAT CONSERVATION GROUP

Private Bag X11
PARKVIEW 2122 Gauteng
Tel: 011 486 1102 • Fax: 011 486 1506
E-mail: bats@ewt.org.za
Website: www.ewt.org.za
Contact Name: Claire Patterson

The Bat Conservation Group, a working group of the EWT, aims to achieve increased public participation in bat conservation through education, communication and support for bat research and conservation projects.

EWT – BLUE SWALLOW WORKING GROUP (EWT–BSWG)

Private Bag X11
PARKVIEW 2122 Gauteng
Tel: 011 486 1102 • Fax: 011 486 1506
Email: blueswallow@ewt.org.za
Website: www.ewt.org.za/blueswallow/index.html
Contact Name: Steven W Evans

The Blue Swallow Working Group is a working group of the Endangered Wildlife Trust dedicated to preventing the Blue Swallow from going extinct throughout its 10 range-state sub-Saharan distribution range.

EWT – THE CARNIVORE CONSERVATION GROUP

Private Bag X11
PARKVIEW 2122 Gauteng
Tel: 011 486 1102 • Fax: 011 486 1102
Email: patfletcherccg@ewt.org.za • Website: www.ewt.org.za
Contact Names: Pat Fletcher, Dr Gus Mills

The Carnivore Conservation Group is a working group of the Endangered Wildlife Trust dedicated to promoting the conservation of carnivores through integrated research on aspects, which will help develop and implement sound management strategies.

EWT – CONSERVATION BREEDING SPECIALIST GROUP (CBSG) SOUTHERN AFRICA

Private Bag X11 PARKVIEW 2122 Gauteng
Tel: 011 486 1506 • Fax: 011 486 1102
Email: cbsgsa@wol.co.za • Website: www.ewt.org.za/cbsg
Contact Names: Yolan Friedmann, Brenda Daly

Conservation Breeding Specialist Group (CBSG) Southern Africa is a regional network of CBSG and is therefore an IUCN / SSC Specialist Group. CBSG Southern Africa is a partnership project with the EWT and aims to catalyse conservation action in southern Africa by assisting in the development of integrated and scientifically sound conservation programmes for species and ecosystems, building capacity in the local conservation community and incorporating practical and globally endorsed tools and processes into current and future conservation programmes.

EWT – THE CONSERVATION LEADERSHIP GROUP

Private Bag X11
PARKVIEW 2122 Gauteng
Tel: 011 486 1102 • Fax: 011 486 1102
Email: leadership@ewt.org.za • Website: www.ewt.org.za
Contact Names: Andre Van Zyl, Edward Farrell

The Conservation Leadership Group is a working group of the Endangered Wildlife Trust dedicated to building capacity by developing conservation ethics for the people of southern Africa through excellence in leadership and role models.

EWT – ESKOM STRATEGIC PARTNERSHIP

Private Bag X11
PARKVIEW 2122 Gauteng
Tel: 011 486 1102 • Fax: 011 486 1506
Email: chrisv@ewt.org.za • Website: www.ewt.org.za/eskom
Contact Name: Chris van Rooyen

Eskom-EWT Strategic Partnership of the Endangered Wildlife Trust is dedicated to implementing an integrated management system to minimise negative interaction between wildlife and electricity structures, in order to improve quality of supply and reduce biodiversity impacts.

EWT – FÓRUM PARA A NATUREZA EM PERIGO

PO Box 4203 MAPUTO
Mozambique
Tel: +258 – 1 308 924
Fax: +258 – 1 308 925
Email: fnp@fnp.org.mz
Website: www.ewt.org.za / www.fnp.org.mz
Contact Name: António Reina

The Fórum para a Natureza em Perigo is Mozambique's regional branch of the Endangered Wildlife Trust dedicated to biodiversity conservation through community based natural resource management.

EWT – ORIBI WORKING GROUP

KZN Wildlife
P O Box 13053 CASCADES 3202 KwaZulu-Natal
Tel: 033 239 1513 • Fax: 033 239 1529
Email: athol@kznwildlife.com
Website: www.kznwildlife.com
Contact Name: Athol Marchant

The Oribi Working Group is a working group of the EWT and aims to promote the long-term survival of Oribi in their natural

grassland habitat through initiating and coordinating provincial conservation programmes. These include: education and awareness, habitat conservation, scientific research and population management.

EWT – THE POISON WORKING GROUP

PO Box 72334
PARKVIEW 2122 Gauteng
Tel: 011 486 1157 • Fax: 011 646 4631
Nasua-PWG Wildlife Poisoning Helpline: 082 446 8946
Email: pwg@ewt.org.za
Website: www.ewt.org.za/poisonworkinggroup/index.html
Contact Names: Prof Gerhard H. Verdoorn, Nicola Van Zyl

The Poison Working Group is a working group of the Endangered Wildlife Trust dedicated to addressing the poisoning of wildlife through data assimilation, dissemination, analysis and investigation on a scientific and interactive basis, and to take the appropriate pro-active education and conservation action for the protection of wildlife and people in southern Africa.

EWT – THE RAPTOR CONSERVATION GROUP

PO Box 72155
PARKVIEW 2122 Gauteng
Tel: 011 646 4629 • Fax: 011 646 4631
Cell: 082 446 8946 • Email: rcg@ewt.co.za
Website: www.ewt.org.za/raptor/index.html
Contact Name: Jenny le Roux

The Raptor Conservation Group is a working group of the Endangered Wildlife Trust dedicated to conserving the living populations of diurnal and nocturnal raptors in southern Africa by the initiation and support of research, conservation and education programmes based on scientific and sound conservation principles.

EWT – RIVERINE RABBIT WORKING GROUP

P.O. Box 172
LOXTON 6985 South Africa
Tel: 053 - 381 31 07 • Fax: 053 - 381 31 07 or 16
E-mail: ahlmann@yebo.co.za • Web: www.riverinerabbit.co.za
Contact Name: Dr. Vicky Ahlmann

The Riverine Rabbit Working Group is a working group of the EWT and aims to ensure the survival of the critically endangered riverine rabbit (Bunolagus monticularis) by establishing and implementing sound conservation programmes for the species and its unique habitat, developing and implementing awareness and education programmes for all levels of society, building capacity in the local communities within the species distribution range and engaging in research programmes on the biology and ecology of the species and on its distribution range (Nama and Succulent Karoo).

EWT – SOUTH AFRICAN CRANE WORKING GROUP

Private Bag X11
PARKVIEW 2122 Gauteng
Tel: 011 486 1102/3 • Fax: 011 486 1506
Email: crane@ewt.org.za
Website: www.ewt.org.za/sacwg/index.html
Contact Names: Kerryn Morrison, Kevin McCann, Kathy Leitch

The South African Crane Working Group is a working group of the Endangered Wildlife Trust and aims to promote the long-term survival of all three species of cranes found in South Africa in their natural and human-modified habitats through initiating and co-ordinating regional and national crane conservation programmes.

EWT – STRATEGIC PARTNERSHIP WITH TRAFFIC EAST/SOUTHERN AFRICA

Private Bag X11
PARKVIEW 2122 Gauteng
Tel: 011 486 1102 • Fax: 011 486 1506
Email: TRAFFICAZA@uskonet.com
Website: www.TRAFFIC.org
Contact Name: David Newton

TRAFFIC, a strategic partner of the Endangered Wildlife Trust supports the work and missions of WWF and IUC, and its purpose is to help ensure that wildlife trade is at sustainable levels and in accordance with domestic and international laws and agreements.

EWT – THE WILDLIFE BIOLOGICAL RESOURCE CENTRE

PO Box 582
PRETORIA 0001 Gauteng
Tel: 012 305 5840 • Fax: 012 305 5840
Email: paulb@wbrc.org.za
Website: www.ewt.org.za/wbrc
Contact Name: Dr. Paul Bartels

The Wildlife Biological Resource Centre is a working group of the Endangered Wildlife Trust dedicated to broadening genetic diversity within Southern African wildlife species through the development of a biological resource centre for wildlife species.

EWT – UNITED STATES OF AMERICA

346 Smith Ridge Road
NEW CANAAN CT 06840 USA
Tel: + 203 966 1981 • Fax: + 203 966 2748
Email: Mdevlin512@aol • Website: www.ewt.org.za
Contact Name: Mike Devlin

The USA Office is a working group of the Endangered Wildlife Trust dedicated to promoting the work of the Endangered Wildlife Trust to audiences in the United States of America, and to procure funds for the Trust's projects in southern Africa.

EWT – VULTURE STUDY GROUP

PO Box 72334 PARKVIEW 2122 Gauteng
Tel: 011 646 8617 • Fax: 011 646 4631
Email: vsg@ewt.co.za
Website: www.ewt.org.za/vulture/index.html
Contact Name: Kerri Wolter

The Vulture Study Group is a working group of the Endangered Wildlife Trust dedicated to furthering the cause of vulture conservation through the facilitation of research, conservation action and communication in southern Africa and internationally.

EWT – WILDCARE STRATEGIC PARTNERSHIP

PO Box 15032 LYNN EAST 0039 Gauteng
Tel: 012 808 1106 • Fax: 012 808 0602
Cell: 083 653 8900
Email: karojay@global.co.za • Website: www.ewt.org.za
Contact Name: Karen Trendler

Wildcare is a strategic partner of the Endangered Wildlife Trust committed to conservation through ethical and responsible wildlife rehabilitation, welfare and crisis response: providing specialist nursing, treatment and care of compromised wildlife, as well as training and information.

ENERGY & DEVELOPMENT GROUP

PO Box 261 NOORDHOEK 7979 W Cape
Tel: 021 702 3622 • Fax: 021 702 3625
Email: admin@edg.co.za • Website: www.edg.co.za
Contact Names: Beverley Duffield or Mark Borchers – Administration

The Energy & Development Group promotes access to energy services for sustainable development in southern Africa, as well as to contribute to the improvement of welfare in developing areas through the provision of affordable energy.

ACTIVITIES:

ENERGY EFFICIENT OPTIONS

PO BOX 316 HOWARD PLACE 7450 W Cape
Tel: 021 534 8640 • Fax: 021 534 8614
Email: judy@energyoptions.co.za
Website: www.energyoptions.co.za
Contact Name: Leon Ravell – Consultant

The owners of Energy Efficient Options, a small family business with 16 years of experience in photovoltaics, have been living solely on solar power for more than nine years.

The family built the house in the Thornton residential area, in Cape Town, independent of the national electricity grid. The equipment needed to live without the grid, i.e. solar modules, regulators, batteries and invertors, came to a cost of R77 000. The option to connect to the national grid, Eskom was only R4 500, but the Ravells decided they would like to "walk their talk" and live in a solar-powered house. After four years they did an upgrade in order to have a bigger fridge and freezer and provide power for two computers for the office. The current system consists of 2 000 watts of solar module, which produces 8 000 watt hours of energy per day. The only other power source they use is gas for cooking and supporting the solar hot water system on cold days.

Photovoltaic technology uses the light properties of the sun, so as long as the sun rises in the morning, there is light being converted into energy. The volume of this charge is however relative to the brightness of the day. In the Ravell's batteries they store 8 days of energy, and have already survived a rainy period as long as 14 days.

The Ravells say "the rule of life is – you will pay"
"the choice in life is – how to pay"

Independence is an option and the road to independence is called efficiency. Solar power systems are not cheaper than coal-fired/nuclear/hydro systems utilised by the national utility to provide electricity in developed urban areas. Solar power is only feasible when one has to pay the national grid R65 000 per kilometer to be extended in your direction, and then to pay for availability and consumption. As consultation and design is free, please evaluate your options.

ACTIVITIES:

ENERGY TECHNOLOGY UNIT – CAPE TECHNIKON

PO Box 652 CAPE TOWN 8000 W Cape
Tel: 021 460 3127 • Fax: 021 460 3887
Email: uken@ctech.ac.za • Website: www.ctech.ac.za
for Energy Technology Unit see: staffnet@ctech.ac.za/uken
Contact Name: Prof Ernst Uken

The Energy Technology Unit (ETU) of the Cape Technikon concentrates on developing affordable, energy-efficient domestic appliances and promoting the use of renewable energy in remote rural areas.

ACTIVITIES:

ENGEN

PO Box 35 CAPE TOWN 8000
Tel: 021 403 5258 • Fax: 021 403 5240
Email: Barbara.Manson@engenoil.com
Website: www.engenoil.com
Contact Name: Barbara Manson – Group Communications Manager

Engen Limited is a holding company with investments in oil and related industries.

Engen Petroleum Limited is a wholly-owned subsidiary of Engen Limited and is the operating company of Engen's downstream activities in South Africa. The core business of the company is the refining of crude oil, the marketing of primary refined petroleum products and the provision of convenience services through its retail network of over 1 250 service stations and over 450 Quickshops.

Engen Petroleum Limited owns and manages a 150 000 barrel per day crude oil refinery and a state-of-the-art lubricants blending plant in Durban.

Engen Holdings Limited is the holding company for operations conducted outside of South Africa. Engen is represented in 13 countries outside South Africa – Namibia, Botswana, Swaziland, Lesotho, Zimbabwe, Zambia, Kenya, Mozambique, Uganda, Tanzania, Burundi, the Democratic Republic of the Congo and Ghana. Engen has about 2 800 employees.

ACTIVITIES:

ENVIRO FRINGE SERVICES

PO Box 176
CRAMER VIEW 2060 Gauteng
Tel: 011 463 4902 • Fax: 011 463 4902
Email: sueb@mweb.co.za
Contact Name: Sue Bellinger – Managing Director

For companies, authorities, communities: awareness raising, practical implementation of waste minimisation/cleaner production, resource use reduction and resulting social development.

ACTIVITIES:

ENVIROCON INSTRUMENTATION

PO Box 2686
NORTHCLIFF 2115 Gauteng
Tel: 011 476 7323 • Fax: 011 476 5995
Email: sales@envirocon.co.za • Website: www.envirocon.co.za
Contact Name: Howard Palmer – Director

Envirocon Instrumentation are involved in sales, marketing, distribution and service of environmental and occupational hygiene monitoring instruments, representing some of the world's leading manufacturers of Occupational Hygiene Equipment.

Envirocon has been in the business for 15 years, supplying equipment to South Africa and many other African Countries, south of the Sahara.

The Company is currently the Sole Appointed Distributors for:
◆ Quest Technologies Inc, Sensidyne Gilian, RAE Systems, T S I, K and M Environmental, Aquaria and Met One Instruments.

These Companies supply, inter alia, instruments for:
◆ Noise, Dust, Gas, Heat Stress, Vibration, Air Flow, Indoor Air Quality, Volatile Organic Compounds, Temperature, Humidity, Pressure, Light Measurement, Weather Monitoring, Environmental Particulate Monitoring.

ACTIVITIES:

ENVIROCOVER (PTY) LTD

PO Box 2756
SAXONWOLD 2132 Gauteng
Tel: 011 325 4089 • Fax: 011 325 4089
Cell: +27 83 308 8600 • Email: eastonj@aforbes.co.za
Contact Name: John Easton

Envirocover was established in 1997 with a view to bringing together the technical and financial skills necessary to manage environmental liabilities for industries and mines. In 2000, Envirocover was acquired by Alexander Forbes Risk Services (ranked as one of the top ten largest professional services organisation of its kind in the world).

Under legislation, including the National Environmental Management Act, National Water Act and Mineral and Petroleum Resources Development Act, companies and individuals are vulnerable to both statutory and civil claims.

Envirocover packages offer:

- environmental risk assessment
- on-site environmental risk management
- environmental risk financing
- environmental insurance

for industry and mining companies during their operational, decommissioning and closure phases achieving:

- smooth transfer of mining* and industrial environmental liabilities
- financing of the foreseen and unforeseen environmental risks, through
  - specialist insurance and risk financing products
  - management of specialist trust funds
- pre-closure and post closure environmental management
- transfer of residual risks to the next land or facility user

* Envirocover is the approved "third party" and "qualified person" in terms of the DME Mine Closure Policy and Regulations under the Mineral and Petroleum Resources Development Act of 2002 and will, through the above packages, enable the issue of unconditional closure certificates for mines.

ACTIVITIES:

ENVIRO-ED CC

PO Box 13067 NELSPRUIT 1200 Mpumalanga
Tel: 013 755 2479 • Fax: 013 755 1324
Email: info@ekukhanyeni.co.za
Website: www.worldnetafrica.com/training/enviroed
Contact Name: Stephany Milne – Director

ACTIVITIES:

ENVIROEDS

24 Michel Walk Park Island
MARINA DA GAMA 7945 WCape
Tel: 021 788 2431 • Fax: 021 788 2431
E-mail: ally@enviroeds.co.za • Website: www.enviroeds.co.za
Contact Name: Ally Ashwell – Consultant

Established in 1999, EnviroEds aims to enable environmental learning and action through the development of educational programmes and materials. EnviroEds also convenes an informal network of environmental educators in Cape Town, known as the EE Friends.
Services offered include:

- Development, management and evaluation of environmental education programmes
- Curriculum and resource material development in environmental education
- Popular writing and editing in the fields of science, environment and environmental education.

ACTIVITIES:

ENVIROLEG

PO Box 1571 BALLITO 4420 KwaZulu-Natal
Tel: 082 490 5369 • Fax: 082 490 5536
Email: queries@enviroleg.co.za
Website: www.enviroleg.co.za
Contact Names: Ian, Magda, John and Lee – Directors

We provide legislation libraries, aspects and legal registers in electronic software and formats. We cover SA and other SADC environmental, health, safety and labour legislation software and other related services.

ACTIVITIES:

ENVIRONMENTAL AND CHEMICAL CONSULTANTS CC

PO Box 2856 CRESTA 2118 Gauteng
Tel: 011 792 1052 • Fax: 011 791 4222
Email: ecconsultants@mweb.co.za
Contact Name: Dr David A Baldwin – Director

Consulting: general, mining, industrial and hazardous waste management; waste classification; waste avoidance, recycling, treatment and disposal; site selection, investigation and remediation; auditing; risk assessment.

ACTIVITIES:

ENVIRONMENTAL & GEOGRAPHICAL SCIENCES DEPARTMENT

University of Cape Town RONDEBOSCH 7701 W Cape
Tel: 021 650 2873 • Fax: 021 650 3791
Email: admin@enviro.uct.ac.za
Website: www.egs.uct.ac.za/engeo/courses/mphil
Contact Name: Sharon Adams

The Master of Philosophy degree programme was started in 1974, and has trained more South African environmental professionals than any other similar programme. The programme aims to produce graduates with a broad understanding of major environmental issues and the ability to effectively analyse and manage environmental problems and conflicts, with emphasis on South Africa. The programme duration is 18 months full-time, starting in February each year:

9 months of course work followed by an 8 month research project applying environmental assessment and management skills and procedures to real life issues. Employment opportunities arise in the public sector; parastatals, the private sector (consulting firms and industry), and in NGOs and CBOs. Scholarships and bursaries are available. Applicants should have a good Honours degree or equivalent 4-year professional degree, preferably with appropriate experience. The closing date for applications is the end of August and the outcome of the selection process is in October.

ACTIVITIES:

ENVIRONMENTAL EDUCATION ASSOCIATION OF SOUTHERN AFRICA (EEASA)

PO Box 394 HOWICK 3290 KwaZulu-Natal
Tel: 033 330 3931 • Fax: 033 330 4576
Email: eeasa@futurenet.co.za
Website: www.info-net.net/eeasa
Contact Name: Tobile Dlamini

EEASA supports environmental education in southern Africa by providing opportunities for the exchange of ideas/opinions through publications, the annual conference and regional working groups.

ACTIVITIES:

ENVIRONMENTAL EDUCATION AND RESOURCES UNIT UNIVERSITY OF WESTERN CAPE

Private Bag X17 BELLVILLE 7535 W Cape
Tel: 021 959 2498 • Fax: 021 959 2484
Email: cklein@uwc.ac.za
Website address: www.botany.uwc.ac.za/eeru
Contact Name: Charmaine Klein – Director

The mission of EERU is to promote environmental education opportunities that will lead to a better understanding of the total environment. The Unit is located on the campus of the University of the Western Cape and functions as a service organisation that provides environmental education support primarily to school and community organisations on the Cape Flats.

Its main activities include Environmental Education, Indigenous Greening and Horticulture as well as Urban Nature Conservation. To achieve this, the Unit's Cape Flats Nature Reserve (a Provincial Heritage site), consisting of Strandveld and coastal Fynbos, its Nursery and EE Resource centre are used. The Indigenous Greening Programme has since 1992 been responsible for the establishment of many indigenous gardens at schools and in communities. Workshops, guided walks and ecological fieldwork are offered.

ACTIVITIES:

ENVIRONMENTAL ETHICS UNIT AT THE UNIVERSITY OF STELLENBOSCH

Private Bag X1
MATIELAND 7602
Tel: 021 808 2988 • Fax: 021 808 3556
Email: jph2@sun.ac.za
Website: www.sun.ac.za/philosophy/cae
Contact Name: Prof J P Hattingh – Head

The Unit for Environmental Ethics is a service and research body that focuses on the independent analysis and critical evaluation of the values informing environmental policy and decision-making in South Africa. It seeks to improve the quality of the environmental decision-making in South Africa. It sets out to achieve this by creating an awareness and critical understanding among leaders in business, non-governmental organisations and government of the values informing environmental policy and decision-making.

As specialists in environmental ethics we offer:
 ◆ Ethical reviews of EIAs
 ◆ Ethical surveys on environmental management practice
 ◆ Facilitation of ethical deliberation in environmental decision-making
 ◆ Ethical analysis of environmental policy
 ◆ Environmental codes of ethics
 ◆ Short courses and workshops

Our unit hosts an environmental ethics e-group on the following website: www.sun.ac.za/philosophy/cae. The e-group operates as a think-tank, bulletin board and advice forum on environmental ethics for its membership. Membership is open to all.

ACTIVITIES:

"Instead of developing a new momentum at the Summit, civil society has had a very hard job of defending the Summit from a complete take-over by the World Trade Organisation (WTO). The WTO says you have to have economic growth and if it negatively affects society and the environment, that's just tough."

~ said at WSSD 2002

ENVIRONMENTAL EVALUATION UNIT
UNIVERSITY OF CAPE TOWN

Private Bag
RONDEBOSCH 7701 W Cape
Tel: 021 650 2866 • Fax: 021 650 3791
Email: sowman@science.uct.ac.za
Website: www.egs.uct.ac.za/eeu
Contact Name: Dr Merle Sowman – Co-Director

The Environmental Evaluation Unit (EEU) is an independent, self-funded environmental consulting, research and training unit, based at the University of Cape Town. Founded in 1985, the EEU has established itself as a centre of excellence in the fields of integrated environmental and coastal management. Over the past 18 years the EEU has undertaken work throughout South Africa and the SADC countries, and has been involved in a wide array of projects ranging from large, complex Environmental and Social Impact Assessments, to the development of the Integrated Environmental Management (IEM) guidelines for South Africa, as well as managing policy formulation processes. In addition to consulting and research, training and capacity building is also a core competency of the EEU. Since its inception the EEU has established itself as a professional training and capacity building unit and has designed and delivered numerous Environmental Assessment and Management training courses, as well as specialised courses, such as Integrated Coastal Management, throughout southern Africa.

ACTIVITIES:

ENVIRONMENTAL JUSTICE NETWORKING FORUM, NATIONAL OFFICE (EJNF)

PO Box 3544 PIETERMARITSBURG 3200 KwaZulu-Natal
Tel: 033 342 4034 • Fax: 033 345 5841
Email: info@ejnf.org.za • Website: www.ejnf.org.za
Contact Name: Dr Amos Dube – National Director

The Environmental Justice Networking Forum is a shared resource established to service the common interests of participating South African non-governmental and community-based organisations on matters concerning environmental justice and sustainable development.

EJNF is a network in the sense that it seeks to improve the strategic information and communications functions of participating organisations. EJNF is also a forum in that it seeks to provide a common structure through which different sectors of South African civil society can explore, strengthen and promote matters of common interest, relating to environmental justice and sustainable development.

ACTIVITIES:

ENVIRONMENTAL LAW ASSOCIATION THE (ELA)

PO Box 82 OBSERVATORY 7935 W Cape
Tel: 021 448 8908 • Fax: 021 448 3422
Email: louis@villiers.co.za
Contact Name: Louis de Villiers

Established in 1991 to promote the enhancement and conservation of the environment and to educate its members and the public in all matters pertaining to the law relating to the environment. Membership of the ELA is open not only to lawyers, but to anyone with an interest in the laws pertaining to the environment.

The objects of the association are:

◆ education and law relating to protection of the environment
◆ encourage collaboration between all persons concerned with the law relating to the environment
◆ disseminate information relating to environmental law
◆ identify, review, advise and comment on issues of environmental law and its application

In furtherance of its aims, the association publishes the South African Journal of Environmental Law and Policy.

ACTIVITIES:

ENVIRONMENTAL MONITORING GROUP

PO Box 13378 MOWBRAY 7705
W Cape
Tel: 021 448 2881 • Fax: 021 448 2922
Email: info@emg.org.za Website : www.emg.org.za
Contact Name: Nontembo Bam – Assistant to the Director

ENVIRONMENTAL
MONITORING
GROUP

EMG aims to strengthen the participation of civil society organisations and community groups in environmental policy and decision-making processes. EMG works in 3 programme areas, namely International Trade & Environmental Governance, Water Justice in Southern Africa, and Rural Resource Management. Within these areas, EMG's focus is on providing information, acting as a process facilitator and demonstrating alternatives.

ACTIVITIES:

ENVIRONMENTAL PLANNING AND RESOURCE MANAGEMENT SERVICES CC

PO Box 4296 DURBANVILLE 7551 W Cape
Tel: 021 975 7396 • Fax: 021 975 1373
Email: eprms@intekom.co.za
Contact Name: Larry Eichstadt – Director

EPRMS is a professional environmental management consultancy that is client focused, transparent and ensures the management of projects in a strategic, sustainable and ethical manner.

The company's mission statement is supported by the working experience (as a regulator and project manager) and knowledge of the director, and sole member Larry Eichstadt, within the water resource, waste management and environmental planning fields. Due to this broad working experience the company provides a professional service in the following areas:

- Project Management – water and waste projects
- Environmental Impact Assessments
- Project Review
- Applications in terms of Section 20, 21 and 26 (Environment Conservation Act)
- Licence and General Authorization applications (National Water Act)
- Environmental Auditing
- Due Diligence and Compliance Audits
- Environmental Education / Capacity Building
- Policy / Guideline Development
- Public participation.

ACTIVITIES:

ENVIRONMENTAL RISK MANAGEMENT CC

PO Box 15051 FARRARMERE 1518 Gauteng
Tel: 011 425 2728 • Fax: 011 425 6894
Email: erm@global.co.za
Contact Name: Mrs Janine Nicholson – Consultant

Specialising in Waste Management, Environmental Impact Assessments, Environmental Auditing, Environmental management systems including ISO 14000, Environmental Training Programmes, Standards Manuals, Environmental Information update services and Health and Safety.

ACTIVITIES:

ENVIRONOMICS CC   ENVIRONOMICS

PO Box 44108 THERESAPARK 0155 Gauteng
Tel: 012 549 5949 • Fax: 012 549 2483
Email: info@environomics.co.za
Contact Name: Debbie Claassen – Partner

Environomics is an environmental consultancy. Established in 1998, we offer extensive experience and professional expertise in all aspects of managing the environmental impacts of development. The principles and procedures of Integrated Environmental Management (IEM) as embodied in national and provincial law, policy and guidelines, form the basis of all work done by Environomics.

Our services include:

- The formulation of Environmental Policy at both organisation and government levels
- Execution of Environmental Impact Assessments including Scoping Studies
- Execution of Strategic Environmental Assessments and the compilation of Environmental Management Frameworks
- Execution of Visual Impact Assessments using the latest technologies
- Development and implementation of Environmental Management Systems (including ISO 14001 certification)

- The design and implementation of Information Management Systems for legal and resource administration
- The development of Environmental Information Systems
- The creation and maintenance of EMFs on a GIS based platform
- The execution of site surveys
- Identification of legal requirements and legal compliance review

ACTIVITIES:

ENVIRO-OPTIONS (PTY) LTD

PO Box 13 Kya Sands RANDBURG 2163 Gauteng
Tel: 011 708 2245 • Fax: 011 708 2180
Email: eloo@mweb.co.za • Website: www.eloo.co.za
Contact Name: Gavin La Trobe – Director

Enviro Options based in Johannesburg, specialises in providing ecological sanitation solutions, utilising the Enviro Loo "waterless" toilet system. Development of the system was initiated in the late eighties and was subjected to extensive field trials. The Enviro Loo is a 100% South African owned and manufactured product.

The system has been implemented in schools, communities, clinics, etc. on a widespread basis within South Africa and Enviro Options is also experiencing a steady growth in the export market.

Authorities are often called upon to use precious financial resources to treat health, environmental and replacement symptoms, brought about by inappropriate sanitation. The Enviro Loo system eradicates these problems!

The Enviro Loo toilet meets the financial, social and environmental criteria necessary for sustainable development. Enviro Options has proven that improved technology principles and design, improves people's standard of living, while conserving the very water and financial resources, which make development possible.

ACTIVITIES:

ENVIROSENSE CC

61 Peninsula Road ZEEKOEVLEI 7945 W Cape
Tel: 021 706 9829 • Fax: 021 706 6622
Email: envirosense@xsinet.co.za
Contact Name: Susanne Dittke – Member

EnviroSense CC develops tailor-made industrial/commercial and residential Resource and Waste Management solutions. Services include baseline assessments, identification of financial saving opportunities, education and training and optimisation of technical/environmental system impacts.

ACTIVITIES:

ENVIROSERV WASTE MANAGEMENT (PTY) LTD

PO Box 2207 BENONI 1500
Gauteng
Tel: 011 422-2560 • Fax: 011 845-1495
Email: info.ho@enviroserv.co.za
Website: www.enviroserv.co.za

EnviroServ is focused on environmentally responsible waste management. Operations are carried out through its primary subsidiary EnviroServ Waste Management (Pty) Ltd and its joint venture empowerment partner, Millennium Waste Management (Pty) Ltd in which it holds a 47% share. EnviroServ Waste Management provides professional, integrated waste management solutions to industry. Five operating divisions specialise in products and services dedicated to meeting the needs of particular markets.

Waste-tech manages, collects, sorts and separates recyclables, transports and disposes of industrial waste. Hazmat offers a 24-hour support service for emergency hazardous material spill response, clean-ups, spillage containment and on-site remediation. Dispose-tech operates permitted hazardous waste disposal sites and incineration facilities and offers expertise in analysis, classification, planning and control of hazardous waste treatment and disposal.Process Management facilitates the development, imple-mentation and operation of process solutions, for the treatment, reduction, recovery, re-use or re-cycling of waste products and effluent. Conquip supplies compaction and earthmoving equipment to the waste, construction and mining industries.

Regional Offices:

Gauteng:	Tel: 011 456-5400 • Fax: 011 453-7583 E-mail: info.gp@enviroserv.co.za
KwaZulu-Natal:	Tel: 031 902-1526 • Fax: 031 902-5778 E-mail: info.kzn@enviroserv.co.za
Western Cape:	Tel: 021 951-8420 • Fax: 021 951-8440 E-mail: info.ct@enviroserv.co.za
Eastern Cape:	Tel: 041 466-2741 • Fax: 041 466-2745 E-mail: info.pe@enviroserv.co.za

ACTIVITIES:

ENVIROWIN CC

146 Boulder Avenue NORTHCLIFFE Ext 18 Gauteng 2195
Tel: 011 478 3366 • Fax: 011 478 3366
Email: ndmaslen@iafrica.com
Contact Name: Rhonwen Maslen – Member

The competencies of the Envirowin partners include:
- ◆ Experience in the compilation of large mining Environmental Impact Assessments (EIA's) and Environmental Management Programme Reports (EMPR's), undertaking such work independently or as part of a team
- ◆ Compilation of mining construction phase Environmental Management Systems (EMSs).
- ◆ Compilation of EMS's for business office environments

- ◆ Compilation of EMS's for mining operational environments
- ◆ Research into various environmental topics

More specifically the products and services that Envirowin offer include:
- ◆ Advice pertaining to the requirements of an EIA/EMPR
- ◆ Provision of pro forma documents to manage and establish long-term systems and processes
- ◆ Workshops (facilitation) – identification of potential environmental impacts and the development of management plans to address these areas
- ◆ Management of the EIA process and compilation of the required documentation
- ◆ Management of the EMPR process and compilation of the required documentation
- ◆ Compilation of EMS documentation
- ◆ Facilitating governmental approval of EIAs and EMPRs
- ◆ Project management/co-ordination/sub contracting (Front and technical)

Involvement within project teams – providing advice and assistance on environmental issues.

ACTIVITIES:

ERA ARCHITECTS (PTY) LTD

3 Garrett Road
PARKTOWN 2193 Gauteng
Tel: 011 482 1994 • Fax: 011 726 5976
Cell: 082 451 8069
Email: kenneth@zanet.co.za
Contact Name: Ken Stucke – Director

This architectural consultancy specialises in sustainable development generally, and low-energy buildings in particular. Specialist expertise in "green" buildings places this practice in an ideal position to help implement projects that have a serious environmental agenda.

ACTIVITIES:

ESI AFRICA

PO Box 321
STEENBERG 7947 W Cape
Tel: 021 700 3508 • Fax: 021 700 3501
Email: Claire@spintelligent.com
Website: www.esi-africa.com
Contact Name: Claire Volkwyn – Publisher

ESI Africa reports on development in the power sector in Africa and provides practical and product related information pertaining to the generation, transmission and distribution of energy on the continent.

ACTIVITIES:

ESKOM
CORPORATE ENVIRONMENTAL AFFAIRS

PO Box 1091 JOHANNESBURG 2000 Gauteng
Tel: 011 800 2792 • Fax: 011 800 5725
Email: annamarie.murray@eskom.co.za
Website: www.eskom.co.za
Contact Name: Annamarie Murray – Account Executive

Eskom is currently one of the top seven utilities in the world in terms of size and sales, and it intends to retain this position by adopting strategies that strengthen its ability to respond to changing requirements, while at the same time embracing flexibility to deal with uncertainty. This position is embodied in Eskom's strategic intent – to be the pre-eminent African energy and related services business, of global stature.

As South Africa's national electricity utility Eskom supplies 95% of the electricity requirements of South Africa, and in the process generates more electricity than all the other suppliers in Africa combined. One of Eskom's key goals is to contribute towards the strengthening of the country.

An integral component of this goal is the provision of affordable energy to that part of the population, which does not have electricity as yet, in a sustainable manner.

Environmental concerns such as the greenhouse effect, ozone depletion and acid rain are also of paramount importance to Eskom, hence it strives to achieve an appropriate balance between environmental quality and the provision of electricity.

Eskom furthermore embraces the future competitive environment by ensuring that it remains at the forefront of research and technology, continually seeking new and innovative ways to expand its operations.

ACTIVITIES:

EURETA ROSENBERG
(Formerly Eureta Janse van Rensburg)

7 Boulder Road
LAKESIDE 7945 W Cape
Tel: 021 788 2238 • Fax: 021 788 4918
Email: eureta@worldonline.co.za
Contact Name: Dr Eureta Rosenberg

Environmental Education and Research

ACTIVITIES:

EWT – SEE ENDANGERED WILDLIFE TRUST (THE)

EZEMVELO AFRICAN WILDERNESS

PO Box 26665 NELSPRUIT
1200 Mpumalanga
Tel: 013 744 9813 • Fax: 013 744 9813
Email: info@africanwild.co.za • Website: www.africanwild.co.za
Contact Name: Mr Les Ashley – Operations Director

Wilderness trails and courses focusing on Leadership; ethics and values; group and interpersonal dynamics; wilderness healing; specialist trails. Operating in Gauteng, Limpopo, Mpumalanga, North West provinces. Malaria free areas.

ACTIVITIES:

EZEMVELO KZN WILDLIFE

PO Box 13053
CASCADES 3202
KwaZulu-Natal
Tel: 033 845 1999 (head office) /
033 845 1000 (reservations)
Fax: 033 845 1299
Email: bovetm@kznwildlife.com
Website: www.kznwildlife.com
Contact Name: Michele Bovet – Group Marketing Manager

KZN Wildlife is the formal nature conservation of the province of KwaZulu-Natal in South Africa and is responsible for the management of 110 protected wildlife areas throughout the province.

These include two World Heritage Sites – the uKhahlamba-Drakensberg Park and the Greater St Lucia Wetland Park. Hluhluwe-Imfolozi Park, established in 1895 and one of the oldest game reserves in Africa, is home of the white rhino and the scene of the internationally renowned Operation Rhino that saw this magnificent animal brought back from the brink of extinction and spread throughout the world in parks, zoos and safari parks.

Parks feature a wide range of rare mammals, fish and reptiles, including white and black rhino, the "Big 5", ice rats on the high Drakensberg; sea-turtles such as the Leatherbacks and Loggerheads that nest on the beaches of the Greater St Lucia Wetland Park; the prehistoric coelacanth – a fish thought to be extinct until rediscovered in 1938 and also found off Sodwana Bay in the Greater St Lucia Wetland Park in 1999.

KZN Wildlife offers a wealth of tourism opportunities throughout the province in parks that have great and diverse natural splendour. Visitors may explore rugged mountains, valley bushveld, grasslands, densely wooded riverine gorges, thornveld, coastal forests and sand dunes and brilliant coral reefs.

There is a range of comfortable and affordable accommodation, and activities to be done that include hiking and climbing, fresh and salt water fishing, boating and water-sports, scuba-diving and snorkelling, mountain biking and 4x4 trails, controlled hunting, game drives and wilderness trails.

ACTIVITIES:

FAIREST CAPE ASSOCIATION (THE)

PO Box 97
CAPE TOWN 8000 W Cape
Tel: 021 462 2040 • Fax: 021 461 9519
Email: faircape@iafrica.com
Website: www.fairestcape.co.za
Contact Name: Rodney Leak – CEO

Aim: The Fairest Cape Association promotes a clean, healthy and sustainable Western Cape, in which everyone demonstrates pride, ownership and respect for the environment.

Services offered: The FCA aims to educate the public about the impact of waste on people and the physical environment; to promote waste stream reduction by encouraging the public to reduce, re-use and recycle; to lobby for appropriate integrated waste legislation, policy and practices; and to assist and encourage individuals, communities, municipalities and companies to take responsibility for the appropriate management of waste. The FCA operates a resource and information centre, assists in the setting up of waste management systems such as recycling depots, and offers education and awareness programmes/workshops for businesses, teachers, pupils, waste managers and the general public. We are specialists in solid waste management courses.

ACTIVITIES:

FERN SOCIETY OF SOUTHERN AFRICA (THE)

PO Box 73125
LYNWOOD RIDGE 0040 Gauteng
Tel: 012 662 1922 • Fax: 012 662 1922
Email: filices@pixie.co.za
Contact Name: Mrs Jolanda Nel

A forum for fern enthusiasts to meet and discuss the cultivation of ferns; with monthly newsletters, quarterly journals and field trips with conservation in mind.

ACTIVITIES:

FOOD AND TREES FOR AFRICA

PO Box 2035
GALLO MANOR 2052 Gauteng
Tel: 011 803 9750 • Fax: 011 803 9604
Email: info@trees.org.za
Website: www.trees.org.za
Contact Name: Jeunesse Park

FTFA works in partnership with government, the private and public sector and civil society to alleviate poverty, develop skills and contribute to food security for the numerous communities that apply to FTFA for assistance. The organisation contributes to the design, implementation and management of sustainable greening and Permaculture projects and aims to create awareness of the benefits of environmental upliftment activities amongst all communities of southern Africa.

FTFA has been greening South Africa since 1990 and has contributed to over 500 community food gardens, greened many suburbs, run environmental education and awareness campaigns, published easy-to-use materials and distributed over 2 million trees to disadvantaged and degraded areas in South Africa. FTFA has won national and international awards and recognition.

With the support of funders and the endorsement and cooperation of government FTFA implements three programmes, each with many projects.

The many companies, aid agencies, organisations, media and individuals who contribute to FTFA are improving the quality of life of impoverished communities by contributing to a sustainable green environment, food security, capacity building and development.

ACTIVITIES:

FOOD GARDENS FOUNDATION

PO Box 41250
CRAIGHALL 2024 Gauteng
Tel: 011 880 5956 • Fax: 011 442 7642
Email: fgf@global.co.za
Contact Name: Joy Niland – Director

The Food Gardens Foundation was established in 1977, as a socio-economic project to teach people to help themselves by growing essential food according to sustainable, organic principles. The vision of FGF is the empowerment of people to overcome malnutrition, hunger and famine.

The small-scale, low-cost and environmentally friendly FGF method is called Food Gardening. Practising this simple skill, people achieve community development and social upliftment – improving their health and quality of life, helping them to escape from the grip of poverty and helplessness, and enabling them to achieve a meaningful level of Household Food Security.

Food Gardening optimises normal domestic and other organic waste to revitalise the soil and feed growing vegetables. It is an appropriate method even for those who have meagre resources, making the best use of limited space and scarce water, restoring life and fertility to poor and arid land.

The FGF motto is MAXIMUM NUTRITION IN MINIMUM SPACE WITH LIMITED WATER AND RESOURCES.

ACTIVITIES: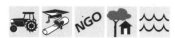

FORD MOTOR COMPANY OF SOUTHERN AFRICA

PO Box 411
PRETORIA 0001 Gauteng
Tel: 012 842 2911 • Fax: 012 803 4404
Email: crouser@ford.co.za • Website: www.mazda.co.za
Contact Name: Rob Crouse – Mazda Brand Manager

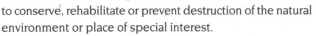

The Mazda Wildlife Fund was formed in April 1990, funded jointly by the Mazda Division of SAMCOR (now Ford Motor Company of Southern Africa) and its nationwide network of Mazda Dealers. The fund has committed an amount in excess of R1 million per annum over the past 10 years for investment in nature conservation.

The Mazda Wildlife Fund currently supports a wide range of projects, covering every aspect of conservation and preservation of our environment. These projects are situated throughout southern Africa and include educational projects and conservation projects. At the last count, 98 projects had already been sponsored by The Mazda Wildlife Fund, with 33 SAMCOR-donated Mazda vehicles presently operating all over South Africa.

The Mazda Wildlife Fund Board comprises senior representatives from the Endangered Wildlife Trust, the World Wide Fund for Nature (WWF-SA), the Wildlife & Environment Society of Southern Africa, as well as Mazda Dealer representatives and Ian Shepherd of Grey Worldwide. The Fund is specifically concerned with a dynamic involvement in activities that will directly provide a better environment for future generations.

ACTIVITIES:

FORUM FOR ECONOMICS AND THE ENVIRONMENT

PO Box 395
PRETORIA 0001
Tel: 012 841 3766 • Fax: 012 841 2689
Email: mdewit@csir.co.za • Website: www.econ4env.co.za
Contact Name: Dr Martin de Wit – Director

The Forum for Economics and Environment is a not-for-profit company focused on promoting the interface between economics and the environment through the dissemination of information, facilitation of open debate and the stimulation of research.

For more information visit our website at www.econ4env.co.za.

ACTIVITIES:

FREEME WILDLIFE REHABILITATION CENTRE

PO Box 1666
CRAMERVIEW 2060 Gauteng
Tel: 011 807 6993 • Fax: 011 706 6563
Email: tremor@icon.co.za • Website: freemewildlife.org.za
Contact Name: Margi Brocklehurst – Director

Freeme is a wildlife rehabilitation centre with an open permit from Gauteng nature conservation to raise, rehabilitate and release all indigenous birds, mammals, reptiles, incorporating environmental education in all spheres.

ACTIVITIES:

FRIENDS GROUPS

Friends are highly focused volunteers who band together to conserve, rehabilitate or prevent destruction of the natural environment or place of special interest.

Friends Groups are affiliated to The Wildlife and Environment Society of South Africa (WESSA) and adhere to the aims and objectives of the Society. Each group has its own constitution and runs quite independently using the conventional committee system. They are responsible for their own finances, do their own fundraising from members, the general public and they may raise sponsorships. Friends work closely with the formal management authorities of their chosen nature area; they help and support them with the aims and objectives for that area.

Some of the activities that the Friends groups around the country do:

- ◆ Run work parties: removing alien plants, repairing fences and various structures in the reserves/natural areas. Rehabilitation of buildings, waterways and wetlands.
- ◆ Identification of indigenous flora and fauna, censussing of birds, helping with reintroduction of additional species, control of problem animals.
- ◆ Fund raising: Acquiring sponsorship for specific projects. Selling merchandise.
- ◆ Education: Arranging conducted outings and courses on various topics including for example birds, spring flowers, trees, dragonflies, frogs, mushrooms, local history and cultural interests.
- ◆ Lobbying the authorities for the preservation of the natural environment and historical remains, arranging exhibitions depicting their special area of concern
- ◆ Improvement in the natural environment: such as planning, developing and building bird hides, trails, walkways, bridges and maintaining them.
- ◆ Collecting, collating information about their area, writing and producing maps, brochures, newsletters for visitors and members.
- ◆ Acting as eyes and ears to combat anti-social behaviour, such as vagrancy, noise pollution, illegal dumping, property encroachments, and lighting of fires other than in designated places. Working directly with the managing authorities on such issues as sewage spills, diverted and or polluted water supplies, poaching.

If you would like to join an existing group – or want to know how to set up a Friends Group – please contact:

The WESSA National Friend's Co-ordinator
Marion Dunkeld-Mengell
Tel/Fax: 012 667 2183 • Cell: 083 455 1736
Email: marion@friendsnylsvley.org.za

ACTIVITIES:

FRIENDS GROUPS:
NORTHERN AREAS:

	Friends of:	Contact Name:	Telephone
Tshwane (Pretoria)	Austin Roberts Bird Sanctuary	Jacques Botha	012 460 0240
Tshwane	Colbyn Valley	Piet-Louis Grundling	012 808 5342
Bronkhorstspruit	Ezemvelo Nature Reserve	Marion Dunkeld-Mengell	012 667 2183
Tshwane	Faerie Glen Nature Reserve	Reuben Heydenrych	012 549 5949
Tshwane	Groenkloof Nature Reserve	Robrecht Tryhou	012 427 8335
Benoni	Korsman's Bird Sanctuary	Ian Dustan	011 849 8574
East Rand	Marivale Nature Reserve	Alan Madden	011 734 4137
Tshwane	Moreletakloof	Uschi Arend	012 997 2347
Tshwane	Moreletaspruit	Barry Puttergill	012 361 4760
Near Modimolle	Nylsvley and Nyl floodplain	Marion Dunkeld	012 667 2183
Tshwane	Rietvlei Dam Nature Reserve	Fred du Ploouy	012 361 7322
Irene	Smuts Foundation	Cheryl Dehning	011 316 1426
Tshwane	Struben Dam	John Clark	012 361 4770
South Johannesburg	Suikerbosrand Nature Reserve	Frances Smith	011 640 2857
Tshwane	Vervet Monkey	Christina Vosloo	083 274 1702
Witbank	Witbank Nature Reserve	Malcolm Suttill	013 656 5932

EASTERN PROVINCE:

Walmer	Schoenmakerkop	Malcolm Smale	041 366 1049
St Francis Bay	Cape St Francis	Ian Benfield	042 294 0319
Greenacres	Groendal (Frog)	Martin de Bruyn	083 463 7598
Linton Grange	Blue Horizon Bay	Jacques Smit	082 928 8146
Grahamstown	Thomas Baines	Annie Hahndick	046 622 4727
Summerstrand	Greater Addo Nat. Park	Valerie Hunt	041 582 2784
Algoa Bay	Algoa Bay	Leon van Zyl	082 632 3057
Jeffreys Bay Rehabilitation Centre	Jeffreys Bay Penguin	Crystal Hartley	042 293 1320

WESTERN CAPE:

Paarden Island, Cape Town	Paarden Island Wetlands	Peter Albert	021 511 7234
Montagu	Montagu Nature Garden	Debbie Murray	023 614 2304
Rondebosch, Cape Town	Friends of Rondebosch Common	Tim Jobson	021 940 5314
Scarborough	Scarborough Conservation Group	Ivan Groenhof	083 2617960
Newlands, Cape Town	Afro Montane Information Forum	Pixie Littlewort	021 686 3964
Paarl	Boland Environment Forum	Sara van Tonder	021 872 7209
Northern suburbs, Cape Town	Durbanville Environment Forum	Johnathan Cartwright	021 976 2959
Newlands, Cape Town	Afro Montane Information Forum	Pixie Littlewort	021 686 3964
Paarl	Boland Environment Forum	Sara van Tonder	021 872 7209
Northern suburbs, Cape Town	Durbanville Environment Forum	Johnathan Cartwright	021 976 2959
Cape of Good Hope area	Friends of the Cape of Good Hope	Dolores Donovan	021 785 2191
Constantia, Cape Town	Friends of the Constantia Valley Green Belts	Brian Ratcliffe	021 794 4620
Camps Bay/Table Mountain area	Friends of the Glen	Sally Forde	021 438 2013
Somerset West	Friends of Helderberg Nature Reserve	A. Groenewald	021 852 8831
Fish Hoek	HOPE Group	Diana Busschau	021 782 7243
Cape Town	Friends of the Liesbeek	Liz Wheeler	021 671 4553
Cape Town/Table Mountain area	Friends of Lion's Head	Cheryl Pocock	021 434 8456
Meadowridge, Cape Town	Friends of Meadowridge Common	Roger Graham	021 715 9206
Newlands, Cape Town	Friends of Newlands Forest	Jeff Goy	021 674 1392
Milnerton, Cape Town	Friends of Rietvlei	Margaret Maciver	021 557 4990
Southern Peninsula	Friends of Silvermine Nature Area	Sandy Barnes	021 785 1477
Simon's Town	Simon's Town Flora Conservation	Peter Salter	021 786 1620
Simon's Town	Friends of Simon's Town Coastline	Rosemary Barker	021 786 2233
Tokai, Cape Town	Friends of Tokai Forest	John Green	021 712 1341
Northern Suburbs, Cape Town	Friends of the Tygerberg Hills	Karen Marais	021 945 2855

FRY'S METALS [A DIVISION OF ZIMCO (PTY) LTD]

PO Box 519 GERMISTON 1400 Gauteng
Tel: 011 827 5413 • Email: guym@frys.co.za
Contact Name: Guy Marshall

Principle products recycled: lead-acid batteries, scrap lead, lead dross and lead oxide.

Principle end products: Battery alloys, antimonial lead alloys, cable sheath alloys, calcium lead alloys and 99.97% refined soft lead.

ACTIVITIES:

FUTURE FOOTPRINTS

PO Box 31152 KYALAMI 1684 Gauteng
Tel: 082 990 7583 • Email: footprint@futurefootprints.co.za
Contact Name: Diane Sheard – Director

Environmental education and animal rehabilitation.

ACTIVITIES:

GAEA PROJECTS

PO Box 30258
MAYVILLE 4058 KwaZulu-Natal
Tel: 031 756 1756 • Fax: 031 765 3172
Email: nicky@gaea.co.za • Website: www.gaea.co.za
Contact Name: Nicky Jones – Office Manager

Gaea Projects is a unique collaboration between an environmentalist, a town and environmental planner, and a landscape architecture specialist. We offer an integrated service in which planning and environmental issues, GIS, and landscape design and architecture complement each other. The advantage being that all elements required by new and existing planning, development and environmental legislation are provided in-house.

ACTIVITIES:

GALAGO VENTURES

PO Box 886 IRENE 0062 Gauteng
Tel: 012 345 5103/4 • Fax: 012 345 2211
Email: rvmarais@lantic.net
Contact Name: Riaan Marais – Project Manager

Galago Ventures specialises in faunal and floral surveys underpinning environmental management objectives. The Galago Ventures team is committed to both the sustainable conservation of South African biodiversity as well as to economic development. These two concepts appear to be in conflict, but we strive to petition creative solutions should irreconcilable positions arise.

Due to the complexity of our biological environment, surveys undertaken by Galago Ventures are done by a number of core specialists which, depending on the specific site, can be expanded.

Our core competencies are:

- ◆ Provide clients with vegetation surveys which, depending on the brief, can be expanded to denote plant communities, medicinal plants and exotic weeds. Vegetation maps or the mapping of ecologically sensitive areas can be provided
- ◆ Provide clients with survey data pertaining to the mammals, birds, reptiles and amphibians of a particular site or region
- ◆ Depending on the assignment, survey findings can be interpreted in the context of habitat classification, and data can be presented to define the faunal and floral components of constituent ecological assemblies
- ◆ Evaluation of the conservation significance of a site, with special emphasis on the presence and status of officially acknowledged threatened species
- ◆ Projections of potential ecological impacts resulting from development
- ◆ Recommendations to mitigate negative and enhance positive environmental impacts, should development be approved
- ◆ Advise on further studies and ecological rehabilitation

ACTIVITIES:

GAME RANGERS' ASSOCIATION OF AFRICA

PO Box 78 ROSETTA 3301
KwaZulu-Natal
Tel: 033 267 7171 • Fax: 033 267 7171
Cell: 082 463 4104 • Email: snowman@ewt.co.za
Contact Name: Tim Snowman

This association, controlled by Professional Game Rangers, is committed to the protection of natural assets, primarily through the maintenance and improvement of field-based training standards.

ACTIVITIES: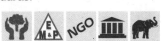

GAUTENG CONSERVANCY ASSOCIATION

PO Box 1552 WALKERVILLE 1876 Gauteng
Tel: 084 5902312
Email: mail@thorntree.co.za
Contact Person: Ivan Parkes – Chairman

The Gauteng Conservancy Association (GCA) was formed in February 2003 to promote conservation on private property in Gauteng and to give "teeth" to efforts to protect Gauteng's fast-disappearing greenbelt areas. It is an independent body, prepared to co-operate with all those who are passionate about conservation and willing to work for it. The GCA represents 27 rural and urban conservancies in Gauteng.

ACTIVITIES:

GEOMEASURE GROUP (PTY) LTD

PO Box 36 Haraldene Road GLENWOOD 4001 KwaZulu-Natal
Tel: 031 205 8624 • Fax: 031 205 4075
Email: theo@geomeasuregroup.co.za
Website: www.geomeasuregroup.co.za
Contact Name: Theo van Niekerk

The Geomeasure Group is represented by a team of hydrogeologists and environmental scientists with extensive experience in the fields of groundwater exploration, wellfield development, aquifer management, regional mapping and resource evaluation, chemical and hydrocarbon groundwater contaminant investigation and remediation, waste and cemetery site identification, investigation and permitting, development of environmental management systems, compliance monitoring and auditing.

ACTIVITIES:

GEOPRECISION SERVICES CC

PO Box 19786 NOORDBRUG 2522
North West
Tel: 018 294 4651 • Fax: 018 294 4651
Email: gpscc1@lantic.net
Contact Name: Mr Pieter Swart – Manager

Each rehabilitation project is unique with its own challenges and impact mitigation strategies. Effective remediation planning implicates practical, cost-effective and innovative solutions and designs. GeoPrecision Services CC specialises in a range of environmental services incorporating these criteria, to ensure minimal environmental risk and liability.

ACTIVITIES:

GET BUSHWISE

PO Box 26361 STEILTES 1213 Nelspruit
Tel: 013 744 0936 • Fax: 013 744 9911
Email: nadine@getbushwise.co.za
Website: www.getbushwise.co.za
Contact Names: Dee Adams / Nadine Clarke

Get Bushwise has brought out a series of fun, environmentally educational books for children (Struik) and a range of kids bush accessories. 10% of the proceeds go towards environmental education.

ACTIVITIES:

GLASS RECYCLING ASSOCIATION (THE)

PO Box 5303 DELMENVILLE 1403 Gauteng
Tel: 011 874 0000 • Fax: 011 827 0210
Email: jhuman@consol.co.za
Website: www.glassrecycling.co.za
Contact Name: Mr J W Human

The Glass Recycling Association represents the glass packaging recovery and recycling interests of both Consol Glass and Metal Box Glass. The Association was formed in 1986 and administers 137 collecting agents throughout southern Africa. Recycled glass is sourced from bottlers, on-premise consumption outlets such as hotels, restaurants and clubs, hawkers and voluntary drop-off points. Presently some 1 500 bottle banks are placed at shopping centres, parking areas and recreation centres for the convenience of the public.

Glass is 100% recyclable and can be re-melted over and over again without losing any of its inherent qualities. Today's new bottles and jars contain in excess of 25% of recycled glass. Glass is made from natural materials. Although these materials are plentiful, there are a number of reasons why glass should be recycled and not simply discarded: it conserves raw materials; it saves space in landfill sites; it saves energy consumption, as recycled glass melts at a lower temperature than virgin raw materials; it reduces air pollution; it reduces the demand for water used in the glass-producing process; it reduces litter.

ACTIVITIES:

GISP (GLOBAL INVASIVE SPECIES PROGRAMME)

C/o NBI Kirstenbosch
CLAREMONT 7735 W Cape
Tel: 021 799 8836 • Fax: 021 797 1561
Email: gisp@nbi.ac.za • Website : www.gisp.org
Contact Name: Kobie Brand

The Global Invasive Species Programme (GISP) was established in 1997 to address global threats caused by IAS and play a leading role towards the implementation of Article 8(h) of the Convention on Biological Diversity (CBD). The GISP mission is to conserve biodiversity and sustain human livelihoods by minimising the spread and impact of invasive alien species.

IAS are those species that become established in a new environment, then proliferate and spread in ways that are destructive to human interests and native biodiversity. IAS are now considered to be one of, if not indeed the greatest threats to the ecological and economic well-being of the planet.

The GISP Secretariat aims to draw upon a global partnership network of organisations, programmes and experts with an interest in IAS issues. The Secretariat will also build on the work achieved and international projects launched during GISP Phase 1. The many innovative and important projects and publications already developed, including the international Toolkit for Invasive Species and a Global Strategy on IAS will be widely promoted.

To mark the Secretariat's launch, GISP has produced a book on the growing danger of IAS in Africa. Entitled 'Africa Invaded', the publication aims to raise awareness in Africa and beyond about some of the most prominent IAS issues facing the continent. Copies are available from the GISP offices.

ACTIVITIES:

GLOBAL OCEAN (PTY) LTD

5 Hazelden House Somerset Business Park
SOMERSET WEST 7129 W Cape
Tel: 021 852 8967 • Fax: 021 852 8017
Email: markm@globalocean.co.za
Website: www.globalocean.co.za
Contact Name: Mark Miles – Sales Director

Development and supply of complete, energy efficient aquaculture systems. Specialising in culture of abalone and other high value marine species. Water is continually cleaned and aerated by in-tank bio-remediation.

ACTIVITIES:

GOLDFIELDS ENVIRONMENTAL EDUCATION CENTRE

NBI Kirstenbosch
Private Bag X7
CLAREMONT 7735 W Cape
Tel: 021 799 8670 • Fax: 021 797 1919
Email: fullard@nbi.ac.za • Website: www.nbi.ac.za
Contact Person: Donavan Fullard – Head of Education

The mission of the Goldfields EE Centre is to use the Kirstenbosch Botanical Gardens and other resources of the NBI to inspire and enable people to take responsibility for the environment. We offer a wide range of environmental education programmes which include:

1. A curriculum linked garden-based school programme.
2. Teacher professional development.
3. Resource materials development.
4. Outreach greening projects with schools.

ACTIVITIES:

GONDWANA ALIVE

Private Bag X101
PRETORIA 0001 Gauteng
Tel: 012 804 3200 • Fax: 012 804 3211
Email: Gondwana@webmail.co.za
Website: www.gondwanaalive.org
Contact Name: Dr John Anderson – Specialist Scientist, National Botanical Institute

Gondwana Alive is an international endeavour to spread awareness of the deepening 6th extinction and to persuade all humans throughout the world to join together in stemming this unacceptable process.

Gondwana Alive publishes:

◆ "Towards Gondwana Alive Vol. 1 Promoting biodiversity and stemming the Sixth Extinction" – a full colour 133 page publication with endorsements from Nelson Mandela, Kofi Annan, the Dali Llama, Sir David Attenborough and others.

◆ "Towards Gondwana Alive Vol. 2 – A set of 100 strategies towards stemming the Sixth Extinction".

"Towards Gondwana Alive Vol. 3 – Biodiversity through deep time & the evolving terrestrial biosphere". This 750-page full colour publication includes contributions from 150 top biological and global scientists.

ACTIVITIES:

GRANT JOHNSTON ASSOCIATES CC

PO Box 1026
PLETTENBERG BAY 6600 W Cape
Tel: 044 533 0728 • Fax: 044 533 2379
Email: dianagrant@mweb.co.za
Contact Name: Di Grant BA (Hons) Landscape Architecture (Heriot-Watt) MLI UK – Principal

Grant Johnston Associates cc is an environmental consultancy with access to GIS expertise via local networking. We specialise in Environmental Impact Assessment (EIA), landscape planning and visual assessment.

ACTIVITIES:

GRASSLAND SOCIETY OF SOUTHERN AFRICA

PO Box 41
HILTON 3245 KwaZulu-Natal
Cell: 083 256 7202 • Fax: 033 390 3113
Email: info@gssa.co.za • Website: www.gssa.co.za
Contact Name: Freyni Killer

Professional association of individuals and institutions working in the field of or interested in rangeland science and forage production. Activities include: dissemination of information through a scientific journal and annual congresses, publication of a bulletin and other special publications; organisation of symposia as an interaction between scientists and producers; and administration of a trust fund to further the objects of the Society.

ACTIVITIES:

'The conservationist's most important task, if we are to save the Earth, is to educate.'

~ Sir Peter Scott,

Founder Chairman of WWF

Printed on Sappi recycled paper

GREEN & GOLD FORUM

GREEN & GOLD
The responsible choice for business & industrial leaders

PO Box 74416 LYNNWOOD RIDGE
0040 Gauteng
Tel: 012 803 9736 • Fax: 012 803 9745
Email: green.gold@pixie.co.za • Website: www.stratek.co.za
Contact Name: Dr Kelvin Kemm – Director

Green and Gold Forum was founded in 1990. The Forum strives to bring a scientifically accurate picture to the public and to decision-makers in both public and private enterprises. Environmental issues are extremely important and can impact considerably on the profitability of a range of projects and endeavours. These issues are at times so emotive that it is all too easy for distorted stories to become established as fact in the public mind. G&G Forum is composed of qualified scientists, economists, lawyers and other professionals who collectively offer a range of consultancy services.

These include: Environmental Impact Assessments, writing proposals and information documents, production of videos, investigations of individual incidents, briefing of companies or government agencies, and international liaison.

Green & Gold Bulletin is the monthly publication of the Green & Gold Forum. G&G has international representation in Washington DC and London, but also has links with many other countries.

ACTIVITIES:

GREEN CLIPPINGS

PO Box 680
STELLENBOSCH 7599 W Cape
Tel: 021 883 3366 • Email: gc@greenclippings.co.za
Website: www.greenclippings.co.za
Contact Name: Richard Weeden

Green Clippings publishes a weekly digest of environmental and conservation news. The service is popular with government, business, educators, environmental organisations and conservation groups, lawyers, students, journalists, health parishioners and the general public.

ACTIVITIES:

GREEN GAIN CONSULTING (PTY) LTD

PO Box 1528
FERNDALE 2160 Gauteng
Tel: 011 792 9129 • Fax: 011 793 2337
Email: info@greengain.co.za
Website: www.greengain.co.za
Contact Name: Adv Nicolai Massyn – Managing Director

GREEN GAIN CONSULTING

Green Gain Consulting provides environmental legal and training services and products to mining, industry and the public sector. Our regular client base consists of most of the larger companies in these sectors. We have offices in Randburg and Richards Bay. Our staff consists of legal and environmental professionals.

Services:
- ◆ Legal compliance – We have developed a turnkey environmental legal compliance system E-LCS. This system creates and integrates environmental legal compliance audits, company activities/aspects/impacts registers, site-specific legal registers, and full text of relevant national, provincial and local legislation and continuous updating of the system with new legal obligations.
- ◆ Risk Management – We provide environmental legal risk assessments, compliance audits and due diligence services.
- ◆ EMS – Organisation conformance to the ISO 14 001 standard. Assistance with licensing and permitting.
- ◆ Environmental Management and Planning – Creation of legislative frameworks for development.
- ◆ Environmental training – We provide cost effective short training courses to all levels of employees in most languages. All trainers and courses are registered and accredited.
- ◆ Consulting – Legal opinions and litigation support.

ACTIVITIES:

GREENHAUS ARCHITECTS

3 Eskol Lane CONSTANTIA 7806 W Cape
Tel: 021 794 4465/6 • Fax: 021 794 0360
Email: greenhaus@icon.co.za
Contact Name: Etienne Bruwer

General practice in organic architecture, focussing on the use of common green/ecological principles and systems that flow naturally out of higher aesthetic and spiritual dimensions.

Pioneered the search in South Africa for context-appropriate practices to transform professional practice and built environment procurement to serve local conditions and realities in the making of sustainable built environments. Interpreted and tailored universal 'green' principles to guide correct ecological approaches to the sustainability triad which reflect a balance of social economic and environmental imperatives.

Activities range from design/supervision, the building of green buildings to general consultancy work, publication in local and foreign 'green' press, occasional academic work, television/radio appearances, co-convening 'strategies for sustainability in the built environment' conference series with UCT / CSIR / Kruger Associates, offering training courses to the NGO sector in green architecture – transforming traditional architecture and deficient R D P housing to better living environments.

ACTIVITIES:

GREENHOUSE PROJECT

PO Box 32025
BRAAMFONTEIN 2017 Gauteng
Tel: 011 720 3773 • Fax: 011 720 3532
Email: info@ghouse.org.za
Website: www.greenhouse.org.za
Contact Name: Dora Lebelo – Information Co-ordinator

The GreenHouse Project runs the GreenHouse People's Environmental Centre in Joubert Park, Johannesburg. The Centre has a demonstration office and gardens, which showcase environmental building methods, energy conservation, water conservation and re-use, waste recycling and organic growing. Visit the Centre every third Saturday of the month. A free information resource on 'Green Living and Development' can be found on the website.

ACTIVITIES:

GREEN INC

PO Box 226 PARKLANDS 2121 Gauteng
Tel: 011 782 2210 • Fax: 011 782 2207
Email: stuart@greeninc.co.za • Website: www.greeninc.co.za
Contact Name: Stuart Glen – Director

At Green Inc, we are sensitive to the context of our sites. This enables us to create landscapes that are both environmentally and architecturally appropriate. We specify only indigenous plant material to help re-establish viable ecosystems on our sites. We establish close working relationships with the members of the professional teams with whom we work. Throughout the process, we remain aware that our projects must be commercially viable.

The dynamic practice provides outstanding graphic communication, innovative construction detailing and is equipped to allow seamless data transfer to other professionals on our teams using CAD.

Services include environmental and open space planning, environmental impact assessments and management plans.

ACTIVITIES:

GREEN TRUST (THE)

Private Bag X2 DIE BOORD 7613 W Cape
Tel: 021 888 2837 • Fax: 021 888 2888
Email: tbrinkcate@wwfsa.org.za
Website: www.panda.org.za
Contact Name: Therese Brinkcate – Manager

The Green Trust was founded in October 1990. It is a subsidiary of WWF South Africa (the conservation organisation) in a mutual benefit partnership with Nedbank. Funds are raised through the use of Nedbank Green products including cheque books, credit cards and savings accounts. The Green Trust funds a broad range of practical and sustainable conservation projects with a significant focus on community-based conservation. Projects range from urban-greening activities, conservation of highly-endangered species to the promotion of sustainable use of natural resources. The Trust aims to protect the unique biodiversity of southern Africa and in so doing to promote an improved quality of life for all South Africans.

Projects funded by The Green Trust:
◆ Urban Greening (Food and Trees For Africa, Abalimi Bezekhaya)

◆ Conservation of Cheetah in Namibia (CCF)
◆ Wildlife Trade Project (TRAFFIC)
◆ Kosi Bay Co-Management Project (sustainable use)
◆ Farm worker awareness programme for Crane Conservation
◆ Special Advisor on Environmental Education to the Minister of Education
◆ Opinion Leader Trails
◆ Itala Co-operative community conservation project
◆ Ground Hornbill Reintroduction project
◆ Ukuvuka: Operation Firestop
◆ Wild Dog Conservation
◆ Conservation of the Knysna Seahorse
◆ Master Farmer Programme – community conservation project
◆ Conservation of the Blue Swallow
◆ Ekangala Grassland Project
◆ Hippo Reintroduction into Rondevlei
◆ Badgers and Beekeepers
◆ Scuba Divers and Ragged Tooth Sharks at Aliwal Shoal
◆ Addressing the Invertebrate extinction crisis
◆ Klip River Wetland Rehabilitation project
◆ Operation Oxpecker
◆ National Biodiversity Strategy and Action Plan – civil society involvement

ACTIVITIES:

GROUND HORNBILL PROJECT

Mabula Game Reserve
Private Bag X1644
WARMBATHS 0480 Mpumalanga
Tel: 014 734 1788 • Fax: 014 734 0013
Email: project@ground-hornbill.org.za
Contact Name: Ann Turner – Co-ordinator

ACTIVITIES:

GROUNDWATER ASSOCIATION OF KWAZULU-NATAL (GAKZN)

Postnet Suite 187,
P/Bag X04
DALBRIDGE 4014 KwaZulu-Natal
Tel: 031 205 8624 • Fax: 031 205 4075
Email: theo@geomeasuregroup.co.za
Website: www.geomeasuregroup.co.za
Contact Name: Theo van Niekerk – Chairman

Members actively involved in groundwater industry, consulting, surveys, environmental review panel, test pumping, pumps, borehole, drilling, soil, water sampling, reticulation equipment materials supplies and services, contracting, project management.

ACTIVITIES:

GROUNDWATER CONSULTING SERVICES

PO Box 2597 RIVONIA 2128 Gauteng
Tel: 011 803 5726 • Fax: 011 803 5745
Email: faniec@gcs-sa.biz
Website: www.gcs-sa.biz
Contact Names: Andrew Johnstone – Senior Partner / Fanie
Coetzee – Manager Environmental Unit

- ◆ Environmental Impact Assessments
- ◆ EMPs and Site Monitoring
- ◆ Environmental Management Programme Reports
- ◆ Rehabilitation and Reclamation Investigations
- ◆ Environmental Auditing and EMS
- ◆ Environmental Risk and Liability
- ◆ Public and Community Participation

ACTIVITIES:

GROUNDWORK

PO Box 2375 PIETERMARITZBURG 3200 KwaZulu-Natal
Tel: 033 342 5662 • Fax: 033 342 5665
Email: gill@groundwork.org.za
Website: www.groundwork.org.za
Contact Name: Gillian Addison – Administrator

groundWork is a non-profit environmental justice service and developmental organisation working primarily in South Africa but increasingly in southern Africa.

groundWork is recognised as the leading South African NGO working on industrial pollution. It focuses on supporting communities faced with environmental threats and aims to build community awareness and solidarity. It supports communities by providing or brokering strategic and technical advice and information. It seeks to build the community voice by facilitating links between communities faced with similar environmental problems, supporting community campaigns by negotiating with industry, accessing government decision-makers, providing press exposure, and linkages with national and international campaigns. Such support to communities contributes to improved democratic environmental governance and poverty alleviation.

groundWork's main project areas are:

- ◆ corporate accountability
- ◆ air quality
- ◆ health care waste and incineration
- ◆ industrial landfill waste, and publications.

ACTIVITIES:

GUY STUBBS PHOTOGRAPHER

PO Box 15 IRENE 0062 Gauteng
Tel: 012 667 3939 • Fax: 086 672 2273
Email: stubbsgr@mweb.co.za
Website: www.guystubbs.co.za
Contact Name: Guy Stubbs

Guy Stubbs is an internationally published, award-winning photographer. Based in the country that inspires him – South Africa – Guy specialises in environmental photography.

He is a fine arts photographer who communicates messages through creative, aesthetically stimulating images. He prefers working on large assignments where he can really get to grips with his subject matter and interpret it fully. Guy works for a variety of clients, which reflects his creative versatility. He has delivered numerous assignments for advertising agencies such as TBWA Hunt Lascaris, social development organisations such as World Bank, international magazines such as ADAC travel magazine in Germany, New York Times Travel magazine in the USA, the Sunday Independent magazine in the UK, and is well known for the eye-catching front cover photography he has done for Out There Magazine in South Africa.

He has worked for graphic design studios such as Wiggins Inc. in the USA and is involved in large-scale projects for big corporates such as Barlow International. Design trends in South Africa traditionally dictate that photographs are dropped into a preconceived concept with little or no consultation with the photographer. Guy's approach is holistic in this sense – he starts with the concept as a whole and concentrates on creating an image that will enhance the message.

Guy likes to be closely involved in the design and publication of his work. To this end he began his own publishing company ten years ago and has built up a network of designers and copywriters who he uses and builds customised teams for his clients' unique communication needs. He has produced upmarket calendars for clients such as the Lesotho Highlands Development Authority, books for large corporations such as Southern Life and international donors such as the European Union, and has developed visual communications programmes for leading development authorities such as the WWF - Green Trust. Guy has a post-graduate degree in photography, having specialised in communicating environmental issues. He has won numerous international photographic awards, produced acclaimed publications, and his creative ideas have influenced many. Because of his experience, enthusiasm and vast knowledge of the people and places in Africa, his concern for the environment and his compassion for the poor, Guy Stubbs is the ideal person to pursue your project.

ACTIVITIES:

HABITAT COUNCIL

1 Bertha Court Exner Avenue
VREDEHOEK 8001 W Cape
Tel: 021 465 3972 • Fax: 021 461 0709
Email: mlroux@new.co.za
Contact Name: Mrs ML Roux – Executive Officer

The Habitat Council, national consultative umbrella NGO, works to prevent environmental harm by promoting ethical environmental management and sound environmental legislation, preserving biodiversity and natural resources, and combating pollution.

ACTIVITIES:

HANGERMAN

PO Box 21
STEENBERG 7947 W Cape
Tel: 021 701 1595 • Fax: 021 701 5053
Email: neil@hangerman.co.za
Contact Name: Neil Weston – Managing Director

Hangerman specialises in the sale of "good as new" plastic clothes hangers, as specified by several of the major national chain stores. The process of "refurbishing" used clothes hangers provides employment to many handicapped people.

ACTIVITIES:

HATCH

Private Bag X20
GALLO MANOR 2052 Gauteng
Tel: 011 239 5577 • Fax: 011 239 5788
Email: praymond@hatch.co.za
Website: www.hatch.co.za
Contact Name: Peter Raymond – Senior Consultant

HATCH is a global engineering company, providing comprehensive consulting project delivery and operational support services to the mining, metallurgical and manufacturing industries.

ACTIVITIES:

HEINRICH BÖLL FOUNDATION

43 Tyrwhitt Avenue
MELROSE 2196 Gauteng
Tel: 011 447 8500 • Fax: 011 447 4418
Email: info@boell.org.za
Website: www.boell.org.za
Contact Name: Dr Stefan Cramer – Regional Director

The Heinrich Böll Foundation is an independent, non-profit and non-governmental foundation from Germany. It has supported and co-operated with civil society initiatives since 1989 through funding, advocacy and networking.

Headquartered in Berlin, Germany, the Heinrich Böll Foundation is a political non-profit Foundation affiliated with the party of Alliance 90/The Greens. Striving to promote democratic ideas, civil society and international understanding, the work of the Heinrich Böll Foundation centres on the core political values of ecology, democracy, solidarity and non-violence.

At the moment, the Foundation is represented by Regional offices in seventeen countries: Belgium, Bosnia-Herzegovina, Brazil, Cambodia, Czech Republic, El Salvador, Israel, Kenya, Nigeria, Pakistan, Palestine, Poland, Russia, South Africa, Thailand, Turkey and USA.

ACTIVITIES:

HEWLETT-PACKARD SOUTH AFRICA (PTY) LTD

12 Autumn Street
RIVONIA 2146 Gauteng
Tel: 011 785 1000 • Fax: 011 785 1500
Email: shane.tyrrell@hp.com
Website: www.hp.com
Contact Name: Shane Tyrrell – Supplies Business Manager

At Hewlett-Packard, our environmental goals are to provide products and services that are environmentally sound throughout their life-cycles and to conduct our operations worldwide in an environmentally responsible manner.

More information at: www.hp.com/go/environment or www.hp.co.za for local information.

ACTIVITIES:

HIKING AFRICA SAFARIS & TOURS

PO Box 76583
WENDYWOOD 2144 Gauteng
Tel: 082 854 5853 • Fax: 011 444 2464
Email: hikingafrica@mweb.co.za
Website: www.hikingafrica.co.za
Contact Name: Milly Jarvis – Managing Member

- ◆ Hike or tour with us and unveil the secrets of South Africa personally. Guided Hiking Trails: 1-10 day guided hiking trails in the Magalies and Drakensberg Mountains, Wildcoast and Namibia.
- ◆ Wildlife Safaris: Pilanesberg National Reserve, Kruger Park, Entebeni, and many other private reserves.
- ◆ Touch Tours: Experience the joy of touching and interacting with young lions, giraffes and feeding young elephants. Experience of a lifetime – not to be missed.
- ◆ City Tours: Experience the history and culture of our three cities in Gauteng:
  - ◆ Soweto: share in Nelson Mandela's life before he went to prison
  - ◆ Johannesburg: where gold was discovered
  - ◆ Pretoria: our beautiful garden city
  - ◆ Cultural Tours: visit a traditional cultural village where five of our tribes celebrate their uniqueness.

Registered DEAT and FGASA Licensed Guides who love their country and are enthusiastic about sharing it with you our visitors.

Pick up and drop off at all local hotels in luxury air-conditioned VW Micro bus. Half day, full day and longer tours available and we will tailor make tours to suit your needs.

ACTIVITIES:

HILLAND ASSOCIATES

PO Box 590
GEORGE 6530 W Cape
Tel: 044 889 0229 • Fax: 044 889 0229
Email: info@hilland.co.za • Website: www.hilland.co.za
Contact Name: Cathy Avierinos – Director

HilLand Associates is a multi-disciplinary consultancy specialising within the disciplines of environmental planning and management. The principles of Integrated Environmental Management are obtained – from project planning, scoping and specialist studies through to project management, environmental control and monitoring. This is ensured through facilitation and co-ordination. Feasibility studies, using ecological surveys and impact analysis, culminate in appropriate decision-making, using GIS applications, spatial analysis, monitoring and auditing.

Research planning and scientific methodology are maintained to the highest standards, through affiliation to SACNASP, SAIE&ES, IAIA-SA and SAAB. Our speciality fields include integration of Golf Estate Planning and Rehabilitation within the natural environment; Impact Assessments and Permaculture project design. We are committed to maintaining a high standard of environmental ethics, to which international standards and norms are adhered.

ACTIVITIES:

HILLSIDE ALUMINIUM

PO Box 897 RICHARDS BAY 3900
KwaZulu-Natal
Tel: 035 908 8854 • Fax: 035 908 8727
Email: hendrik.louw@bhpbilliton.com
Website: www.hillside.co.za
Contact Name: Hendrik Louw – Environmental Specialist

The Hillside Aluminium Smelter is situated in the coastal city of Richards Bay, uMhlathuze, in northern KwaZulu-Natal and is South Africa's major producer of primary aluminium. 500 000 tons of high quality aluminium is produced on an annual basis for the export market. The current expansion of the smelter, following the International Association for Impact Assessment SA (IAIAsa) award winning EIA, will result in an annual production of 640 000 tons. Aluminium is the "metal for the future" in line with sustainability, especially with regards to its impact on recycling and indirect effect on energy consumption, i.e. lighter motorcars for better fuel efficiency. An accredited Environmental Management System (ISO 14001) has been implemented in order to manage and minimise all associated environmental impacts. Biannual Interested and Affected Party meetings are being held for feedback on environmental progress. A sustainability report is also produced in accordance with the Global Reporting Initiative (GRI). Hillside Aluminium sees itself as a leader in the quest towards sustainability.

ACTIVITIES:

HOEDSPRUIT ENDANGERED SPECIES CENTRE

Po Box 912031
SILVERTON 0127 Gauteng
Tel: 012 804 4840 • Fax: 012 804 3984
Email: info@cheeetahresearch.co.za /
Ina.michau@gentour.co.za
Website: www.cheetahresearch.co.za
Contact Name: Ina Michau – Marketing Manager

Established in 1990 as a breeding programme for cheetah, the Centre now also breeds and conserves other African species for reintroduction into the wild.

Great care is taken to ensure suitable breeding environments for all animal inhabitants, in some cases in co-operation with breeding programmes of the National Zoological Gardens in Pretoria and the Zoology Department of the University of Pretoria. The cheetah rescue unit was also established at the centre to assist farmers with problem animals.

Operating as a non-profit organisation, the Centre is funded from gate entrance fees and sales from the curio shop on the premises. The objective of the Centre is to make an important contribution to the conservation of Africa's endangered wildlife for posterity.

ACTIVITIES:

HOLCIM (SOUTH AFRICA)
(Previously Alpha (Pty) Ltd)

PO Box 6367
WELTEVREDEN PARK 1715
Gauteng
Tel: 011 670 5614 • Fax: 011 670 5790
Email: acluett@alpha.co.za • Website: www.alpha.co.za
Contact Name: A Cluett – Environmental Consultant

Holcim's Environmental Policy is based on the recognition that OUR ACTIONS TODAY MOULD THE FUTURE. Holcim endorses the universal right of present and future generations to an environment that is not harmful to human health or well-being. We believe the best mechanism to ensure that this right is achieved is through the process of Sustainable Development within the framework of the principles of Integrated Environmental Management. The Holcim Environmental Policy, first published in 1994, was revised in June 2000 to reflect the commitment of Holcim to the South African Constitution and to our goal of continual environmental performance improvement.

Our strategy to achieve policy objectives includes: compliance with environmental legislation; open participation in environmentally related dialogue with interested and affected parties; optimal utilisation of resources; monitoring and auditing of environmental performance; and, on closure of an operation, the rehabilitation of the site to a self-sustainable or positively usable landform.

ACTIVITIES:

HOLGATE AND ASSOCIATES

PO Box 97497 PETERVALE 2151 Gauteng
Tel: 011 486 2567 • Fax: 011 486 2567
Email: holgate@worldonline.co.za
Contact Name: Claudia Holgate – Director

Holgate and Associates – Environmental Management Services Provide professional environmental services in the following fields:

◆ Environmental Impact Assessments (EIAs) and EIA Review
◆ Environmental Management Plans (EMPs)
◆ Strategic Environmental Assessments
◆ Visual Assessments
◆ Faunal and Floral Surveys (Incl. Avian surveys)
◆ Catchment Management and Biomonitoring
◆ Wetland Rehabilitation
◆ Environmental Research and Training
◆ Policy Development
◆ Project Management

ACTIVITIES:

HOTBAG PROJECT

PO Box 651794 BENMORE 2010 Gauteng
Tel: 011 792-8675 / 083-539 5192 • Fax: 086 672-2910
Email: hotbag@mweb.co.za
Contact Name: Wendy Chandler

The HotBag is a thermal bag which holds a cooking pot and can save up to 75% of fuel when cooking with retained heat. Because the pot of food only needs to remain on the stove for about a third of the normal cooking time, it takes the stress out of cooking and frees one to use only one heat source for more than one pot of food. This neat, light-weight and washable bag is easy to use at home and away. It cooks food to perfection in normal time and there is no moisture loss or chance of food burning. Food stays hot for 3-5 hours and longer. It can also be used as a cooler bag. The HotBag Project was started in 2001 as a job-creation initiative in Gauteng when Wendy Chandler began to develop and promote this unique design on the hay-box concept to become recognised as an essential item in every household.

ACTIVITIES:

HOUT BAY AND LLANDUDNO HERITAGE TRUST

PO Box 27091 HOUT BAY 7872 W Cape
Tel: 021 790 2008
Email: hb.heritage@zsd.co.za
Website: www.zsd.co.za/~houtbay
Contact Name: Dave Cowley

Mission is to protect and preserve the Architectural, Historical, Environmental and Cultural Heritage of the Community. Projects include the Hout Bay River and the military history sites of the area.

ACTIVITIES: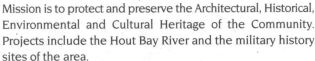

HRH THE PRINCE OF WALES'S BUSINESS & THE ENVIRONMENT PROGRAMME

PO Box 2264 CLAREINCH 7740 W Cape
Tel: 021 671 8803 Fax: 021 674 4334
Email: peter.willis@cpi.cam.ac.uk
Website: www.cpi.cam.ac.uk/bep
Contact Name: Peter Willis

The Business & the Environment Programme is an initiative of HRH the Prince of Wales and is developed and run by the University of Cambridge Programme for Industry. Established in 1994, it is designed as a way for senior executives to get rapidly up to speed on the complexities of the sustainable development debate.

Week-long seminars are held each year in Cambridge (April), Salzburg (September), Washington (October) and, since 2003, Cape Town (February). The seminars bring together around 45 leaders, mainly from across different industry sectors but with a number of senior government and civil society representatives as well. In the company of their peers and a Core Faculty made up of international experts and facilitators, delegates are given a rare chance to consider deeply the challenges and opportunities posed by the rapidly changing social and environmental context for business, both in Africa and globally. The seminars are generally regarded as the best of their kind worldwide.

After the seminar delegates become part of the global Alumni Network, which now comprises more than 1 000 leaders from across the business spectrum. Through the network alumni are able to access support in transforming their businesses and attend further meetings, workshops and lectures where they can deepen their learning and forge active partnerships and alliances for change.

ACTIVITIES:

IBA ENVIRONNEMENT SA (PTY) LTD

PO Box 100 SIR
LOWRY'S PASS 7133 W Cape
Tel: 021 856 4103 • Fax: 021 856 1736
Email: ambico@iafrica.com
Contact Name: Barbara Court – Managing Director

IBA Biological Products & Services
Bioremediation of pollution, in soil and water, rebalancing of disturbed ecosystems using the latest French Biotechnology and Bioremediation Products. IBA also provides economic, biodegradable, non-toxic solutions for industrial applications and biological solutions for a wide range of problem areas encountered by Local, Regional and National Authorities in the provision of services to communities in urban, rural and informal settlements, including purification and recycling of water, treatment of sewage and solid waste.

IBA Bioremediation products are formulated using NATURAL BACTERIA and NATURAL ENZYMES which are not genetically modified, non pathogenic and not harmful to humans, flora and fauna.

Our solutions enable: Production without pollution; Cleansing and purification; Maintenance, protection and rebalancing of ecosystems.

The IBA Product Range:
- SERIES 1400 Treatment of Flowing Water
- SERIES 1500 Agro-Industry & Sanitation
- SERIES 1600 Industrial Bacteria
- SERIES 1700 Cleaning & Odour
- SERIES 1800 Cleaning & Disinfecting

ACTIVITIES:

ICANDO

PO Box 115 LINK HILLS 3652
KwaZulu-Natal
Tel: 031 763 3222 • Fax: 031 7633041
Email: june@icando.co.za • Website: www.icando.co.za
Contact Name: June Lombard – Senior Member

ICANDO'S Integrated Environmental Management and Training service offers:
- Compliance with legal requirements
- Independent EIA team
- Experienced environmental scientists and educators
- Empowered I&APs who can participate effectively
- Innovative and sustainable ways to accommodate I&AP interests
- Reduced levels of conflict
- Learner-centred environmental education and training
- Competent learners who have the confidence to apply what they have learned
- Access to our extensive network of associates
- Environmental management
- Public participation at all levels
- Project management
- Environmental education and training
- Skills development and capacity building in the public and private sector

Outcomes-based short courses in waste and environmental management:
- Introductory waste management course for decision-makers
- Advanced waste management courses:

Waste minimisation; Integrated waste management planning; Hazardous waste management; Landfill site establishment and operation; Public participation processes; Health care waste management; Municipal services partnerships; The practicalities of recycling; maximising business opportunities in waste.

ACTIVITIES:

ILITHA RISCOM (PTY) LTD

PO Box 336 WIERDA PARK 0149 Pretoria
Tel: 012 668 1075 • Fax: 012 668 1828
Email: mikeo@ilitha.com
Website: www.ilitha.com
Contact Name: Michael Oberholzer

Risk Management and Assessment (including Insurance and Legal Services).

ACTIVITIES:

IMAGIS

PO Box 100941
SCOTTSVILLE 3209 KwaZulu-Natal
Tel: 033 346 1575 • Fax: 033 346 1575
Email: jwhyte@mweb.co.za
Contact Name: Chris Whyte – Managing Director

Professional consultancy services in spatial planning and information management, specialising in natural resource management, conservation, environmental applications, sustainable development and infrastructure planning. IMAGIS provides more than simple GIS, providing integrated input to multidisciplinary and multinational projects with experience and extensive databases on many African countries. Work experience with public and private entities and foreign agencies such as World Bank, GEF, European Commission, PHRD, USAID, DfID, AfDB, DBSA and IDA. Facilitation of design, implementation, management and monitoring of small local or large international planning projects through appropriate usage of spatial information management systems. Particular focus is given to spatial management and analysis of data for large-scale natural resource management projects. In addition, extensive experience in the development and application of GIS-based planning tools for infrastructural requirements, environmental monitoring, resource management, facilities management, town planning and strategic economic planning.

ACTIVITIES:

IMBEWU ENVIRO-LEGAL SPECIALISTS (PTY) LTD

1st Floor, Block 6, Albury Park
SANDTON 2146 Gauteng
Tel: 011 325 4928 • Fax: 011 325 4901
Email: admin@imbewu.co.za • Website: www.imbewu.co.za
Contact Name: Catherine Warburton – Managing Director

"Imbewu" means "seed" in Zulu and Xhosa. It illustrates our goal of providing fruitful and practical legal advice on Sustainable Development issues that gives rise to tangible results within the African context. We aim to facilitate the continual improvement of safety, health and environmental management issues by our clients and in particular the achievement of legal compliance in order to limit and manage negative impacts on the environment, on employees and communities. Partnerships are developed with our clients to address legal compliance issues in the most appropriate way. By being involved at the initiation of our clients' projects we are able to deal with legal and substantive requirements practically and efficiently. This kind of intervention permits us to achieve time and cost effective results.

Our core team of 5 lawyers has specialist expertise in the following areas:

- Mining, hazardous substances and waste management
- Legal audits and electronic SHE legal registers as a component of EMSs
- General legal advice and opinions on all areas of environmental and health and safety law
- EIAs – process advice and legal input
- Environmental management plans
- Environmental legal risk assessments
- Legal awareness training
- Assistance with public consultation processes
- Advice on climate change, SHE policy and sustainable development issues
- Conservation, biodiversity, marine law and heritage resources law

Litigious matters are referred to our associated law firm, Warburton Attorneys.

ACTIVITIES:

INCE (PTY) LTD

7 Ravenscraig Road WOODSTOCK 7925 W Cape
Tel: 021 440 7400 • Fax: 021 448 7581
Email: vinceb@incecape.co.za • Website: www.ince.co.za
Contact Name: Vince van der Bijl – Sales & Marketing Director, Western Cape Region

The Environment Diary, published by Ince (Pty) Ltd, empowers the child to embrace the environment to improve quality of life. It has a school sponsorship programme endorsed by the Department of Education, the Department of Environmental Affairs & Tourism, the Department of Water Affairs and WWF SA, which enables 50 000 diaries to go out to poorer schools. The diaries support Food & Trees for Africa and the Wilderness Foundation. Ince firmly believes that it is through sustainable development that the economic and social health and well being of South Africa is going to be assured. The diary underpins this principle to ensure the next generation is environmentally literate.

ACTIVITIES:

INDEPENDENT QUALITY SERVICES

PO Box 89 KLOOF 3640 KwaZulu-Natal
Tel: 031 767 1825 • Fax: 031 767 0017
Email: qmatters@mweb.co.za • Website: www.iqs.co.za
Contact Name: Hugh Davies – Member

IQS provides workshops, coaching and consulting for companies striving to achieve ISO 14001 certification.

We provide multilingual training from boardroom to shopfloor, backed up with motivational posters and booklets.

ACTIVITIES:

INSPIRE!

Postnet Suite 505
Private Bag X4
SUN VALLEY 7985 W Cape
Tel: 021 780 9169 • Fax: 021 80 9169
Email: seton.bailey@inspia.org
Website: www.inspire.gb.com
Contact Name: Seton Bailey

The INSPIRE! aim is to promote personal leadership and teamwork through positive participation – tackling real environmental challenges on missions in Antarctica and around the World. Our aim is to inspire business and industry to engage environmental challenges proactively. Our target market is global business, commerce and industry.

ACTIVITIES:

INSTITUTE OF DIRECTORS (IOD)

PO Box 908
PARKLANDS 2121 Gauteng
Tel: 011 643 8086 • Fax: 011 484 1416
Email: iod@icon.co.za
Website: www.iodsa.co.za
Contact Name: A D Tony Dixon – Executive Director

Vision: An Institute of Directors for Directors and leaders of the business professions. A prime and dynamic body promoting the highest ethical and professional standards in business through education, sharing of knowledge, leadership and information communication.

Motto: Integrity and Enterprise

History: The Institute of Directors was founded in London in 1903 and the South African division was formed in 1960. Autonomy was established by starting a separate Institute of Directors in Southern Africa, through an association not for gain, in 1985.

Environment: The IoD has a Portfolio Committee, which concentrates on Integrated Sustainability, Safety, Health and Environment (SHE) issues as these affect a Director and business enterprises. Social and Ethical accounting is encouraged through the application of good corporate governance practices. Directors of companies are seen as "Stewards of the Earth".

ACTIVITIES:

INSTITUTE OF LANDSCAPE ARCHITECTS OF SOUTH AFRICA

PO Box 78 GROENKLOOF 0027 Gauteng
Tel: 012 347 2371 • Fax: 012 347 2371
Email: ilasa@ilasa.co.za • Website: www.ilasa.co.za
Contact Name: Barend Smit – President

ACTIVITIES:

INSTITUTE OF NATURAL RESOURCES (INR)

Private Bag X01 SCOTTSVILLE 3209
KwaZulu-Natal
Tel: 033 346 0796 • Fax: 033 346 0895
Email: inr@nu.ac.za • Website:
www.inr.unp.ac.za
Contact Name: Ms Jennifer Mander – CEO

The Institute of Natural Resources aims for excellence in the delivery of innovative and integrated natural resource management solutions to Africa's development challenges. The Institute was formed 22 years ago and is now a public benefit, Section 21 company and an Associate of the University of Natal. The service and work of the INR ultimately aims to make a substantial contribution to poverty alleviation, sustainable livelihoods, development initiatives and wise use of natural resources. Our services are based on participative, multi-disciplinary and integrated approaches to projects. We aim to bridge the gap between science and application, with emphasis on practical management systems. Our solutions aim for innovation in integrating the needs of People, Environment and Development.

ACTIVITIES:

INSTITUTE OF SAFETY MANAGEMENT (IOSM)

PO Box 2098 BEACON BAY 5205 Eastern Cape
Tel: 043 722 3128 • Cell: 073 130 1515
Email: dalene.atterbury@afrox.boc.com
Contact Name: Dalene Atterbury – Chairperson

The IOSM promotes safety, health and the environment in the Border region. We run projects and training on these issues.

ACTIVITIES:

INSTITUTE OF WASTE MANAGEMENT – SOUTHERN AFRICA (IWM-SA)

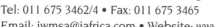

PO Box 79 ALLENS NEK 1737 Gauteng
Tel: 011 675 3462/4 • Fax: 011 675 3465
Email: iwmsa@iafrica.com • Website: www.iwmsa.co.za
Contact Name: Gail Smit – Office Co-ordinator

The Institute of Waste Management of Southern Africa (IWMSA) is a professional organisation whose main aim is to promote the science and practice of waste management. It is a non-aligned body committed to protecting the environment and people of southern Africa from the adverse effects of poor waste management. The Institute has Branches and Interest

Groups as well as Chapters in Botswana, Zambia and Zimbabwe. Membership is spread throughout the SADC region and is made up from both the public and private sectors.

Main objectives:
- to promote the science and practice of responsible waste management
- to promote sustainable living by advocating cleaner production technologies that avoid waste, then reduce, reuse, and recycle waste before choosing the best environmentally acceptable, cost effective options to treat and dispose of waste.

IWMSA is committed to moving away from uncoordinated 'end of pipe' methods which deal with waste only once it is generated towards consideration of the entire life cycle of waste, integrating the management of waste to minimise its impacts on the air, water and land.

IWMSA supports appropriate waste management policy, plans, programmes and technologies in its implementation of the following:
- Pro-active pollution prevention
- Integrated environmental management
- Consultation with and involvement of stakeholders
- The best practicable environmental option
- Poverty alleviation and creation of employment opportunities by extending waste management services into formalizing communities

The Institute strongly promotes:
- Research and development
- Duty of care
- Channelling of all waste into the formal waste stream

It achieves its objectives through:
- Capacity building and awareness raising
- Training and education, participating in NQF standards setting and course development
- Technology transfer
- International technology bridges

IWMSA supports cooperative partnerships with any organisation or individual with complementary or similar objectives throughout the SADC Region.

ACTIVITIES:

INTERIOR PLANTSCAPERS' ASSOCIATION OF SOUTH AFRICA (IPSA)

PO Box 472 MODDERFONTEIN 1645 Gauteng
Tel: 011 606 3156 • Fax: 011 606 2895
Email: val@sali.co.za
Contact Name: Val Wamsteker – Director of Operations

IPSA represents the interests of the fastest growing sector of the horticultural industry – the supply and maintenance of indoor office plants and the creation of interior landscapes.

ACTIVITIES:

INTERNATIONAL ASSOCIATION FOR IMPACT ASSESSMENT (IAIA)

South African Affiliate (IAIA-SA)
PO Box 599 ONRUSRIVIER 7201 W Cape
Tel: 028 316 2905 • Fax: 028 316 4658
Email: kruger@jaywalk.com • Website: www.iaia.za.org
Contact Name: Mrs Glaudin Kruger – Executive Officer Secretariat

IAIAsa is a forum for advancing innovative, development and communication of best practice in impact assessment. Its international membership promotes development of local and global capacity for the application of environmental assessment in which sound science and full public participation provide a foundation for equitable and sustainable development. Our members are from consulting, government, business and industry, NGO and academic sectors. IAIAsa has branches in most of the provinces. Membership benefits include access to the IAIAsa annual conference, newsletters, seminars, meetings, training courses, local, national, sub regional and international networking.

ACTIVITIES:

INTERNATIONAL FUND FOR ANIMAL WELFARE (IFAW)

PO Box 16497
VLAEBERG 8018 W Cape
Tel: 021 424 2086 • Fax: 021 424 2427
Email: cpretorius@ifaw.org
Website: www.ifaw.org
Contact Name: Christina Pretorius – Public Affairs Manager

The International Fund for Animal Welfare (IFAW) has worked in Southern Africa since 1985, to protect numerous species including elephants, seals, whales, penguins and sharks. IFAW has also contributed significantly to habitat protection and emergency relief efforts in the region.

Our pragmatic approach to diverse issues – from providing primary health care to companion animals in impoverished communities to campaigning against the ivory trade – have established IFAW as one of the most important animal welfare organisations in southern African.

IFAW supports projects as varied as Community Led Animal Welfare (CLAW), a primary health care provider to domestic animals within the informal settlements of Gauteng Province; Wildcare Africa, a wild animal rehabilitation centre specialising in rhinoceros near Pretoria; and C.A.R.E., a primate rehabilitation centre. Generous donations from our many supporters have allowed us to help SANParks purchase land to extend elephant habitat at Marakele National Park and Addo Elephant National Park. In Malawi, IFAW supports the rangers of Liwonde National Park in protecting the animals in their care and, in Zambia, IFAW is part of a team working to research elephant populations in Kafue National Park. We are committed to addressing the critical balance between human and animal needs in the developing world and to promoting the well-being of both wild and domestic animals globally.

IFAW's dedicated team of employees employs sound science, innovative educational programmes, and effective advocacy tools to ease the plight of animals in Southern Africa.

ACTIVITIES:

INTERNATIONAL INSTITUTE FOR ENERGY CONSERVATION – AFRICA (IIEC-AFRICA)

62A 5th Avenue MELVILLE 2092 Gauteng
Tel: 011 482 5990 • Fax: 011 482 4723
Email: iiec@iiec.org.za • Websit: www.iiec.org
Contact Name: Jason Schaffler – Project Manager Climate Change and Energy Efficiency

The International Institute for Energy Conservation's works to accelerate the global adoption of energy efficiency and renewable energy policies, technologies and practices in developing countries.

ACTIVITIES:

INTERNATIONAL OCEAN INSTITUTE – SOUTH AFRICA (IOI-SA)

Department of Botany University of the Western Cape
Private Bag X17 BELLVILLE 7535 W Cape
Tel: 021 959 2594 • Fax: 021 959 1213
Email: kprochazka@uwc.ac.za
Website: www.ioinst.org/ioisa/
Contact Name: Kim Prochazka

IOI-SA promotes the peaceful and sustainable use of ocean space and resources in a southern African context by facilitating sustainable livelihoods of coastal peoples through awareness creation, education, information dissemination, research, and community initiatives. This is achieved through four work programmes, viz: people, oceans and coasts, biodiversity and informatics, education through technology and IOI-SA Online Services.

ACTIVITIES:

INTERNATIONAL PLANT GENETIC RESOURCES INSTITUTE (IPGRI)

PO Box 30677 NAIROBI Kenya
Tel: +254 2524 501 • Fax: +254 2524 501
Email: e.obel-lawson@cgiar.org / ipgri-kenya@cgiar.org
Website: www.ipgri.cgiar.org
Contact Name: Elizabeth Obel-Lawson – Scientific Writer

The goal of IPGRI in sub-Saharan Africa is to support countries in the region to develop capacities to conserve and use their crop and forestry genetic resources through effective, co-ordinated national programmes.

ACTIVITIES:

INTERNATIONAL SOLAR ENERGY SOCIETY (ISES AFRICA)

Constituency: Regional Office 5 Guild House 239
Bronkhorst Street
NEW MUCKLENEUK 0181 Pretoria
Fax: 012 420 3837
E-mail: dieterholm@worldonline.co.za
Website: www.ises.org
Contact Name: Dieter Holm – President

Aims of the Society:

◆ To encourage the use and acceptance of
Renewable Energy technologies

◆ To realise a global community of industry,
individuals and institutions in support of
Renewable Energy

◆ To create international structures to facilitate
cooperation and exchange

◆ To create and distribute publications for various
target groups to support the dissemination of
renewable energy technologies

◆ To bring together industry, science and politics in
workshops, conferences and summits on
Renewable Energy

◆ To advise governments and organisations in
policy, implementation and sustainability of
Renewable Energy activities world wide

ACTIVITIES:

IRVIN & JOHNSON (I&J)

PO Box 1628
CAPE TOWN 8000 W Cape
Tel: 021 402 9212 • Fax: 021 402 9378
Email: sharonm@ij.co.za
Website: www.ij.co.za
Contact Name: Sharon Mattinson – Corporate Affairs
Manager

Irvin & Johnson (I&J) has long been active in conserving and
protecting the environments in which it operates, in
particular the marine environment and its precious
resources. I&J is South Africa's major fishing company and
has a conservation ethic entrenched throughout its
Trawling Division. Marine pollution is rigorously controlled
with its fleet of 42 trawlers, returning its substantial non-
biodegradable waste to shore for proper disposal. Crew
incentives are in place to encourage strict adherence to the
company's environmental procedures.

I&J sponsors many diverse marine conservation projects,
ranging from the Department of Environmental Affairs &
Tourism's National Marine Day to providing food for the
marine life at Cape Town's Two Oceans Aquarium. All
sponsorships are aimed at creating and influencing the
general public's awareness of the need to protect the
country's marine resources for future generations.

ACTIVITIES:

ISHECON

PO Box 320
MODDERFONTEIN 1645 Gauteng
Tel: 011 605 2129 • Fax: 011 608 2000
Email: conin@ishecon.co.za • Website: www.ishecon.co.za
Contact Name: Nigel Coni – Member

Ishecon specialise in providing integrated safety, health
and environmental consulting services to the chemical and
process industries, using the following techniques: process
hazard identification; risk assessment; environmental
impact assessment and environmental management
systems.

ACTIVITIES:

ISKHUS POWER (PTY) LTD

PO Box 6807 HOMESTEAD
Germiston 1412 Gauteng
Tel: 011 392 4838 • Fax: 011 392 5751
Email: info@iskhus.co.za • Website: www.iskus.co.za
Contact Name: Sheridan Fritz – Project Administration
Manager

Iskhus Power is a black-owned energy service company
based in Johannesburg. Our key business focus is to
optimise our customer's energy procurement process
through holistic energy management, thereby leading to
energy efficiency improvements and cost reductions. In
practising holistic energy management we take four
elements into consideration when working together with our
customers, these being Organisational, Facility, Financial
and Operational elements. Only when one takes a holistic
approach to energy management is one assured of
sustainable benefits.

Our service of managing the Energy Procurement process
of our customers is based on five pillars:

Qualifying: We determine what type of useful energy
is being utilised by the customer

Quantifying: We determine the quantity of useful
energy being utilised by the customer

Optimising: We optimise the process and energy
utilisation thereby improving efficiency

Procuring: We obtain the most competitive pricing
for the customer and then

Managing: Manage the entire process for them

Iskhus Power's products cover a wide range of energy
related services such as energy audits, energy flow studies,
energy waste evaluation, tariff studies, equipment
retrofitting, project management, design consulting and
management.

ACTIVITIES:

IUCN – SOUTH AFRICA COUNTRY OFFICE (IUCN-SA)

IUCN
The World Conservation Union

PO Box 11536
HATFIELD 0028 Pretoria
Tel: 012 342 8304 • Fax: 012 342 8289
Email: iucnsa@iucn.org • Website: www.iucnsa.org.za
Contact Name: Saliem Fakir – Director

IUCN-SA is a leading partner for conservation and development in South Africa. We provide strategic direction for development through the use of natural resources. To deal with challenges, we assist communities, government bodies and corporations to create and implement their own processes towards the development of new policies and strategies. We develop linkages between science and governance, economics, social equity and access to natural resources, protection of resources and sustainable use.

Since 1998, we have worked to create unique solutions to development and conservation challenges in collaboration with a wide range of experts and stakeholders from the private, public and non-profit sectors. We are able to achieve our goals through the generation, transformation and dissemination of scientific knowledge and tools. By providing the necessary planning strategies and support, we are able to effectively aid in the interaction between the common boundaries of conservation and development.

ACTIVITIES:

J MITCHELL AND ASSOCIATES ENVIRONMENTAL SERVICES

PO Box 398 DUNDEE 3000 KwaZulu-Natal
Tel: 034 212 2983 • Fax: 034 212 2933
Email: enviro@dundeekzn.co.za
Contact Name: J Mitchell

Scoping, Impact Assessment, Environmental Management Plans. Hazardous Materials. We specialise in all aspects of environmental management for developments in KwaZulu-Natal.

ACTIVITIES:

JACANA MEDIA

PO Box 2004
HOUGHTON 2041 Gauteng
Tel: 011 648 1157 • Fax: 011 648 5516
Email: marketing@jacana.co.za
Website: www.jacana.co.za
Contact Name: Angela McClelland – Marketing & Sales Manager

Jacana Media is an independent South African publisher, known as the experts in developing educational works about eco-tourism and the environment, life skills, medical and primary healthcare and contract publishing.

ACTIVITIES:

JARROD BALL & ASSOCIATES CC

JARROD BALL & ASSOCIATES
WASTE MANAGEMENT CONSULTANTS

9A Louis Road
ORCHARDS 2192 Gauteng
Tel: 011 485 1391 • Fax: 011 640 2463
Email: info@jbawaste.co.za • Contact Name: Jarrod Ball

Jarrod Ball and Associates (established in 1985) is a specialist waste management consultancy committed to providing waste management services of a high professional and technical standard.

Specific expertise includes:

◆ Integrated Waste Management Plans (status quo analyses, comparison of alternatives, strategy formulation and implementation)

◆ Waste reduction and treatment (minimisation, separation, resource recovery, recycling, treatment and hazard rating)

◆ Waste disposal strategies (negative mapping, modelling, landfill site identification, investigation, EIA, design, and permitting)

◆ Landfill upgrading (auditing, pollution monitoring, problem identification, operating plans, remedial design, rehabilitation and closure)

◆ Waste management training (developing tailored courses and documentation for all levels)

◆ Miscellaneous (project management, documentation, guidelines, manuals and codes of practice, review consulting, research and due diligence)

Specialising in waste management in developing countries.

ACTIVITIES:

JOULE OIL (PTY) LTD

PO Box 1189 BRACKPAN 1540 Gauteng
Tel: 011 738 2824 • Fax: 011 738 1240
Email: asinenergy@iafrica.com
Contact Name: Cecil van de Venter

Principle products recycled: Used vegetable cooking oil, used vegetable fats. Principle Products Produced: Putty oil used in the manufacture of putty, lubricants and soaps.

ACTIVITIES:

The best and most beautiful things in the world cannot be seen or even touched – they must be felt with the heart.

~ Helen Keller

KALAHARI CONSERVATION SOCIETY (KCS)

PO Box 859
GABORONE Botswana
Tel: +267 397 4557
Fax: +267 314 259
Email: admin@kcs.org.bw
Contact Name: Felix Monggae – CEO

Kalahari Conservation Society was formed in 1982 when some concerned individuals realised that even though Botswana was renowned for her rich wildlife species, few people realised that fact. There were concerns that through ignorance or misguidance wildlife could suffer, and the irreplaceable resources be neglected and unwittingly destroyed, hence the formation of the society. The Society's mission is to promote knowledge of Botswana's rich wildlife resource through education and publicity. It also encourages research into issues affecting these resources and their conservation. KCS also promotes and supports policies of conservation towards wildlife and its habitat.

ACTIVITIES:

KANTEY & TEMPLER CONSULTING ENGINEERS

PO Box 3132
CAPE TOWN 8000 W Cape
Tel: 021 405 9600 • Fax: 021 419 6774
Website: www.kanteys.co.za

Johannesburg:
Johan Kriek – Environmental Geologist
E-mail: jhb@kanteys.co.za
Tel: (011) 447 2981 • Fax: (011) 447 2989

Cape Town:
Willie van der Merwe - Director
E-mail: info@kanteys.co.za
Paul Aucamp – Engineering Geologist
Craig Stevenson – Environmental Consultant
Tel: (021) 405 9600 • Fax: (021) 419 6774

Durban:
Dion van Schalkwyk – Executive Associate
E-mail: ktdbn@kanteys.co.za
Tel: (031) 266 6535 • Fax: (031) 266 5786

Independent Geo-environmental engineers providing specialist services in:

◆ Environmental Impact Assessments
◆ Environmental Management Systems
◆ Environmental Management Reports
◆ Environmental Auditing
◆ Contamination Assessments
◆ Risk based corrective action
◆ Design and implementation of remediation

ACTIVITIES:

KAREN SUSKIN INTERIORS ARCHITECTURE AND DESIGN

Essential Design for Life
PO Box 672 NOORDHOEK W Cape
Tel: 021 789 2697 • Cell: 084 824 9592
Email: tipi@icon.co.za • Contact Name: Karen Suskin

In a time of growing environmental awareness there is a need to deepen the essential link between ones home and the larger habitat of the Earth. Innovative and contemporary advice and services in design, planning, finishes, custom designed and built furniture and cabinetry, low impact lighting, soft furnishings and landscapes. Our emphasis is on a natural, non-toxic environment, drawing on a network of "green" specialists and products. Residential and corporate.

ACTIVITIES:

KEEP DURBAN METRO BEAUTIFUL ASSOCIATION

PO Box 1535 DURBAN 4000 KwaZulu-Natal
Tel: 031 303 1665 • Fax: 031 3033969
Email: robertab@dmws.durban.gov.za
Contact Name: Robert Abbu – Business Manager

The Association is responsible for enabling and empowering all sectors of the eThekwini Unicity to make informed and responsible decisions regarding the way they and others manage the waste which they produce.

ACTIVITIES:

KEEP PIETERMARITZBURG CLEAN ASSOCIATION

Private Bag 321 PIETERMARITZBURG 3200 KwaZulu-Natal
Tel: 033 395 1201 • Fax: 033 345 7558
Email: kpca@pmbcc.gov.za
Contact Name: Cherie Pascoe – Campaign Director

A non-profit organisation conducting waste management education for schools, individuals, communities, business and industry. KPCA promotes the Integrated Waste Management process, and assists communities to seek solutions to their waste management challenges.

ACTIVITIES:

KEN SMITH ENVIRONMENTALISTS

PO Box 1297 PIET RETIEF 2380
Mpumalanga
Tel: 017 826 1427/1434 • Fax: 017 826 0238
Email: kse@ksenviro.com • Website: www.ksenviro.com
Contact Names: Ken Smith / Graham Shand – Manager

KSE offers a service in the field of environmental management in the coal-mining and forestry industries and hazardous landfills, and specialises in Public Participation and Risk Communication in the Mpumalanga, Gauteng and KwaZulu-Natal provinces.

ACTIVITIES:

KHAMA RHINO SANCTUARY TRUST

PO Box 10 SEROWE Botswana
Tel +267 430 713 • Fax: +267 435 808
Email: krst@botsnet.bw
Website: www.khamarhinosanctuary.org
Contact Name: Frederik Schutyser –
Development Officer

Khama Rhino Sanctuary Trust is a community trust protecting the white rhino in Botswana. It is home to the country's single largest rhino population, as well as to many other mammal species. KRST also has an environmental education and conference centre.

ACTIVITIES:

KNYSNA ESTUARINE AQUARIUM

Postnet Suite 24 Private Bag X31 KNYSNA 6570 W Cape
Tel: 083 400 3266 • Fax: 044 382 6302
Email: knysnaseahorse@cyberperk.co.za
Contact Name: Jim Morel – Director

The main objective of the Knysna Estuarine Aquarium & Educational Centre is to raise awareness and educate the general public of the uniqueness of estuarine biodiversity and the importance of ensuring its continued survival. The sustainable use of the estuary can only be achieved when all parties are brought together and recognise the significance of this goal.

ACTIVITIES:

KOEBERG NUCLEAR POWER STATION

Private Bag X10 KERNKRAG 7440 W Cape
Tel: 021 550 4021 • Fax: 021 550 5120
Email: marina.jenkins@eskom.co.za
Contact Name: Marina Jenkins – Head of Koeberg Visitors Centre

ACTIVITIES:

KPMG
SUSTAINABILITY SERVICES

Private Bag X9 PARKVIEW 2122 Gauteng
Tel: 011 647 6202 • Fax: 011 484 0534
Email: shireen.naidoo@kpmg.co.za
Website: www.kpmg.co.za
Contact Name: Shireen Naidoo – Director

KPMG is recognised as a leading global service provider of assurance, tax and legal, and financial advisory services. As a result of the expanding KPMG Global Sustainability Network, the firm is also now recognised as a leader in the field of Sustainability Services. Today, the KPMG Global Sustainability Network consists of more than 350 professionals in 27 countries, enabling us to apply a global approach to service delivery and respond to clients' complex business challenges with services that span industry sectors and national boundaries. In South Africa, KPMG Sustainability Services provides assurance and advisory services to many leading national and international companies. Our approach is focused on relating environmental, social, ethical, economic and financial issues to the core strategy, management objectives and operational processes of any organisation. With 10 experienced professionals and offices in Cape Town, Durban, and Johannesburg, KPMG's Sustainability Services in South Africa provides core Assurance and Advisory Services in the following areas:

- ◆ Environmental Accounting
- ◆ Environmental Due Diligence
- ◆ Environmental Legal Compliance
- ◆ Corporate Governance Assessment
- ◆ Sustainability Reporting & Assurance
- ◆ Sustainability Research & Benchmarking
- ◆ Organisational Integrity & Business Ethics Tools
- ◆ Social Accountability & Stakeholder Engagement
- ◆ Global Climate Change Strategy & Greenhouse Gas Assessment
- ◆ Management System Assessment & Certification (ISO 14001, etc.)

To learn more about how KPMG Sustainability Services can assist your organisation, please contact Shireen Naidoo, Director at 011 647 6202.

ACTIVITIES:

KRISTINA GUBIC T/A
THE AFRICAN SCRIBE

Tel: 083 651 7087 • Fax: 011 616 7521
Email: africanscribe@yebo.co.za
Website: www.africanscribe.co.za
Contact Name: Kristina Gubic – Freelance Environmental Writer

I am an established freelance writer and amateur photographer focusing on issues that affect our environment. Continuously travelling east and southern Africa to keep up to date with developments – I am renowned for my meticulous research and personalised approach to my stories and my dedication to delivering objective and factual editorial. My integrated approach encompasses academic training in Zoology and Environmental Management, field experience in both rural and corporate surroundings and my commitment to preserving our planet through sustainable development.

Please contact me with your writing requirements for:
- ◆ Environmental impact issues
- ◆ Cultural heritage
- ◆ Community upliftment projects
- ◆ Ecotourism
- ◆ Conservation of biodiversity and natural resources
- ◆ Resource management

editorial – scriptwriting – website copy – outdoor travel features – news – research

ACTIVITIES:

KRUGER & ASSOCIATES

PO Box 599 ONRUSRIVIER 7201 W Cape
Tel: 028 316 2905 • Fax: 028 316 4658
Email: kruger@jaywalk.com
Website: www.kruger-associates.com
Contact Name: Glaudin Kruger – Sole Proprietor

Kruger & Associates have planned and managed successful conferences since 1982. We provide conference services to associations, corporations and institutes, and manage conferences on the natural / built environment and technology.

We co-ordinate all aspects of meetings, including:

♦ Administration
♦ Financial management
♦ Marketing, promotion and publicity
♦ Processing of abstracts and papers
♦ Programmes
♦ Publication
♦ Social programmes
♦ Travel, transport and accommodation
♦ Exhibitions and displays
♦ Registration

ACTIVITIES:

KUMBA RESOURCES

PO Box 9229 PRETORIA 0001
Tel: 012 307 4561 • Fax: 012 307 4080
Email: mitzi.duplessis@kumbaresources.com
Website: www.kumbaresources.com
Contact Name: Mitzi du Plessis
– Corporate Relations Manager

KUMBA RESOURCES, one of the largest South African-domiciled mining groups, is a focused metals and mining company with a diverse commodity portfolio consisting of iron ore, heavy minerals, coal, base metals and industrial minerals. The company's strategic intent is to optimise its existing assets and to increase the size of its businesses through organic growth, acquisitions and mergers.

Proud of its South African roots, Kumba also has substantial links and growth prospects abroad and is widely recognised as one of South Africa's major earners of foreign exchange.

The company boasts a substantial knowledge and skills base vested in carefully selected, demographically representative employees and is building a credible empowerment base by aggressively searching for local empowerment partners.

Kumba recently joined the ranks of the Peace Park Foundation's Club 21, aimed at facilitating the establishment of transfrontier conservation areas.

For further information and contact details, visit our website on www.kumbaresources.com

ACTIVITIES:

KZN CRANE FOUNDATION

PO Box 115 MOOI RIVER 3300
KwaZulu-Natal
Tel: 033 266 6268 • Fax: 033 266 6268
Email: crane@futurenet.co.za
Website: www.kzncrane.co.za
Contact Name: Brent Coverdale

Conserving the Wattled, Blue and Grey-Crowned Cranes and their wetland and grassland habitats through education/extension; research/monitoring; habitat protection/restoration and captive breeding.

ACTIVITIES:

LANDSCAPE ARCHITECTS UYS & WHITE (PTY) LTD

PO Box 7001 CENTURION 0046 Gauteng
Tel: 012 663 1045 • Fax: 012 663 5907
Contact Names: Ms Linda Kuhn
– Environmental Manager

Gauteng (Established 1964)

Tel: (012) 663-1045/6/7 • Fax: (012) 663-5907
e-mail: luw@iafrica.com
Neville White, Danie Rebel (Landscape Architecture)
Kim Read, Linda Kühn (Environmental Facilitation)

KwaZulu-Natal

Tel: (031) 562-0620 • Fax: (031) 562-0625
E-mail: luwkzn@iafrica.com
Contact: Lucas Uys

ENVIRONMENTAL FACILITATION
Environmental Impact Assessments; Environmental Management Programmes; Conservation and Ecological Planning; Landscape Rehabilitation; End-use Planning of Landfills.

LANDSCAPE ARCHITECTURE
Planning and detail design of urban, office, residential, industrial, commercial and institutional landscapes; Recreation, entertainment, tourism and sport facilities; Parks and open space planning; Land-use master planning; Cemeteries; and Eco-tourism planning.

ACTIVITIES:

LANDSCAPE IRRIGATION ASSOCIATION (LIA)

PO Box 472 MODDERFONTEIN 1645 Gauteng
Tel: 011 606 3156 • Fax: 011 606 2895
Email: val@sali.co.za
Contact Name: Val Wamsteker – Director of Operations

The LIA is a National organisation representing irrigation suppliers and contractors. The aims of the Association include maintaining high standards of workmanship through training and certification of management and employees.

ACTIVITIES:

LC CONSULTING (CAPE) PTY LTD

Suite 111
Stadium on Main Main Road
CLAREMONT W Cape
Tel: 021 670 9800 • Fax: 021 670 9810
Email: ahakin@lcct.co.za
Contact Name: Andrew Hakin – Managing Director

LC Consulting services include the basic structural and civil engineering services to the built environment and extend through their façade engineering work into: ·

1. Passive building envelope design including solar shading, insulation and daylighting analysis.
2. Preliminary building energy efficiency design including thermal mass, building envelope performance, usage profiling and low energy systems.
3. Building emissions calculations from energy usage calculations.

ACTIVITIES:

LGWSETA

PO Box 1964 BEDFORDVIEW 2008 Gauteng
Tel: 011 456 8579 • Fax: 011 450 4948
Email: info@lgwseta.co.za
Website: www.lgwseta.co.za
Contact Name: Nonhlanhla Dube – Water Chamber Manager

Local Government water and related services SETA is an education training qualification authority for local government, water, sanitation and waste management.

ACTIVITIES:

LILLIPUT™ BY "AFRICA WATER & WASTE (PTY) LTD"

Lilliput™
SEWAGE EFFLUENT TREATMENT PLANTS
PATENT PENDING NO: 2002/0614
AFRICA WATER & WASTE (Pty)Ltd
" IF IT DOES NOT SAY LILLIPUT ™ IT IS NOT A LILLIPUT ™ "

PO Box 1413 HILLCREST 3650
KwaZulu-Natal
Tel: 031 783 4276 • Cell: 082 808 9344
Email: mross@mweb.co.za • Website: www.aww.co.za
Contact Name: Mark Ross – Director

The LILLIPUT™ is a treatment SYSTEM for the purification of organic wastes, especially domestic sewage, to permit safe discharge to the environment or re-use such as re-flushing or irrigation.

It can be sized to serve a range of duties, down to single households. Treatment is effected through the use of a combination of proven technologies successfully improved and blended with process chains and proprietary techniques developed by Africa Water and Waste.

The reception pre-digester contains the anaerobic phase of the process. Here the organic solids are degraded by microbial digestion, converting complex organic compounds to simple soluble organic compounds. These are further degraded under the aerobic conditions in the fixed-film bioreactor.

The bioreactor is an upflow submerged fixed media environment designed for a specific hydraulic residence, rise rate and COD loading rate. A proprietary random-packed media occupies the majority of the reactor, onto which the biofilm adhere, creating a three-dimensional environment that facilitates contact between the biomass and the carbo-naceous material. Carbonaceous degradation and nitrification are achieved through the controlled input of dissolved oxygen.

Plants are modular and can be designed for almost any site-specific condition, such as containerisation for temporary applications or in modular form to suit topographical and/or aesthetic requirements.

Plants can be modified to conform to stricter environmental requirements where necessary such as phosphate removal. The combination of technologies, process flow, proprietary componentry configurations and maximisation of variable hydraulic implications to minute applications, places the Lilliput™ technologies in a unique situation.

ACTIVITIES:

LIZ ANDERSON & ASSOCIATES

Quality, Environmental & Integrated Management Systems Consulting Services

PO Box 776 UMHLALI 4390 KwaZulu-Natal
Tel: 032 942 8367 • Fax: 032 942 8328
Email: lizanderson@mjvn.co.za
Website: www.mjvn.co.za/lizanderson
Contact Name: Liz Anderson – Director

Assisting Industry to achieve competitive advantage through incorporating Environment, Social, Health & Safety issues as:

- ◆ Key to Business Strategy
- ◆ Part of the holistic Business Management System
- ◆ Providing Quality and Environmental Consulting, Training, Communications & Public Reporting Solutions tailored to individual companies needs.

ACTIVITIES:

LIZ KNEALE COMMUNICATIONS CC

PO Box 1960 JUKSKEI PARK 2153 Gauteng
Tel: 011 793 6911 • Fax: 011 793 3639
Email: lizcomm@netactive.co.za
Contact Name: Liz Kneale

Environmental communications and education consultancy (waste management) offering education/training, public relations, project management, exhibitions, writing services (news releases, brochures, newsletters), public participation, community development.

ACTIVITIES:

LOMBARD & ASSOCIATES

PO Box 115 LINK HILLS 3652
KwaZulu-Natal
Tel: 031 763 3222 • Fax: 031 763 3041
Email: ray@rlombard.co.za
Contact Name: Ray Lombard – Senior Member

Lombard & Associates was established in 1985 to offer environmentally-appropriate, cost-effective solutions to waste management and environmental pollution problems.

SPECIFIC AREAS OF EXPERTISE INCLUDE:

- ◆ Regional and industry specific waste management strategies and appropriate waste management systems
- ◆ Environmental impact assessments, motivation reports and permit applications for waste management facilities
- ◆ Selection and design of new waste management and processing facilities, modification/rehabilitation of existing facilities by applying integrated environmental management and waste minimisation
- ◆ Hazardous waste management, treatment plants, process control and ultimate disposal facilities
- ◆ Operating guidelines for sanitary landfill and hazardous waste disposal facility management
- ◆ Leachate treatment and landfill gas control
- ◆ Capacity building, education and training in waste management
- ◆ Bioremediation of contaminated industrial sites and chemical spill clean-ups
- ◆ Environmental audits, pollution monitoring and risk assessments project management

ACTIVITIES:

LOUIS DE VILLIERS

PO Box 82 OBSERVATORY 7935 W Cape
Tel: 021 448 8908 • Fax: 021 448 3422
Email: louis@villiers.co.za
Contact Name: Louis de Villiers – Enviro-Law Consultant

ACTIVITIES:

MASTER FARMER PROGRAMME

Umzimvubu Nursery
PO Box 70 PORT ST JOHNS 5120 Eastern Cape
Tel: 083 700 8612 • Fax: 044 534 8827
Email: bolus@global.co.za
Contact Name: Richard Bolus – Co-Ordinator

The Master Farmer Programme works from the Umzimvubu Nursery and resource centre at Port St. Johns, Eastern Cape. It has been operating in communities close to the forests, 20km east and west of the town for the past 7 years. It gives technical and organisational support to farmers by assisting with inputs, and providing advice and training. Emphasis of the technical training is the development of farming systems that are essentially organic and self sustaining. Efficient water use is being promoted to intensify land use, increase out-of-season production and lower the need to open lands

in the forests. Central to the work is the promotion of the status of farmers, by giving them a platform to express themselves at Farmers' Days and Agricultural Shows and rewarding achievements. Apart from its focus on conservation farming the programmes is actively promoting land use planning. It is funded by the Green Trust and WWF-SA.

ACTIVITIES:

MAXLITE

PO Box 460 BEDFORDVIEW 2008 Gauteng
Tel: 011 622 2827 • Fax: 011 615 1516
Email: trevor@maxlite.co.za
Website: www.maxlite.co.za
Contact Name: Trevor van der Vyver

Supply and manufacturer of energy efficient lighting. Focus on solar or D.C. voltages. Range includes solar powered street lights. Distributor/importer of electronic ballasts and LEDs.

ACTIVITIES:

MBB CONSULTING ENGINEERS INCORPORATED

PO Box 3011 MATIELAND 7602 W Cape
Tel: 021 887 1026 • Fax: 021 883 8514
Email: johanvh@mbb.co.za • Website: www.mbb.co.za
Contact Name: Johan van Huyssteen – Director

- ◆ Resource recycling facilities
- ◆ Composting of garden refuse and/or the organic component of domestic waste
- ◆ Composting of sewage sludge
- ◆ Operating guidelines for landfills
- ◆ Landfill auditing

ACTIVITIES:

METAGO ENVIRONMENTAL ENGINEERS (PTY) LTD

PO Box 1596 CRAMERVIÉW 2060 Gauteng
Tel: 011 789 8785 • Fax: 011 789 8788
Email: general@metago.co.za • Website: www.metago.com
Contact Name: Mr Paul de Kock – Director

Metago provides environmental engineering services internationally through offices in Johannesburg, South Africa, and Perth, Western Australia. Our specialist skills include:

Environmental Science

- ◆ Implementation of Management Systems (ISO 14001, OHSAS18001, ISO9000)
- ◆ Environmental Management Programme Reports (EMPRs)
- ◆ Environmental Impact Assessments (EIAs)
- ◆ Environmental Due Diligence Assessments and Audits.
- ◆ Corporate Environmental Reports
- ◆ Environmental Risk Assessment
- ◆ Environmental Monitoring

Environmental Law
◆ Opinions
◆ Risk Assessments
◆ Licensing
◆ Legal registers
◆ Raining

Business
◆ Closure planning / business risk management
◆ Quantification of liabilities and obligations
◆ Structuring rehabilitation programmes
◆ Supply and installation of management system software for ISO 14001, OHSAS 18001, ISO 9000

Engineering
◆ Tailings and waste rock engineering
◆ Acid mine drainage assessment and control
◆ Heap leach engineering
◆ Geotechnical engineering
◆ Water and salt balance investigations & assessments
◆ Risk assessment and risk-based design
◆ Liquefaction and flow slide assessment
◆ Closure planning and site remediation

ACTIVITIES:

METRIX SOFTWARE SOLUTIONS

Liberty Gardens, 10 South Boulevard,
BRUMA 2198 Gauteng
Tel: 011 635 9311 • Fax: 011 622 0124
Email: info@isometrix.co.za
Website: www.isometrix.co.za
Contact Name: Lucille Hallowes – Product Manager

IsoMetrix, the latest generation software from Metrix Software Solutions, will help you manage environmental, health, safety and quality risk across your organisation. Developed to meet the needs of users, IsoMetrix has been designed based on the structure of the ISO 14000, ISO 9000 and OHSAS series of standards. IsoMetrix operates with intuitive logic and functionality, helping companies identify, assess and prioritise issues, plan and follow up on management requirements and, critically, measure performance. IsoMetrix includes comprehensive reporting, analysis, tracking, follow-up and monitoring tools to assist in highlighting trends and potential areas of concern.

Features of IsoMetrix:
◆ Feature rich
◆ Web based
◆ Multi-site functionality
◆ Solid database
◆ Strong support
◆ 60 sites locally and internationally

IsoMetrix can be cost effectively deployed across your entire organisation, allowing you to not only manage SHEQ issues but also promote your drive toward continual improvement.

ACTIVITIES:

MILLENNIUM WASTE MANAGEMENT (PTY) LTD

PO Box 232 BEDFORDVIEW 8002 Gauteng
Tel: 011 456 5597/8 • Fax: 011 453 2241
Email: info.gp@millenniumwaste.co.za
Website: www.millenniumwaste.co.za

Millennium Waste Management (Pty) Ltd is focused on providing professional waste management services to satisfy specific needs of government, local authorities and the private sector. Millennium has four operating divisions: SanuMed provides services for the safe and responsible containerisation, collection, transport and disposal of healthcare waste. Wade Refuse is a domestic waste service, offering door-to-door collections, street cleaning, litter picking and cleaning of illegal general waste dumping. Landfill Management manages general waste disposal sites and facilities. Sediba provides total sewage, water effluent and potable water management solutions.

Regional Offices Gauteng:
Tel: (011) 456 5597/8 • Fax: (011) 453 2241
Email: info.gp@millenniumwaste.co.za
KwaZulu-Natal:
Tel: (031) 700 3921 • Fax: (031) 700 3208
Email: info.kzn@millenniumwaste.co.za
Western Cape:
Tel: (021) 951 5383/9 • Fax: (021) 951 8440
Email: info.ct@millenniumwaste.co.za
Eastern Cape:
Tel: (041) 466-2741 • Fax: (041) 467-1578

ACTIVITIES:

MINERALS & ENERGY EDUCATION TRAINING INSTITUTE (MEETI)

PO Box 599 RANDBURG 2125 Gauteng
Tel: (011) 709 4311 • Fax: 011 709 4657
Email: info@meeti.org.za • Website: www.meeti.org.za
Contact Name: Dr Olga Svoboda – Director

MEETI offers certificate courses and short seminars on the current developments in the implementation and management of minerals and energy policy and industry as well as on the environmental management in mining and energy sector. Environmental courses include: Environmental Policy and Management in Mining, Multi-Stakeholder Negotiations – Key to Sustainable Success, Climate Change and Environmental Policy in Energy. Specialised courses on the petroleum industry and economics, electricity industry, commercialisation of natural gas, petroleum products, mining legislation, Mine Health and Safety Act, and opportunities for women and entrepreneurs in mining and energy are also offered. Strong focus of the environmental training is on sustainable development and social environmental issues. Courses are accredited with the University of the Witwatersrand and lectured by experts from industry, academia, government, independent consultants and lawyers. Participants from government, industry, unions, NGOs and community are encouraged to attend.

ACTIVITIES:

MINING REVIEW AFRICA

Po Box 321 STEENBERG 7947 W Cape
Tel: 021 700 3511 • Fax: 021 700 3501
Email: jonathan@spintelligent.com
Website: www.miningreview.com
Contact Name: Jonathan Spencer Jones – Publisher

Mining Review Africa supports the sustainable development of Africa's mineral resources, by promoting human and institutional capacity, liaison between majors and juniors, and the introduction of suppliers to mines.

ACTIVITIES:

MINTEK

Private Bag X3015
RANDBURG 2125 Gauteng
Tel: 011 709 4061
Fax: 011 709 4684
Email: cobusk@mintek.co.za • Website: www.mintek.co.za
Contact Name: Dr R J (Cobus) Kriek – Environmental Specialist

Mintek is a world-renowned specialist in the field of mineral and metallurgical technologies. In conjunction with our core activities, Mintek is in the position to provide industry with specialist processes, advice and services. Amongst others, these include the treatment of metal-based effluents resulting from processes such as metals refining and electro-winning, acid mine drainage solutions, TCLP analysis, as well as air monitoring. In our research and development, we have committed ourselves to be environmentally conscious and to provide clients throughout the world with mineral and metal processing solutions that meet the demands of modern environ-mental legislation. Our process engineers are engaged in developing processes involving the above fields, and are well aware of and able to meet the environmental demands of the present and the future.

The implementation of both the ISO 14001 environmental management system and the ISO 9001 quality management system form an integral part of the above services we provide to industry.

ACTIVITIES:

MJ MOUNTAIN & PARTNERS

4 Castleview Road CONSTANTIA 7806 W Cape
Tel: 021 7155273 / 851 2010 • Fax: 021 852 2055
Cell: 082 658 4528 • Email: mjmandp@mweb.co.za
Contact Name: Richard Galliers – Managing Partner

M J Mountain & Partners have undertaken regional and site-specific geological investigations for environmental and development projects since 1968 for government, NGOs and private organisations.

ACTIVITIES:

MONDI WETLANDS PROJECT

PO Box 338 IRENE Pretoria 0062
Tel: 012 667 6597
Fax: 012 667 5720
Email: info@wetland.org.za
Website: www.wetland.org.za
Contact Name: David Lindley – Manager

The Mondi Wetlands Project (previously called the Rennies Wetlands Project) has changed the face of wetland conservation in South Africa forever, by pioneering the conservation of wetlands outside protected areas. Over the past decade the Mondi Wetlands Project (MWP) has also moved wetlands conservation from a side issue to centre stage and even inspired central government to pledge millions for wetland conservation.

MWP works nationally to conserve wetlands outside declared nature reserves. Most of the country's wetlands in fact lie outside protected areas so MWP's interventions are of vital importance. MWP promotes the wise use, rehabilitation and sustainable management of palustrine wetlands – these are predominantly wet meadows, marshes and floodplain wetlands. MWP's interventions take place at both grassroots and political decision-making levels: it works with commercial farmers, government agricultural and conservation extension services, historically disadvantaged rural communities, and key decision makers on a national basis.

ACTIVITIES:

MONKEYLAND PRIMATE SANCTUARY

THE CRAGS 6602 E Cape
Tel: 044 534 8906 • Fax: 044 534 8907
Email: monkeys@global.co.za
Website: www.monkeyland.co.za
Contact Names: Lara Mostert – Marketing
Tony Blignaut – CEO

Primate sanctuary and tourist attraction.

ACTIVITIES:

MORRIS ENVIRONMENTAL & GROUNDWATER ALLIANCES (MEGA)

PO Box 26870
HOUT·BAY 7872 W Cape
Tel: 021 790 5793 • Fax: 021 790 5793
Email: info@megateam.co.za
Website: www.megateam.co.za
Contact Names: Ritchie Morris – Managing Member and Environmental Hydrologist
Mary-Jane Morris – Environmental Management Specialist

Morris Environmental & Groundwater Alliances (MEGA) bring together project teams with the appropriate mix of environmental skills, technical expertise and social objectives. The MEGAteam has over 17 years of experience throughout southern Africa in all aspects of the environmental and groundwater fields. MEGA work with large and small corporations, local and central government and other consulting groups (www.megateam.co.za). We are service and solution driven and it is our philosophy to work with our clients to provide innovative, practical and cost effective solutions. MEGA promote employment equity, empowerment and capacity building within our project partnerships. Within the MEGAteam is the specialist subsidiary, Industrial Groundwater Monitoring Services (IGMS) – see www.megateam.co.za/igms.htm. Our joint partner in this venture is Mr Deven Naidoo BSc Hons of Geosure PTY Ltd, based in Durban 031 266 0458. We hold Professional Indemnity Insurance through Manwood Underwriters and are registered with the Department of Labour and Workman's Compensation Commission.

ACTIVITIES:

MOUNTAIN CLUB OF
SOUTH AFRICA (THE)

97 Hatfield Street CAPE TOWN 8001 W Cape
Tel: 021 465 3412 • Fax: 021 461 8456
Email: mcsacc@iafrica.com
Website: www.mcsa.org.za
Contact Name: Fran Hunziker – Honorary Secretary

The MCSA's objectives are to further the interests of mountaineering, conserve and protect mountains, access to mountain areas, fauna, flora, historical and archaeological sites. It also promotes safety and organises mountain search and rescue.

ACTIVITIES:

MTN WHALE ROUTE

PO Box 2791 KNYSNA 6570 W Cape
Tel: 044 382 1436 • Fax: 044 382 7078
Email: whalemaster@mweb.co.za
Website: www.cape-whaleroute.co.za
Contact Name: Greg Vogt – CEO

With the MTN Whale Route, whale tourism, whale education and a healthy marine environment are priority number one. For more information, please visit our website or call Whale & Flower Hotline on 083-910-1028.

ACTIVITIES:

MUSEUM PARK ENVIRO CENTRE

PO Box 413 PRETORIA 0001
Tel: 012 326 3279 • Fax: 012 326 4067
Email: sraath@mweb.co.za
Website: www.museumpark.co.za
Contact Name: Dr Schalk Rath – Manager

Museum Park is a non-governmental organisation that markets the heritage activities of museums and historical sites

in Pretoria. It supports and facilitates environmental awareness on a partnership basis.

ACTIVITIES:

NAARTJIE CLOTHING

PO Box 13050
WOODSTOCK 7915 W Cape
Tel: 021 448 3502
Fax: 021 448 3609
Email: headoffice@naartjie.co.za
Website: www.naartjie.com
Contact Names: M Elliott and Steve Eales – Directors

As a leading South African kidswear brand, Naartjie Clothing tries to draw attention to environmental issues such as global warming, endangered species, etc. through prints, swingtags, and window displays. With Naartjie's US retail expansion already at three stores in San Francisco and more to follow, this will help raise awareness of these issues with kids and their parents beyond our shores too.

ACTIVITIES:

NADEET
(NAMIB DESERT ENVIRONMENTAL EDUCATION TRUST)

PO Box 31017 WINDHOEK Namibia
Tel: +264 63 693 012 • Fax: +264 63 693 012
Email: nadeet@iway.na • Website: www.nadeet.org
Contact Name: Viktoria Paulick – Environmental Education Director

The Namib Desert Environmental Education Trust (NaDEET) is a non-profit organisation aimed to empower Namibians to make decisions for a sustainable future. Located 100 km south of Sesriem on the NamibRand Nature Reserve, NaDEET provides an outdoor, hands-on, experiential learning opportunity. It was started because there was an urgent need to provide an environmental education (EE) opportunity in the South of Namibia as it was identified that there was a dire lack of outdoor, educational facilities.

ACTIVITIES:

Flowers are the poetry of earth,
as stars are the poetry of heaven.

~ Unknown

NAMIBIA NATURE FOUNDATION

PO Box 245 WINDHOEK Namibia
Tel: 09 264 61 248 345
Fax: 09 264 61 248 344
Email: cb@nnf.org.na
Website: www.nnf.org.na
Contact Name: Dr C J Brown – Director

The Namibia Nature Foundation is a non-governmental organisation, the primary aims of which are to promote sustainable development, to conserve biological diversity and natural ecosystems, and to promote the wise and ethical use of natural resources for the benefit of all Namibians, both present and future.

Objectives

In order to fulfil this mission, the following objectives are upheld:

- raise and administer funds for conservation and environmental initiatives in support of the mission statement
- initiate, support and promote activities that conserve Namibia's environment, protect biological diversity and promote the wise use of natural resources
- plan, develop, implement, manage and administer projects
- establish and maintain good communications with local, national and international partner organisations
- provide grants in order to support and promote initiatives that strengthen Namibian institutions to better understand and manage natural resources
- encourage and support community-based natural resource management initiatives
- raise awareness and educate the public on Namibia's environment and its natural resources

Services

- Excellent financial services and accountability in the management and administration of funds
- A broad and detailed knowledge of the Namibian environment, biodiversity conservation and sustainable development issues
- A good overview of the conservation and environmental work being carried out in the country
- Extensive experience of project planning, development of proposals, project management and administration
- Good networking with local, regional and international organisations
- Vast experience with donors from all over the world

ACTIVITIES:

NAMIBIAN ENVIRONMENTAL EDUCATION NETWORK (NEEN)

C/o CCF: PO Box 1755
OTJIWARONGO Namibia
Tel: 09264 67 306 225 • Fax: 09264 67 306 247
Email: neen@iafrica.com.na
Website: www.natmus.cul.na/neen/
Contact Name: Graeme Wilson

NEEN endeavours to maintain and increase the popularity of environmental education in Namibia.

NEEN networks and links environmental education related efforts throughout Namibia and the SADC region.

ACTIVITIES:

NAMPAK RECYCLING

Private Bag X85
BRYANSTON 2021
Gauteng
Tel: 011 799 7258 • Fax: 011 799 7297
Email: recycling@nampak.co.za • Website: www.nampak.co.za
Contact Name: Srini Naidoo – Commercial Manager

Nampak Paper Recycling's function is to procure waste paper for all Nampak Group paper mills. The Paper Mills are situated in Bellville, Kliprivier, Verulam and Rosslyn, producing kraft and tissue products.

Nampak Recycling is one of the largest buyers of secondary fibre in South Africa, actively promoting the recovery of recyclable papers.

Branches are situated in Pretoria, Johannesburg, Cape Town and Durban, with a supporting infrastructure of appointed agents to provide national coverage.

Amongst the range of services offered are confidentiality, compaction equipment, recycling aids, assistance with recycling programmes and small business development.

ACTIVITIES:

NATAL SHARKS BOARD

Private Bag 2
UMHLANGA 4320
KwaZulu-Natal
Tel: 031 566 0400 • Fax: 031 566 0499
Email: hargreaves@shark.co.za
Website: www.shark.co.za
Contact Name: Mrs Debbie Hargreaves – PR Manager

A visit to the Natal Sharks Board offers an audio-visual presentation and shark dissection, a truly unique insight into the world of the shark.

ACTIVITIES:

NATIONAL AIR POLLUTION ASSESSMENT SERVICES

10 Alexandra Street FLORIDA 1711 Gauteng
Tel: 011 674 2080 • Fax: 011 674 1575
Email: co-napas@iafrica.com
Contact Name: Christo Oberholzer – Manager

National Air Pollution Assessment Services is structured to provide a wide range of Environmental and Occupational Hygiene services to the mining and local industries. Environmental Impact Assessments, monitoring of air quality and personal exposures to airborne pollutants are receiving increased attention. Napas assists management to provide a safe and healthy working environment. Personal and public exposure assessments are carried out using scientific methods recognised by authorities across the world. Air filtration systems are designed and installed. Mine management is assisted in the ventilation planning, with specific attention being given to safety and health. The dilution of diesel exhaust emissions and the comfort of workers in the underground environment are of paramount importance when planning a mine. Mine ventilation systems are designed with the aid of computer programmes. Napas is the local agent for the CIRRUS range, state of the art, noise measuring instruments as well as other hygiene equipment.

ACTIVITIES:

NATIONAL ASSOCIATION FOR CLEAN AIR (NACA)

PO Box 2036 PARKLANDS 2121 Gauteng
Tel: 012 543 0151 • Fax: 012 543 0151
Email: naca@mweb.co.za
Contact Name: Piet Odendaal – Administrator

Since 1969, NACA has championed the cause of clean air, creating a forum of broad expertise, sharing and disseminating knowledge and sponsoring training and promoting research.

ACTIVITIES:

NATIONAL BOTANICAL INSTITUTE (NBI)

Private Bag X7
CLAREMONT 7735 W Cape
Tel: 021 799 8680 • Fax: 021 797 8390
Email: cole-rous@nbict.nbi.ac.za
Website: www.nbi.ac.za
Contact Name: Jenny Cole-Rous – Media Officer

The National Botanical Institute (NBI) of South Africa, with its Head Office at Kirstenbosch National Botanical Garden in Cape Town, manages a network of eight National Botanical Gardens. These are strategically situated throughout the country in five provinces for the propagation and display of South Africa's unique and diverse floral wealth.

All National Botanical Gardens have a selection of visitor/tourist facilities such as restaurants, coffee shops, curio/gift shops, display gardens (medicinal, herbal, plants used in perfumes, etc.) and waterwise gardens. Depending on location, a Xhosa herb garden, a Nama kookskerm, a tradi-tional Zulu hut and a Ndebele homestead with a medicinal plant garden are also on display at some of them. Activities and events include theme-guided walks, concerts, conferences, craft markets and art exhibitions.

The Gardens are situated in Cape Town, Betty's Bay, Worcester, Bloemfontein, Pietermaritzburg, Pretoria, Roodepoort/Krugersdorp and Nelspruit.

ACTIVITIES:

NATIONAL BUSINESS INITIATIVE

PO Box 294 AUCKLAND PARK 2006 Gauteng
Tel: 011 482 5100 • Fax: 011 482 5508
Email: charlotte@nbi.org.za
Website: www.nbi.org.za
Contact Name: Charlotte Middleton – Manager

Since its inception in 1995, the National Business Initiative (NBI) has gained the respect and recognition as a leading non-profit organisation using business leadership and resources to meet the challenges of South Africa. Through the mandate of 150 member companies, representing South Africa's top business leadership, the NBI acts at the intersection of the private and the public sector to contribute to political and economic stability and to enhance the country's competitiveness as a key to sustained growth.

The NBI's strategic direction is firmly based on the belief that investment, job creation, economic growth and sustainable development are the critical challenges facing South Africa over the next decade. The NBI has already made progress in this regard through flagship programmes which have for a number of years successfully targeted areas that promote job creation and skills development (see www.nbi.org.za). The NBI is also the managing agency of the Business Trust, widely recognised as one of the most successful structured partnerships between government and business for social progress in the world.

In ensuring a balance between catalytic and implementation capabilities, the NBI has shown the ability to demonstrate practical impact on the ground and to elevate the lessons learned to the strategic level. Although a fairly new focus area, the NBI is convinced its corporate citizenship programme will build on the solid foundations of the NBI and the rich history of corporate social responsibility amongst the South African business sector. The recent incorporation of the South African Business Council for Sustainable Development (BCSD: SA) into the NBI holds great promise in this regard.

ACTIVITIES:

NATIONAL CLEANER PRODUCTION CENTRE

PO Box 395 CSIR
PRETORIA 0001
Tel: 012 841 3665 • Fax: 012 841 2135
Email: cmasuku@csir.co.za
Website: ncpcsa.co.za
Contact Name: Dr CM Masuku – Director

The South African National Cleaner Production Centre (NCPC) aims to enhance the competitiveness and productive capacity of the national industry through cleaner production (CP) techniques, and the transfer and development of environmentally friendly technologies. The NCPC plays an important catalytic and coordinating role in fostering the promotion of sustainable industrial development in South Africa by building national capacity in CP and fostering multi-stakeholder policy dialogues in CP. The main activities of the NCPC cover in-plant assessments; promotion of CP technologies and investments; training and information dissemination.

The NCPC was launched during the World Summit on Sustainable Development in Johannesburg in 2002. This is a cooperation programme between South Africa and the United Nations Industrial Development Organisation (UNIDO) with financial assistance from the dti and the governments of Austria and Switzerland. The NCPC is hosted at the CSIR in the Manufacturing and Materials Division, Pretoria, with Regional Focal Points in Cape Town and Durban. The work of the NCPC is guided by an Executive Board, involving the main national and international stakeholders of the Centre. The NCPC has a memorandum of understanding with the Kingdom of Denmark covering the activities of the Cleaner Textile Linkage Centre in Cape Town.

CP is an integrated preventive strategy applied to processes, products (life cycle)and services to eliminate or at least minimize the production of waste and pollution, and to optimise the consumption of resources (energy, water). CP activities generally improve an organisation's bottom line and invariably improve productivity and competitiveness.

ACTIVITIES:

NATURAL STEP (THE)

The Learning Centre
Dreyersdal Farm
BERGVLIET 7945 W Cape
Tel: 021 7150526/5 • Fax: 021 7150325
Email: sjakes@iafrica.com
Website: www.thenaturalstep.org
Contact Name: Stephen Jacobs – Executive Director

the NATURAL STEP

The Natural Step (TNS) is an international non-profit organisation founded in Sweden in 1989. It develops and shares a scientifically sound conceptual framework that helps people, particularly in business and government, to understand sustainability and move towards it. Its focus is on the "Why?" of sustainable development, coupled with some highly action-oriented tools, which are accessible to people at all levels.

Besides Sweden, it is now operating in the UK, USA, Canada, Japan, Australia, New Zealand, Israel, Brazil and, since 1999, South Africa. Many multi-national corporations (e.g. Electrolux, Nike, IKEA), smaller businesses and around seventy municipalities use the TNS framework as a strategic planning and environmental management tool. It provides a powerful foundation to any management system (such as ISO 14001) and the basis for any environmental education programme.

Southern African clients include Woolworths, GSK, Triangle Sugar, Bokomo Foods, Lanzerac Manor Hotel, Eskom.

The TNS framework is made available in workshops, seminars, training programmes, consultancy, books and electronic materials. TNS in southern Africa is funded mainly through fees charged for the above services.

ACTIVITIES:

NEW AFRICA SKILLS DEVELOPMENT (PTY) LTD

PO Box 278 MERRIVALE 3291
KwaZulu-Natal
Tel: 033 330 7002 • Fax: 033 330 7005
Email: info@nasd.co.za
Website: www.newafricaskills.co.za
Contact Name: Mark Dicks – National Training Manager

We are a FIETA accredited training company specialising in herbicide courses. Courses are offered to clients in the forestry, rangeland, conservation and industrial sectors.

ACTIVITIES:

NEWTOWN LANDSCAPE ARCHITECTS

PO Box 36
FOURWAYS 2055 Gauteng
Tel: 011 462 6967 • Fax: 011 462 9285
Cell: 082 442 6114 or 082 462 1491
Email: graham@newla.co.za / johan@newla.co.za
Website: www.newla.co.za
Contact Names: Graham Young, Johan Barnard

NLA strive for sustainable environments that have cultural, artistic and ecological merit. In our designs we recognise the realities of the contemporary situation as well as the influences that gave uniqueness to place. This approach is evident in the many different projects completed by the firm.

NLA provide well-managed design, development and environmental plans and offer an array of services and expertise that can add long-term value to a project.

Services include:

- ◆ Landscape and urban design
- ◆ Ecological planning and design
- ◆ Environmental planning including environmental impact assessments and environmental management programmes
- ◆ Visual impact assessments and computer generated photo-simulation
- ◆ Open space planning and detail park design
- ◆ Site and landscape design for domestic, commercial and industrial sites

NLA are committed to quality of design and service. Drawing on the partners' extensive experience we are able to respond quickly and effectively to the specific needs of our clients.

ACTIVITIES:

NEWTOWN LANDSCAPE ARCHITECTS LIMPOPO CC

No 6 Pierre Street Hampton Court
POLOKWANE 0700 Limpopo
Tel: 015 296 4452 • Fax: 015 296 4453
Email: newtown@telkomsa.net • Website: www.newla.co.za
Contact Name: Cobus Scheepers – Managing Member

NLA Limpopo CC provide a comprehensive service in the fields of Landscape Architecture, Environmental Planning and Environmental Assessment in the Limpopo Province. We have successfully completed more than 30 Landscaping and Environmental Assessment projects ranging from small to large scales. Our client base includes Municipalities, Developers, Provincial Government and private individuals.

Project types include:

- ◆ Recreation planning and detail design
- ◆ Site Masterplanning / detail design
- ◆ Institutional campuses
- ◆ Residential landscaping
- ◆ Commercial Developments and offices
- ◆ Site re-vegetation and stabilisation
- ◆ Environmental Reports
- ◆ Environmental Management Plans

ACTIVITIES:

NINHAM SHAND – CONSULTING ENGINEERS

81 Church Street
CAPE TOWN 8001 W Cape
Tel: 021 424 5544 • Fax: 021 424 5588
Email: enviro@shands.co.za • Website: www.shands.co.za
Contact Name: Mike Luger – Environmental Division Head

Ninham Shand Environmental Section forms part of a multidisciplinary engineering consultancy providing a range of services supported by the belief that development should be accompanied by a commitment to environmental responsibility, for the benefit of society. We have specialist knowledge and expertise in Cape Town and Centurion, in the fields of:

- ◆ Environmental Impact Assessment Processes
- ◆ Environmental Management Plans & Systems
- ◆ Environmental Management Programme Reports
- ◆ Water Resource Development and Catchment Management
- ◆ Institutional & Policy Development and Professional Review Services
- ◆ Specialist Facilitation, Public Processes and Training
- ◆ Industrial/Construction & Mining
- ◆ Environmental Management
- ◆ Environmental Planning & Design, Visual Impact Assessment and Landscape Architecture.

We are familiar with the planning, engineering and environmental processes and are thus readily able to balance the environmental and social considerations and engineering and authority requirements of any project.

ACTIVITIES:

NOORDHOEK ENVIRONMENTAL ACTION GROUP (NEAG)

PO Box 222
NOORDHOEK 7979 W Cape
Tel: 021 789 1751 • Email: ekogaia@iafrica.com
Contact Name: Glenn Ashton

ACTIVITIES:

NORTH WEST UNIVERSITY

Centre for Environmental Management (CEM)
Potchefstroom Campus
Private Bag X6001 POTCHEFSTROOM 2250 North West
Tel: 018 299 2725 • Fax: 018 299 2726
Email: aokml@puknet.puk.ac.za
Website: www.cem.puk.ac.za
Contact Name: Madel Lottering – Public Relations Officer

The Centre for Environmental Management (CEM) is positioned to deliver high-quality training programmes and finding innovative solutions to environmental challenges that facilitate change in your organisation on virtually any topic related to the environment and environmental management. An extensive network of supporting specialists, state of the art facilities and equipment as well as dedicated, professional staff, ensure high-quality service delivery.

ACTIVITIES:

NORTHERN CAPE ENVIRONMENTAL EDUCATION DEPARTMENT

Department of Agriculture, Land Reform,
Environment and Conservation
Private Bag X5018
KIMBERLEY 8300 Northern Cape
Tel: 053 832 2143 • Fax: 053 831 3530
Email: elameyer@grand.gov.ncape.za

The Environmental Education section imparts knowledge and awareness mainly due to the people of the Northern Cape. The resource centre is available to the public. Resource development is an ongoing process. Enviro club workshops are held to demonstrate to youth how to start and maintain clubs. Field excursions for school children are undertaken to Goegap and Rolfontein Nature Reserves, as well as to Rooipoort Game Farm. Environmental days are celebrated. Talks, especially on request, are offered to the public.

ACTIVITIES:

NOSA

PO Box 26434 ARCADIA 0007 Pretoria
Tel: 012 303 9762 • Fax: 012 303 9855
Email: debbyp@nosa.co.za • Website: www.nosa.co.za
Contact Name: Debby Parsonson – MC Senior Administrator

NOSA
★★★★★

NOSA's holistic approach to the demands of occupational risk management is developed around the nature and the needs of multi-national organisations on a global basis. Our offering ranges from services and products for SMEs (Small and Medium Enterprises) to multi-national global enterprises. We support companies with the following operational and strategic undertakings:

- ◆ Corporate governance linked to your corporate standards, linked to global standards, suitable for reporting to stakeholders and ensuring that no impacts are passed over
- ◆ Sustainable development and triple bottom line reporting
- ◆ Integration of management systems
- ◆ Third party audits, driven by client requirements
- ◆ Corporate identity: SH&E creditworthiness in the market
- ◆ Corporate frameworks to manage corporate risks, including aspects such as legal registers and international standards
- ◆ Implementation philosophy of operational risk underpinned by relevant corporate, international and SH&E standards, all focused on ultimately enhancing corporate reputation

The holistic approach is also the framework of NOSA's diversified services: a full spectrum of consultative implementation services to the management process, focused on aspects of SHEQ (Safety, Health, Environment, Quality) expertise.

ACTIVITIES:

NURSERY ASSOCIATION OF KWAZULU-NATAL

PO Box 11636 DORPSPRUIT 3206 KwaZulu-Natal
Tel: 033 342 5779 • Fax: 033 394 4842
Email: sc@futurenet.co.za
Contact Name: Lolly Stuart – Director of Operations

ACTIVITIES:

OCEAN BLUE ADVENTURES

OCEAN BLUE ADVENTURES
plettenberg bay ~ south africa

PO Box 1812
PLETTENBERG BAY 6600
W Cape
Tel: 044 533 5083 • Fax: 044 533 5083
Email: info@oceanadventures.co.za
Website: www.oceanadventures.co.za
Contact Name: Linda Packwood – Marketing

Winner of the Green Trust Award 2001, Ocean Adventures is a company founded in 1995 primarily for the purpose of developing and constructing professional, marine eco-tours, based on the non-consumptive, non-invasive sustainable utilisation of marine resources.

The main focus is on dolphins, whales and other marine mammals but includes as much of the whole marine environment as possible. Our aim is to stimulate guests' awareness of and interest in the marine environment and the realistic, holistic conservation thereof. Simultaneously we aim to ensure that from the fast-developing tourism industry all local inhabitants benefit both directly and indirectly. We believe that socio-economic upliftment and realistic conservation must and can only work hand in hand.

ACTIVITIES:

OELSNER GROUP

PO Box 13 DARLING 7345 W Cape
Tel: 022 492 3095 • Fax: 022 429 3095
Email: oelsnergrp@wcaccess.co.za
Website: www.ruethlein.com
Contact Name: Dr Herman Oelsner

The Oelsner Group has much experience in implementing renewable projects and is the main shareholder of the Darling IPP (DARLIPP), which will implement the first wind farm in South Africa. The wind farm was declared as a National Demonstration Project for WSSD 2002 by the Minister of Minerals and Energy.

Other skills within the Oelsner group include:
- ◆ Industrial Representatives
- ◆ Transfer of Technologies
- ◆ Facilitating South African – European Technology Cooperation
- ◆ Industrial, Commercial and Agricultural Properties
- ◆ Relocation of Industries

ACTIVITIES:

OIL INDUSTRY ENVIRONMENT COMMITTEE

PO Box 7082 ROGGEBAAI 8012 W Cape
Tel: 021 419 8054 • Fax: 021 419 8058
Email: amoldan@icon.co.za • Website: www.sapia.co.za
Contact Name: Anton Moldan – Environmental Advisor

The OIEC co-ordinates the South African oil industry efforts to effectively manage the environmental issues relating to its business activities as well as to promote sustainability within the industry.

ACTIVITIES:

OPEN AFRICA

PO Box 44814 CLAREMONT 7735 W Cape
Tel: 021 683 9639 • Fax: 021 683 9639
Email: admin@openafrica.org
Website: www.africandream.org
Contact Name: Carolyn Kinsey – Administration Manager

Open Africa optimises the synergies between tourism, job creation and conservation by promoting Afrikatourism through the integration of GIS technology with the Internet. For route maps and information, visit www.africandream.org

ACTIVITIES:

ORCA FOUNDATION (THE)
(OCEAN RESEARCH & CONSERVATION AFRICA)

PO Box 1812 PLETTENBERG BAY 6600 W Cape
Tel: 044 533 4897 • Cell: 082 880 2604
Email: info@orcafoundation.com
Website: www.orcafoundation.com
Contact Name: Sam Kaine – Project Co-Ordinator

ORCA is the first privately-funded initiative to support, fund and facilitate the protection of Plettenberg Bay through creating a multi-user approach to managing the Bay. The main focus is on on Research, Education and Monitoring. Together with local business, Ocean Blue Adventures and the community (Qolweni Community Development Trust), ORCA has taken the lead in marine tourism and conservation by implementing partnerships between government, the private sector and NGOs. We at ORCA believe that this is the future for our country, its people and the environment.

ACTIVITIES:

ORYX ENVIRONMENTAL

6th Floor Everite House 20 De Korte Street
BRAAMFONTEIN 2001 Gauteng
Tel: 011 403 2889 • Fax: 011 403 2887
Email: oryx.env@global.co.za
Contact Name: Andrew Duthie

Oryx Environmental provides project management and specialist expertise for environmental assessments of mining, industrial, tourism and infrastructure projects. Our approach ensures early integration of environmental issues in project decision making.

ACTIVITIES:

OVP ASSOCIATES CC

PO Box 15208 VLAEBERG 8018 W Cape
Tel: 021 462 1262 • Fax: 021 461 6162 • Email: doc@ovp.co.za
Contact Name: Johan van Papendorp

Established in 1983, OVP offers full consulting services in landscape architecture, environmental planning and urban design, serving public authorities, private development and community-based organisations.

ACTIVITIES:

PACKAGING COUNCIL OF SOUTH AFRICA

PO Box 782205 SANDTON 2146 Gauteng
Tel: 011 783 4782 • Fax: 011 883 7170
Email: packagec@cis.co.za • Website: packagingsa.co.za
Contact Name: Owen C Bruyns – Executive Director

The Packaging Council of South Africa (PCSA) is a national, voluntary association of raw material suppliers, packaging manufacturers, packaging users (fillers), retailers, packaging designers, consultants and other organisations with similar aims.

ACTIVITIES:

PALMER DEVELOPMENT CONSULTING

PO Box 11906 QUEENSWOOD 0012 Pretoria
Tel: 012 349 1901 • Fax: 012 349 1913
Email: marlett@pdcl.co.za • Website: www.pdc1.co.za
Contact Name: Marlett Wentzel

PDC specialises in renewable and household energy research, capacity building and training.

ACTIVITIES:

PAPER MAN CC

PO Box 5216 GREENFIELDS 5218 E Cape
Tel: 043 736 1016 • Cell 082 929 6353
Contact Name: Lynne Cockcroft

Principle Product: Recycled Cardboard, Paper, Magazines, Newspaper, Plastic. We collect these products, press them into bales and transport them to the mill by road, where they get pulped and recycled.

ACTIVITIES:

You must be the change you wish to see in the world.

~ Mahatma Gandhi

PEACE PARKS FOUNDATION

PO Box 12743 DIE BOORD
Stellenbosch 7613 W Cape
Tel: 021 887 6188 • Fax: 021 887 6189
Email: parks@ppf.org.za
Website: www.peaceparks.org
Contact Name: Prof W. van Riet – Executive Vice Chairman

The Peace Parks Foundation facilitates the establishment of transfrontier conservation areas or Peace Parks in Africa, supporting sustainable economic development, the conservation of biodiversity and regional peace and stability.

ACTIVITIES:

PENINSULA MOUNTAIN FORUM (PMF)

PO Box 30145
TOKAI 7966 W Cape
Tel: 021 448 8908 • Fax: 021 448 3422
Email: louis@villiers.co.za
Contact Name: Louis de Villiers – Chairman

A network of NGOs and CBOs that are concerned about the long-term protection and sustainable use of the Cape Peninsula Mountain Chain and its river systems.

ACTIVITIES:

PENINSULA PERMITS CC /ENVIROCENTRIC

PO Box 30223 TOKAI 7966 W Cape
Tel: 021 715 0011 • Fax: 021 712 9928
Email: penperm@iafrica.com
Website: www.peninsulapermits.co.za
Contact Name: Leigh Ann Ferreira /
James Jackelman – Owners

 Founded in 1999, Peninsula Permits cc has over 30 trained and qualified staff at its service, covering a wide range of expertise, including but not limited to: environmental scientists, environmental lawyers, zoologists, botanists; nature conservationists, horticulturists, foresters and law enforcement officers. Peninsula Permits provides specialist in situ operational services in the mitigation of environmental impacts as a result of major events, functions and infrastructural developments. Peninsula Permits is a registered member of IAIA(SA).

Envirocentric

Founded in 2002, Envirocentric provides specialist strategic planning and management expertise to protected areas and institutions, and specialist environmental audit and project management services in mitigating the impacts of events, functions and developments on sensitive species, habitats and ecosystems. Specialist environmental, legal, socio-economic, communications and public participation services are provided though strategic alliances and partnerships.

Key Services Offered

The key service offered is the provision of professional environmental site/control officers for events, functions, industrial developments, commercial developments, tourism developments and public service infrastructure developments to ensure the effective management, and mitigation of, environmental impacts as a result of these activities. This service includes:

◆ Development of Environmental Management Plans (EMP)

◆ Independent review of EMP's for events, functions, commercial developments & public services

◆ Implementation of on site environmental quality control, and auditing, of EMP's

◆ Project-based environmental education and awareness programs for contractors and staff

◆ Project-based environmental reviews and audits of method statements

◆ Project-based environmental advice and support services

◆ Project-based environmental public relations

◆ Representation of environmental issues at project meetings

◆ Project-based environmental reporting and communications

◆ Implementation of project-based environmental mitigatory measures

ACTIVITIES:

PEOPLE'S DISPENSARY FOR SICK ANIMALS

PO Box 29003
MELVILLE 2109 Gauteng
Tel: 011 726 6100 • Fax: 011 726 8513
Email: marilyne@netactive.co.za
Contact Name: Marilyn Eaton – Admin Officer

ACTIVITIES:

PERMACORE PERMACULTURE FOUNDATION OF THE WESTERN CAPE

PO Box 13024 MOWBRAY 7705 W Cape
Tel: 021 789 2494 • Fax: 021 789 2496
Email: info@permacore.org.za
Contact Names: Beau Horgan or
after hours Lisa Kriel 021 447 0331

Permaculture (PERMAnent AgriCULTURE) supports environmentally responsible human subsistence through design that integrates landscape, water, plants animals, energy, buildings, and people. PERMACORE promotes Permaculture in South Africa.

ACTIVITIES:

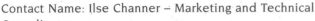

PLAASKEM (PTY) LTD

PO Box 14418 WITFIELD 1467 Gauteng
Tel: 011 823 8000 • Fax: 011 826 7241
Email: research@plaaskem.co.za
Website: www.plaaskem.co.za
Contact Name: Ilse Channer – Marketing and Technical
Co-ordinator

The business of Plaaskem (a South African company established in 1972), was acquired by Chemical Services Limited (CSL), on 1 March 2001. The businesses of Plaaskem and Applied Chemical Products were merged to capitalise on synergies between the two operations. Plaaskem now manufactures a range of products serving the specialised agricultural, animal health and foundry industries. The merger was successful and resulted in a highly motivated and focused company, which exceeded all its financial targets. The Agricultural division performed well and this division's product range was further expanded following its acquisition of Kynochem's nitrates business. Plaaskem successfully entered the fertigation market through its new joint venture, Fertiplant. Plaaskem's agricultural outlook is positive. A significant portion of the company's product range has a high local manufacture content. Plans include consolidation of the production units, expansion of existing plant capacities, increasing exports, overall growth in productivity and continued attention to safety, health and quality issues.

ACTIVITIES:

PLANTLIFE

PO Box 50030 MUSGRAVE 4062 KwaZulu-Natal
Cell: 083 360 8173 • Email: tritonia@telkomsa.net
Contact Name: David Styles – Editor

PlantLife is a non-profit journal committed to promoting interest in and conservation of southern African plants. Subject matter includes horticulture, features on particular species, taxonomy, botanical explorations and collecting expeditions.

ACTIVITIES:

PLASTICS FEDERATION OF SOUTHERN AFRICA

 The Plastics **e** nviromark

Private Bag X68
HALFWAYHOUSE 1685 Gauteng
Tel: 011 314 4021 • Fax: 011 314 3765
Email: dsteyn@plasfed.co.za • Website: www.plasfed.co.za
Contact Name: Douw Steyn – Environmental Manager

Plastics have revolutionised many aspects of daily life. The use of this highly versatile range of raw materials, in applications as varied as food packaging and artificial heart valves, has had an impact on what we eat, what we wear, how we travel, how we communicate and our general quality of life. Most plastics are derived from petrochemical feedstocks, which in turn originate from oil, natural gas and, in South Africa, from coal. Despite the widespread use of plastics worldwide, less than 4% of these energy resources worldwide are used in the manufacture of plastics.

Plastic bags have been hammered by environmentalists for a long time, suggesting they are environmentally unfriendly. Post-consumer waste of all types is a major environmental problem. This is not, however, the fault of the materials, but rather the result of irresponsible social attitudes. The same properties, which make plastics so useful to man – low mass to strength ratio and durability – also cause plastics to be highly visible in the environment. In 1997 the Plastics Federation of South Africa embarked on a campaign to help address the problem of plastics littering in our country, through the Plastics Environmental Initiative – the Plastics Enviromark.

ACTIVITIES:

POLYTECHNIC OF NAMIBIA

Department of Natural Resource Management and Tourism
Private Bag 13388 WINDHOEK Namibia
Tel: 09264 62 207 2146 • Fax: 09264 62 207 2196
Email: mdeklerk@polytechnic.edu.na
Contact Name: Marietjie de Klerk

ACTIVITIES:

PPC
(PRETORIA PORTLAND CEMENT CO LTD)

PO Box 3811 GAUTENG 2000
Tel: 011 488 1700 • Fax: 011 488 1905
Email: bharris@ppc.co.za
Website: www.ppc.co.za
Contact: Beth Harris – Public Relations Manager

PPC is South Africa's leading cement and lime producer and is a subsidiary of the global conglomerate, Barloworld. PPC boasts seven cement manufacturing facilities in South Africa, with functioning plants in Botswana and Zimbabwe. Its interests in mining is in the extraction of limestone and dolomite materials, which are used in the cement manufacturing industry as well as a host of other related industries. PPC is committed to the communities in which it operates and commits a large amount of funding to worthy community upliftment projects as well as arts and culture and environmental projects.

Of key interest is its involvement with the SA Parks Board's Game Capture Unit, based in Skukuza. PPC has been instrumental in the development of a customised transport trailer, which is able to transport two adult elephant bulls to parks and reserves in South and southern Africa. PPC has furthered this commitment by supporting research projects in the Pilanesberg and Hluhluwe-Umfolozi Parks, which concentrated on the reintroduction of adult elephant bulls into the existing and immature herds of these parks. Based on the success of these projects, PPC continues to support the SA Parks Board's elephant relocation programme in the interests of conserving this magnificent species.

ACTIVITIES:

PRICEWATERHOUSECOOPERS

2 Elgin Road SUNNINGHILL
2157 Gauteng
Tel: 011 797 4769 • Fax: 011 209 4769
Email: andrew.j.smith@za.pwc.com
Website: www.pwc.com/za
Contact Name: Andrew J Smith – Associate Director,
Sustainable Business Solutions

Good corporate citizenship is an integral aspect of good governance practice. PricewaterhouseCoopers offers a range of advisory and assurance services designed to help our clients operate and be recognised as socially reponsible corporate citizens. Our focus areas include:

Corporate governance and ethics
- Best practice reviews and gap analysis
- Framework design and development
- Policy and code development
- Training and capacity-building
- Design and development of stakeholder engagement programmes

Sustainable management practices
- Life cycle assessments
- Climate change
- HIV/AIDS risk management
- Ethical supply chain management
- Transformation
- Corporate social investment
- Environmental, health and safety reviews and due diligence
- ISO 14001 certification

Non-financial reporting and assurance
- Report development
- Report assurance

ACTIVITIES:

PROGRESS MAGAZINE

PO Box 182 PINEGOWRIE 2123 Gauteng
Tel: 011 280 5362 • Fax: 011 787 4725
Email: jasperr@jpl.co.za
Contact Name: Jasper Raats – Editor:
Tracking Development in Africa

The latest edition to the Johnnic business publishing stable, Progress is a quarterly magazine focusing on social, economic and environmental issues in southern Africa and the rest of the continent. It strives to be both topical in terms of current affairs, but also serve as a watchdog, by questioning the sustainability of development in its three spheres of coverage. While Progress is aimed at both public and private sector decision makers, it provides content in a manner that would appeal to all people in the rainbow nation and their neighbours, keeping them informed how decisions and actions on a national level will impact on their lives.

ACTIVITIES:

QUALITY TRACK

PO Box 17023 CONGELLA 4013 KwaZulu-Natal
Tel: 031 261 7228 • Fax: 031 261 3141
Email: info@qualtrak.co.za
Website: www.qualitytrack.co.za
Contact Name: Ric Neville

ACTIVITIES:

RAND AFRIKAANS UNIVERSITY (RAU)

PO Box 524
AUCKLAND PARK 2006 Gauteng

Department of Geography and Environmental Management
Tel: 011 489 2433 • Fax: 011 489 2430
Email: ggf@na.rau.ac.za
Contact Name: Dr June Meeuwis

Rauecon-Department of Zoology
Tel: 011 489 2445/2441 • Fax: 011 489 2951
Email: pk@na.rau.ac.za / jhvv@na.rau.ac.za
Contact Names: Dr Pieter Kotze or Prof Johan J. van Vuren

RAUECON is a group of environmental specialists, associated with the Science Faculty at the Rand Afrikaans University, committed to maintaining aquatic health and environmental protection.

ACTIVITIES:

RAND WATER

PO Box 1127
Johannesburg 2000 Gauteng
Tel: 011 682 0911 • Fax: 011 682 0444
Email: lwagner@randwater.co.za
Website: www.randwater.co.za
Contact Name: Lenise Wagner – Marketing Brand Manager: Water Wise

Rand Water was appointed as the sole bulk water supplier to Gauteng in 1903, and has never failed to supply its customers with clean, healthy drinking water. As times have changed, Rand Water has evolved to meet the progressive requirements of a growing population and expanding economy. It has been a key partner in the development of Gauteng and is well positioned to play an even more strategic role in the future. Rand Water's business is in every way tied to the environment. Water is a scarce natural resource and Rand Water gives equal weighting to the legal requirements and ethical obligations posed by Environmental Management. The "Water Wise" brand aims at instilling in Rand Water's customers the value of water and the need to respect it, through environmental education and community involvement (i.e. the Water Wise Education Team; Working for Water Project; Catchment Management; Water Wise Gardening; etc.)

ACTIVITIES:

RECYCLING FORUM (THE NATIONAL)

PO Box 79 ALLENS NEK 1737 Gauteng
Tel: 011 675 3462/4 • Fax: 011 675 3465
Email: iwmsa@iafrica.com
Website: www.recycling.co.za
Contact Name: Gail Smit – Secretary

The National Recycling Forum is a non-profit organisation that was created to further the interest of the formal recycling industries in South Africa.

ACTIVITIES:

RECYCLING PROJECTS (PTY) LTD

PO Box 346 RIVER CLUB 2149 Gauteng
Tel: 011 783 9794 • Fax: 011 783 9794
Email: recyclep@global.co.za
Contact Name: John des Ligneris

Total recycling of domestic waste with over 50% reduction goal, MRF sorting (jobs), composting all organics and garden waste. Milling, turning, screening equipment, builder's rubble (bricks and houses).

ACTIVITIES:

RENEWABLE ENERGY AND ENERGY EFFICIENCY PARTNERSHIP (REEEP)

Regional coordinator for Southern Africa
- AGAMA Energy
PO Box 606 CONSTANTIA 7848 W Cape
Tel: 021 701 7052 • Fax: 021 701 7056
Email: glynn@agama.co.za • Website: www.reeep.org
Contact Name: Glynn Morris

REEEP is a coalition of progressive governments, businesses and organisations committed to accelerating the development of renewable and energy efficient systems. Initiated at the Johannesburg World Summit on Sustainable Development (WSSD) in August 2002 by the UK Government, the REEEP provides an open and flexible framework within which governments work together to meet their own sustainable energy objectives according to their own timetables.

ACTIVITIES:

RESOURCE RECOVERY SYSTEMS (PTY) LTD

PO Box 2645 HALFWAY HOUSE 1685
Tel: 011 805 1066 • Fax: 011 805 0979
Email: recovery@icon.co.za
Contact Name: Mr Rob Heather – Managing Director

A chemical engineering-based company supplying services and capital plant and equipment for the recovery of solvents, acids, alkalis, base and precious metals, and organics.

ACTIVITIES:

RESPONSIBLE CONTAINER MANAGEMENT ASSOCIATION OF SOUTHERN AFRICA (RCMASA)

PO Box 894 UMHLALI 4390 KwaZulu-Natal
Tel: 032 942 8256 • Fax: 032 942 8328
Email: liz@rcmasa.org.za
Website: www.rcmasa.co.za
Contact Name: Liz Anderson – Executive Manager

Steel and plastic drums and containers of all sizes are used for an amazing variety of products such as:

- ◆ petrochemicals, paraffin, oils & paints
- ◆ agricultural & industrial chemicals
- ◆ pharmaceuticals & household products
- ◆ foodstuffs, fruit juices & beverages

All pose significant pollution, health, safety and environmental risks if not handled responsibly.

The Association promotes a life cycle and value chain approach to sustainable, safe, efficient and environmentally responsible Manufacture, Fill & Handling, Transport & Distribution, Reconditioning, Recycling, Reuse, and Final Disposal of drums and containers, whether 250ml, 210l or 1000l. It is issues driven cutting across industry sectors and silos, breaking down barriers and improving communication for more effective solutions.

RCMASA provides a focal point for all organisations involved in the life cycle of such packaging, promotes Responsible Container Management, and provides networking opportunities, information, awareness raising workshops, a voice to and liaison with government and authorities on new legislation and regulations, and coordination with international bodies to promote improved standards in our region.

RCMASA encourages industry's commitment to the public to continuously improve its health, safety and environmental practice and performance, to demonstrate transparency, and to be accountable for its action. In line with International trends, membership is open to ALL members in the life cycle and value chain of such packaging.

ACTIVITIES:

"... Pollution doesn't carry a passport..."

~ Thomas McMillan,
Canadian Environment Minister

RHODES UNIVERSITY

Environmental Education Unit (RUEEU)
PO Box 94 GRAHAMSTOWN 6140 E Cape
Tel: 046 603 8389 • Fax: 046 636 1495
Email: hlotz@ru.ac.za
Website:www.ru.ac.za/academic/departments/education/ee
unit education/eeunit
Contact Name: Prof Heila Lotz-Sisitka

Both the Murray & Roberts Chair of Environmental Education, the only one of its kind in Africa, and the Gold Fields Environmental Education Service Centre are located within the Department of Education on the Grahamstown campus of Rhodes University. The function of the Chair is to promote environmental education through research, post-graduate teaching and community empowerment. The primary purpose of the Centre is to service the growing environmental education needs of business, industry, agriculture and the formal education sector. The Murray & Roberts Chair, the Gold Fields Centre and those elements of the Faculty of Education directly involved with environmental education collectively constitute the Rhodes University Environmental Education Unit, which apart from teaching and research, takes on consultative work. The Unit also interacts closely with NGOs and community projects all over southern Africa.

ACTIVITIES:

RICHARDS BAY CLEAN AIR ASSOCIATION

P.O.Box 10299
RICHARDS BAY 3901 Kwazulu-Natal
Tel: 035 901 5340 • Fax: 035 901 5342
Email: camminga@iafrica.com • Website: www.rbcaa.co.za
Contact Name: Sandy Camminga – Public Officer

RBCAA was established in 1996 as a result of public concerns regarding levels of pollution. Its main business is to monitor and predict air quality.

ACTIVITIES:

RICHARDS BAY MINERALS

P O Box 401 RICHARDS BAY 3900
KwaZulu-Natal
Tel: 035 901 3441 • Fax: 035 901 3442
Email: jeanette.small@rbm.co.za
Website: www.rbm.co.za
Contact Name: Jeanette Small – Head of PR

RBM

Reaching towards sustainability
Since its inception 26 years ago Richards Bay Minerals (RBM) has recognised that it faces many challenges regarding the effect its various operations have on the environment and surrounding communities. Accordingly, the company implements a comprehensive Social, Environmental, Health and Safety management system of international standard. RBM was one of the first mining companies to have been awarded an ISO 14001 listing for its environmental management system. In recognition of its performance RBM also achieved the EMEM Award from the Department of Minerals and Energy for Excellence in Mining Environmental Management (National). Furthermore, the company's extensive social investment programme, which has improved the lives of thousands of people, received the Shell International Award for Sustainable Development at the Worldaware Business Awards in recognition of its Mbonambi rural development programme. Through these initiatives, RBM is contributing to the global transition to sustainable development. For a copy of the 2002 Sustainable Development Report, phone Richard O'Brien on (035) 9013153 or visit www.rbm.co.za.

ACTIVITIES:

RIVIERA WETLANDS AND AFRICAN BIRD SANCTUARY

PO Box 263327 THREE RIVERS 1935 Gauteng
Cell: 082 9000 656 • Fax: 016 423 2225
Email: treehaven@pixie.co.za
Website: treehaven.co.za
Contact Name: Heidi Weingartz – Curator

Our aim is to educate the public about African birds and their habitats. We are open to the public Wednesdays to Sundays, 9am to 5pm.

ACTIVITIES:

RONDEVLEI NATURE RESERVE

Fisherman's Walk
ZEEKOEVLEI 7941 W Cape
Tel: 021 706 2404 • Fax: 021 706 2405
Email: rondevlei@sybaweb.co.za
Website: www.rondevlei.co.za
Contact Name: Mr D J Gibbs – Manager

Rondevlei is a small 50 ha inland water body and 240 ha of surrounding land on the Cape Flats that has been protected as a nature reserve since 1952.

ACTIVITIES:

ROSE FOUNDATION

Suite A9 Waverley Court
7 Kotzee Street
MOWBRAY 7700 W Cape
Tel: 021 448 7492 • Fax: 021 448 7563
Email: usedoil@iafrica.com
Website: www.rosefoundation.org.za
Contact Name: Jeannette Rustenberg

The Rose Foundation manages the environmentally acceptable collection, storage and recycling of used lubricating oil throughout South Africa. For used oil collections call toll free 0800 107 107.

ACTIVITIES:

RURAL SUPPORT SERVICES

6 Muir St Southern Wood EAST LONDON 5213 E Cape
Tel: 043 743 0051 • Fax: 043 743 0051
Email: rss@rss.co.za
Contact Name: Ms Mihlali Fimukonda

RSS is an Eastern Cape NGO, which supports integrated development for rural communities through gender equity; health promotion and facilitation of water and sanitation supply.

ACTIVITIES:

SANTAM/CAPE ARGUS UKUVUKA OPERATION FIRESTOP

Goldfields Education Centre
Kirstenbosch Botanical Gardens
NEWLANDS 7700 W Cape
Tel: 021 762 7474 • Fax: 021 762 8337
Email: sandra@ukuvuka.org.za • Website: www.ukuvuka.org.za
Contact Name: Sandra Fowkes – Campaign Manager

The Santam/Cape Argus Ukuvuka Campaign was initiated after the devastating fires in January 2000, which destroyed vast areas in the Western Cape and 30% (8 370 hectares) of the natural area of the Cape Peninsula as well as 8 homes and damaged over 50 houses and buildings. The fires were a wake-up call that a significant change in the behaviour of citizens, both individual and institutional, was needed in addition to the implementation of an integrated fire management plan, if a similar disaster was to be prevented from occurring again.

The overriding aim of the Campaign and its implementing partners is to significantly reduce the damage and danger from uncontrolled fires in the mountains and informal settlements of the Cape Peninsula. In addition, the Campaign seeks to contribute to poverty relief through creating training and employment opportunities, and lessening the vulnerability of communities to uncontrolled fire. The Campaign further wishes to address institutional issues by supporting the formation of a Fire Protection Association and developing a strategy and tools for the local authority to better manage the natural-urban interface. Finally the Campaign is creating a role model of a government/ business/ Community partnership that can be replicated in other areas in South Africa as well as internationally.

In case of fire call the City of Cape Town Fire and Emergency Services on 107 or 021 480 7700

ACTIVITIES:

SAPLER POPULATION TRUST

PO Box 51446 RAEDENE 2092 Gauteng
Tel: 011 477 6447 • Fax: 011 477 6468
Email: info@population.org.za
Website: www.population.org.za
Contact Name: Mr David Hirsch – Executive Director

The SAPLER population trust advocates limiting population size through friendly, accessible, family planning. It currently provides schools' sexuality education by community youth in schools in the North West Province.

ACTIVITIES:

SAPPI LIMITED

PO Box 31560
BRAAMFONTEIN 2017 Gauteng
Tel: 011 407 8111 • Fax: 011 403 8236
Email: corporateaffairs@sappi.com
Website: www.sappi.com
Contact: Corporate Affairs and Communications Manager

Sappi Limited, South Africa's global pulp and paper group, is the world's leading producer of coated fine paper and dissolving pulp. The company recognises that responsible management of natural resources, linked with social responsibility and sound economic performance, is a prerequisite for sustainable development. We strive for continual improvement in the mitigation of our impact on the environment through the use of recognised management systems by which we also monitor our progress. Sappi has chosen ISO 14001 as the preferred environmental management system.

More than 25% of the land on our tree farms is especially managed for the conservation of water, flora and fauna. We are a key participant in the Department of Water Affairs and Forestry's 'Working for Water' programme, and we continue to invest in projects in our mills to increase the use of recycled water and reduce waste, emissions and energy use.

Sappi has initiated many environmentally-based projects focused on uplifting rural communities and enhancing environmental awareness. One of these is the R10 million 'TreeRoutes' initiative, launched in 1998, for developing viable eco-tourism routes through some of South Africa's most sensitive indigenous forests and wetlands. This initiative was honoured by the World Wide Fund For Nature (WWF) as a 'Gift to the Earth' in 2000.

ACTIVITIES:

It we are not doing what needs to be done, it is certainly not for lack of knowledge. Since Rio, we have shared more knowledge of what is right and wrong than ever before.

~ Poul Nyrup Rasmussen,
Prime Minister of Denmark.

SAPPI FINE PAPER

sappi

P O Box 32706
BRAAMFONTEIN 2017 Gauteng
Tel: 011 407 8111• Fax: 011 339 8022
Website: www.sappi.com
Contact:
Adamas mill – Environmental Manager • Tel: 041 408 4111
Stanger mill – Environmental Manager • Tel: 032 437 2222
Enstra mill – Environmental Manager • Tel: 011 360 0000

Headquartered in London, Sappi Fine Paper leads the fast growing coated fine paper market with a 15% share of the world market. Coated fine paper is used in high quality publications such as annual reports, coffee table books, calendars, catalogues, brochures and magazines. Sappi's Adamas, Enstra and Stanger mills, situated in South Africa, form part of Sappi Fine Paper.

All Fine Paper mills in South Africa are certified to the international recognised ISO 14001 environmental management system, while all the European mills have both ISO 14001 and European standard, EMAS (Environmental Management and Audit System) certification.

SAPPI FOREST PRODUCTS

sappi

PO Box 32706
BRAAMFONTEIN 2017 Gauteng
Tel: 011 407 8111 • Fax: 011 403 1885
Website: www.sappi.com
Contact: Group Environmental Manager

Sappi Forest Products, headquartered in Johannesburg, is Africa's largest forest products business. It comprises three business units: Sappi Forests, Sappi Kraft and Sappi Saiccor.

SAPPI FORESTS

sappi

P O Box 32706
BRAAMFONTEIN 2017 Gauteng
Tel: 011 407 8111 • Fax: 011 403 1885
P O Box 13124 Cascades 3202
Tel: 033 347 6600 • Fax: 033 347 6790
Website: www.sappi.com
Contact: Environmental Manager

Sappi Forests owns and manages approximately 550 000 hectares of land in southern Africa. In 1999, Sappi Forests received the international environmental management system ISO 14001 certification, the first multi-site operation in Africa, and one of only a few in the world, to achieve this international environmental standard. In addition, all Sappi lands are certified to the Forest Stewardship Council standard for well managed forests. The company voluntarily pledges to the South African Natural Heritage Programme and has 18 Natural Heritage Sites on approximately 6 500 hectares and 28 Sites of Conservation Significance covering 4 000 hectares.

SAPPI KRAFT

sappi

P O Box 32706
BRAAMFONTEIN 2017 Gauteng
Tel: 011 407 8111 • Fax: 011 339 6929
Website: www.sappi.com
Contact:
Ngodwana mill – Environmental Manager • Tel: 013 734 6111
Tugela mill – Environmental Manager • Tel: 032 456 1111
Cape Kraft mill – Environmental Manager • Tel: 021 552 2127
Usutu mill – Environmental Manager • Tel: 09628 452 6010

Sappi's Kraft division is South Africa's largest producer of packaging papers and newsprint. Three of the division's mills are based in South Africa and produce papers and boards used in the manufacture of corrugated boxes, paper sacks and bags. Sappi Kraft is a major exporter. The Sapoxyl oxygen bleaching process for pulp, developed at Ngodwana mill, has now become a worldwide industry standard. Two of the mills, Ngodwana and Tugela, have been certificated to the ISO 14001 environmental management system standard.

SAPPI WASTE PAPER

sappi

P O Box 32706
BRAAMFONTEIN 2017 Gauteng
Tel: 011 407 8111• Fax: 011 339 6929
Website: www.sappi.com • Contact: General Manager

Sappi Waste Paper, a division of Sappi Kraft, drives the procurement of waste paper through outsourced national on-site waste management and recycling programmes. While the company contracts numerous recycling operations throughout the country, primarily for collection of larger volumes of paper from businesses and schools, it also utilises an extensive network of recycling centres, which purchase waste paper from street traders. The key business objective of Sappi's recycling programme is to supply waste paper to its South African Kraft operations. One of these operations, Cape Kraft, operates on 100% recycled paper. Sappi's War on Waste (WOW) paper recycling campaign is an initiative aimed at leveraging individual responsibility – to change South Africa's 'throw away' culture to one that endorses recycling.

SAPPI SAICCOR

sappi

P O Box 62
UMKOMAAS 4170 KwaZulu-Natal
Tel: 039 973 8911 • Fax: 039 973 8054
Website: www.sappi.com
Contact: Environmental Manager

Sappi Saiccor is the world's largest producer of dissolving pulp, used in the manufacture of products such as viscose, rayon and cellophane. Effluent generated is discharged into the sea via a 6.5km pipeline. Compared to the previous 3km pipeline, this has substantially reduced the impact of the effluent on the beach. In addition, the lignosulphonate plant, LignoTech SA, which uses some of the waste from the Saiccor plant, has recently been expanded. This has resulted in a further reduction to the amount of effluent discharged into the sea.

SCHOOL ENVIRONMENTAL EDUCATION AND DEVELOPMENT (S.E.E.D)

PO Box 44
OBSERVATORY 7935 W Cape
Tel: 021 447 7686 • Fax: 021 447 7686
Email: seed@myafrica.com
Contact Name: Leigh Brown

S.E.E.D is facilitating a broad-based teacher and school driven movement, which results in Permaculture being taught across all learning areas in the RNCS.

ACTIVITIES:

SEAL ALERT-SA

PO Box 221 Postnet
HOUT BAY 7872 W Cape
Tel: 021 790 8774 • Cell: 072 579 3154
Email: sealalert@arcticonline.net / sasealion@wam.co.za
Website: www.sealalert.co.za
Contact Name: Francois Hugo – Founder

Seal Alert-SA is involved in addressing historical imbalances, caused by exploitation, disturbance and commercial sealing. It is opposed to sealing, seal tagging or disturbance of seal colonies in any way. It also opposes unnatural mainland seal colonies. It supports the development of natural environment seal rescue and rehabilitation. It is instrumental in developing improved conservation, protection, welfare and awareness of all species of seal, found along the coastlines of southern Africa.

ACTIVITIES:

SEAWATCH

PO Box 457 BETTY'S BAY 7141 W Cape
Tel: 028 272 9532 • Fax: 028 272 9532
Email: richards@globalocean.co.za
Contact Name: Mike Tannett – Coordinator

Seawatch is concerned with the conservation of marine and coastal habitats, monitoring and reporting of suspected marine crimes to police and sea fisheries inspectors. They have been instrumental in assisting with the confiscation of 55000 shell fish, and the recovery of 300 birds. They also provide environmental education and training for S.A.P, schools and fishermen.

ACTIVITIES:

SEEDLING GROWERS' ASSOCIATION OF SOUTH AFRICA

Postnet Suite 178 PB X9118
PIETERMARITZBURG 3200 KwaZulu-Natal
Tel: 033 342 5779 • Fax: 033 394 4842
Email: sc@futurenet.co.za
Website: www.seedlinggrowers.co.za
Contact Name: Lolly Stuart – Director of Operations

This is a national organisation comprising 140 members. Focuses on providing latest technical information on seedling growing to its members by an on-going intensive research programme through universities and supplier companies. The Seedling Growers Association holds an annual symposium and field day focusing on research and best business practice.

ACTIVITIES:

SEMODISE SE KOPANE CORPORATE FARMERS

PO Box 5552
MMABATHO 2735 North West
Tel: 018 384 0970
Email: imodise@tsa.ac.za
Contact Name: Mr Itumeleng Modise – CEO
Emerging Farmers' Association.

ACTIVITIES:

SETTLEMENT PLANNING SERVICES (SETPLAN)

PO Box 3405
CAPE TOWN 8000 W Cape
Tel: 021 422 1946
Fax: 021 424 3490
Email: rod@setplan.com
Contact Name: Rod Cronwright – Director

SetPlan is a land use and environmental planning & management consultancy. The practice has a 23 year southern African track record, and operates out of offices in the centres of Cape Town, Johannesburg, East London, Port Elizabeth, Kimberley and George.

SetPlan has extensive experience in the planning and setting up of protected areas, with particular attention to their regional integration and delivering socio-economic benefits to surrounding communities. For their work in this field on the Lesotho Highlands Water Project, the practice was awarded the South African Planning Institute's Bi-annual Merit Award. Similar work in this field has been undertaken for the Western Cape Nature Conservation Board and for SANParks in the recently established Cape Peninsula National Park and the Agulhas National Park.

In addition to providing planning services, the practice also provides land use and environmental management services related to project implementation and operation.

ACTIVITIES:

To attack nature is to attack mankind.

~ Jacques Chirac, President of France

SHARE-NET

WESSA PO Box 394
HOWICK 3290 KwaZulu-Natal
Tel: 033 330 3931 • Fax: 033 330 4576
Email: jt@futurenet.co.za
Contact Name: Dr Jim Taylor – Director:
SADC Regional EE Support

Contact Share-net for inexpensive resource materials to support environmental processes in South Africa. Our resources are copyright free and cover a very wide range of topics such as:

- The "Hand-On" series for coastal, inland and urban environments
- Action orientated "How to" activities
- Indigenous Knowledge series
- Wetland and water testing packs
- Bird resource Pack
- A series of Enviro fact sheets
- School Environmental Policy pack
- E-info libraries
- An interactive Environmental learning CD.

ACTIVITIES:

SI ANALYTICS

PO Box 154
MAGALIES VIEW 2067 Gauteng
Tel: 011 465 6004 • Fax: 011 465 6139
Email: mturnbull@icon.co.za
Website: www.sianalytics.co.za
Contact Name: Mike Turnbull

S I Analytics, founded in 1976, specialises in Ambient Air Quality and Continuous Emission Monitoring Systems. We design, manufacture, supply, install, commission and provide after sales service to comprehensive systems or individual analysers in sub-Saharan Africa. Our sole suppliers are world leaders in their fields:

Ambient Air Quality

- Rupprecht and Patashnick – manufactures a range of Ambient Air Quality Instruments certified as Reference or Equivalent methods by US EPA and German TUV.

Real time ambient continuous mass measurement, size selective (PM10, PM2.5, PM1.0 & TSP) Particulate monitors – TEOM and it's various derivatives

- Sampling platforms for particulate collection – Partisol range
- Speciation Samplers
- Bioaerosol Samplers
- Personal real time Dust Monitors

Sulphate, Nitrate and Carbon Monitors

- Monitor Europe – manufactures a range of Ambient Air Quality Instruments certified as Reference or Equivalent methods by US EPA, German TUV, and UK MCerts.

Gas analysers for SO_2, NOx, O_3, H_2S, NH_3, CO, CO_2

- Baseline Mocon – Hydrocarbon Analysers – Heated Total Hydrocarbon (THC), non Methane Hydrocarbon (NMHC) and BTEX (Benzene, Toluene, Xylene, Ethyl Benzene).
- Sabio Engineering – Dynamic Gas Dilution and Permeation Bench Calibrators
- Envitech – Data Acquisition and Reporting Software package systems – Tabulated, graphical, wind & pollution Roses, GIS, maintenance software etc.

Continuous Emission Monitoring

- Procal Analytics – In-situ continuous stack gas analysers – Infra red and Ultra Violet

Local and remote access to analyser data and maintenance control parameters

- MIP Electronics Oy – Opacity/Particulates Monitors with the unique continuous ability to monitor the Zero/Calibration function
- Ecochem – Extractive high sensitivity gas analyser system (Hot Wet Sampling) e.g. Incinerator Exhaust Gas Monitoring

ACTIVITIES:

SOCIETY FOR ANIMALS IN DISTRESS

PO Box 391164 BRAMLEY 2018 Gauteng
Tel: 011 466 0261 • Fax: 011 466 0262
Email: sanzio@icon.co.za
Contact Name: Peter Lake – Executive General Manager

The Society for Animals in Distress provides veterinary care to large and small animals in townships and informal settlements. We also offer education on basic animal care in disadvantaged schools.

ACTIVITIES:

SOCIETY OF SOUTH AFRICAN GEOGRAPHERS

Department of Geography
Rhodes University
PO Box 94 GRAHAMSTOWN 6140
Tel: 046 603 8319 • Fax: 046 636 1199
Email: R.Fox@ru.ac.za
Website: www.egs.uct.ac.za/ssag
Contact Name: Prof R.C. Fox – President

The prime focus of the SSAG is the promotion of geographical and environmental research, education and communication in southern Africa. The Society promotes interaction between organisations and individuals interested in Geography and related disciplines.

These aims are achieved by:

- Publication of the South African Geographical Journal at least twice annually
- Publication of regular Newsletters and other occasional publications for members

◆ Arrangement and organisation of a biennial conference

◆ Encouragement of scientists of international standing to visit southern Africa

◆ Organisation of workshops, lectures and other appropriate activities at national and regional level

◆ Representation of geographical and environmental educators through engaging with government bodies over policy formulation.

ACTIVITIES:

SOFTCHEM

PO Box 1525
NORTH RIDING 2162 Gauteng
Tel: 082 554 8900 • Fax: 011 462 2985
Email: francois@softchem.co.za
Website: www.softchem.co.za
Contact Name: Francois Friend – Managing Member

Softchem provides engineering consulting services specialising in environmental management (including implementation of ISO 14001 environmental management systems, environmental impact assessments and environmental management programme reports – EMPRs), water management (including water balances), functional specifications and operating procedures for water treatment plants, dedicated software for the water and environmental industry, technical and environmental auditing and the implementation of waste management systems (including dangerous substance registers).

ACTIVITIES:

SOLAR COOKER PROJEKT (SOLCO)

PO Box 13732 PRETORIA 0028
Tel: 012 342 0181 • Fax: 012 342 0185
Email: david.hancock@gtz.de
Website: www.solarcookers.co.za
Contact Name: David Hancock

Since 1996, the Department of Minerals and Energy (DME) and the German Technical Corporation (GTZ) have cooperated in introducing solar cookers to the South African market. Besides undertaking field studies and demonstrations concerning use and acceptance of solar energy in rural areas, the mission of GTZ's Solar Cooker Project is to open up a whole new industrial sector, constructing and selling cookers everywhere in South Africa.

The vision of Solco is to turn solar energy into an equal partner of other forms of energy such as electricity, gas, paraffin, or wood. South African partners of Solco are Vesto, SunStove and Hotbag.

ACTIVITIES:

SOLARDOME SA

18 Stoffelsmit Street
PLANKENBRUG 7600, W Cape
Tel: 021 886 6321
Fax: 021 886 5121
Email: sunlight@solardome.co.za
Website: www.solardome.co.za
Contact Name: Ryan Dearlove – Sales

Solardome SA cc was established in 1965 as Mikado (Pty) Ltd, later changing its name to Solardome SA cc.

As the most established manufacturer of solar heating systems in South Africa, we can draw on 37 years of experience in the field.

Solardome SA cc offers the following:

◆ Solar water heating systems for domestic, and commercial applications

◆ Solar collector panels and solar geysers for customised heating systems

◆ Solar electric systems (photovoltaic) for remote power applications, both large and small, including pumps, chargers etc.

◆ Accessories and support for all solar driven systems.

We are able to supply nationally, as well as internationally, and can arrange delivery to your door. Our confidence in the quality of our products is backed by a 3-year warranty on any goods manufactured in our factory.

ACTIVITIES:

SOLAR ENGINEERING SERVICES

PO Box 628 KLOOF 3640
KwaZulu-Natal
Tel: 031 764 6292
Fax: 031 764 1266
Email: soleng@solarengineering.co.za
Website: www.solarengineering.co.za
Contact Name: Will Cawood – Director

Solar Engineering Services design and implement sustainable rural development, income generation and labour saving strategies. These involve the integrated use of renewable energy technologies. Urban solar water heating is also a focal activity.

While some of these strategies may be sustainable individually, in our experience they are most likely to endure when implemented as a cohesive interdependent and integrated programme.

Cognisance must be taken of gender issues to ensure programme sustainability.

ACTIVITIES:

SOLAR HEAT EXCHANGERS

Postnet Suite 292 Private Bag X3 NORTH RIDING 2162 Gauteng
Tel: 011 462 0024 • Fax: 011 704 6570
Email: info@solarheat.co.za • Website: www.solarheat.co.za
Contact Name: Kathy Oliver – Office Manager

Manufacture, design, sales and installation of solar hot water systems.

ACTIVITIES:

SOUTH AFRICAN ASSOCIATION AGAINST PAINFUL EXPERIMENTS ON ANIMALS (SAAAPEA)

PO Box 85228 EMMARENTIA 2029 Gauteng
Tel: 011 646 5956 • Fax: 011 646 5956
Email: saaapea@mweb.co.za
Website: www.geocities.com/ PetburghHaven/1496
Contact Name: Rae Pearl – Secretary-General

SAAAPEA was founded on 10 April 1976. Our main objective is to promote by all possible means the phasing out of animal-based research.

ACTIVITIES:

SOUTH AFRICAN ASSOCIATION FOR MARINE BIOLOGICAL RESEARCH (SAAMBR)

PO Box 10712 MARINE PARADE 4056
KwaZulu-Natal
Tel: 031 337 3536 • Fax: 031 337 2132
Email: info@saambr.org.za • Website: www.saambr.org.za
Contact Names: Dr Mark Penning – Director, Ann Kunz – Marketing Manager

SAAMBR was founded in 1951, as a unique centre for marine science, conservation and education. It is also one of Durban's foremost tourist attractions, now located at uShaka Marine World.

SAAMBR is an NGO that stimulates community awareness of the marine environment through education and promotes wise, sustainable use of marine resources through scientific investigation. It fulfils its mission by operating three integrated divisions:

- ◆ Oceanographic Research Institute (ORI) – Scientific information provided by ORI's experienced marine scientists contributes towards ensuring sustainable resources for the future
- ◆ Sea World – which provides visitors with a glimpse of the rich diversity of life in the western Indian Ocean and promotes the need for marine conservation
- ◆ Sea World Education Centre – with its extensive marine education programmes, reaches learners, teachers and resource users from all sectors of the community.

ACTIVITIES:

SOUTH AFRICAN CLIMATE ACTION NETWORK (SACAN)

PO Box 11383 JOHANNESBURG 2000 Gauteng
Tel: 011 339 3660 • Fax : 011 339 3270
Email: elin@earthlife.org.za
Website: www.climatenetwork.org
Contact Nam : Elin Lorimer

The secretariat is provided by the SECCP project of Earthlife Africa Jhb.

SACAN is a network established to facilitate civil society participation in responding to climate change and promote government, industry and individual action to limit human-induced climate change. See: www.earthlife.org.za

SACAN and the regional SARCAN are driven by membership, which is open to citizen-based non profit-organisations and individuals that subscribe to the goals of the global Climate Action Network (CAN) and the CAN Charter: www.climatenetwork.org.

ACTIVITIES:

SOUTH AFRICAN DARK SKY ASSOCIATION

20 Nerine Avenue PINELANDS 7405 WCape
Tel: 021 531 5250 • Email: cliffturk@yebo.co.za
Contact Name: Cliff Turk – Acting Director

SADSA was formed under the sponsorship of The Astronomical Society of southern Africa, and consists of individuals, organisations and businesses interested in avoiding pollution by invasive and wasteful lighting.

ACTIVITIES:

SOUTH AFRICAN FREEZE ALLIANCE ON GENETIC ENGINEERING (SAFeAGE)

PO Box 13477, MOWBRAY 7704 W Cape
Tel: 021 447 8445 • Fax: 021 447 5974
Email: safeage@mweb.co.za
Contact Name: Karen Kallman – Co-ordinator

SAFeAGE is an alliance of organisations and individuals concerned about the effects of genetic engineering on our health, environment, farmers and economy. We are campaigning for a minimum five-year freeze/moratorium on genetic engineering and patenting on food and farming in South Africa. During those five years we would like the following to be developed:

- ◆ Labelling in order that people can exercise their democratic right to choose products free of genetically engineered (GE) ingredients and derivatives
- ◆ A comprehensive review of government policy and legislation concerning GE
- ◆ Public participation in decision-making including civil society representation on the Advisory Committee to the Executive Council of Genetically Modified Organisms (GMOs)

- Independent assessment of social and economic impacts of GMOs on farmers
- A system whereby genetic contamination of the environment can be prevented
- Independent assessment of the implications of patenting genetic resources.

ACTIVITIES:

SOUTH AFRICAN INSTITUTE OF ECOLOGISTS AND ENVIRONMENTAL SCIENTISTS

PO Box 1749 NOORDHOEK 7979 W Cape
Tel: 021 789 1385 • Fax: 021 789 1385
Email: saie-es@intekom.co.za • Website: www.saie-es.za.org
Contact Name: Ms Jeanette Walker – Secretary

The Southern African Institute of Ecologists and Environmental Scientists is a professional body set up to advance the science and uphold the standards of professional practice for ecology and environmental science throughout southern Africa.

ACTIVITIES:

SOUTH AFRICAN INSTITUTE OF MINING & METALLURGY (SAIMM)

PO Box 61127 MARSHALLTOWN 2107 Gauteng
Tel: 011 834 1273 • Fax: 011 838 5923
Email: sam@saimm.co.za • Website: www.saimm.co.za
Contact Name: Ms Sam Moodley – Manager

Since its inception in 1894, The South African Institute of Mining & Metallurgy has been a dynamic organisation attending professionally to all the needs of its members. There are approximately 2 900 members and the Institute's membership shows a steady growth.

The key objective of the Institute is to identify the needs of its members and to initiate and give effect to the transfer of technology and scientific knowledge relevant to the sustainable development of the minerals and metals section of the South African economy to its stakeholders; and furthermore to represent and promote the interests of its members.

ACTIVITIES:

SOUTH AFRICAN LANDSCAPERS' INSTITUTE (SALI)

PO Box 472 MODDERFONTEIN 1645 Gauteng
Tel: 011 606 3156 • Fax: 011 606 2895
Email: val@sali.co.za
Contact Name: Val Wamsteker – Director of Operations

ACTIVITIES:

SOUTH AFRICAN NATIONAL PARKS HEAD OFFICE

PO Box 787 PRETORIA 0001
Tel: 012 426 5000 • Fax: 012 343 4605
Website: www.parks-sa.co.za
Email: reservations@parks-sa.co.za
Contact Name: – Ms Wanda Mkutshulwa

See Page 362 for full list of Parks

ACTIVITIES:

SOUTH AFRICAN NEW ECONOMICS FOUNDATION (SANE)

PO Box 44928
CLAREMONT 7735 W Cape
Tel: 021 689 6892
Fax: 021 686 1560
Email: sane@iafrica.com • Website: www.sane.org.za
Contact Name: Laura Bishop – Administrator

A network and research foundation offering and investigating alternatives to modern free-market economics. The alternatives offer an end to both poverty and environmental degradation.

ACTIVITIES:

SOUTH AFRICAN NURSERY ASSOCIATION (SANA)

PO Box 514
HALFWAY HOUSE 1685 Gauteng
Tel: 011 464 1098 • Fax: 011 464 1099
Email: info@sana.com • Contact Name: Debbie Parker

The Green Industry is the umbrella body for the horticultural industry in South Africa and covers nurseries, landscapers, irrigation specialists and flower growers.

ACTIVITIES:

SOUTH AFRICAN SOCIETY OF OCCUPATIONAL HEALTH NURSING PRACTITIONERS (SASOHN)

PO Box 18793
SUNWARD PARK 1470 Gauteng
Tel: 011 892 3174 • Fax: 011 892 5355
Email: sasohnoffice@mweb.co.za
Website: www.sasohn.org.za
Contact Name: Karen Michel – President

SASOHN is a non-profit professional organisation whose goal is to promote occupational health nursing practice. Provides a supportive network for the almost 1 000 full and associate members. An environment is provided for development of professional capacity and excellence of practice.

ACTIVITIES:

SOUTH AFRICAN NATIONAL PARKS

PARK NAME	LOCATION	TELEPHONE	FAX
Addo Elephant National Park Park Manager: Lucius Moolman	ADDO E Cape	042 233 0556	042 233 0196
Agulhas National Park Park Manager: Etienne Fourie	L'Agulhas W Cape	028 435 6222	028 435 6225
Augrabies National Park Park Manager: Jan van Deventer	AUGRABIES N Cape	054 452 9200	054 451 5003
Bontebok National Park Park Manager: Jorum Mkosana	SWELLENDAM W Cape	028 514 2735	028 514 2646
Golden Gate Highlands National Park Park Manager: Johan Taljaard	CLARENS Free State	058 255 0012	058 255 0022
Karoo National Park Park Manager: Norman Johnson	BEAUFORT WEST W Cape	023 415 2828	023 415 1671
Kgalagadi Transfrontier National Park Park Manager: Steve Smith	UPINGTON N Cape	054 561 2000	054 561 2005
Knysna National Lake Area National Park Park Manager: Roy Ernstzen	KNYSNA W Cape	044 382 2095	044 382 5801
Kruger National Park Park Manager: Josias Chabani	SKUKUZA Mpumalanga	013 735 4000	013 735 4054
Marakele National Park Park Manager: Nicholas Funda	THABAZIMBI Limpopo	014 7771745	014 777 1866
Mountain Zebra National Park Park Manager: Phumla Mzazi	CRADOCK E Cape	048 881 2427	048 881 3943
Namaqua National Park Park Manager: Matthew Norval	KAMIESKROON N Cape	027 672 1948	027 672 1015
Richtersveld National Park Park Manager: Willem Louw	ALEXANDER BAY N Cape	027 831 1506	027 831 1175
Table Mountain National Park Park Manager: Brett Myrdal	CAPE TOWN W Cape	021 701 8692	021 701 4261
Tankwa Karoo National Park Park Manager: Conrad Strauss	CALVINIA N Cape	0273 412 389	0273 412 389
Tsitsikamma National Park Park Manager: Nico van der Walt	STORMS RIVER E Cape	042 281 1607	042 281 1629
Vaalbos National Park Park Manager: Deon Joubert	BARKLEY WEST N Cape	053 561 0088	053 561 0080
Vhembe Dongola National Park Park Manager: Bernard van Lente	MESSINA Limpopo	015 534 0102	
West Coast National Park Park Manager: Gary de Kok	LANGEBAAN W Cape	022 772 2144	022 772 2607
Wilderness National Park Park Manager: Roy Ernstzen	WILDERNESS W Cape	044 877 0046	044 877 0111

SOUTH AFRICAN SUGAR ASSOCIATION (SASA)

PO Box 700
MOUNT EDGECOMBE 4300 KwaZulu-Natal
Tel: 031 508 7024 • Fax: 031 508 7191
Email: rory.lynsky@sasa.org.za
Website: www.sasa.org.za
Contact Name: Rory Lynsky – Environment Liaison Manager

The South African sugar industry, recognising that its constituent members, the sugar millers and cane growers, are major custodians of natural resources, encourages its members to promote and take responsibility for the resources which are used within the sugar industry.

ACTIVITIES:

SOUTH AFRICAN WHITE SHARK RESEARCH INSTITUTE

PO Box 50775
V&A Waterfront
CAPE TOWN 8002 W Cape
Tel: 021 552 9794 • Fax: 021 552 9795
Email: whiteshk@iafrica.com • Website: www.whiteshark.co.za
Contact Name: Jytte Ferreira

ACTIVITIES:

SOUTHERN AFRICA ENVIRONMENT PROJECT (SAEP)

10 Surbiton Road ROSEBANK 7700 W Cape
Tel: 021 689 2020 • Fax: 021 689 2020
Email: saep@worldonline.co.za • Website: www.saep.org
Contact Name: Norton Tennille – Executive Director

"Green Your Mind, Green Your Soul, Green Your Body, Green Your World." Our educational and youth development activities offer a holistic approach to personal development and constructive involvement in the human environment of South Africa's townships and the natural environment of the Western Cape.

ACTIVITIES:

SOUTHERN AFRICAN FOUNDATION FOR THE CONSERVATION OF COASTAL BIRDS (THE) (SANCCOB)

PO Box 11116
BLOUBERGRANT 7443 W Cape
Tel: 021 557 6155 • Fax: 021 557 8804
Email: info@sanccob.co.za
Website: www.sanccob.co.za
Contact Name: Alan Jardine – Chief Executive Officer

SANCCOB was established in 1968 in response to the increasing numbers of oiled sea birds in the Western Cape. Throughout the last 35 years, SANCCOB has responded to every major oil spill along the South Africa coast and attended to the chronic influx of ill, injured and oiled sea birds every day of every year. Since its inception, the centre has helped to treat over 81,000 seabirds, in particular the African Penguin, which is classified as "Vulnerable" in the Red Data Book. Recent research proves that the African Penguin population is 19% higher today than it would have been in the absence of SANCCOB's efforts. Educationals and tours are offered by appointment only.

ACTIVITIES:

SOUTHERN AFRICAN WILDLIFE COLLEGE (SAWC)

Private Bag X3015
HOEDSPRUIT 1380 Mpumalanga
Tel: 015 793 7300 • Fax: 015 793 7314
Email: info@sawe.org.za
Website: www.wildlifecollege.org.za
Contact Name: Fanie Greyling – Executive Director

The Southern African Wildlife College began as a vision of a centre of excellence in conservation education and training. WWF-SA, with funding from the German Government (KfW), made this vision a reality, and the College opened its doors in 1997 to its students from southern Africa. The College provides natural heritage managers from all over Africa, in cooperation with stakeholders, with the motivation and relevant skills to manage their areas and associated wildlife populations sustainably, and in a culturally acceptable manner. The College has an extraordinary attitude to training, aimed at providing practical skills which supplement the theoretical knowledge embedded in the core of the College's modules. In addition, the College's strict selection criteria ensure a situation where students also learn from each other through shared experience. All trainers are contracted in to ensure we offer the latest training from the coalface of conservation.

ACTIVITIES:

SOUTHERN CROSS SCHOOLS

PO Box 116
HOEDSPRUIT 1380 Mpumalanga
Tel: 015 793 0590/1
Fax: 015 793 0557
Email: raptor@scschools.co.za
Contact Name: Jumbo Williams – Headmaster

Southern Cross Schools is a unique, value-based and co-educational institution. It is based on 'Raptors' View', a wildlife estate on the outskirts of Hoedspruit, in Limpopo Province. Its ethos is entrenched in an environmental code, which serves as its vehicle for instruction. The methodology through which learning is facilitated, is learner-centred and outcomes-based motivated.

ACTIVITIES:

SOUTHERN HEALTH & ECOLOGY INSTITUTE

PO Box 67608
BRYANSTON 2021 Gauteng
Tel: 073 197 9094 • Fax: 073 197 9094
Email: shae@worldonline.co.za
Contact Name: Nicole Venter – Programme Development

The Southern Health & Ecology Institute specialises in educating people about macrobiotic impacts of their personal actions. We provide full courses in Macrobiotics, Personal Health, Deep Ecology and we do experiential workshops, host seminars and camps or tours for nature integration and environmental awareness. We publish manuals, media and newsletters. As activists we produce campaign materials and initiate community action.

ACTIVITIES:

SOUTHSOUTHNORTH PROJECT (THE)

South Africa Office
138 Waterkant Street GREEN POINT 8001 W Cape
Tel: 021 425 1465 • Fax: 021 425 1463
Email: lester@southsouthnorth.org
Website: www.southsouthnorth.org
Contact Name: Lester Malgas – Country Liaison Officer

Our Website, www.southsouthnorth.org includes up to the minute news on all SSN activities and projects in our different regions around the world. It functions as an interface for linking with our partners and team members. Our library section contains easily downloadable templates and reports, which are based on SSN project experience. It is designed to be useful for everyone who is interested in Clean Development Mechanism (CDM) and Climate Change.

ACTIVITIES:

SPACE FOR ELEPHANTS FOUNDATION

PO Box 1020 PONGOLA
3170 KwaZulu-Natal
Tel: 034 413 2477 • Fax: 034 413 2478
Email: info@space4elephants.co.za
Website: www.space4elephants.co.za
Contact Name: Dr Heinz Kohrs – Founding Trustee

The foundation aims to find a long term solution to challenges facing elephants in a fenced and fragmented environment, through partnerships, knowledge sharing and educating local communities about the importance of sustainable wildlife, conservation and tourism for social and economic upliftment.

ACTIVITIES:

SPCA (THE NATIONAL COUNCIL OF SPCAS)

PO Box 1320
ALBERTON 1450 Gauteng
Tel: 011 907 3590 • Fax: 011 907 4013
Email: spca@global.co.za
Website: www.nspca.org.za
Contact Name: Christine Kuch – PR Officer

The SPCA movement is the only animal welfare organisation in South Africa with statutory rights to deal with all animals – not only domestic animals or wildlife. Falling directly under the auspices of the NSPCA are three specialised national units: The National Farm Animal Unit, The National Wildlife Unit and The National Special Projects Unit.

There are over 100 SPCAs in South Africa, each of which is registered as a charitable organisation. Each SPCA is an autonomous, independent society with its own fundraising/non-profit organisations registration as issued by the Department of Welfare in Pretoria.

The NSPCA is the governing body of all SPCAs in the country. It co-ordinates the efforts of all SPCAs and provides professional assistance, training, education and specialist services. There is an annually elected Board of Directors and one Executive Director.

The NSPCA constantly monitors and motivates changes and improvements to all legislation relating to animal welfare in South Africa. It is affiliated to the RSPCA, Humane Society of the United States and the World Society for the Protection of Animals.

ACTIVITIES:

SPEROSENS

Private Bag X115
CENTURION 0046 Gauteng
Tel: 012 665 0317 • Fax: 012 665 0337
Email: info@spero.co.za
Contact Name: Johan van Tonder

Spero Sensors and Instruments In-Stack Dust Monitoring.

ACTIVITIES:

SPOORNET

Private Bag X47
JOHANNESBURG 2000 Gauteng
Tel: 011 773 9279 • Fax: 011 774 4432
Email: annemaries@transnet.co.za
Website: www.spoornet.co.za
Contact Name: Annemarie Strydom – Manager Corporate Communications

ACTIVITIES:

SRK CONSULTING ENGINEERS AND SCIENTISTS

Website: www.srk.co.za

Contact Names:

Eastern Cape:	Rob Gardiner	041 581 1911
Gauteng:	Allison Burger-Pinter	011 441 1111
KwaZulu-Natal:	Nick Holdcroft	031 312 1355
W Cape:	Christopher Dalgliesh	021 409 2400

SRK Consulting is a South African founded group of consulting engineers and scientists providing a comprehensive range of services to the natural resource industries. Our professional team provides environmental management consulting services to stakeholders in the sustainable utilisation of human and natural resources, such as:

- ◆ Environmental, economic and social assessments and systems
- ◆ Sustainability management, assessment and reporting
- ◆ Auditing, monitoring and evaluation
- ◆ Environmental and social costing
- ◆ Multi-stakeholder management and needs analysis
- ◆ Facilitation
- ◆ Public involvement and conflict management
- ◆ Training and capacity building
- ◆ Mine closure planning
- ◆ Organisational development
- ◆ Social and labour planning

ACTIVITIES:

STRATEK – BUSINESS STRATEGY CONSULTANTS

PO Box 74416

LYNWOOD RIDGE 0040 Pretoria

Tel: 012 803 9736 • Fax: 012 803 9745

Email: stratek@pixie.co.za • Website:www.stratek.co.za

Contact Name: Dr Kelvin Kemm – Director

Management Consultants in the fields of Strategic Plan Development and Implementation: as Stratek carries out scenario development in conjunction with the client. Strategy developments include the formulation of probable paths of development and market strategies. Stratek also engages in general problem solving.

Management Consulting:

Particularly appropriate for Stratek are projects requiring a complex multidisciplinary approach. Stratek can move between different professional groups and individuals to maintain all in harmony. The Stratek Associates have qualifications in a wide variety of disciplines including: nuclear physics, chemistry, economics, law, management and policy design.

Training:

Most training needs are company specific and, as a result, most of Stratek's training products are developed for each client's needs. Other established courses are also available.

Project Communications:

Stratek is skilled in converting complex language into layman's language. Services include the production of corporate and training videos, and the writing of documents for marketing or public information purposes.

International:

Stratek maintains a wide network of international links, and can operate in other countries, or rapidly link other countries to local requirements.

ACTIVITIES: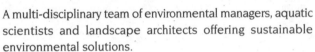

STRATEGIC ENVIRONMENTAL FOCUS

Tel: 012 349 1307 • Fax: 012 349 1229

Cell: 083 442 5417 – Dave Rudolph

Cell: 083 305 9597 – Ainsley Simpson

Email: sef@sefsa.co.za • Website: www.sefsa.co.za

A multi-disciplinary team of environmental managers, aquatic scientists and landscape architects offering sustainable environmental solutions.

SERVICES:

Environmental Management Unit:

- ◆ Environmental Impact Assessments
- ◆ Scoping Reports
- ◆ Environmental Management Plans
- ◆ Environmental Management Systems
- ◆ ISO 14000 Implementation
- ◆ Environmental Auditing
- ◆ Strategic Planning
- ◆ State of the Environment Reports
- ◆ Environmental Management Frameworks
- ◆ Strategic Development Initiatives

Mining & Industrial Unit:

- ◆ Environmental Management Programmes
- ◆ Environmental Risk Analysis

Communication & Training Unit:

- ◆ Complementary & independent public participation
- ◆ Training & awareness raising programmes

Specialist Unit:

- ◆ Floral & Faunal assessments
- ◆ Wetland assessments
- ◆ Artificial wetland design
- ◆ Water license applications
- ◆ Social Impact Assessments

Landscape Architecture & Graphic Design Unit:

- ◆ Visual Impact Assessments
- ◆ Landscape Planning & Design

SEFGIS:

- ◆ Spatial data modelling & analyses
- ◆ Database design, maintenance & management
- ◆ User training for GIS tools and products
- ◆ Tools for decision making & strategic planning

ACTIVITIES:

SUNSTOVE ORGANISATION

PO Box 21960 CRYSTAL PARK 1315 Gauteng
Tel: 011 969 2818 • Fax: 011 969 5110
Email: sunstove@iafrica.com • Website: www.sungravity.com
Contact Name: Mrs M R Bennett – Director

Promotion, production and distribution of a low-cost, user-friendly solar cooker.

ACTIVITIES:

SUSTAINABILITY INSTITUTE

Lynedoch Road LYNEDOCH 7603 W Cape
Tel: 021 881 3196 • Fax: 021 881 3294
Email: info@sustainabilityinstitute.net
Website: www.sustainabilityinstitute.net
Contact Name: Eve Annecke – Director

International living and learning centre for studies in ecology, community and spirit. Delivers a Masters Programme in Sustainable Development in partnership with the University of Stellenbosch, School of Public Management and Planning. Forms part of the larger Lynedoch EcoVillage which is a socially mixed ecologically designed multi-use development that includes housing, schools, NGOs, businesses and farming projects.

ACTIVITIES:

SUSTAINABILITY UNITED (SUN)

PO Box 428 UMZINTO 4200 KwaZulu-Natal
Tel: 039 974 1769 • Fax: 031 710 5255
Email: sustainu@ananzi.co.za
Contact Name: Troy Govender – Chairperson

An environmental network for children, youth, women and NGOs.

- ◆ Environmental capacity building
- ◆ Environmental networking
- ◆ Pilot environmental projects
- ◆ Distribution of environmental information and opportunities from international processes to local levels.

ACTIVITIES:

SUSTAINABLE ENERGY AFRICA

PO Box 261 NOORDHOEK 7979 W Cape
Tel: 021 702 3622 • Fax: 021 702 3625
Email: info@sustainable.org.za
Website: www.sustainable.org.za
Contact Names: Leila Mahomed / Sarah Ward – Directors

Sustainable Energy Africa, promotes sustainable energy approaches and practices in the development of South Africa through research, capacity building, appropriate demonstration, information dissemination, practical project development & management, lobbying, networking, policy & strategy development.

SEA's activities include:

- ◆ Co-ordinating the Sustainable Energy for Environment and Development (SEED) Programme. This programme focuses on making sustainable energy approaches and practices part of the core functions of its partner institutions.
- ◆ Working with Local Authorities to develop local sustainable energy strategies and projects.
- ◆ Assisting in achieving the milestones of the Cities for Climate Protection campaign in Cape Town such as the greenhouse gas emission inventory.
- ◆ Assisting the City of Cape Town with the development of a local sustainable energy strategy. The State of Energy Report, has been completed, and comprises the first comprehensive city-wide energy assessment on the African continent.
- ◆ Assisting in demonstrating energy efficiency and carbon dioxide emission reductions potential by implementing projects such as the energy efficiency measures in the Tygerberg Municipal Building in Parow.

ACTIVITIES:

SUSTAINABLE ENERGY SOCIETY OF SOUTHERN AFRICA (SESSA)

PO Box 152 LA MONTAGNE 0184 Gauteng
Email: info@sessa.org.za • Website: www.sessa.org.za
Contact Name: Trevor van der Vyver

SESSA is a 30-year old multi-disciplined NGO dedicated to the practical use of energy efficiency. Members include academics, students, consultants and the RE industry. Divisions include architecture, solar water heating, photo-voltaics and components, solar water pumping.

ACTIVITIES:

SUSTAINABLE LIVING CENTRE

The Green Building Bell Crescent Close
Westlake Business Park TOKAI 7945
Tel: 021 702 3622 • Fax: 021 702 3625
Email: info@sustainable.co.za
Website: www.sustainable.co.za
Contact Name: Mark Borchers – Director

The Sustainable Living Centre (SLC):

- ◆ supplies a wide range of goods which help people live more sustainably
- ◆ provides information which alerts people to the local and global environmental situation
- ◆ provides practical guidance on measures to live more sustainably
- ◆ provides information on organizations that provide services linked to sustainable living, such as environmental consultancies and renewable energy expertise

It is fast becoming apparent that the planet cannot sustain our current lifestyles. We need to make a change. The SLC aims to make it easy for the average person or household to improve their environmental 'footprint' through the provision of online information and goods. To browse the information and products available, or to order products, go to www.sustainable.co.za.

Goods Supplied: House & Garden; Renewable Energy; Energy Efficiency; Water Saving; Recycled Products; Green Buildings; General (books etc).

ACTIVITIES:

TECHNIKON PRETORIA

Private Bag X680 PRETORIA 0001
Tel: 012 318 6115 • Fax: 012 318 6354
Email: mohrm@techpta.ac.za
Contact Name: M Mohr-Swart – Head of Department

Department of Environmental Sciences

Undergraduate and Post graduate programmes offered in Geology and Environmental Sciences.
Short courses and correspondence courses also available.
Contact person: Geology: Prof Coetzee • Tel: 012 318 6279.
Environmental Sciences: Ms Maryna Mohr Swart
Tel: 012 318 6115

Environmental Law and Management Clinic

The department also offers a community service. The Environmental Law and Management Clinic advises indigent persons on environmental management matters free of charge.
Contact person: Ms Catherine Haffer • Tel: 012 318 6226

ACTIVITIES:

TERRATEST INC

5 Cascades Crescent Montrose
PIETERMARITZBURG 3201
KwaZulu-Natal
Tel: 033 347 2992 • Fax: 033 347 1845
Email: norrisj@terratest.co.za
Contact Name: Jan Norris

Established in 1990, Terratest is now a PDI company operating throughout southern Africa. Our team is made up of geotechnical engineers, engineering geologists, geophysicists and environmental scientists. We provide services to the civil engineering and mining industries, infrastructure developers, government departments, municipalities and NGOs.

At Terratest we take pride in our ability to work closely with our clients, always taking time to understand their needs so that we can offer practical, cost-effective and innovative solutions.

We provide both geotechnical and environmental services as follows:

Geotechnical Services:
- Geotechnical investigations
- Construction materials investigations
- Geotechnical design
- Geophysics
- Site supervision & monitoring

Environmental Services:
- Waste Management
- EIAs
- EMS
- EMPs (mining)
- Catchment management
- Environmental auditing
- Environmental planning and policy

ACTIVITIES:

THORNTREE CONSERVANCY

PO Box 1552 WALKERVILLE 1876 Gauteng
Tel: 016 590 2312 • Fax: 016 590 2312
Email: mail@thorntree.co.za • Website: www.thorntree.co.za
Contact Name: Ivan Parkes – Chairman

Thorntree Conservancy is a rural conservancy which was established April 1999 in Walker's Fruit Farms, northwest of Meyerton. The area encompasses agricultural holdings around a range of unspoilt koppies. The region falls within the Bankenfeld grassland type and consequently the koppies are covered in a huge variety of grasses, wildflowers, proteas and indigenous trees.

ACTIVITIES:

TIGER CHEMICAL TREATMENT (PTY) LTD

PO Box 6674 HOMESTEAD 1412 Gauteng
Tel: 011 828 5652 • Fax: 011 828 4741
Email: tctrcem@icon.co.za
Contact Name: Joe Gruber – General Manager

Waste treatment and reclamation of metals such as precious metals, zinc, copper, cobalt and nickel from metal finishing industries and refineries. Upgrading of residues.

ACTIVITIES:

"God is not making any more water."

~ Said at the World Summit on Sustainable Development 2002

TIMBER PLASTICS (PTY) LTD

186 Second Avenue Unit 54 Joanick Park
ALBERTON 1449 Gauteng
Tel: 011 907 5727 • Fax: 011 907 5783
Email: del@timberplastics.co.za
Website: www.timberplastics.co.za
Contact Name: Del Kotze – Owner / Director

TIMBER PLASTICS established in 1991 is a leader in recycling plastic.

The recycled plastic is used to produce environmental-friendly maintenance free products for a wide spectrum of uses in outdoor entertainment and industry.

The product widely known as Polywood is well established in:
◆ Agriculture
◆ Marine Engineering
◆ Civil Engineering
◆ Landscaping and Horticulture
◆ Recreational
◆ Industrial

The Polywood product has a unique resemblance to timber but unlike timber it does not rot, fade or splinter.

The life span of Polywood is far superior to timber and also:
◆ withstands humidity
◆ is not affected by chemicals and solutes
◆ is not a conductor of electricity
◆ is weather resistant
◆ is insect resistant

Polywood's unique characteristics makes it a product of the future and is supported by the Wild Life Society of South Africa.

ACTIVITIES:

TLHOLEGO DEVELOPMENT PROJECT

PO Box 1668 RUSTENBURG 0300 Mpumalanga
Tel: 014 592 7090 • Fax: 014 592 7090
Email: tlholego@iafrica.com
Contact Name: Mr Paul Cohen - Director

The Tlholego Ecovillage, over the past 13 years, has provided a rural educational environment based on ecological principles with the purpose of training people in the design and implementation of sustainable land use and village settlements. Tlholego is a Tswana word meaning "creation from natural origins".

ACTIVITIES:

TOYOTA S.A.

PO Box 481
BERGVLEI 2012 Gauteng
Tel: 011 809 2232 • Fax: 011 444 8245
Email: leades@tsb.toyota.co.za • Website: www.toyota.co.za
Contact Name: Lee-Anne Eades – Senior Media Relations Officer

Toyota South Africa is the country's leading motor manu-facturer, having been market leader for the past 23 consecutive years. The company manufacturers passenger cars, light commercial vehicles and trucks in a facility at Prospecton, near Durban and distributes vehicles imported from Toyota companies in Japan and Australia and a number of countries in Africa. Toyota SA's shareholding is held 75% by Toyota Motor Corporation, Japan, and 25% by Wesco, a South African company listed on the JSE Securities Exchange. The company was established in 1961.

ACTIVITIES:

TRANSVAAL MUSEUM

PO Box 413 PRETORIA 0001
Tel: 012 322 7632 • Fax: 012 322 7939
Email: malherbe@nfi.co.za
Contact Name: Carina Malherbe – Ecologist

Biodiversity surveys of southern African fauna including small mammals, bats, birds, herpetofauna, insects and spiders.

ACTIVITIES:

TREE SOCIETY OF SOUTHERN AFRICA

PO Box 70720 BRYANSTON 2021 Gauteng
Tel: 011 465 6045 • Fax: 011 465 6045
Email: dehning@mweb.co.za / walterb@icon.co.za
Contact Name: Walter Barker – Chairman

Since 1946, the TSSA has been promoting knowledge and appreciation of our valuable indigenous flora, mainly trees, with emphasis on field activities.

ACTIVITIES:

TREEHAVEN WATERFOWL TRUST

PO Box 263327 THREE RIVERS 1935 Gauteng
Tel: 082 900 0656 • Fax: 016 423 2 225
Email: treehaven@pixie.co.za
Website: www.treehaven.co.za
Contact Name: Heidi Weingartz – Curator

Thirty acres, dedicated to the world's waterfowl, TWT is the largest collection of waterfowl in the world. We participate in the Wattled Crane Recovery Programme.

ACTIVITIES:

TRIBAL CONSULTANTS CC

PO Box 77448
ELDOGLEN 0171 Gauteng
Tel: 012 658 5579
Fax: 082 131 3765 933
Email: tribal@lantic.net
Contact Name: Riaan Robbeson – Director

Tribal Consultants offer proficient floristic investigations, providing solutions for the interface between environment and development. We provide a comprehensive range of vegetation related services to our clients in the compilation of EIAs, EMRs, Environmental Management Systems, Strategic Environmental Assessments, Ecological Surveys and Conservation Planning. Our base of clients includes academic institutions, mining companies, landscape architects, environmental managers and developers, conservation reserves and government councils.

Our services include:

◆ Floristic Investigations, Descriptions & Mapping
◆ Pre-development Floristic Investigations & Process Recommendations
◆ Red Data Plant Species Investigations & Management Recommendations
◆ Plant Species Diversity Investigations
◆ Medicinal Plant Species Investigations
◆ Integrated Sensitivity Analysis
◆ Ecological Interpretations specific to EIAs and EMPRs
◆ Environmental Baseline Studies

Post-graduate degrees and affiliation with the South African Council for Natural Scientific Professions guarantee a professional and scientific approach, while years of experience ensure a thorough business approach. We strive to apply a multi-disciplinary and holistic approach for each project, promoting optimal solutions to environmental problems.

ACTIVITIES:

TSOGA ENVIRONMENTAL RESOURCE CENTRE (TERC)

PO Box 254
LANGA 7456 W Cape
Tel: 021 694 0004 • Fax: 021 694 9813
Email: nosipdil@iafraica.com
Contact Name: Mr Sonwabo Ndandani – Acting Director

To promote a greater public awareness of environmental issues; inform and educate communities of their environmental rights; promote community participation and involvement in environmental policy formulation; build a community-accessible resource centre; and facilitate information sharing locally and internationally.

ACTIVITIES:

TWO OCEANS AQUARIUM

Dock Rd Victoria & Albert
Waterfront Cape Town
PO Box 50603
WATERFRONT 8002 W Cape
Tel: 021 418 3823/4 • Fax: 021 418 3952
Email: aquarium@aquarium.co.za
Website: www.aquarium.co.za
Contact Name: Helen Lockhart – Communications Manager

The Two Oceans Aquarium is a window on the oceans, offering glimpses of the diverse life found off the South African coastline. Over 3 000 living animals, including fishes, invertebrates, mammals, reptiles, birds and plants can be seen in this spectacular underwater nature reserve.

The Aquarium offers unique opportunities such as diving with the sharks and copper hat diving, sleepovers for children and facilities for conferences and functions. Music concerts enchant audiences against the spectacular backdrop of the I&J Predator Exhibit.

The Aquarium's Education Centre offers a variety of educational programmes and facilities. These include a range of theme-based lessons which are presented by the Aquarium's educators to visiting school groups; teacher enrichment workshops and open-days; a mobile teaching unit and Wet Laboratory.

Training courses are also offered several times a year to people who are interested in volunteering as front-of-house guides or behind the scenes.

ACTIVITIES:

TYRRELL ASSOCIATES

PO Box 2341
CAPE TOWN 8000
Tel: 021 511 8108 • Fax: 021 510 0093
Email: tyrrell@iafrica.com • Website: www.tyrrell.co.za

Media communications, marketing, promotions, awareness-raising, strategy and implementation.

ACTIVITIES:

*"Only after the last tree
has been cut down,
Only after the last river
has been poisoned,
Only after the last fish
has been caught
Only then will you find
that money cannot be eaten."*

~ Cree Indian Prophecy

UBUNGANI WILDERNESS EXPERIENCE

PO Box 1610 NIGEL 1490 Gauteng
Tel: 011 454 1414 • Fax: 011 454 1414
Email: ubungani@global.co.za Website:
www.ubungani.org.za
Contact Name: Hanneke van der Merwe – Field Director

We are a non-profit organisation providing environmental education directed at urban and rural communities of all ages, creeds and cultures with the view to improving understanding, standard of living, confidence and self worth of participants. We strive to develop educators with the necessary skills to facilitate integration of Environmental Education into the formal sector. Ubungani offers a well-balanced and established core educational programme focused on individual growth. Our field staff are known and respected and have experience in teaching both local and international students. Our participants are exposed to interpretive walks in a natural environment, case studies and interactive activities.

Our key aims are to:

◆ Create an awareness and appreciation of the natural environment and the bio-diversity of the fauna and flora
◆ Share knowledge and to foster an understanding of the ecological principles and the importance of our natural resources
◆ Create an awareness of local and global environmental issues
◆ Provide the skills to address local and global environmental issues.

ACTIVITIES:

UMGENI WATER

PO Box 9 PIETERMARITZBURG 3200
KwaZulu-Natal
Tel: 033 341 1152 • Fax: 033 341 1501
Email: martine.hedley@umgeni.co.za
Website: www.umgeni.co.za
Contact Name: Mr Pumezo Jonas – General Manager

Umgeni Water is one of the largest water producers and service providers in Africa. The organisation won the recent World Wide Fund for Nature Southern Africa Environmental Reporting Award. Umgeni Water provides African solutions to global water services challenges through commercialisation of its services into Africa. Services include: water resource planning; catchment management; operation/maintenance of large and small water and wastewater works; water, wastewater and industrial effluent control; pollution prevention and control; process design, construction, commissioning of works and quality management systems; health, safety and environmental services; human resource management and professional staff training that can be customised for the client. These activities are supported by state-of-the-art ISO accredited laboratory services.

ACTIVITIES:

UMGENI WATER – EXTERNAL EDUCATION SERVICES

PO Box 9 PIETERMARITZBURG
3200 KwaZulu-Natal
Tel: 031 268 7160 • Fax: 03031 262 8543
Email: sunitha.doodhnath@umgeni.co.za
Website: www.umgeni.co.za
Contact Name: Sunitha Doodhnath

Established in 1987 EES offers full consulting services in water-related education, training and resource development ranging from secondary to tertiary level including ABE. Services the public, schools, universities, NGOs and CBOs. A free water education mail order catalogue is also available.

ACTIVITIES:

UNILEVER CENTRE FOR ENVIRONMENTAL WATER QUALITY

PO Box 94
GRAHAMSTOWN 6140 Eastern Cape
(c/o Institute for Water Research)
Tel: 046 622 2428 • Fax: 046 622 9427
Email: ucewq@iwr.ru.ac.za
Website: www.ru.ac.za/departments/iwr/ucewq
Contact Name: Prof Carolyn Palmer – Director

UCEW-QIWR contributes to water resource management through:

◆ Research into the toxicity of pollutants on aquatic ecosystems
◆ Consulting services around water quality issues
◆ Training
◆ Contributions to national policy development and implementation.

ACTIVITIES:

UNISA

Department of Anthropology, Archaeology, Geography and Environmental Studies, Unisa,

PO Box 392
UNISA 0003 Gauteng
Tel: 012 429 6013 • Fax: 012 429 6868
Email: steenhe@unisa.ac.za
Website: www.unisa.ac.za/dept/geo
Contact Name: Dr Sux Zietsman – Programme Co-ordinator

Undergraduate degree programmes with core Geography modules: BA (Environmental Management) and BSc (Environmental Management) with options in Chemistry, Botany and Zoology.

ACTIVITIES:

UNIVERSITY OF CAPE TOWN

 **UNIVERSITY OF CAPE TOWN**

RONDEBOSCH 7701
W Cape
Tel: 021 650 2873 • Fax: 021 650 3791
Email: admin@enviro.uct.ac.za
Website: www.egs.uct.ac.za/engeo/courses/mphil
Contact Name: Sharon Adams

The Master of Philosophy degree programme was started in 1974, and has trained more South African environmental professionals than any other similar programme. The programme aims to produce graduates with a broad understanding of major environmental issues and the ability to effectively analyse and manage environmental problems and conflicts, with emphasis on South Africa. The programme duration is 18 months full-time, starting in February each year: 9 months of course work followed by an 8 month research project applying environmental assessment and management skills and procedures to real life issues. Employment opportunities arise in the public sector, parastatals, the private sector (consulting firms and industry), and in NGOs and CBOs. Scholarships and bursaries are available. Applicants should have a good Honours degree or equivalent 4-year professional degree, preferably with appropriate experience. The closing date for applications is the end of August and the outcome of the selection process is in October.

Energy & Development Research Centre
Tel: 021 650 3230
Email: harold@energetic.uct.ac.za
Contact Name: Harold Winkler

Environmental Evaluation Unit
Tel: 021 650 2866
Email: sowman@science.uct.ac.za
Contact Name: Dr Merle Sowman

Freshwater Research Unit
Tel: 021 650 3301
Email: jday@botzoo.uct.ac.za
Contact Name: Dr Jenny Day

Industrial Health Research Group
Tel: 021 650 3508
Email: ihnick@protem.uct.ac.za
Contact Name: Mr N. Henwood

Marine Biology Research Institute
Tel: 021 650 3610
Email: clgriff@botzoo.uct.ac.za
Contact Name: Prof Charlie Griffith

Molecular and Cell Biology
Tel: 021 650 3256/67
Contact Name: Professor Jennifer A Thomson
Schools Development Unit

Schools Development Unit
Tel: 021 650 3276
Email: gs@humanities.uct.ac.za
Contact Name: Gail September

South African Bird Ringing Unit
SAFRING Avian Demography Unit
Tel: 021 650 2421
Email: safring@maths.uct.ac.za
Contact Name: Dieter Oschadleus

Zoology Department
Tel: 021 650 3604
Email: jday@botzoo.uct.ac.za
Contact Name: Dr Jenny Day

UNIVERSITY OF NATAL – DURBAN

King George V Ave Durban 4041
Private Bag X01 SCOTSVILLE 3209 KwaZulu-Natal
Websitewww.nu.ac.za – for full a profile of all environmentally-related departments.

Department of Geographical & Environmental Science
Tel: 031 260 2283
Contact Name: Prof Robert Preston-Whyte

Pollution Research Group
Tel: 031 260 3375
Email: buckley@nu.ac.za
Contact Name: Prof Chris Buckley

School of Applied Environmental Sciences
Tel: 033 260 5514
Contact Name: Prof MJ Savage

School of Botany & Zoology
Tel: 033 260 5104
Email: brothers@nu.ac.za
Contact Name: Prof Dennis Brothers

School of Life and Environmental Sciences
Tel: 031 260 3197
Email: berjak@nu.ac.za
Contact Name: Prof Pat Berjak

UNIVERSITY OF ORANGE FREE STATE

PO Box 339 BLOEMFONTEIN 9300 Free State
Tel: 051 401 9111 / 401 2112
Website: www.uovs.ac.za/faculties/nat/zent/sob for a full profile of all environmentally-related departments.

Centre for Environmental Management
Tel: 051 401 2863
Email: seamanmt@sci.uovs.ac.za
Contact Name: Maitland Seaman

Department of Microbiology & Biochemistry
Tel: 051 401 2396
Email: dpreezjc@sci.uovs.ac.za
Contact Name: Prof J du Preez

Department of Plant Sciences & Genetics
Tel: 051 401 2514
Email: grobbeju@sci.mail.uovs.ac.za
Contact Name: Prof Johan Grobbelaar

Department of Zoology & Entomology
Tel: 051 401 2427
Email: vanasjg@dre.nw.uovs.ac.za
Contact Name: Prof Jo van As

Institute for Groundwater Studies
Tel: 051 401 2394
Email: frank@igs.uovs.ac.za
Contact Name: Prof FDI Hodgson

UNIVERSITY OF PORT ELIZABETH

PO Box 1600
PORT ELIZABETH 6000 E Cape
Tel: 041 504 2111/ 504 2313
Website: www.upe.ac.za – for a full profile
of all environmentally-related departments.

Department of Zoology

Tel: 041 504 2341
Email: zladdb@upe.ac.za
Contact Name: Prof D Baird

Terrestrial Ecology Research Unit

Tel: 041 504 23 08
Email: teru@upe.ac.za
Contact Name: Prof G Kerley

NORTH WEST UNIVERSITY – SEE UNDER "N"

UNIVERSITY OF PRETORIA

PRETORIA 0002
Tel: 012 420 2489
Website: www.up.ac.za/ – for a full
profile of all environmentally-
related departments.

University of Pretoria

Mammal Research Institute

Department of Zoology & Entomology
Tel: 012 420 2066 • Fax: 012 420 2534

Centre for Africa -Tourism

Department of Tourism Management
Tel: 012 420 4073
Contact Name: Prof GHD Wilson

Centre for Wildlife Management

Hatfield Experimental Farm, South Street
HATFIELD 0002 Pretoria
Tel: 012 420 2627 • Fax : 012 420 6096
Email: lvessen@wildlife.up.ac.za
Website : www.wildlife.up.ac.za/centre
Contact Name: Mr L D van Essen – Lecturer

RAND AFRIKAANS UNIVERSITY – SEE UNDER "R"

RHODES UNIVERSITY – SEE UNDER "R"

UNIVERSITY OF STELLENBOSCH

Private Bag X1
MATIELAND 7602 W Cape
Tel: 021 808 4515
Website: www.sun.ac.za – for a full profile
of all environmentally-related
departments.
Tel: 021 808 3064

UNIVERSITEIT
STELLENBOSCH
UNIVERSITY

Department of Botany

Email: fcb@sun.ac.za
Contact Name: Prof FC Botha

Department of Geography and Environmental Studies

Tel: 021 808 3103
Email: hlz@sun.ac.za
Contact Name: Prof HL Zietsman

Unit for Environmental Ethics

Tel: 021 808 2988 • Fax: 021 808 3556
Email: jph2@sun.ac.za
Website: www.sun.ac.za/philosophy/cae
Contact Name: Prof J P Hattingh

UNIVERSITY OF THE WESTERN CAPE (UWC)

Welcome to the University of the Western Cape

Private Bag X17
BELLVILLE 7535 W Cape
Tel: 021 959 2911

Department of Geography and Environmental Studies

Tel: 021 959 2421
Email: mdysel@uwc.ac.za
Contact Name: Michael Dyssel

Environmental Education and Resources Unit

Tel: 021 959 2489 • Fax: 021 959 2484
Email: cklein@uwc.ac.za
Website: www.botany.uwc.ac.za/eeru
Contact Name: Charmaine Klein – Director

The mission of EERU is to promote environmental education opportunities that will lead to a better understanding of the total environment. The Unit is located on the campus of the University of the Western Cape and functions as a service organisation that provides environmental education support primarily to school and community organisations on the Cape Flats.

Its main activities include Environmental Education, Indigenous Greening and Horticulture as well as Urban Nature Conservation. To achieve this, the Unit's Cape Flats Nature Reserve (a Provincial Heritage site), consisting of Strandveld and coastal Fynbos, its Nursery and EE Resource centre are used. The Indigenous Greening Programme has since 1992 been responsible for the establishment of many indigenous gardens at schools and in communities. Workshops, guided walks and ecological fieldwork are offered.

UNIVERSITY OF THE WITWATERSRAND

Private Bag 3 WITS 2050 BRAAMFONTEIN
Gauteng
Tel: 011 717 1000 / 717 1010
Website: www.wits.ac.za – for a full profile of all environmentally-related departments.

University of the
Witwatersrand,
Johannesburg

Sasol Centre for Innovative Environmental Management

Tel: 011 717 7103
Email: banem@civil.wits.ac.za
Contact Name: Prof P Marjanovic

SASOL Centre for Innovative Environmental Management is engaged in research in the field of Industrial Ecology and seeks partners and cooperation to move the Industrial Ecology Agenda forward.

Wits School for the Environment (WISE)
Tel: 011 717 6551
Email: developmentstudies@social.wits.ac.za
Contact Name: Prof H Annegarn

School of Animal Plant and Environmental Sciences
Tel: 011 717 6406
Email: shirley@gecko.biol.wits.ac.za
Contact Name: Prof S Hanrahan

School of Geography Archeology and Environmental Science
Tel: 011 717 6521
Email: moonb@geoarc.wits.ac.za
Contact Name: Prof B Moon

Water in the Environment Research Group
Tel: 011 717 6424
Contact Name: Prof KH Rogers

URBAN GREEN FILE

Published by
Brooke Pattrick (Pty) Ltd
PO Box 422
BEDFORDVIEW 2008 Gauteng
Tel: 011 622 4666 • Fax: 011 616 7196
Email: carolknoll@brookepattrick.co.za
Contact Name: Carol Knoll – Editor

Urban Green File is a multidisciplinary, professional, trade and technical journal which deals with environmental issues, largely in the urban environment. These include, amongst many others, coverage of 'green buildings'; the ecological and social dimensions of horticulture and landscaping; conservation and restoration of rivers and wetlands; rehabilitation of mines, road embankments and degraded sites; Environmental Impact Assessments; and social projects aimed at community development and poverty alleviation.

ACTIVITIES:

VAN RIET & LOUW LANDSCAPE ARCHITECTS

PO Box 36723 MENLO PARK 0102 Gauteng
Tel: 012 346 1289 • Fax: 012 346 1289
Email: vrl@vrl.co.za
Contact Names: Johan Louw, Peter Velcich

- ◆ Environmental planning and management
- ◆ Tourism planning
- ◆ Landscape design
- ◆ Landscape rehabilitation
- ◆ Urban design
- ◆ Architectural design

ACTIVITIES:

VERVET MONKEY FOUNDATION

PO Box 458
BANBURY CROSS 2164 Gauteng
Tel: 012 205 1206 – Gauteng
Tel: 082 954 3200 – Rehabilitation Centre – Tzaneen
Email: vervets@enviro.co.za • Website. www.enviro.co.za
Contact Name: Arthur Hunt

In 1991, Dave du Toit, Alma du Toit, and Arthur Hunt became involved in saving orphaned vervet monkeys. These monkeys had no place to go and when confiscated were killed by the authorities. The vervet monkey species had been declared vermin and did not have the right to life in our country. In 1993, they formed the Vervet Monkey Foundation to fight for the monkeys' rights and to develop a rehabilitation programme for orphaned vervet monkeys. The Foundation has become the advocates for this species and is managed by a group of people who volunteer their services free of charge. Today it operates from Tzaneen in the Limpopo Province and Isobel Hits runs a halfway house in Gauteng. To protect this species it has more than one hundred monkey watchers throughout South Africa who report indiscriminate killing or poisoning of monkeys.

ACTIVITIES:

VIRIDUS TECHNOLOGIES (PTY) LTD

T/a Eko Rehab P O Box 19752 NOORDBRUG 2522 North West
Tel: 018 297 7320 • Fax: 018 293 2258
Email: ekojb@puknet.puk.ac.za
Website: www.ecorehab.co.za
Contact Name: Dr Johann Booysen – Managing Director

Eko Rehab delivers the following professional and research-based services at a moderate tariff:

Rehabilitation of disturbed areas, i.e.
- ◆ Open cast mining
- ◆ Mine dumps
- ◆ Tailings
- ◆ Industrial sites
- ◆ Roadsides
- ◆ Polluted areas
- ◆ Degraded land
- ◆ Computerised surveying and mapping
- ◆ Detail rehabilitation designs
- ◆ Physical implementation of rehabilitation
- ◆ Ecological-, plant-, climate- and soil surveys
- ◆ Plant stress and mineral toxicity assessments
- ◆ Soil and Leaf analyses and recommendations
- ◆ Soil amelioration
- ◆ Revegetation
- ◆ Species selection for specific locality revegetation
- ◆ Seed for revegetation
- ◆ Harvesting of in situ plant material for revegetation
- ◆ Monitoring services

The objective of the eko rehab concerning rehabilitation and environmental management, is to create a one-stop service of high quality to the clients at all times.

ACTIVITIES:

VULA ENVIRONMENTAL SERVICES (PTY) LTD

P O Box 858
VREDENBURG 7380 W Cape
Tel: 082 564 5748 • Fax: 022 766 1106
Email: dve@iafrica.com
Contact Name: Deon van Eeden – Managing Director

We, as winners of the 2002 SA Landscape Institute Award of the Best Project in SA, The Rand Water Board Award for the Best Water Wise Project in SA, are proud to offer world-class environmental rehabilitation solutions.

We are experienced in mine, dune and quarry rehabilitation and our knowledge, experience and people make us the leaders in the field. We offer a range of services from conceptual planning to contracting, from plant search and rescue to comprehensive rehabilitation.

Vula is half-owned by a workers trust ensuring commitment of all staff to all our projects, large or small.

Our experience in rehabilitation and plant production is translated into our outstanding appropriate gardens that we design and construct. These are ideal for local conditions and are truly water-wise.

ACTIVITIES:

WALES: ENVIRONMENTAL PARTNERSHIPS

Representative in South Africa:
Africa Business Direct
PO Box 413586 CRAIGHALL 2024
Gauteng
Tel: 011 445 9460 • Fax: 011 445 9459
Email: msudarkasa@africabusinessdirect.com
Website Wales Trade International: www.walestrade.com
Website Welsh Development Agency: www.wda.co.uk
Contact Name: Michael Sudarkasa – CEO

Sustainable development is a priority area and high on the Welsh Assembly Government's agenda. This commitment is reinforced by the fact that the Assembly is one of only three regional governments in the world that has a statutory duty to promote sustainable development. Our First Minister chaired the meeting in Johannesburg in September 2002 which led to the Gauteng Declaration. (www.nrg4sd.net)

Strong potential trade partnerships with South Africa have been identified in environment – a common language, legal system and long historical association underline this.

Benefits of two way partnerships:
- Opportunities for environmental companies in both countries
- Technology transfer / capacity building
- Improved environmental protection / resource management.

The industrial revolution left Wales with a damaged environment. However, regeneration programmes since the 1960s have developed a wealth of experience and technological know-how. Wales has over 1 000 companies specialising in:
- Water supply / treatment & waste water treatment.
- Waste management / recycling
- Land remediation & landscape creation / management
- Energy efficiency & renewables
- Environmental instrumentation / monitoring / analysis
- Environmental consultancy
- Air pollution control
- Marine pollution control
- Noise & vibration control

Wales is at the forefront in the UK of supporting Environmental Goods and Services through WalesTrade International and Welsh Development Agency. We are helping Welsh companies to form fruitful partnerships with organisations and companies in South Africa.

ACTIVITIES:

WATER INSTITUTE OF SOUTHERN AFRICA (WISA)

PO Box 6011 HALFWAY HOUSE
1685 Gauteng
Tel: 011 805 3537
Fax: 011 315 1258
Email: wisa@wisa.co.za • Website: www.wisa.co.za
Contact Name: Colette Joubert – Office Manager

The Water Institute of Southern Africa (WISA) provides a forum for the exchange of information and views to improve water resource management in southern Africa. WISA has been in operation since 1937 and is the only indigenous South African organisation for professionals in the water industry. WISA has four Branches, plus thirteen specifically focused Divisions.

Every two years WISA hosts a Biennial Conference, which is viewed as a premier event on the South African water calendar. WISA also endorses the Afriwater Exhibition held in non-conference years. Additionally, WISA hosts the Stander Evening Lectures for academics. Annually, WISA organises nearly one hundred meetings, which are open to all, as well as publishing related text books.

The WISA website is serviced virtually daily and hosts a number of informational materials and links, as well as a Members Only section.

WISA was actively involved with the World Summit on Sustainable Development (WSSD) and formulated an input concerning the South African position paper on water. Furthermore, WISA was also the only water related organisation worldwide to be recognised by the UN as official observers to this event.

WISA is affiliated to a number of similar organisations abroad and in Africa, and has close ties with the Department of Water

Affairs and Forestry. WISA is also a reference point for the Parliamentary Portfolio Committee on Water and coordinates briefings for the Committee and serves them with information on specialised issues.

Please contact Head Office for information regarding new individual or company membership of WISA.

ACTIVITIES:

WEB ENVIRONMENTAL CONSULTANCY

P.O. Box 1425
NORTH RIDING 2162 Gauteng
Tel: 011 708 2626 • Fax: 011 708 3222
Email: kesnest@mweb.co.za
Contact Name: Bronwen Griffiths –
Consultant

WEB Environmental is a personalised service oriented environmental consultancy that has been in operation since 1999. Our focus is trained on:

◆ Environmental Impact Assessment projects including the preparation of:
  ◆ Exemption applications
  ◆ "Mini EIAs", Scoping Reports, and, Environmental Impact Reports
  ◆ Environmental Management Programmes
  ◆ Environmental reviews forming part of the Townplanning process
  ◆ Biological specialist reviews
  ◆ Peer review of EIA projects.

Additionally, WEB Environmental has been involved in the development of filtering / review systems for environmentally related applications (EIA/townplanning), including the database control systems.

ACTIVITIES:

WELEDA HEALTHCARE CENTRES

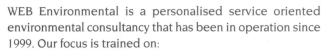

PO Box 98426
SLOANE PARK 2152 Gauteng
Tel: 011 463 3604 • Fax: 011 4638523
Email: weleda@global.co.za
Contact Name: Chamilla Sanua

Weleda pharmacies and Natural Remedy centres are accessible to most of the population in South Africa. The culture in these shops is one of caring, nurturing and fostering the culture of taking responsibility for one's own health through an extensive education programme. All members of the staff are qualified to give expert advice on the alternative approach to health. You can have on-line appointments and the medication will be sent to you via the post. The Help-Line number is 082 456 1897. We have helped many people onto the road of health and we would be delighted to start you on yours. It is never too late to start. Our way of life respects nature, the human body and the fact that we are the creator of our reality. We help you to achieve a healthy balance

between work and play in a happy, relaxed environment. This will be conducive to growth mentally – emotionally and physically.

ACTIVITIES:

WESSA – SEE WILDLIFE AND ENVIRONMENT SOCIETY OF SOUTH AFRICA

WESTERN CAPE ENVIRONMENTAL ETHICS FORUM

Dept of Philosophy Private Bag X1
STELLENBOSCH 7602 W Cape
Tel: 021 808 2058 • Fax: 021 808 3556
Email: jph@sun.ac.za
Website: www.sun.ac.za/philosophy/wceef
Contact Name: Johan Hattingh – Chairman

The Western Cape Environmental Ethics Forum is an issues-driven forum to promote more accountable and informed decision-making in the Western Cape. It aims to achieve this through the systematic analysis of the arguments being used to justify actions in current environmental case studies.

Ethical case studies:
The Forum creates academic working groups who analyse concrete case studies with definable parameters. Using factual research and informed academic debate, these working groups identify the values and principles involved in these specific cases, evaluate the current practice and then finally propose alternatives or interventions where appropriate.

Membership:
Forum membership is open to academics, business people, environmental protection agencies, all tiers of government, environmental activist organisations, consultants and concerned individuals. Fees are R20 for individual membership and R100 for corporate membership. For more details on the organisation you are welcome to visit our website.

ACTIVITIES:

WESTERN CAPE NATURE CONSERVATION BOARD

Private Bag X100
CAPE TOWN 8000 W Cape
Tel: 021 483 4615 • Fax: 021 483 3500
Email: nrockman@pgwc.gov.za
Website: www.capenature.org.za
Contact Name: Natasha Rockman – Communication Officer

Cape Nature Conservation protects the unique bio-diversity of the Cape Floristic Kingdom and surroundings by managing conservation areas, promoting eco-tourism, administering environmental law and ecological planning in the Western Cape to benefit the environment and society.

ACTIVITIES:

WHITE ELEPHANT CC

PO Box 792
PONGOLA 3170 KwaZulu-Natal
Tel: 034 413 2489 • Fax: 034 413 2499
Email: info@whiteelephant.co.za
Website: www.whiteelephant.co.za
Contact Name: Dr Heinz Kohrs – Member

White Elephant Safari Lodge: Set above the mirrored surface of Lake Jozini at the foot of the Lebombo mountains, White Elephant Safari Lodge captures the history and romance of the colonial era. Eight luxurious safari tents, each with ball and claw bath and outdoor canvas shower, provide a secluded retreat in the unspoilt Pongola Game Reserve wilderness. Stylish dining, sophisticated simplicity, attention to detail, sweeping views and a myriad of bush and water-based game activities promise a unique African experience.

White Elephant Bush Camp: Situated in the beautiful Pongola Game Reserve, this unique bush camp provides a rich experience. Thatched chalets, positioned between Knobthorn trees, are generously spacious with private verandahs and spectacular views. Magical moon rises, Africa's hues and uncomplicated dining add to the serenity and harmony with nature. Venture into the wilderness on guided walks, game drives or 'elephanting', cast the waters for tigerfish or unwind on a gentle boat cruise.

ACTIVITIES:

WIECHERS ENVIRONMENTAL CONSULTANCY CC

PO Box 197
FERNDALE 2160 Gauteng
Tel: 011 886 5709 • Fax: 011 787 6853
Email: wieenv@mweb.co.za
Website: www.home.mweb.co.za/wi/wieenv/
Contact Name: Dr Herman Wiechers

Wiechers Environmental Consultancy cc is a small South African enterprise that provides specialist environmental and project management services both locally and internationally. It has access to, and association with, a number of local and international environmental specialists, including ecologists, social impact practitioners, public participation facilitators, liminologists and hydro-geologists. By utilising teams of specialists with appropriate expertise, provides a comprehensive environmental service. Its clients include government, business, commerce, industry, agriculture and mining, as well as labour, community-based organisations, and non-governmental organisations. The services offered by Wiechers Environmental Consultancy include the following: Project Management, Workshop and Stakeholder Facilitation, Water, Wastewater and Waste Management, Environmental Policy, Strategy and Action Plans and Aquatic Chemistry.

ACTIVITIES:

WILBRINK & ASSOCIATES

PO Box 119 WESTVILLE 3630 KwaZulu-Natal
Tel: 031 266 9035 • Fax: 031 266 5613
Email: wilbrink@icon.co.za
Contact Name: Frans Wilbrink – Owner

Wilbrink & Associates, a Durban based technical management consulting firm, operating nationally, was founded in 1990. During the past thirteen years, the firm has experienced a steady growth. Some of the contracts completed successfully and to the full satisfaction of our clients are:

◆ Conducting Risk Assessments as per Regulation No.5 of the Hazardous Chemical Substances Regulations of the Occupational Health & Safety Act (85 of 1993)
◆ Conducting Risk Assessments as per Regulation No.5 of the Major Hazard Installation Regulations of the Occupational Health & Safety Act (85 of 1993)
◆ Conducting environmental monitoring programmes such as air monitoring, noise surveys, heat stress monitoring, ventilation surveys, stack emission-monitoring, etc.
◆ Conducting Environmental Impact Assessments
◆ Assisting operations (production) in plant optimisation / yield improvement, and working off redundant stock, etc.
◆ Improving safety through safety audits, training and by compiling safety manuals and safety talks
◆ Conducting accident / incident investigations
◆ Classification of Hazardous Locations of Factories for determining the type of equipment required, e.g flameproof, explosion proof, intrinsically safe, etc.
◆ Acting as safety consultant for small-to-medium-size companies
◆ Assisting companies complying with the various Acts and Regulations such as the OH&SA, Road Transportation Act, Atmospheric Pollution Prevention Act, etc.
◆ Presenting various training courses as per the OH&S Act
◆ Acting as an Arbitrator in disputes involving chemicals.

ACTIVITIES:

WILDERNESS ACTION GROUP OF SOUTHERN AFRICA

PO Box 1098 WESTVILLE 3630 KwaZulu-Natal
Tel: 031 267 5300 • Fax: 031 267 1838
Email: densham@sai.co.za
Contact Name: WD Densham – Chairman

ORGANISATIONAL INFORMATION

The Wilderness Action Group is a voluntary non-government organisation (NGO) and was formed in 1983 under Section 21 of the RSA Companies Act, that is, a company not-for-gain. The Group is based in Durban, KwaZulu-Natal.

HISTORY AND MISSION

The Group was formed after the 3rd World Wilderness Congress in Scotland in 1983. Originally it was comprised of a body of concerned people, who had attended this Congress. They were concerned about wilderness conservation in

South Africa and wanted to form a Non-Governmental Organisation (NGO) dedicated to promote the interests of wilderness conservation in southern Africa. Subsequently, directors were appointed according to their experience and expertise in wilderness conservation, which was not paralleled elsewhere in the NGO movement of the sub-continent. This was meaningfully employed in a number of successful endeavours, which included preparation of detailed submissions to the national government on wilderness related issues.

Membership of the Group

The Memorandum and Articles of Association of the Group makes provision for both a Board of Directors, members and for the appointment of Honorary Members ("persons whom the Group considers appropriate and to dispense with any membership fees payable by such persons"). The Directors and Members have interests and professional experience in a wide range of aspects related to Nature Conservation with a deep concern for and commitment to wilderness conservation. These include environmental science, environmental law, wilderness management, and psychology.

Vision

The Wilderness Action Group fosters the enduring wilderness of southern Africa and elsewhere for its value to the present and future generations together with people.

Mission Statement: The mission of the Wilderness Action Group is to promote and advocate the wilderness concept throughout southern Africa and elsewhere, on public, communal and private land through a range of activities, which may include amongst other things, the following:

- Promoting the concept of wilderness in an African context through the recognition of indigenous knowledge and the preservation of natural and cultural heritage
- The provision of advice on the professional and technical aspects of the conservation, protection and management of wilderness areas
- Making submissions to promote the development of appropriate legislation and policy
- Offering professional training courses and awareness seminars to bridge knowledge and information gaps in wilderness conservation, protection and management
- Monitoring the state of conservation and management of existing wilderness areas
- Promoting the designation of candidate wilderness areas
- Acting in partnership and networking with individuals and other organisations with similar aims and objectives
- Publishing material in the popular media and scientific journals, in an ethical, credible and effective manner, as experts, advisors, educators, facilitators, trainers, and leaders in the sector.

ACTIVITIES:

WILDERNESS FOUNDATION

PO Box 12509
CENTRAHIL 6006 Port Elizabeth
Tel: 041 582 1885 • Fax: 041 582 1905
Email: info@sa.wild.org • Website: www.wild.org
Contact Name: Andrew Muir – Executive Director

The Wilderness Foundation (SA) – founded by Dr Ian Player in 1972 – has been a pioneer in using the wilderness as a positive force for social change in South Africa by bringing historically disadvantaged youth, as well as political and community leaders on trails to experience wild nature, often for the very first time.

It is a not-for-profit, non-governmental organisation (NGO) working in Southern Africa to protect and sustain wilderness, wildlife and wildlands by providing environmental education, experience and training to all contemporary and indigenous communities; and to further human understanding and cooperation for the conservation of wild habitats.

The Wilderness Foundation (SA) is part of an international network of similar organisations, and accomplishes its mission of the protection of wilderness areas through public awareness programs and campaigns; by promoting wilderness as a resource for all South Africans; by monitoring wilderness and planning for new wilderness designations in South Africa; by assisting with the management of existing wilderness areas under private and public ownership; and by advocating for enlightened policy and research that sustains wilderness and wildlands.

ACTIVITIES:

WILDERNESS LEADERSHIP SCHOOL TRUST

PO Box 53058
YELLOWWOOD PARK 4011 KwaZulu-Natal
Tel: 031 462 8642 • Fax: 031 462 8675
Email: wilderness.trails@eastcoast.co.za
Website: www.ref.org.za/wildernessleadership/trails/
Contact Name: Paul Cryer – Manager Trails

We strive to restore a balanced relationship between nature and humanity by providing a direct experience of wilderness.

ACTIVITIES:

An explorer is someone who goes to the edge of knowledge and brings back something new.

~ John Hemming
Royal Geographical Society

WILDLIFE AND ENVIRONMENT SOCIETY OF SOUTH AFRICA

PO Box 394
HOWICK 3290
KwaZulu-Natal
Tel: 033 330 3931 • Fax: 033 330 4576
Email: alisonk@futurenet.co.za
Website: www.wildlifesociety.org.za
Contact Name: Alison Kelley

People Caring for the Earth.

The Wildlife and Environment Society of South Africa (WESSA), one of South Africa's leading environmental NGOs, has an impressive track record of environmental action over the past 75 years. The Society has been a motivating force behind many significant conservation actions and environmental decisions.

The Society believes that the most pressing environmental issues of the next decade will be health-related concerns (HIV/Aids and waterborne diseases), as well as the consequences of global climate change, landscape changes, the alleviation of poverty, and the threat of increasing, unchecked development with its impacts on natural resources.

The vision of WESSA is to achieve a country, which is wisely managed by all to ensure long-term environmental sustainability. The strength of the Society is its membership, volunteers who take action at a local level. WESSA calls on all South Africans to work together to address environmental concerns and promote sustainable development in this region.

ACTIVITIES:

WILDLIFE TRANSLOCATION ASSOCIATION

PO Box 12452
ONDERSTEPOORT 0110 Gauteng
Tel: 012 565 4939
Email: info@wta.org.za • Website: www.wta.org.za
Contact Name: Elise Berning – Secretary

Professional association of game capture operators and associated role players in the wildlife industry.

ACTIVITIES:

WILDNET AFRICA PROPERTIES CC

PO Box 68289
BRYANSTON 2021 Gauteng
Tel: 011 468 1303 • Fax: 011 468 1303
Email: craig@wildnetafrica.com
Website: www.gameranch.wildnetafrica.com
Contact Name: Craig McKenzie – MD

Specialising in the sale of Wildlife related properties around southern Africa including Game Farms, Eco-tourism, Big 5 Reserves and Leisure Properties. For more information please see our Website.

ACTIVITIES:

WINSTANLEY SMITH & CULLINAN INC

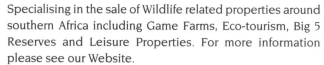

76 Strand Street, CAPE TOWN 8001W Cape
Tel: 021 425 7068 • Fax 021 425 7065
Email : win@law.co.za wscinc@law.co.za
Contact Name : Terry Winstanley

Winstanley, Smith & Cullinan Inc. ("WSC Inc.") is a specialist environmental law firm, based in Cape Town, but operating throughout South Africa as well as in other African countries.

ACTIVITIES:

WIRED COMMUNICATIONS

PO Box 928
GREEN POINT 8051 W Cape
Tel: 021 439 4975 • Fax: 021 439 0386
Email: ke@wiredcommunications.co.za
Contact Name: Karey Evett

Communications and PR Consultancy with experience in social responsibility and sustainable development. Recent publications include *Our Coast for Life*, a book on sustainable coastal development in southern Africa.

ACTIVITIES:

WORKING FOR WATER

Private Bag X4390
CAPE TOWN 8000 W Cape
Tel: 021 405 2200 • Fax: 021 425 7880
Email: noemdos@dwaf.gov.za
Website: www.dwaf.gov.za/wfw/
Contact Name: Simone Noemdoe – Communications Manager

The Working for Water Programme will contribute to the sustainable control of invasive alien plants, and thereby the optimising of the conservation and use of natural resources. In doing so, it will address poverty relief and promote economic empowerment and transformation within a public works framework.

Objectives:

Hydrological:To enhance water security. To promote equity, efficacy and sustainability in the supply and use of water.

Ecological: To improve the ecological integrity of our natural systems. To protect and restore biodiversity.

Social: To invest in the marginalised sectors of society and to enhance their quality of life.

Natural resources: To restore the productive potential of the land, in partnership with the Landcare initiative. To promote the sustainable use of natural resources.

Economic: To develop the economic benefits (from land, water and people) from clearing these plants by facilitating training, economic empowerment and the development of secondary industries.

ACTIVITIES:

WORLD OF BIRDS WILDLIFE SANCTUARY & MONKEY PARK

Valley Road
HOUT BAY 7806 W Cape
Tel: 021 790 2730 • Fax: 021 790 4839
Email: worldofbirds@mweb.co.za
Website: www.worldofbirds.org.za
Contact Name: Walter Mangold – Director

A walk through Hout Bay's World of Birds Wildlife Sanctuary and Monkey Park is truly a walk through nature – a fascinating glimpse into the private lives of about 400 different species of birds, small mammals and monkeys in walk-through enclosures. It is one of the largest bird parks in the world. It also serves as a hospital, orphanage and breeding centre, voluntarily caring for injured wild birds and animals, and for the breeding of threatened species wherever possible.

ACTIVITIES:

WSP WALMSLEY

PO Box 5384 RIVONIA 2128 Gauteng
Tel: 011 233 7880 • Fax: 011 807 1362
Email: bw@wspgroup.co.za
Website: www.wspgroup.co.za
Contact Name: B Walmsley – Managing Director

WSP Walmsley (Pty) Ltd is a subsidiary of WSP Environment Ltd, which is part of the London Stock Exchange listed company, WSP Group plc, an interdisciplinary business providing a range of consultancy services through the world.

WSP Walmsley brings to the South African market place a dynamic blend of local expertise and global cutting-edge technology. We have a well-established team of environmental scientists on staff and an extensive network of specialists in southern Africa. Being part of a large global company, we are also able to draw on considerable international resources and expertise accumulated over many years.

Our vision is to provide an independent, innovative and professional service whereby we strive to achieve a balance between environmental protection, social desirability and economic development.

Our project-specific teams can provide a complete range of environmental solutions to businesses in the following fields:

- Integrated environmental management
- Sustainable development
- Environmental management systems and plans
- Environmental auditing
- Due diligence, compliance and liability assessments
- Remediation, rehabilitation and revegetation programmes
- Coastal zone management
- Aquatic resources management
- Environmental procedures and guidelines
- Environmental legal advice
- Environmental and financial risk and scenario assessment.

ACTIVITIES:

WWF-SOUTH AFRICA

Private Bag X2 DIE BOORD 7613 W Cape
Tel: 021 888 2855 • Fax: 021 888 2888
Email: ctreasure@wwfsa.org.za
Website: www.panda.org.za
Contact Name: Cathryn Treasure – Marketing Manager

WWF-SA is a non-governmental organisation which facilitates environmental and biodiversity conservation. This is achieved through fundraising for priority projects, and not by acting as a conservation implementing agent. Our driving force is to access funds and to use these optimally for conservation.

WWF-SA has initiated new projects every year since it was founded. Many of these projects run for several years and a few have run for as long as WWF-SA has existed. At present fewer projects are being initiated each year than was the case in the early 1990s, and this trend is expected to continue, with larger, more effective projects being selected ever more carefully. WWF-SA is currently supporting some 150 projects which are organised into 10 programmes that focus on international and national priorities. These are Marine, Freshwater, Grasslands, Fynbos, Conservation Education, Succulent Karoo, Forests, Species and Climate Change, and Toxics.

Applications for funding are carefully screened and projects are scientifically evaluated by experts in the respective field. As the scale of environmental demands is enormous, projects which have the greatest ripple effect, and those which have long-term sustainability will be favoured. In considering projects for support, WWF-SA gives particular preference to exemplary, innovative, catalytic projects in order to optimise its limited resources.

ACTIVITIES:

ZANDVLEI ESTUARY NATURE RESERVE

PO Box 30028 TOKAI 7966 W Cape
Tel: 021 701 7542 • Fax: 021 701 7542
Email: spmzandvlei@sybaweb.co.za
Website: www.zandvleitrust.org.za
Contact Name: Cliff Dorse – Reserve Manager

The only functioning estuary along the False Bay coastline, Zandvlei is ranked 46th out of 250 estuaries in South Africa in terms of conservation importance. It is rich in plant, marine, bird and animal life.

ACTIVITIES:

ZANDVLEI TRUST (THE)

PO Box 30017 TOKAI 7966 W Cape
Tel: 021 788 3011
Fax: 021 788 5909
Email: vincentm@iafrica.com
Website: www.zandvleitrust.org.za
Contact Name: Vincent Marincowitz – Chair

The Zandvlei Trust aims to conserve the indigenous fauna and flora of the Zandvlei and to enhance this natural resource for the benefit of all. Refer to our website for further details.

ACTIVITIES:

ZEEKOEVLEI ENVIRONMENTAL EDUCATIONAL PROGRAMME

Peninsula Road
ZEEKOEVLEI 7941 W Cape
Tel: 021 706 8523 • Fax 021 706 2405
Email: zeep@telkomsa.net
Contact Name: Bevan Lill – Project Manager

An NPO Trust that runs high quality, low cost, two day /three night education camps based at Zeekoevlei/Rondevlei. Excellent facilities and programme – dedicated, trained field rangers.

ACTIVITIES:

ZERI-SOUTH AFRICA

4 Stuartfield
Avenue Upper Trovato Estate
WYNBERG 7800 W Cape
Tel: 021 762 1228
Email: nnair@iafrica.com • Website: www.zeri.org
Contact Name: Nirmala Nair – Coordinator

ZERI-SA assists organisations to work with natural systems and develop capacity enabling in job creation and meeting basic needs. ZERI engages in research and development appropriate to the South/Southern African context.

ACTIVITIES:

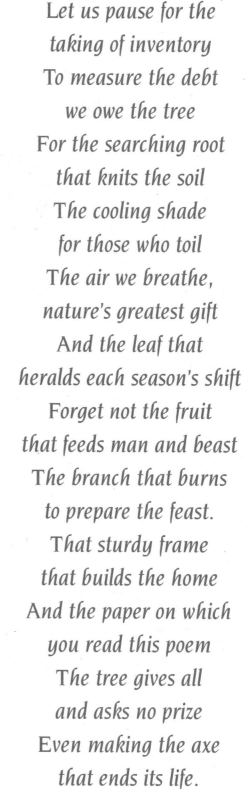

Let us pause for the
taking of inventory
To measure the debt
we owe the tree
For the searching root
that knits the soil
The cooling shade
for those who toil
The air we breathe,
nature's greatest gift
And the leaf that
heralds each season's shift
Forget not the fruit
that feeds man and beast
The branch that burns
to prepare the feast.
That sturdy frame
that builds the home
And the paper on which
you read this poem
The tree gives all
and asks no prize
Even making the axe
that ends its life.

~ Anon

INDEX
SECTION

INDEX I:

TOPICS, ARTICLES AND RESOURCES

INDEX 2:
ORGANISATIONS LISTED IN THE DIRECTORY SECTION

AGRICULTURE

ANIMAL

AVIAN

BOTANICAL

CERTIFICATION ✔

COMMERCE, TRADE AND INDUSTRY 🏭

CONSERVATION 🤲

CONSULTANCY SERVICES

INDEX 3

ORGANISATONS LISTED BY CATEGORY OF ACTIVITY

ECO-TOURISM

ENVIRONMENTAL EDUCATION & TRAINING

ENVIRONMENTAL MANAGEMENT & PLANNING

ENVIRONMENTAL MEDIA

ENVIRONMENTAL NETWORKING

ENVIRONMENTAL RESEARCH & DEVELOPMENT

R&D

ENVIRONMENTALLY FRIENDLY PRODUCTS

ENVIRONMENTAL ETHICS

GOVERNMENTAL BODIES*

* See pages 409—410 for further government details.

HEALTHCARE

PROFESSIONAL ASSOCIATIONS & INSTITUTIONS

ENVIRONMENTAL RISK OR LIABILITY MANAGEMENT

WATER

INDEX 3
ORGANISATONS LISTED BY CATEGORY OF ACTIVITY

REFERENCE

&

RESOURCE

SECTION

ENVIRONMENTAL PHOTOGRAPHY COMPETITION

The *Enviropaedia* is on the lookout for photographs to illustrate Topics & Issues.

Photos that:

◆ Clearly depict the Topic (without the need for explanatory notes)

◆ Are visually bold and colourful

◆ Are dramatic or interesting and unusual

◆ Wherever possible, include human activity (action orientated)

For every photo that we accept for use in The *Enviropaedia*, we will:

◆ Send you a free copy of The *Enviropaedia* (*You can choose to receive a copy of the current edition or to wait for the new version, which includes your photo*)

◆ Credit the name of the photographer and / or the organisation that submitted the photo.
(*This is a very useful form of PR and publicity – and we are particularly keen to use material submitted by our "Listings" in the Directory Section.*)

Here's what to do …

1. Identify a specific Topic from the Encyclopaedia Section.

2. Take a photograph, which you think would clearly depict that Topic
(remembering the type of photo that we prefer – as outlined above).

3. Email a digital copy (in either JPG or TIF format ONLY) to The *Enviropaedia* at this email address: enviropaedia@iafrica.com. Please include your name and contact details.

OR

4. Post a full-colour photograph to the following address: The *Enviropaedia*, PO Box 425 Simonstown 7995 W Cape. Please include your name and contact details.

* Please note: 1. The Editor's decision is final. 2. Photographs cannot be returned once submitted. 3. No correspondence will be entered into regarding the merits of photographs, if not chosen.

ENVIRONMENTAL NEWS & MEDIA

PUBLICATIONS:

Africa Birds and Birding	Bi-monthly magazine. Offers a balanced status of the continent's abundant avifauna. Editor: Peter Borchert	Tel: 021 762 2180 Email: wildmags@blackeaglemedia.co.za
Africa Geographic	Monthly magazine. Wildlife, nature, conservation and travel. Editor: Peter Borchert	Tel: 021 762 2180 Email: wildmags@blackeaglemedia.co.za
Africa Wild	Monthly magazine. Eco-tourism and adventure travel. Editor: Cheri-Ann Potgieter	Tel: 011 454 0535 Email: editor@africawildgroup.co.za
African Wildlife	Quarterly journal for members of the Wildlife and Environment Society of SA - "People caring for the earth." Editor: Sandie Anderson	Tel: 021 535 1818 Email: wildmag@yebo.co.za
Eagle Bulletin	Bi-monthly bulletin. Provides concise focused environmental information for environmental professionals. Editor: Arend Hoogervorst	Tel: 031 767 0244 Email: arend@eagleenv.co.za
Earthyear Environmental	Quarterly magazine. Focuses on sustainable development and enlightened environmentalism. Editor: Fiona McCleod	Tel: 011 727 7000 Email: earthyear@mg.co.za
Endangered Wildlife	Quarterly journal for members of the Endangered Wildlife Trust. Editor: David Holt-Biddle	Tel: 011 486 1102 Email: ewt@ewt.org.za
Enviro Fact Sheets	Environmental education resource materials in loose leaf format (available free from Pick 'n Pay branches.)	Tel: 033 330 3931 Email: sharenet@futurenet.co.za
Enviropaedia (The)	Annually updated environmental encyclopaedia and networking directory of southern Africa. Editor: David Parry-Davies	Tel: 021 786 4311 Email: enviropaedia@iafrica.com
ESI Africa	Quarterly magazine. Reports on development in the power (energy) sector. Editor: Jonathan Spencer Jones	Tel: 021 700 3500 Email: info@spintelligent.com
Getaway Magazine	Monthly travel magazine with major focus on environmental issues and eco-tourism. Editor: David Bristow (Sub Editor: Don Pinnock)	Tel: 021 530 3100 Email: getaway@rsp.co.za
Mining Review Africa	Bi-monthly magazine. Supports the sustainable development of Africa's mineral resources. Editor: Jonathan Spencer Jones	Tel: 021 700 3500 Email: jonathan@spintelligent.com
Plantlife	Quarterly non-profit journal committed to promoting interest in and conservation of southern African plants. Editor: David Styles	Tel: 083 360 8173 Email: tritonia@telkomsa.net
Progress Magazine	Quarterly magazine. Tracking sustainable development in Africa. Editor: Jasper Raatz	Tel: 011 280 5362 Email: jasperr@jpl.co.za
Travel Africa	Quarterly magazine. Focusing on Eco-Tourism. Editor: Craig Rix	Tel: (+263-4) 331 801 (Zimbabwe) Email: editor@travelafricamag.com
Urban Green File	Bi-monthly journal. A professional, trade and technical journal dealing with environmental issues. Editor: Carol Knoll	Tel: 011 268 6723 Email: carolknoll@brookepattrick.co.za
Veld & Flora	Quarterly journal for members of the SA Botanical Society. Editor: Caroline Voget	Tel: 021 797 2090 Email: info@botanicalsociety.org.za

TV & RADIO:

50/50	SABC2: Environmental and nature conservation programme. Broadcast 5.30 pm on Sundays. Executive producer: Danie van der Walt	Tel: 011 714 6621 Email: 5050@sabc.co.za
Fokus	SABC2: Focus on current environmental and social issues. Broadcast 7.30 pm on Sundays. Executive Producer: Alet Joubert	Tel: 011 714 6569 Email: focus@sabc.co.za
Ecowatch	SAFM Ecowatch reflects contemporary environmental concerns and issues. Broadcast: 6pm Sundays and 5am Tuesdays. Presented by John Richards	Tel: 021 430 8182 Email: richardsja@safm.co.za
Sappi Nature Journal	702 Wildlife programme, 4pm Sunday. Cape Talk, 4 pm Sunday. Eco, conservation and environmental issues. Presented by Tim Neary	Tel: 011 704 1009 Email: tneary@intekom.co.za

INTERNET ENVIRONMENTAL NEWS SERVICES:

Green Clippings	A weekly report on the latest environmental and conservation news delivered to subscribers by email.	Tel: 021 883 3366 Email: info@greenclippings.co.za
Earthwire Africa	The UN environmental news clippings website for southern Africa.	Tel: N/A www.earthwire.org

PRINT MEDIA JOURNALISTS:

The *Enviropaedia* regrets that due to space limitations, it has not been possible to include a comprehensive listing of environmental journalists. The following therefore represents only a short list of additional environmental journalists not represented in the media information above.

A longer list of environmental journalists is available from Tyrrell Associates:
Tel: 021 511 8108 • Email: tyrrell@iafrica.com
(Co-ordinators of the annual Environmental Journalists Conference)

Addison, Graeme	Writer, researcher	mediaman@worldonline.co.za • 056 818 1814
Berruti, Sharon	BirdLife South Africa	berruit@iafrica.com • 011 789 1122
Bonthuys, Jorisna	Die Burger	jbonthuy@dieburger.com
Booyens, Bun	Wegbreek	bboyens@wegbreek.co.za • 021 417 1111
Carnie, Tony	Mercury and Daily News	carnie@popserver.nn.independent.co.za • 031 308 2314
Dalgliesh, Geoff	Drive Out	geoffd@netactive.co.za
Darroll, Leigh	Urban Green File	ldarroll@mweb.co.za
Derwent, Sue	Freelance journalist	sued@saol.com • 031 916 7618
Du Toit, Julienne	Freelance writer	features@global.co.za • 011 802 3262
Elias, Lew	Daily Dispatch	lewe@dispatch.co.za • 043 702 2000
Ferris, Melanie-Anne	The Star	mre@argus.co.za
Gosling, Melanie	The Cape Times	melanieg@ctn.independent.co.za
Gowans, Jill	Sunday Tribune	gowans@klington.nn.independent.co.za • 031 308 2394
Hanks, Karoline	Freelance writer	karabos@telkomsa.net
Harvey, Ebraham	Freelance writer	eharvey@telkomsa.net • 011 673 5340
Hetherington, Alex	Freelance writer	alex@ahmedia.co.zan • 021 789 2858
Koch, Eddie	Freelance writer	eddiek@icon.co.za • 011 454 3775
Marshall, Leon	Independent Newspapers	lmar@star.co.za • 011 633 2145
Michler, Ian	Africa Geographic	ianmichler@mweb.co.za
Molefe, Russell	Sowetan	molefer@sowetan.co.za • 011 471 4161
Mphaki, Ali	City Press	mali@citypress.co.za • 011 713 9001
Nel, Dirk	Freelance writer	djnews@absamail.co.za • 082 821 6456
Nel, Michelle	Freelance writer	michelle.nel@iafrica.com
Padayachee, Nicki	Sunday Times	padayacheen@sundaytimes.co.za • 011 280 5150
Rogers, Guy	Eastern Province Herald	grogers@johnnicec.co.za
Wassterfall, Margaret	Country Life	margaretw@dbn.caxton.co.za • 031 910 5713
Yeld, John	Cape Argus	johny@ctn.independent.co.za • 021 488 4156

The ENVIRONMENTAL NEWS AND MEDIA REFERENCE SECTION *has been compiled with the assistance of Tyrrell Associates (environmental media and communications consultants) and has been kindly sponsored by South African Breweries.*

ENVIRONMENTAL WEBSITES

A SELECTION OF USEFUL WEBSITES – REPRESENTING (MOSTLY) INTERNATIONAL ORGANISATIONS NOT LISTED IN THE "SOUTHERN AFRICAN" DIRECTORY SECTION OF THE ENVIROPAEDIA

INTERNATIONAL ENVIRONMENTAL ORGANISATIONS

Australian Greenhouse Office: www.greenhouse.gov.au – Australia's leading environmental action group.

African Elephant Trust: www.africanelephanttrust.org – Assisting the elephant habitat in sub-Saharan Africa.

African Lion Working Group: www.african-lion.org – Dedicated to African lion conservation.

African Wildlife Foundation: www.awf.org – Works to ensure the wildlife and wild lands of Africa will endure.

Africa Water: www.africanwater.org – Dedicated to the promotion of sustainable water resources management.

BirdLife International: www.birdlife.net – Working to improve the quality of life for birds, people and wildlife.

Centre for Environmental Citizenship: www.envirocitizen.org – Networking young leaders to protect the environment.

Climate Action Network (CAN): www.climatenetwork.org – Working to promote government and individual action to limit human-induced climate change.

Conservation International: www.conservation.org – International conservation network.

Earth Charter Campaign: www.earthcharter.org – Promoting values and principles for a sustainable future.

Earth Share: www.earthshare.org – An American nationwide network of the country's most respected and responsible non-profit environmental and conservation organisations.

Earth Trends: www.earthtrends.wri.org – Information on environmental, social, and economic trends.

English Nature Conservation: www.englishnature.org.uk – Conservation of the English countryside and wildlife.

Environment Agency of England and Wales: www.environment-agency.gov.uk – Responsible for protecting and improving the environment in England and Wales.

Environment Canada: www.ec.gc.ca/regeng.html – Protecting and improving the environment in Canada.

Environment Network: www.unep.net – UN Environmental information.

Environmental Business: www.greenbiz.com – Environmental business strategy.

Environmental Education: www.eelink.net – Environmental education resources on the Internet.

Environmental Web Directory: www.webdirectory.com – Links to environmental sites.

Environmental Sustainability Information: www.environmentalsustainability.info – Environmental sustainability information.

European Environment Agency: www.eea.eu.int – Information for improving Europe's environment.

Five E's Unlimited: www.eeeee.net – Specialising in environmental sustainability, strengthened economies and social equity.

Friends of the Earth (US): www.foe.org – One of the world's largest networks of environmental groups.

Global Environmental Information Centre: www.geic.or.jp – Asian environmental information.

Greenpeace International: www.greenpeace.org – Global environmental watchdog.

IISD Linkages: www.iisd.ca – A multimedia resource for environment and development policy makers.

International Corporate Environmental Reporting: www.enviroreporting.com – Environmental and sustainable business reporting.

International Development Research Centre: www.idrc.ca – Helps communities in the developing world find solutions to social, economic and environmental problems.

International Federation of Organic Agriculture Movements (IFOAM): www.ifoam.org – Ecologically, socially and economically sound systems that are based on the principles of Organic Agriculture.

International Fund for Animal Welfare (IFAW): www.ifaw.org – Animal rights protection group.

International Institute for Environment and Development: www.iied.org – Works on trade, livelihoods and sustainable development.

International Institute for Sustainable Development (IISD): www.iisd.org – Economic development and the well being of all people.

IUCN The World Conservation Union (IUCN): www.iucn.org – World's largest Union for the conservation of nature and natural resources (with representatives from Governments, NGOs, science and society).

Jane Goodall Institute: www.janegoodall.org – Advances the power of individuals to take informed and compassionate action to improve the environment for all living things.

National Kilowatt Count of Household Energy Use (Australia): www.erin.gov.au – Australian Government Department of Environment and Heritage site.

Natural Resource Defence Council: www.nrdc.org – US-based environmental action organisation with environmental topics and news.

Organisation for Economic Cooperation and Development (OECD): www.oecd.org/env – Building sustainable partnerships for economic progress.

Rainforest Alliance: www.rainforest-alliance.org – To protect ecosystems, and wildlife that live within them.

Southern Africa Environment Project (SAEP): www.saep.org
– Environmental educational organisation.

Stockholm Environment Institute: www.sei.se – Specialising in sustainable development and environment issues.

TRAFFIC: www.traffic.org – Ensuring that trade in wild plants and animals is not a threat to conservation.

UK Department of the Environment: www.detr.gov.uk – UK Government Department for environment.

UNEP World Conservation Monitoring Centre: www.unep-wcmc.org – The United Nations Conservation website.

United Nations Development Programme (UNDP): www.undp.org – United Nations programme for development.

United Nations Environment Programme (UNEP): www.unep.org – United Nations programme for environment.

United Nations Population Fund (UNFPA): www.unfpa.org – United Nations programme for population.

USAID: www.usaid.gov/environment – United States programme for environmental aid.

US Environmental Protection Agency (EPA): www.epa.gov – United States Government environmental protection agency website.

Water Environment Federation: www.wef.org – A not-for-profit technical and educational organisation for the preservation and enhancement of the global water environment.

Wildlife and Environment Zimbabwe: www.zimwild.co.zw – Zimbabwe wildlife conservation group.

World Conservation Monitoring Centre: www.wcmc.org.uk – Global conservation monitoring.

World Resources Institute: www.wri.org – An independent non-profit organisation with a staff of more than a hundred scientists, economists, policy experts, business analysts, statistical analysts, mapmakers, and communicators working to protect the Earth and improve people's lives.

WWF Global Network: www.panda.org – The World Wildlife Fund's official website.

WWW Virtual Library: www.earthsystems.org/environment.shtml – Advancement of environmental information and education to the world community.

Women's Environment and Development Organisation (WEDO): www.wedo.org – An international advocacy organisation that seeks to increase the power of women worldwide.

PRACTICAL INFORMATION AND NEWS

Care2Company: www.care2.com – US-based website presenting practical information and resources for sustainable living.

Centre for International Environmental Law: www.ciel.org – International environmental law.

Centre for Science and Environment: www.cseindia.org – Newsletters, news releases.

Centre for Sustainable Design: www.cfsd.org.uk – Facilitates discussion and research on eco-design.

Cities Environment Reports on the Internet (CEROI): www.ceroi.net – Works within the framework of Local Agenda 21 to facilitate access to environmental information.

Cities (SA) for climate protection: www.iclei.org/africa/ccp – Global warming and climate change information.

Community Based Natural Resource Management in Southern Africa: www.cbnrm.uwc.ac.za – Community-based approaches to natural resource management.

Compendium of Sustainable Development Indicators: www.iisd.ca/measure/compindex – A worldwide directory of who is doing what in the field of sustainability indicators.

Delta Environmental Centre: www.deltaenviro.org.za – South African environmental education and resource centre.

Dow Jones Sustainability Indexes: www.sustainability-index.com – Global index tracking of sustainability driven companies.

Earth Charter Initiative, The: www.earthcharter.org – A global environmental initiative of values, principles and aspirations that are widely shared by growing numbers of individuals and organisations.

Earth Council: www.ecouncil.ac.cr – Together with partner organisations, has undertaken the Earth Charter Millennium Campaign to support and empower people to build a secure, equitable and sustainable future.

Earthwire Africa: www.earthwire.org.com – Global environmental news clippings site (daily update).

EcoLink: www.lowveld.com/ecolink – South African based environmental education through community outreach.

Ecosystem Valuation: www.ecosystemvaluation.org – How economists value ecosystem conservation.

Environmental Goods and Services: www.sustainableabc.com – USA sustainability products and services site.

Environmental News Network: www.enn.com – US-based environmental news and views, global reporting and information network.

Environmental Sustainability Index: www.ciesin.org/indicators/ESI – Global index tracking of sustainability driven companies.

Fostering Sustainable Behaviour: www.cbsm.com – Numerous initiatives to reduce waste and pollution, increase water and energy efficiency.

Global Environmental Management Initiative (GEMI): www.gemi.org – Organisation of leading companies dedicated to fostering environmental, health and safety excellence and corporate citizenship worldwide.

Green Clippings: www.greenclippings.com – South Africa's environmental news clippings (weekly internet delivery service).

Hydro Power in South Africa: www.microhydropower.net/rsa – South Africa's hydro power sites including maps and technical information.

International Chamber of Commerce: www.iccwbo.org/index_sdcharter.asp – UK-based business action for sustainable development.

MELISSA: www.melissa.org – Reporting on information regarding the environment in sub-Saharan Africa.

Permaculture: www.permaculture.net – International Institute for Ecological Agriculture.

Second Nature: www.secondnature.org – Transformation in higher education. To assist colleges and universities in their quest to integrate sustainability.

South African Association for Energy Efficiency: www.saee.co.za – Renewable energy association.

Sustainable Living Centre: www.sustainable.co.za – Practical information on how to live a sustainable lifestyle.

Thermal Insulation Society of South Africa: www.tiasa.orga.za – Renewable resource association.

Urban Greenfile: www.urbangreen.co.za – South African industrial/environmental magazine.

World Watch Institute: www.worldwatch.org – Leading environmental think-tank.

SUSTAINABLE DEVELOPMENT

Care2Company: www.care2.com – US-based website presenting practical information and resources for sustainable living.

Centre for Sustainable Design: www.cfsd.org.uk – Facilitates discussion and research on eco-design.

Compendium of Sustainable Development Indicators: www.iisd.ca/measure/compindex.asp – A worldwide directory of who is doing what in the field of sustainability indicators.

Dow Jones Sustainability Indexes: www.sustainability-index.com – Global index tracking of sustainability driven companies.

Earth Charter Initiative, The: www.earthcharter.org – The Earth Charter is a global environmental initiative of values, principles, and aspirations that are widely shared and supported by growing numbers of individuals and organisations.

Environmental Goods and Services: www.sustainableabc.com – USA sustainable products and services.

Environmental Sustainability Index: www.ciesin.org/indicators/ESI – Global index tracking of sustainability driven companies.

Environmental Sustainability Information: www.environmentalsustainability.info – An information gateway empowering the movement for environmental sustainability.

Five E's Unlimited: www.eeeee.net – Specialising in environmental sustainability, strengthened economies and social equity.

Fostering Sustainable Behaviour: www.cbsm.com – Numerous initiatives to reduce waste and pollution, increase water and energy efficiency.

Heinrich Boell Foundation: www.boell.org.za – Cutting edge sustainable development thinking, plus information on their foundation work and partnerships. Provides funding for environmental and civil society initiatives.

International Chamber of Commerce: www.iccwbo.org/index_sdcharter.asp – UK-based business action for sustainable development.

International Corporate Environmental Reporting: www.enviroreporting.com – Environmental and sustainable business reporting.

International Institute for Environment and Development: www.iied.org – Works on trade, livelihoods and sustainable development.

International Institute for Sustainable Development (IISD): www.iisd.org – An institute that promotes economic development and the well-being of all people.

Norway – Department for International Co-operation: www.odin.dep.no/md/norsk/dep/statsraad_a/022001-990332 – Ministry of the Environment (Minister Brende) and new Chairman of the UN Sustainable Development Council.

Stockholm Environment Institute: www.sei.se – Specialising in sustainable development and environment issues from Sweden.

Sustainable Development: www.rfweston.com/sd/welcome.htm – Sustainable development case studies involving attempts to pursue sustainable strategies and links to other sustainable development sites.

United Nations Development Programme (UNDP): www.undp.org – United Nations programme for development.

Sustainable Living Centre: www.sustainable.co.za – How to live a sustainable lifestyle.

Sustainable Development Communications Network in Canada: www.sdcn.org – Civil society seeking to accelerate sustainable development.

United Nations Environment Programme (UNEP): www.unep.org – The United Nations programme for environment and conservation.

United Nations Sustainable Development: www.un.org/esa/sustdev – United Nations website on sustainable development.

Unit for Sustainable Development Environment: www.oas.org/usde – The USDE is the principal technical arm of the OAS General Secretariat responsible for meeting the needs of the member states in matters of sustainable economic development.

United Nations Commission for Sustainable Development: (CSD) www.un.org/esa/sustdev/csd/about_csd.html – The CSD is the appointed UN body to facilitate the implementation of sustainable development targets agreed at WSSD.

World Bank Sustainable Development: www-esd.worldbank.org – Explains their three objectives of sustainable development: improving the quality of life; improving the quality of growth; protecting the quality of the regional and global commons.

World Business Council for Sustainable Development: www.wbcsd.org – Is a coalition of 170 international companies united by a shared commitment to sustainable development via the three pillars of economic growth, ecological balance and social progress.

World Sustainability Hearing: www.earthisland.org/wosh – Rio+10 initiative on sustainable development. A Hearing at which ordinary people from around the world testified on their experiences of how the promise and practice of "sustainable development" has affected their lives over the past ten years.

Although national government carries the co-ordinating function of environmental management in South Africa, the practical implementation and enforcement of most environmental regulation rests with the provinces. Guidance on environmental issues, particularly those that affect local matters, should be acquired from provincial authorities as the first stage of investigation. The provinces have organised their environmental responsibilities in different ways and therefore responsibility for "the environment" falls under different departments in different provinces. You should be aware that the following list of officials and contact details are constantly changing and the details may not necessarily remain accurate, even in the short term.

Should you find the listed information to have changed, you should contact Government Communications Information Services (GCIS) who can provide the updated information and also provide contact details of other Departments not listed here.
* GCIS General Enquiries Tel: 012 314 2211 www.gcis.gov.za

NATIONAL DEPARTMENTS

DEPARTMENT OF AGRICULTURE

Website: www.nda.agric.za
Director-General: Ms Bongiwe Njobe
Tel: (012) 319 6000 • Email: dg@nda.agric.za

Directorate: Agricultural Land Resource Management
Tel: (012) 319 7685 • Fax: (012) 329 5938
Contact Name: Mr B Msomi

Directorate: Agricultural Water Use Management
Tel: (012) 842 4279 • Fax: (012) 804 3048
Contact Name: Mr A T van Coller

Directorate: Genetic Resources
Tel: (012) 319 6329 • Fax: (012) 319 7279
Contact Name: Dr S Moephuli

Directorate: Food & Quarantine Inspection Services
Tel: (012) 319 6502 • Fax: (012) 326 5606
Contact Name: Mr E Rademeyer

Directorate: Agricultural Production Inputs
Tel: (012) 319 7000 • Fax: (012) 319 7179
Contact Name: Miss Ratebe

DEPARTMENT OF EDUCATION

Website: www.education.pwv.gov.za
Director-General: Mr Thamsanqa Dennis Mseleku
Tel: (012) 312 5911

National Environmental Education Programme (NEEP)
Tel: (012) 312 5210 • Email: kuhles.r@doe.gov.za
Contact Name: Mr Reinhard Kuhles – Deputy Director – School Education.

DEPARTMENT OF ENVIRONMENTAL AFFAIRS AND TOURISM

Website: www.environment.gov.za
Director-General: Dr Chrispian Olver • Tel: (012) 310 3911

Sub-Directorate: Air Quality Management
Tel: (012) 3103911 • Fax: (012) 320 0488
Contact Name: Dr Matjiola

Directorate: Biodiversity Management
Tel: (012) 310-3911 Fax: (012) 322 2682
Contact Name: Ms Maria Mbengashe

Directorate: Cultural & Local Natural Resources Management
Tel: (012) 310 3911 • Fax: (012) 322 2682
Contact Name: Ms Maria Mbengashe

Directorate: Coastal & Inshore Resource Management
Tel: (021) 402 3911 • Fax: (021) 402 3360
Contact Name: Dr J van Zyl

Directorate: Information and Reporting
Tel: (012) 310 3693 • Fax: (012) 322 6287

Directorate: Integrated Pollution Prevention & Waste Management
Tel: (012) 310 3654 • Fax: (012) 322 2682
Contact Name: Mr S Gamede

Chief Air Pollution Control Officer
Tel: (012) 310 3911 • Fax: (012) 322 2682
Contact Name: Dr Matjiola

Directorate: Environmental Planning & Impact Management
Tel: (012) 310 3911 • Fax: (012) 322 2682
Contact Name: Mr W Fourie

Chief Directorate: Weather Bureau
Tel: (012) 309 3911 • Fax: (012) 309 3127
Contact Name: Mr G Schulze

Directorate: Climatology
Tel: (012) 309 3911 • Fax: (012) 309 3127
Contact Name: Mr M Laing

Sub-Directorate: Montreal Protocol
Tel: (012) 312 0215 • Fax: (012) 322 2682
Contact Name: Ms M Schuurman

DEPARTMENT OF HEALTH

Website: www.doh.gov.za
Director-General: Dr N K Matsau
Head of Communications: Ms Jo-Anne Collinge
Tel: (012) 312 0000 • Email: collij@health.gov.za

DEPARTMENT OF MINERALS & ENERGY

Website: www.dme.gov.za
Director-General: Adv Sandile Nogxina
Communications Officer: Mr Andre Eager
Tel: (012) 317 9000 • Email: esther@mepta.pwv.gov.za

Directorate: Renewable Energy
PRO: Mr Andre Otto • Tel: (012) 317 9225

DEPARTMENT OF TRADE & INDUSTRY (DTI)

Website: www.thedti.gov.za
Director-General: Dr Alistair Ruiters

Directorate: Standards & Environment
Deputy Director Environment: Mrs M J Visagie
Tel: (012) 310 9822

DEPARTMENT OF WATER AFFAIRS AND FORESTRY

Website: www.dwaf.gov.za
Director-General: Mr Arnold Mike Muller
Head of Communications: Mr Babs Naidoo
Tel: (012) 336 8726

Directorate: Catchment Management
Tel: (012) 336 8829 • Fax: (012) 326 2630
Contact Name: Mr H Karar

Directorate: Conservation Forestry
Tel: (012) 336 7742 • Fax: (012) 336 8942
Contact Name: Mr M Teter

Directorate: Information Services
Tel: (012) 336 8774 • Fax: (012) 332 1532
Contact Name: Mr G Manyatche

Directorate: Social & Ecological Services
Contact Name: Mr M Msiza
Tel: (012) 336 7255 • Fax: (012) 326 2715

Directorate: Water Conservation
Tel: (012) 336 8818 • Fax: (012) 326 2630
Contact Name: Mr C Ruiters

Directorate: Water Resources Planning
Tel: (012) 336 8814 • Fax: (012) 332 1532
Contact Name: Mr J van Rooyen

Endangered Species Protection Unit
Tel: (012) 803 9900 • Fax: (012) 803 8379
Contact Name: Mrs B Benson

PROVINCIAL DEPARTMENTS

EASTERN CAPE

MEC for Education: Ms Nomsa Jajula
Tel: (040) 608 4202 • Email: beryl@edumtzs.ecape.gov.za

MEC for Environment & Tourism: Mr Enoch Godongwane
Tel: (040) 609 4889 • Email:godongw@eetmind.escape.gov.za

FREE STATE

MEC for Agriculture: Mr Mann Oelrich
Tel: (051) 861 1000 • Email: oelrich@glen.agric.za

MEC for Education: Mr Diratsagae Alfred Kganare
Tel: (051) 404 8411 • Email: meceguc@majuba.ofs.gov.za

MEC for Tourism Environmental and Economic Affairs:
Mr Mann Oelrich
Tel: (051) 861 1000 • Email: oelrich@glen.agric.za

GAUTENG

MEC for Agriculture, Conservation, Environment and Land Affairs: Ms Mary Metcalfe

KWAZULU-NATAL

MEC for Agriculture and Environmental Affairs:
Mr Dumisani Makhaye
Tel: (035) 874 3291

Directorate: Environmental Impact Assessment
Tel: (033) 355 9120 • Contact Name: Ms Sarah Allan

LIMPOPO

MEC for Agriculture and Environmental Affairs:
Dr Pakishe Aaron Motsoaledi • Tel: (015) 295 7023

MPUMALANGA

MEC for Agriculture, Conservation and Environmental:
Ms Kwati Candith Mashego-Dlamini
Tel: (013) 766 6074 • Email: mec@nelagri.agric.za

NORTH WEST

MEC for Agriculture, Conservation and Environment:
Ms Bomo Edith Molewa
Tel: (018) 389 5688 • Email: molewab@nwpg.org.za

NORTHERN CAPE

MEC for Agriculture, Land Reform, Environment and Conservation: Mr David Rooi • Tel: (053) 831 4049

WESTERN CAPE

MEC for Agriculture, Environmental Affairs and Development Planning: Mr Johan Gelderblom
Tel: (021) 483 4700

Director: Environmental Management
Tel: (021) 483 4093 • Contact Name: Mrs D Alfred

USEFUL GOVERNMENT BODIES, COMMISSIONS AND TASK GROUPS

Agricultural Research Council (ARC) *
Contact Name: Ms Mariana Purnell
Tel: (012) 427 9700 • Email: mpurnell@arc.agric.za

Central Energy Fund (CEF)
Mrs M Moll: PRO
Tel: (011) 535 7000 • Email: mollm@cef.org.za

Council for Geoscience *
Dr P K Zawada: PRO
Tel: (012) 841 1911 • Email: pzawada@geoscience.org.za

CSIR Environment Technology *
Dr Pat Manders – Business Development
Tel: (012) 841 3680 • Email: pmanders@csir.co.za

ESKOM *
Annamarie Murray – Account Executive
Tel: (011) 800 2792 • Email: annamarie.murray@eskom.co.za

Human Sciences Research Council
CEO: Dr Mark Orkin • Tel: (012) 302 2999
Email: rpcf@hsrc.ac.za

Mintek *
Dr R J (Cobus) Kriek – Environmental Specialist
Tel: (011) 709 4061 • Email: cobusk@mintek.co.za

National Cleaner Production Centre *
Dr C M Masuku: Director
Tel: (012) 841 3665 • Email: cmasuku@csir.co.za

National Nuclear Regulator
PRO: Mr P Nkhwashu • Tel: (012) 674 7191

Rand Water *
Lenise Wagner Marketing Brand Manager – Water Wise
Tel: (011) 682 0911 • Email: lwagner@randwater.co.za
Email: customerservice@randwater.co.za

South African Forestry Company (Ltd) (SAFCOL)
Mrs E. Cornelius: PRO
Tel: (012) 481 3500 • Email: annemie@safcol.co.za

South African Nuclear Energy Corporation (NECSA)
Nomsa Sitole: Information Officer
Tel: (012) 305 5750 • Email: mailto:nsithole@necsa.co.za

South African Veterinary Council
Ms Hanri Kruger: Information Officer
Tel: (012) 324 2392 • Email: savc@intekom.co.za

Water Research Commission (WRC)
Mr Yuven Gounden – PRO
Tel: (012) 330 0340 • Email: yuveng@wrc.org.za

These organisations are also listed in the "Directory" Section of The Enviropaedia

Printed on Sappi recycled paper

USING THE ENVIROPAEDIA FOR ENVIRONMENTAL EDUCATION

Minister Asmal was instrumental in setting up a National Environmental Education Project for the General Education and Training band of schooling (NEEP-GET). He also ensured that when the curriculum was revised, A Healthy Environment was addressed in every Learning Area.

Every teacher from Grade R to Grade 9 (GET) now shares responsibility for promoting A Healthy Environment. But many lack a good grasp of environmental issues, and do not always know how best to use available resources to support environmental learning. The Enviropaedia is a comprehensive resource on environmental matters, which can be used to support teachers and learners for school-based curriculum planning and delivery.

This is how! Five Steps to Planning Environmental Learning using *The Enviropaedia*

STEP 1. RECOGNISE THE ENVIRONMENTAL PRINCIPLES IN THE CURRICULUM

GET teachers are currently delivering one of two versions of Curriculum 2005. There are technical differences between the two versions, but the environment is prominent in both.

In the first version of C2005 'environment' is one of the phase organisers, that is, a key idea used to organise teaching and learning in a particular phase, across all the Learning Areas.

To work with environment as a phase organiser teachers need to recognise that environmental issues (and solutions) have social, cultural, economic, scientific, technological, lifestyle, management and even political dimensions which are relevant across the curriculum. Enviropaedia articles illustrate this. For example, the article on Energy higlights the different technologies used to generate energy, the socio-ecconomic and political factors which determine people's access to energy service and lifestyle and management solutions to the environmental problems associated with unsustainable energy generation and use.

The new version of C2005 is the Revised National Curriculum Statement (RNCS). This streamlined and strengthened framework does not have phase organisers. A Healthy Environment is one of four values which form the First Principle of the RNCS, and which are addressed in all Learning Areas, through various Learning Outcomes.

For the RNCS, teachers need to understand the principles and the practicalities of A Healthy Environment. Enviropaedia entries and articles explain how A Healthy Environment is threatened by factors like pollution, climate change and excessive consumption, and what we can do about it (e.g. sustainable development). Its articles are also useful for exploring how A Healthy Environment links to the other values in the RNCS first principle: Human Rights, Social Justice and Inclusivity. For example, the Energy article explains that some options for generating energy have greater impacts on people and the environment and threaten human rights, while other options have fewer negative impacts, create more jobs and make energy services more inclusive.

The NEEP-GET says:

"Given the socio-political history of our country, which promoted unjust conservation laws, this RNCS principle prepares learners to deal meaningfully with the socio-ecological issues of the past, present and future. Within the framework of this principle, teachers have unlimited opportunities for environmental learning related to the South African context. The Enviropaedia and many other resources can be used to enable a better understanding of this principle."

STEP 2. LOCATE ENVIRONMENTAL LEARNING IN THE LEARNING AREAS

Read the relevant curriculum policy to see how the Learning Area(s) that you teach, support environmental learning. In the RNCS, this would be through one or more of the following:

◆ Particular Learning Area principles (as in Mathematics, Languages);

◆ Key concepts and knowledge – content (in Social Sciences and Natural Sciences);

◆ Particular Learning Outcomes (e.g. LO3 in Natural Sciences, Technology; LO2 & LO3 in Social Sciences; LO1 in Life Orientation; LO2 in Economics & Management Sciences);

◆ Associated Assessment Standards (see examples below) – all of which work towards the cross-curricular value of A Healthy Environment.

> Some of the Assessment Standards, which GET teachers will use to assess if the required Learning Outcomes have been achieved.

Learners must be able to:

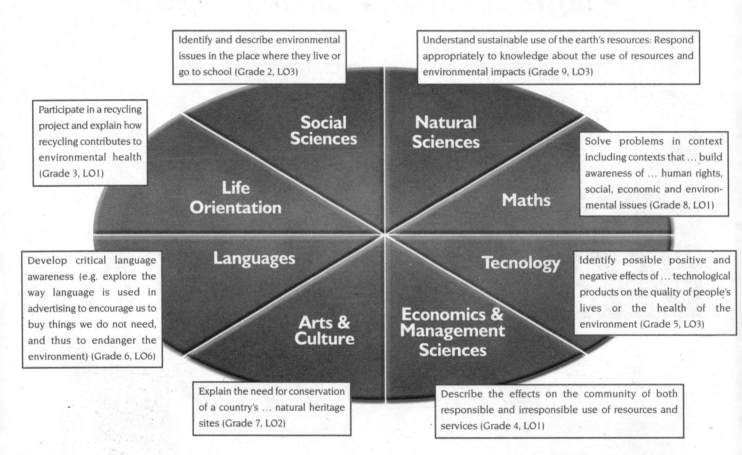

Identify and describe environmental issues in the place where they live or go to school (Grade 2, LO3)

Understand sustainable use of the earth's resources: Respond appropriately to knowledge about the use of resources and environmental impacts (Grade 9, LO3)

Participate in a recycling project and explain how recycling contributes to environmental health (Grade 3, LO1)

Social Sciences

Natural Sciences

Solve problems in context including contexts that ... build awareness of ... human rights, social, economic and environmental issues (Grade 8, LO1)

Life Orientation

Maths

Develop critical language awareness (e.g. explore the way language is used in advertising to encourage us to buy things we do not need, and thus to endanger the environment) (Grade 6, LO6)

Languages

Tecnology

Identify possible positive and negative effects of ... technological products on the quality of people's lives - or the health of the environment (Grade 5, LO3)

Arts & Culture

Economics & Management Sciences

Explain the need for conservation of a country's ... natural heritage sites (Grade 7, LO2)

Describe the effects on the community of both responsible and irresponsible use of resources and services (Grade 4, LO1)

STEP 3. DRAW UP A LEARNING PROGRAMME AND LESSON PLANS

To deliver these and other environment-related curriculum outcomes, the Learning Programmes that districts or schools develop for each Grade must include an environmental focus. The Enviropaedia gives school and departmental staff the broad background on "A healthy environment" as a cross-curricular concern, and can therefore inform district and school-based curriculum planning.

Classroom-based planning of individual teacher's Lesson Plans also requires a sound understanding of environmental matters. For example, if a Grade 6 Language teacher plans a lesson on advertising and the dangers to the environment, the teacher needs to understand how the excessive consumption of goods threatens A Healthy Environment. The Enviropaedia topic Consumerism will provide the relevant background information.

STEP 4. USE TEACHING AND LEARNING SUPPORT MATERIALS EFFECTIVELY

During teaching, *The Enviropaedia* can be used to help learners "tune into" environmental issues, research and take appropriate action. Teachers can, for example:

◆ Use Encyclopaedia Topics, Issues and Articles as stimuli for art; for Language comprehension; or to "tune in" at the start of inquiries into local environmental issues.

◆ Use Encyclopaedia Topics, Issues and Articles to prepare worksheets and questionnaires that learners use as part of fieldwork on local environmental issues.

◆ Deepen learners' understanding of local issues by providing new information from the Articles and Related Topics given.

◆ Provide definitions of key concepts and issues, using the Encyclopaedia Section.

◆ Use the Networking Directory, Related Organisations and Advertorials to request further materials; to organise resource persons or field trips to particular sites, etc.

◆ Refer learners to the Networking Directory, to support their project work.

◆ Encourage learners to take action, e.g. writing to parties listed in the Directory or Advertorials, to lobby for better environmental management.

STEP 5. ASSESS ENVIRONMENTAL LEARNING!

Remember – in C2005 and the RNCS environmental learning is part of the formal curriculum and needs to be assessed.

The curriculum requires the assessment of Knowledge, Skills, Values and Attitudes. Teachers know how to assess Knowledge (e.g. the ability to identify and describe environmental issues) and Skills (e.g. the ability to do a calculation related to water use). It is harder to assess Values and Attitudes. Use criteria for assessing participation (e.g. in recycling) or "appropriate responses to knowledge about environmental impacts" which reflect appropriate values – including the values of A Healthy Environment, Social Justice and Human Rights.

> For more resources, contact Share-Net or the Environmental Education Association of Southern Africa (details in the Directory section of *Enviropaedia*) or the environmental education co-ordinators in the provincial education departments (NEEP GET's national office can refer you). Or try the new Portal to electronically available environmental resources, at:
> www.eviroLEARN.org.sa.

Draw on balanced and comprehensive resources like *The Enviropaedia* to prepare learners adequately, and make sure you do justice to the environment in your Learning Area(s).

Guest author:
Dr Eureta Rosenberg

More About The NEEP-GET:

The Department of Education's National Environmental Education Project for the General Education and Training band (the NEEP-GET) has to ensure that current curriculum policy implementation and educator training processes successfully incorporate environmental learning. Since its establishment in 2001 NEEP-GET has played a key role in policy development (including the development of the RNCS). It also facilitates a professional development programme and activities focused on environmental learning in schools. The project is led in each province by a provincial co-ordinator who works closely with official curriculum support staff. The main pillars of the NEEP-GET are:

◆ **Professional development** of educators, curriculum support staff and provincial environmental education co-ordinators, based on comprehensive professional development programmes.

◆ **Materials development** involving accessible policy guideline documents for educators, and support for resource developers to produce relevant learning and teaching support materials.

◆ **Curriculum development** contributing to curriculum policy development and implementation processes at national, provincial, district and school levels.

For further information or help,
Contact the NEEP-GET on 021-312 5122, Olivier.C@doe.gov.za or Neluvhalani.E@doe.gov.za

GREEN CONSUMER GUIDE
From
The Greenhouse Project

Every purchase you make has either a direct or indirect effect on the environment. When you exercise your power by choosing where and what you buy, you help change the world for the better. What and how much we decide to buy also determines what and how much we throw away. Careless shopping causes waste. The Greenhouse Project, GLAD Files (see our listing in the Directory Section) can provide a detailed guide to help you find environmentally-friendly Products, Services, Help, Working examples and Resources. The Glad Files also provide a wide range of consumer shopping tips. Some extracts and abbreviated examples of these tips include:

◆ Avoid using disposable plastic bags given out by shops – it is better to acquire a reusable shopping bag or basket.

◆ Avoid buying goods with unnecessary packaging – excess packaging not only cheats you if the box is only partially filled – but it also creates unnecessary waste material that has to be disposed of.

◆ Wherever possible, choose natural, environmentally-friendly and certified "green products" that have less harmful chemical content for your home and garden.

◆ Buy organic products that are free from toxic chemicals, which can have a serious impact on the environment and your health.

◆ Be a natural beauty – buy products, which are cruelty free to the planet, people and animals too. Many products (especially cosmetics) have been tested painfully on animals. Many of the chemicals in toiletries and cosmetics are absorbed through your skin and can be harmful to your health.

◆ Use natural soaps instead of commercial soaps, which often contain strong degreasing agents like sodium laurel sulphate or sodium laureth sulphate, a potent chemical that can destroy delicate tissues in the eye and skin.

◆ Never use an aerosol – although they no longer contain ozone-depleting CFCs, many still contain hydrocarbon propellants that contribute to air pollution and when inhaled, irritate the lungs.

◆ Avoid products containing chlorine. Chlorine is highly corrosive and harmful to both people and the environment.

◆ Buy low-phosphate washing-up liquid and washing powder. Phosphates are used as water softeners, which when discharged into the water supply, stimulate excessive algae growth. These algae starve water of oxygen, killing plant and fish life, and disrupt the sewage treatment process.

◆ Buy bicarbonate of soda to clear your drains instead of abrasive powders. Many commercial drain cleaners contain corrosive and toxic products such as sodium or potassium hydrochloric acid and petroleum distillates that kill aquatic life and make water even more expensive to treat.

◆ Never use optical brighteners to wash your clothes – they disrupt the ecosystems in the rivers because they cannot be broken down.

◆ Avoid using chemical insect sprays – rather alter your local environment to make it less attractive to pests. For example, clean and wash out rubbish bins with hot water and make sure that there is no decaying food lying around. You could use natural deterrents like lemon, cloves, pine and cedar oils.

◆ Use natural methods to control weeds. The damaging effects of pesticides are multiplied as they are washed by rainwater into sewers and then into waterways. Even weed killers that claim to be safe have damaging effects.

◆ Buy energy-efficient white goods like washing machines, tumble dryers and dishwashers.

◆ Never buy wooden furniture, which comes from unmanaged forests.

◆ Avoid chipboard and MDF (Medium density fibreboard) – they have a high formaldehyde content, which gases out of the board over time. This is a recognised carcinogen.

These and many more can be found at website: www.greenhouse.org.za
or call Dorah Lebelo on 011 720 3773

Printed on Sappi recycled paper

Sustainable
Living Centre

P O Box 261 Noordhoek 7979 Cape Town
Tel (021) 789-2920 Fax (021) 789-2954
info@sustainable.co.za web www.sustainable.co.za

Products that help you live more sustainably   "BUY WITH A GREEN CONSCIENCE"

PRODUCT	BRIEF DESCRIPTION
ENERGY EFFICIENCY	
Compact fluorescent lights	Energy (and money) saving 12W, 16W and 20W Compact Fluorescent light. Uses a quarter of the energy of a normal incandescent 'bulb' light for the same light output (e.g. 16W = 60W bulb), lasts 8 to 10 times longer. Saves money and the environment.
Hotbox	Once boiled, simply place the pot between the two insulated cushions and it keeps on cooking without using any more energy. Simple, and remarkably effective.
Geyser blanket	A bubble wrap blanket that cuts geyser heat losses by up to 50%. Quickly pays for itself.
Hot water pipe insulation	Polyethylene lagging cuts hot water pipe heat losses by up to 50%.
Geyser timer	Installing a timer, which regulates when the geyser water is heated can reduce electricity used to heat water by 30 to 40%. Pays for itself in a few months.
RENEWABLE ENERGY	
Solar photovoltaic panels (PV) and systems	High quality solar PV panels from 10 watt to 120 watt capacity. 20 year guarantee on many panels. Other solar electric system components also available (regulators, inverters, batteries).
AirX wind charger – 400W rated	400 watt nominal output at 12.5m/s wind speed. Aero wind charger. Built-in regulation, few moving parts. Easy installation.
Sunstove solar cooker	Cost-effective locally made solar cooker. Can take two pots at a time.
Parabolic dish solar cooker	Powerful parabolic solar cooker.
Solar water heaters	The sun's energy is directly used to heat water, typically saving at least half of your water heating electricity consumption. (SLC does not supply solar water heaters directly. Contact us and we'll get a recommended supplier to call you.)
WATER EFFICIENCY & sanitation	
Low-flow showerhead range	A range of different low-flow showerheads are available for different water pressures and tastes. Typically they save between 50 and 75% of shower water with little or no comfort sacrifice.
Tap aerator (kitchen)	This easy to fit aerator can save 75% of kitchen tap water use.
Multi-flush toilet system	A simple system that lets you control the flush volume. Can save 20% of your total water consumption. (SLC does not supply these systems directly. Contact us and we'll get the recommended supplier to call you.)
Greywater recycling system	This system recycles grey water to the garden – typically reducing total water consumption by 35%. (SLC does not supply these systems directly. Contact us and we'll get the recommended supplier to call you.)
Other water saving systems	There are systems available to redirect rainwater to swimming pools, redirect grey water to the toilet cistern and re-use backwashed pool water. (SLC does not supply these systems directly. Contact us and we'll get the recommended supplier to call you, or visit www.water-rhapsody.co.za)
Drip irrigation	Drip irrigation reduces garden water use dramatically – by as much as 90%. (SLC does not supply these systems directly. Contact us and we'll get the supplier to call you.)
Composting toilet	Self-contained biological toilet system (SLC does not supply these systems directly. Contact us and we'll get the suppliers to call you.)
GREEN BUILDING	
Enviro touch paints	A range of non-toxic, environmentally-friendly wood finishes, wood care products, wall finishes and more. Visit www.envirotouch.com (the Enviro Touch website) for more info, or call us.
Breathecoat paints	Non-toxic, durable paints which are easy to apply and allow walls to 'breathe' while remaining waterproof. Visit www.breathecoat.co.za for more info, or call us.
GARDEN & HOME	
Counter-top composter	The convenient counter-top composter stores your organic waste while absorbing odours.
Worm bins for home & garden	The earthworm bin converts household and garden organic waste to hi-quality compost in 4 to 6 weeks.
Composter (tumbler)	This rotating compost bin speeds up the composting process.
Composter (thermo)	Well-designed composter. Made from recycled plastics.
Biogrow gardening products	Biogrow products include a range of environmentally friendly, certified organic pesticides and fungicides.

More environmentally sound products being sourced continually.

THE ECO-LOGICAL LIFESTYLE GUIDE

(Be part of the solution – no change happens without people, like us, changing our own behaviour)

USE YOUR CONSUMER POWER
– BE A "GREEN CONSUMER"

You can make use of your "consumer power" on a daily basis to improve our environment! Every purchase that you make has an impact on the environment in many ways. By selecting products and services from organisations that demonstrate that they care for the environment, we can send a direct message to manufacturers and suppliers, that their "care for our environment" will be rewarded by increased sales. Your spending decisions can motivate them to:

◆ Supply us with more environmentally-friendly products (See: Green Consumer Guide in the Resource Section)

◆ Operate their business in a more environmentally-sustainable manner.

Organisations can also demonstrate their "environmental care" in many other ways. Here are some examples:

◆ Motorcars – Mazda have created – and financially support – the Mazda Wildlife Fund.

◆ Office electronics – Hewlett Packard and Canon SA have projects to collect and recycle used toner cartridges.

◆ Clothing – Woolworths support an initiative (Hangerman project) to recycle the clothes hangers used in their shops.

◆ Pick 'n Pay and Spar stores sell environmentally-friendly shopping bags to replace disposable plastic bags.

◆ Many corporates are members of – and annually donate money to – environmental organisations such as: The Botanical Society (BOTSOC); Endangered Wildlife Trust (EWT); Wildlife and Environmental Society (WESSA); World Wildlife Fund (WWFSA) and many other worthy organisations.

◆ Many organisations are improving the way in which they produce products or deliver services and are reducing their environmental impact, by choosing to implement an ISO series accreditation.

JOIN AN ENVIRONMENTAL ORGANISATION

◆ Not many of us have the time or knowledge to go out saving whales or chopping down alien invasive plants – but by paying a subscription to an environmental organisation, you are paying them to do the job of protecting and conserving our environment for you (in the same way that you might pay an accountant to do your accounts for you). Some organisations may also send you a member's magazine – which keeps you informed about what is going on in the area of your interest.

◆ If you do have the time, you might also have a pleasant surprise in meeting other members who may share similar interests and values. Some organisations (particularly a local "FRIENDS GROUP") can help you conserve and protect your own local area from insensitive developments or environmental damage.

◆ Choose to join one of the organisations listed in *The Enviropaedia* – **your support can make a world of difference.**

SAVE ENERGY AND NATURAL RESOURCES

◆ Turn off lights and electrical appliances when not needed.

◆ Use low energy light bulbs wherever possible – they cost more initially, but repay themselves many times over.

◆ When you make a cup of tea or coffee, boil only the amount of water you need.

◆ Consider using renewable energy sources in your home – which don't cost the earth. They include solar panels, photo-voltaics, and wind generators. They can pay you back in a few years, and thereafter save you money.

◆ Keep your car properly tuned and maintained – this prevents air pollution and saves you money.

◆ Reduce the amount of water that you use: Showering uses much less water than a bath; Reduce the water level if you do bath and use the bath water to water plants; Put a water-filled plastic tub or brick in your toilet cistern – this will save many litres of water with every flush; Use short bursts of water from the tap when brushing your teeth; Collect and use rainwater that runs off your roof to water plants; Only use a washing machine when you have a full load; Wash the car with a few buckets of water instead of the hose; Do not leave taps running unless you are using every drop of water and never leave taps dripping.

LOOK AFTER YOUR OWN LOCAL ENVIRONMENT

◆ Be responsible for your own litter. Put it in a litterbin if one is available, or take it away to dispose of it at home.

◆ Help keep our natural areas clean and free of rubbish or alien vegetation, especially: bushveld, beaches, rivers and mountains.

◆ Prevent unnecessary fires and never burn materials that give off polluting fumes and gasses – especially rubber and plastics.

CUT DOWN ON WASTE

Re-use bottles and containers or take them to recycling depots. Contact the following numbers for information on collection points in your area: (Plastic containers) Plastics Federation 011 314 4021; (Cans and tins) Collect-a-Can 011 466 2939; (Glass) Consol Glass 011 874 2010; (Motor and cooking oils) Oilkol 011 762 5506; (Paper) Nampak 0800 018 818. Recycling means that we do not waste resources. Recycling can also create employment opportunities. Use a re-usable shopping bag or basket. Avoid buying goods with unnecessary packaging (like those big boxes of breakfast cereals that are only half-filled with the product).

Whenever possible, repair things instead of throwing them away or give them to other people who could use them.

Grow your own organic vegetables and herbs; it's a lot cheaper and saves on the hidden costs of commercial foods such as packaging, transport, and fertiliser inputs.

Instead of throwing leftovers and food waste away, create a compost heap. It offers a great way to reduce and recycle waste, and a natural way of fertilising your garden.